CLINICAL
NEUROPHYSIOLOGY

CONTEMPORARY NEUROLOGY SERIES AVAILABLE:

CLINICAL NEUROPHYSIOLOGY

Edited by Jasper R. Daube, M.D.
Chair, Department of Neurology
Mayo Clinic and Mayo Foundation
Professor of Neurology, Mayo Medical School
Rochester, Minnesota

with contributors

 F. A. DAVIS COMPANY • Philadelphia

F. A. Davis Company
1915 Arch Street
Philadelphia, PA 19103

Printed in the United States of America

Last digit indicates print number: 10 9 8 7 6 5 4 3 2

Medical Editor: Robert W. Reinhardt
Medical Developmental Editor: Bernice M. Wissler
Production Editor: Glenn L. Fechner
Cover Designer: Steven R. Morrone

As new scientific information becomes available through basic and clinical research, recommended treatments and drug therapies undergo changes. The author and publisher have done everything possible to make this book accurate, up to date, and in accord with accepted standards at the time of publication. The author, editors, and publisher are not responsible for errors or omissions or for consequences from application of the book, and make no warranty, expressed or implied, in regard to the contents of the book. Any practice described in this book should be applied by the reader in accordance with professional standards of care used in regard to the unique circumstances that may apply in each situation. The reader is advised always to check product information (package inserts) for changes and new information regarding dose and contraindications before administering any drug. Caution is especially urged when using new or infrequently ordered drugs. Nothing in this publication implies that Mayo Foundation endorses the products or equipment mentioned in this book.

Library of Congress Cataloging in Publication Data

Clinical neurophysiology/edited by Jasper R. Daube; with
    contributors.
        p.   cm.—(Contemporary neurology series ; 46)
    Includes bibliographical references and index.
    ISBN 0-8036-0073-9
    1. Electroencephalography.   2. Electromyography.   3. Evoked
potentials (Electrophysiology)   4.Neurophysiology.   5. Nervous
system—Diseases—Diagnosis.   I. Daube, Jasper R.   II. Series.
    [DNLM: 1. Nervous System Diseases—diagnosis.   2. Nervous System
Diseases—therapy.   3. Neurophysiology.   4. Electroencephalography.
5. Electromyography.   6. Evoked Potentials.   W1 CO769N v. 46 1996/
WL 140 C641 1996]
    RC386.6.E43C585   1996
    616.8′047547—dc20
    DNLM/DLC
    for Library of Congress   95-31428

FOREWORD

Activity is booming in clinical neurophysiology. The American Board of Psychiatry and Neurology now offers a certificate for Added Qualifications in Clinical Neurophysiology. Because this process was approved by the American Board of Medical Specialties, clinical neurophysiology is clearly recognized as a subspecialty in the practice of medicine. A Section on Clinical Neurophysiology has been formed at the American Academy of Neurology. Numerous societies in the United States and throughout the world are devoted to clinical neurophysiology, and their membership is growing.

Clinical neurophysiology is a large field. Like neurology, it encompasses a wide variety of issues and illnesses, ranging from the peripheral nervous system to the central nervous system. Like neurology, it is difficult to master all of clinical neurophysiology, and most practitioners have a focused interest in only part of the field. Also, as in neurology, there is an essential unity to clinical neurophysiology. There is a physiologic approach to problems with methods that measure the electrical activity of the nervous system. It is useful for practitioners to have a view of the whole field even if they practice only a part of it.

The Mayo Clinic has been a central force in the United States in many areas of clinical neurophysiology. In the area of electromyography, Dr. Edward Lambert, a pioneer in the field, has made many basic observations that still guide current practice, and, of course, he identified an illness that now bears his name. He has trained many leaders of modern electromyography in the United States. In electroencephalography, Dr. Reginald Bickford was also a pioneer and was active in many areas, including evoked potentials and even early attempts at magnetic stimulation of the brain. There have been many other leaders in electroencephalography at the Mayo Clinic, and four of them, in addition to Dr. Bickford, have been presidents of the American EEG Society.

Thus, when Mayo Clinic neurologists speak about clinical neurophysiology, they speak with special authority. No one is better suited to orchestrate the writing of a textbook on clinical neurophysiology than Dr. Jasper Daube, who is currently Chair, Department of Neurology, at the Mayo Clinic. Dr. Daube is well recognized internationally as an expert in electromyography; he is very knowledgeable about all areas of the subject, basic and applied. He is an outstanding leader with a gift for organization. "Organization" characterized his presidency of the American Association of Electrodiagnostic Medicine, his role in promoting the certificate for Added Qualifications in Clinical Neurophysiology, and his founding of the Section on Clinical Neurophysiology at the American Academy of Neurology.

What about this textbook on clinical neurophysiology? Its many chapters cover the field of clinical neurophysiology. The first several chapters discuss the basic issues of neuronal generators, biologic electricity, and measurement techniques central to all areas of clinical neurophysiology. Next, the individual areas of the field are discussed: areas including classic electromyography, electroencephalography, and evoked potentials and extending to autonomic nervous system testing, sleep, surgical monitoring, motor control, vestibular testing, and magnetic stimulation. The text is organized for physicians who want to know how to make an assessment of a particular symptom, of a particular system, or for a particular disease. There is valuable information on the use of clinical neurophysiologic testing in a practical setting. This is a welcome new textbook that should serve both students and practitioners.

Mark Hallett, M.D.
Clinical Director, National Institute of
  Neurological Diseases and Stroke
Bethesda, Maryland

PREFACE

Clinical Neurophysiology is the result of more than 50 years of experience at the Mayo Clinic in training clinicians in the neurophysiologic methods for assessing diseases of the central and peripheral nervous systems. The lectures and handouts that were developed initially by Drs. Reginald Bickford and Edward Lambert in electroencephalography and electromyography, respectively, were the seeds of what has grown into a far-reaching but still unified field of endeavor in clinical neurophysiology at the Mayo Clinic. As the base of knowledge in neurophysiology and its clinical applications and as the range of techniques available for diagnosing and treating diseases of the central and peripheral nervous systems have expanded during the past 20 years, the teaching program at the Mayo Clinic has evolved into a formal, unified, 2-month-long course in clinical neurophysiology. It serves as an introduction to clinical neurophysiology for residents, fellows, and other trainees. The course includes lectures, small group seminars, practical workshops, and clinical experience in each of the areas of clinical neurophysiology. The faculty for the course consists entirely of Mayo Clinic staff members. These staff are among the authors of the chapters in this textbook.

Over the years, the material for the clinical neurophysiology course has been consolidated from individual lecture handouts into more organized (and thick) manuals, which have been requested with increasing frequency by persons outside out institution who have learned about them mainly by word of mouth. The positive responses expressed by these people and by our trainees have prompted us to write *Clinical Neurophysiology*. We hope that it meets the needs of trainees in the various areas of clinical neurophysiology at other medical institutions and that it provides a good background and review of the field for those already active in clinical neurophysiology.

The organization of our textbook is unique: it is built around the concept of testing systems within the nervous system. The book consists of three major sections. The first division is a review of the basics of clinical neurophysiology, that is, knowledge that is common to each of the various areas in clinical neurophysiology. The second section considers the assessment of diseases by techniques. Thus, methods for assessing the motor system are grouped together, followed by those for assessing the sensory system, the higher cortical functions, and the autonomic nervous system. The third section is an attempt to synthesize how the different techniques are used in the clinical approach to diseases of the nervous system.

ACKNOWLEDGMENTS

The privilege of editing this textbook is akin to that of being the chair of the Department of Neurology at the Mayo Clinic. It provides the opportunity and honor to work with national leaders in each of the areas of clinical neurophysiology. All the authors bring unique knowledge and skills to bear in the chapters they have written, making the task of the editor a remarkably easy and straight forward one of organizing and coordinating the material.

It is a pleasure to recognize the skill and professionalism of the people in the Sections of Publications and Visual Information at the Mayo Clinic; they have had an integral part in the development of this textbook. Leaders at the Mayo Clinic who have encouraged the development of clinical neurophysiology at the Clinic, including the formal establishment of the Division of Neurophysiology more than 10 years ago, helped foster this textbook. Although all the staff in the Division of Neurophysiology have contributed in a major way to the clinical neurophysiology course on which this textbook is based, the directors of the individual laboratories should be singled out: Dr. Phillip Low, chair of the Division of Clinical Neurophysiology and director of the Autonomic Reflex Laboratory and the Nerve Physiology Laboratory; Dr. William Litchy, director of the Electromyography Laboratory; Dr. Frank Sharbrough, director of the Electroencephalographic Laboratory; Dr. Peter Hauri, director of the Sleep Laboratory; Dr. Robert Brey, director of the Posturography Laboratory; Dr. Philip McManis, director of the Movement Disorder Laboratory; Dr. Robert Fealey, director of the Thermoregulatory Sweat Laboratory; Dr. Peter J. Dyck, director of the Morphology Laboratory and the Peripheral Neuropathy and Research Laboratory; and Dr. Andrew Engel, director of the Muscle Laboratory. Special thanks must be given to Drs. Barbara Westmoreland and Robert C. Hermann, who are the major organizers of the clinical neurophysiology course; they also provide much of the teaching for the trainees participating in the course.

The support of the Mayo Foundation has been critical in the development of new directions and unique training programs in clinical neurophysiology. We acknowledge not only this support but also the help given by many others: the trainees who have participated in our clinical neurophysiology program and the students in our courses in continuing medical education who have given us feedback on our teaching material, the technicians who have been a major part of our teaching program and who have provided a helpful critique of our activities, the secretarial staff who has worked diligently to make corrections and revisions long after we said that we were finished, and other physicians at our institution who have found our help in clinical neurophysiology useful in the care of their patients. Particular recognition must be given to the dedication, commitment, and skill of our secretaries, Ms. Rita Loftus and Jeanette Connaughty, who were involved in organizing the original syllabus material and bringing it together in this textbook.

Jasper Daube, M.D.

LIST OF CONTRIBUTORS

Christopher D. Bauch, Ph.D.
Consultant, Section of Audiology, Mayo Clinic and Mayo Foundation; Associate Professor of Audiology, Mayo Medical School; Rochester, Minnesota

Eduardo E. Benarroch, M.D.
Consultant, Department of Neurology, Mayo Clinic and Mayo Foundation; Associate Professor of Neurology, Mayo Medical School; Rochester, Minnesota

Robert H. Brey, Ph.D.
Consultant, Section of Audiology, Mayo Clinic and Mayo Foundation; Professor of Audiology, Mayo Medical School; Rochester, Minnesota

Gregory D. Cascino, M.D.
Consultant, Department of Neurology, Mayo Clinic and Mayo Foundation; Associate Professor of Neurology, Mayo Medical School; Rochester, Minnesota

Bruce A. Evans, M.D.
Consultant, Department of Neurology, Mayo Clinic and Mayo Foundation; Assistant Professor of Neurology, Mayo Medical School; Rochester, Minnesota

Robert D. Fealey, M.D.
Consultant, Department of Neurology, Mayo Clinic and Mayo Foundation; Assistant Professor of Neurology, Mayo Medical School; Rochester, Minnesota

C. Michel Harper, Jr., M.D.
Consultant, Department of Neurology, Mayo Clinic and Mayo Foundation; Associate Professor of Neurology, Mayo Medical School; Rochester, Minnesota

Cameron D. Harris, B.S., R.PSG.T.
Associate, Sleep Disorders Center, Mayo Clinic and Mayo Foundation; Instructor in Medicine, Mayo Medical School; Rochester, Minnesota

Peter J. Hauri, Ph.D.
Consultant, Sleep Disorders Center, and Consultant, Section of Psychology, Mayo Clinic and Mayo Foundation; Professor of Psychology, Mayo Medical School; Rochester, Minnesota

Robert C. Hermann, Jr., M.D.
Consultant, Department of Neurology, Mayo Clinic and Mayo Foundation; Assistant Professor of Neurology, Mayo Medical School; Rochester, Minnesota

Kathryn A. Hirschorn, M.D.
Consultant, Department of Neurology, Mayo Clinic Scottsdale, Scottsdale, Arizona; Assistant Professor of Neurology, Mayo Medical School; Rochester, Minnesota

Donald W. Klass, M.D.
Emeritus Member, Section of Electroencephalography, Mayo Clinic and Mayo Foundation; Emeritus Professor of Neurology, Mayo Medical School; Rochester, Minnesota

Terrence D. Lagerlund, M.D., Ph.D.
Consultant, Department of Neurology, Mayo Clinic and Mayo Foundation; Associate Professor of Neurology, Mayo Medical School; Rochester, Minnesota

William J. Litchy, M.D.
Consultant, Department of Neurology, Mayo Clinic and Mayo Foundation; Associate Professor of Neurology, Mayo Medical School; Rochester, Minnesota

Phillip A. Low, M.D.
Consultant, Department of Neurology, Mayo Clinic and Mayo Foundation; Professor of Neurology, Mayo Medical School; Rochester, Minnesota

Joseph Y. Matsumoto, M.D.
Consultant, Department of Neurology, Mayo Clinic and Mayo Foundation; Assistant

Professor of Neurology, Mayo Medical School; Rochester, Minnesota

Wayne O. Olsen, Ph.D.
Consultant, Section of Audiology, Mayo Clinic and Mayo Foundation; Professor of Audiology, Mayo Medical School; Rochester, Minnesota

Frank W. Sharbrough, M.D.
Head, Section of Electroencephalography, Mayo Clinic and Mayo Foundation; Professor of Neurology, Mayo Medical School; Rochester, Minnesota

Benn E. Smith, M.D.
Consultant, Department of Neurology, Mayo Clinic Scottsdale, Scottsdale, Arizona; Assistant Professor of Neurology, Mayo Medical School; Rochester, Minnesota

J. Clarke Stevens, M.D.
Consultant, Department of Neurology, Mayo Clinic Scottsdale, Scottsdale, Arizona; Associate Professor of Neurology, Mayo Medical School; Rochester, Minnesota

Kathryn A. Stolp-Smith, M.D.
Consultant, Department of Physical Medicine and Rehabilitation, Mayo Clinic and Mayo Foundation; Assistant Professor of Physical Medicine and Rehabilitation, Mayo Medical School; Rochester, Minnesota

Barbara F. Westmoreland, M.D.
Consultant, Section of Electroencephalography, Mayo Clinic and Mayo Foundation; Professor of Neurology, Mayo Medical School; Rochester, Minnesota

CONTENTS

SECTION 1

Analysis of Electro-physiologic Waveforms

Clinical neurophysiology is an area of medical practice focused primarily on measuring function in the central and peripheral nervous systems, including the autonomic nervous system, and in the muscles. The specialty identifies and characterizes diseases of these areas, understands their pathophysiology, and, to a limited extent, treats them. Clinical neurophysiology relies entirely on the measurement of ongoing function—either spontaneous or in response to a defined stimulus—in a patient. Generally, the techniques used in clinical neurophysiology are noninvasive and do not require the removal of tissue for testing. Each of the clinical neurophysiologic methods measures function by recording alterations in physiology as manifested by changes in electrical waveforms, electromagnetic fields, force, or secretory activities. Each of these variables is measured as a waveform that changes over time. Electrical measurements in which the voltage or current flow associated with a neural activity is plotted on a temporal basis are the most common measurements used in clinical neurophysiology. Knowledge of the generation, recording, measurement, and analysis of such waveforms is critical in learning how to perform and apply clinical neurophysiologic methods in the study of disease.

The first section of this introductory textbook on clinical neurophysiology reviews the generation, recording, and analysis of the waveforms needed for the practice of clinical neurophysiology. The principles of electricity and electronics needed to make the recordings are reviewed in Chapter 1. To make the appropriate measurements, clinical neurophysiologists rely on equipment with technical specifications; this requires that they understand the principles reviewed in Chapter 1. All electrical stimulation and recording methods require applying electrical connections that pass currents (usually

1

of a small amount) through human tissue. Although the risks of harm from this current flow are small, they must be understood. The principles of electrical safety necessary to minimize, reduce, or eliminate any risk are discussed in Chapter 2. These principles are of critical importance to the practice of clinical neurophysiology.

The familiar forms of electricity—as used in the home and in business—are conducted by wires or electromagnetic fields that can be specified in the simple rules reviewed in Chapter 1. Electrical recordings made from human tissue are distinctly different because the electrical currents are carried by charged ions present throughout the tissue. Therefore, electrical currents in the human body flow through all parts of the tissue, and the density of current flow varies with the resistance and capacitance of the different types of tissue. This widespread flow of electricity is referred to as *volume conduction*. Volume conduction produces the unique aspects of the generation and recording of those physiologic waveforms recorded from human tissue. The principles of volume conduction described in Chapter 3 are applicable to the many forms of electrical recording used in clinical neurophysiology, whether the waveforms are recorded from the head (electroencephalography), nerves (nerve conduction studies), muscles (electromyography), or skin (autonomic function testing).

Measurement of current flow, or potential differences, between areas of the body and its fluctuation over time was first made using analog electronic devices. However, in the past 15 years, digital electronic devices have become an integral part of the measurement of electrical activity in humans. Although a small number of recordings are still made with analog devices, most use digital techniques for selecting, displaying, and storing the waveforms. These techniques are described in Chapter 4.

Virtually all tissues (and the cells composing them) in the human body have electrical potentials associated with their activity. These potentials are much larger for nerve and muscle tissue and can easily be recorded for analyzing the function of these tissues and their alteration with disease. The range of generators of electricity and the waveforms associated with their activity are described in Chapter 5. Each of these generators gives rise to many different electrical signals that are measured to characterize normal and abnormal activity.

All electrical waveforms, whatever their source, have several features in common. These are summarized in Chapter 6. The range of alterations that occur in these waveforms in disease and the electrical artifacts that occur in association with the physiologic waveforms are reviewed in Chapter 7.

CHAPTER 1

ELECTRICITY AND ELECTRONICS IN CLINICAL NEUROPHYSIOLOGY

Terrence D. Lagerlund, M.D., Ph.D.

PRINCIPLES AND DEFINITIONS IN ELECTRICITY

Electric Charges and Force

The two types of electric charges are designated *positive* and *negative*. Charges exert electric forces on each other: like charges repel, opposite charges attract. An electric charge can be thought of as generating an electric field in its surrounding space. The field around a positive charge points radially outward in all directions from that charge, whereas the field around a negative charge points radially inward. Numerically, the electric force F on charge q in a region of space in which there is an electric field E is given by $F = qE$. Thus, the electric field can be thought of as the electric force per unit charge ($E = F/q$).

Ordinary matter consists of atoms, which

3

contain a nucleus composed of positively charged protons and uncharged neutrons. Negatively charged electrons occupy the space around the nucleus, to which they are normally bound by the attractive electric force between them and the nucleus. In un-ionized atoms, the net charge of the electrons is equal and opposite to the charge of the nucleus so that the atom as a whole is electrically neutral. The charge carried by 6.25×10^{18} protons is one coulomb (C) (the unit of electric charge in the International System of Units [SI]).

Electric Potential

Energy (work) is required to move a charge in an electric field because of the electric force acting on that charge. The energy required, U, is proportional to the charge q; thus, it makes sense to talk about the energy per unit charge. This quantity is called the *electric potential* ($V = U/q$) and is measured in volts (V). One volt is one joule of energy per coulomb of charge. The energy required to move a charge in a uniform electric field is also proportional to the distance moved. It can be shown that the difference in electric potential between two points a distance l apart in a region of space containing an electric field E is given by $V = El$.

To have continuous movement of charges, as in an electric circuit, energy must be continuously supplied by a device such as an electrochemical cell, or battery of such cells (which convert chemical energy to electric energy), or an electric generator (which converts mechanical energy to electric energy). Such a device is called a *seat of electromotive force* (EMF). The EMF of a battery or a generator is equal to the energy supplied per unit of charge and is measured in volts. The rate at which energy U is supplied is called *power* ($P = dU/dt$); it is measured in watts (W). One watt is one joule per second.

Electric Current and Resistors

The movement of electric charges is *electric current*. The current i is numerically equal to the rate of flow of charge q ($i = dq/dt$) and

is measured in amperes (A). One ampere is one coulomb per second. A *conductor* is a substance that has free charges that can be induced to move when an electric field is applied. For example, a salt solution contains sodium (Na^+) and chloride (Cl^-) ions; when this solution is immersed in an electric field, the Na^+ ions move in the direction of the field, while the Cl^- ions move in the opposite direction. The direction of flow of current is determined by the movement of positive charges and, hence, is in the direction of the applied electric field. A metal contains free electrons; when an electric field is applied, the electrons (being negatively charged) move in the direction opposite to the electric field, but by convention, the current is still taken to be in the direction of the field (i.e., opposite to the direction of charge movement). The current flowing in a conductor, for example, a wire, divided by the cross-sectional area A of that conductor, is called the *current density* (J), that is, $J = i/A$.

Movement of charges in an ordinary conductor is not completely free; there is friction, which is called *resistance*. Many conductors are *linear*, that is, the electric field that causes current flow is proportional to the current density in the conductor. The *resistivity* (ρ) of a substance determines how much current it will conduct for a given applied electric field and is numerically equal to the ratio of the electric field to the current density ($\rho = E/J$). Resistivity is a constant for any given substance. In contrast, the *resistance* R of an individual conductor (also called a *resistor* in this context), which is equal to the ratio of the potential difference (V) across the resistor to the current flow i in the resistor ($R = V/i$), depends on the geometry of the resistor as well as on the material from which it is made. Resistance is measured in volts per ampere, or *ohms* (Ω). The resistance R of a long cylindrical conductor of length l, cross-sectional area A, and resistivity ρ is given by $R = \rho l/A$.

Capacitors

A *capacitor* is a device for storing electric charge; it generally consists of two charged conductors separated by a dielectric (insulator). A capacitor has the property that the charge q stored is proportional to the potential difference V across the capacitor; $q = CV$, where C, the charge per unit potential, is called the *capacitance*. Capacitance is measured in coulombs per volt, or *farads* (F).

Coils (Inductors)

A *coil* (also known as an *inductor* or *electromagnet*) is a device that generates a magnetic field when a current flows in it. Physically, it consists of a coil of wire wrapped around a magnetic core, for example, a ferromagnetic substance such as iron, cobalt, or nickel. A coil has a property, called *inductance*, which is analogous to the mechanical property of inertia; it resists any change (either increase or decrease) in current flow. More precisely, a coil is capable of generating an EMF in response to any *change* in the current flowing in it; the direction of the EMF is always such as to oppose the current change, and numerically the potential difference V across a coil is proportional to the rate of change of the current i in the coil: $V = -L(di/dt)$, where L is the inductance of the coil and the minus sign indicates that the potential is in the direction opposite to the current change. Inductance is measured in volt-seconds per ampere, or *henrys* (H).

CIRCUIT ANALYSIS

A *circuit* is a closed loop or series of loops composed of circuit elements connected by conducting wires. Circuit elements include a source of EMF (i.e., power supply), resistors, capacitors, inductors, and transistors.

Kirchhoff's First Law

For any loop of a circuit, the energy per unit charge, or potential, imparted to the loop must equal the energy per unit charge dissipated; this follows the principle of the conservation of energy. Stated another way, if electric potential is to have any meaning, a given point can have only one value of potential at any given time. If we start at any point in a circuit and go around the circuit in either direction and algebraically add up the changes in potential that we encounter, we must arrive at the same potential when we return to the starting point. In other words, the algebraic sum of the changes in potential encountered in a complete loop of a circuit must be zero. This is Kirchhoff's first law.

Rules for Seats of Electromotive Force, Resistors, Capacitors, and Inductors

To apply Kirchhoff's first law to a circuit, the following rules must be used to determine the algebraic signs of the potentials across circuit components:

1. If a seat of EMF ε is traversed in the direction of the EMF (i.e., from the negative to the positive terminal), the change in potential is $+\varepsilon$; in the opposite direction, it is $-\varepsilon$.
2. If a resistor is traversed in the direction of the current, the change in potential is $-iR$; in the opposite direction, it is $+iR$.
3. If a capacitor is traversed in the direction of its positively charged plate to its negatively charged plate, the change in potential is $-q/C$; in the opposite direction, it is $+q/C$.
4. If an inductor is traversed in the direction of the current, the change in potential is $-L(di/dt)$; in the opposite direction, it is $+L(di/dt)$.

Figure 1–1 shows a simple circuit containing a source of EMF and a resistor, to which Kirchhoff's first law may be applied to determine the current flow, as follows:

$$-iR + \varepsilon = 0$$
$$\varepsilon = iR$$
$$i = \frac{\varepsilon}{R}$$

Thus, the current is given by the EMF divided by the resistance. A similar analysis can be made when a circuit contains multiple sources of EMF, or multiple resistors, connected in series. The effective net EMF is the algebraic sum of the individual EMFs. The

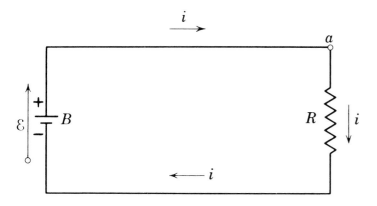

Figure 1–1. A simple electric circuit containing an electromotive force, ε, and a resistor, R. a = starting location for application of Kirchhoff's first law to the circuit; B = battery (seat of electromotive force); i = current. (From Halliday, D and Resnick, R: Physics, Part II, rev. ed 2. John Wiley & Sons, New York, 1962, p 790, with permission.)

effective net resistance is the sum of the individual resistances.

Kirchhoff's Second Law

A *node* is a junction point of two or more conductors in a circuit. The node cannot act as a repository of electric charge; therefore, the sum of all the current flowing into a node must equal the sum of all the current flowing out of the node (otherwise, charge would be constantly accumulating at the node). This is Kirchhoff's second law. By using this law for each node in the circuit together with the first law for each loop in the circuit, any arbitrary circuit problem, no matter how complex, can be expressed as a mathematical equation. Whether the equations can be solved and how difficult it is to

obtain a solution depend on the nature of the circuit and the elements it contains and on the mathematical skills available for the task.

Figure 1–2 shows a circuit containing a single source of EMF connected to three resistors in parallel. Kirchhoff's second law is applied to the junction of the three resistors and Kirchhoff's first law is applied to each branch of the circuit to give four independent equations in the four current variables i, i_1, i_2, and i_3, as follows:

$$i = i_1 + i_2 + i_3$$
$$-i_1 R_1 + \varepsilon = 0$$
$$-i_2 R_2 + \varepsilon = 0$$
$$-i_3 R_3 + \varepsilon = 0$$

These equations are solved for the four currents as follows:

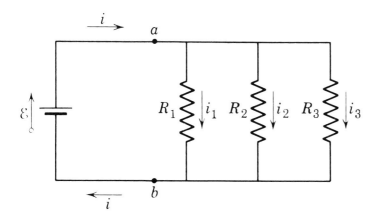

Figure 1–2. A circuit containing an EMF, ε, and three resistors (R_1, R_2, R_3) in parallel. a = junction point (node); b = junction point (node); i = current. (From Halliday, D and Resnick, R: Physics, Part II, rev. ed 2. John Wiley & Sons, New York, 1962, p 800, with permission.)

$$\varepsilon = i_1R_1 = i_2R_2 = i_3R_3$$

$$i_1 = \frac{\varepsilon}{R_1}$$

$$i_2 = \frac{\varepsilon}{R_2}$$

$$i_3 = \frac{\varepsilon}{R_3}$$

$$i = i_1 + i_2 + i_3$$

$$= \varepsilon\left(\frac{1}{R_1} + \frac{1}{R_2} + \frac{1}{R_3}\right)$$

Equivalent resistance R

$$= \frac{1}{1/R_1 + 1/R_2 + 1/R_3}$$

The net current i in the circuit can be calculated from Ohm's law, using an effective net resistance given as the reciprocal of the sum of the reciprocals of the three resistances in parallel.

RESISTIVE-CAPACITIVE AND RESISTIVE-INDUCTIVE CIRCUITS

Resistive-Capacitive Circuits and Time Constant

Figure 1–3 shows a resistive-capacitive (*RC*) circuit containing a single source of EMF connected to a resistor and capacitor in series. When the switch is placed in position *a*, the current flows in such a way as to charge the capacitor. Because of the presence of the resistor, the capacitor is not charged all at once but gradually over time. When Kirchhoff's first law is applied to this circuit and use is made of the fact that the current is charging the capacitor at the rate

$i = dq/dt$, a differential equation results. Its solution is an exponential curve, as follows:

$$\varepsilon - iR - \frac{q}{C} = 0$$

$$\varepsilon = Ri + \left(\frac{1}{C}\right)q$$

$$\varepsilon = R\left(\frac{dq}{dt}\right) + \left(\frac{1}{C}\right)q$$

$$q = C\varepsilon(1 - e^{-t/RC})$$

$$i = \left(\frac{\varepsilon}{R}\right)e^{-t/RC}$$

Figure 1–4A shows the exponential rise in the charge q on the capacitor, and Figure 1–4B shows that the current i (which is the slope of the curve representing q against time) falls exponentially to zero as the capacitor becomes fully charged. The *time constant* of the *RC* circuit is the time required for the current to fall to $1/e$ (37%) of its initial value; it is also the time needed for the charge to increase to $1 - 1/e$ (63%) of its final value; e is the base of the natural logarithm. The time constant is equal to *RC*.

When the switch shown in Figure 1–3 is placed in position *b*, the direction of current flow is reversed and the capacitor is discharged. Application of Kirchhoff's first law to this circuit yields a differential equation whose solution is an exponential, as follows:

$$-iR - \frac{q}{C} = 0$$

$$0 = Ri + \left(\frac{1}{C}\right)q$$

$$0 = R\left(\frac{dq}{dt}\right) + \left(\frac{1}{C}\right)q$$

$$q = q_0 e^{-t/RC}$$

$$i = -(q_0/RC)e^{-t/RC}$$

Figure 1–3. A circuit containing an EMF, ε, a resistor, R, and a capacitor, C. With the switch, S, in position a, the capacitor is charged. In position b, it is discharged. x = starting location for application of Kirchhoff's first law to the circuit. (From Halliday, D and Resnick, R: Physics, Part II, rev. ed 2. John Wiley & Sons, New York, 1962, p 802, with permission.)

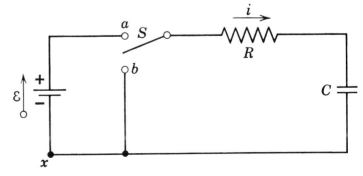

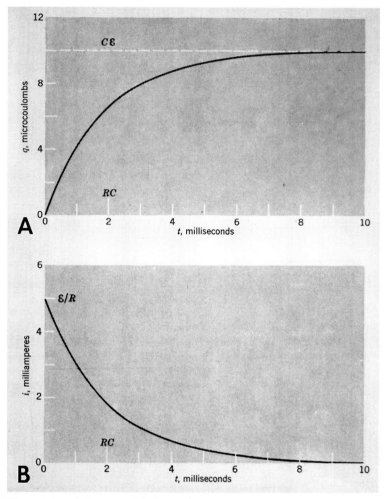

Figure 1–4. (*A*) The variation of charge, *q*, with time, *t*, during the charging process. (*B*) The variation of current, *i*, with time, *t*. For $R = 2,000 \ \Omega$, $C = 1.0 \ \mu f$, and $\varepsilon = 10$ V. (From Halliday, D and Resnick, R: Physics, Part II, rev. ed 2. John Wiley & Sons, New York, 1962, p 804, with permission.)

In this situation, both the charge on the capacitor and the current fall exponentially with time with the same time constant, *RC*. The current is negative in this case (i.e., it flows counterclockwise in the circuit as the capacitor is discharged).

Resistive-Inductive Circuits and Time Constant

A circuit containing a resistor *R* and an inductor *L* (an *RL* circuit) as well as a source

of EMF can be studied by similar methods. Because of the inductance, the current flow in this circuit does not rise immediately to its eventual value when a switch is closed but rather rises exponentially with a time constant (time to reach 63% of final value) given by *L/R*. This is due to the "inertial" effect of the inductor. Similarly, when a switch bypasses the source of EMF in this circuit, the current will not immediately drop to zero but will fall off exponentially in time with time constant *L/R* because of the effect of the inductor.

CIRCUITS CONTAINING INDUCTORS AND CAPACITORS

Inductive-Capacitive Circuits

Figure 1–5 shows an ideal circuit containing a capacitor and an inductor, an *inductive-capacitive, or LC, circuit.* The circuit is "ideal" because it contains no resistance. In stage (*a*), the capacitor is fully charged and there is no current flow. All the energy present in the circuit is stored as electric energy (U_E) in the capacitor. By stage (*b*), a current flow has partially discharged the capacitor but at the same time has created a magnetic field in the inductor. The total energy is split between electric energy in the capacitor and magnetic energy in the inductor (U_B). At stage (*c*), the current flow has reached its maximum and the capacitor is fully discharged. However, the inductance effect causes the current to continue to flow, now charging the capacitor in the opposite direction, as shown in stages (*d*) and (*e*). In stages (*f*) through (*h*), the scenario is repeated, with the capacitor discharging and the current flowing in the opposite direction. Finally, stage (*a*) is reached again and the entire cycle repeats. Thus, an *LC* circuit is an oscillator, and the charge on the capacitor is a cosine function of time; similarly, the current in the circuit is a sine function of time. This is, in fact, the solution to the equations that result from applying Kirchhoff's first law to the circuit, as follows:

$$\frac{-q}{C} - L\left(\frac{di}{dt}\right) = 0$$

$$L\left(\frac{d^2q}{dt^2}\right) + \left(\frac{1}{C}\right)q = 0$$

$$q = q_0 \cos{(2\pi ft + \phi)}$$

$$i = -2\pi f q_0 \sin{(2\pi ft + \phi)}$$

Figure 1–5. A simple LC circuit showing eight stages (*a* to *h*) in one cycle of oscillation. The bar graphs under each stage show the relative amounts of magnetic (U_B) and electric (U_E) energy stored in the circuit at any time. (From Halliday, D and Resnick, R: Physics, Part II, rev. ed 2. John Wiley & Sons, New York, 1962, p 944, with permission.)

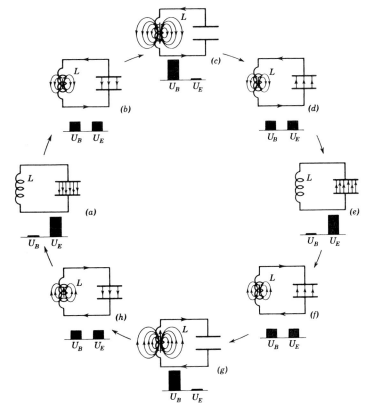

where the frequency f of oscillations is given by

$$f = \frac{1}{2\pi\sqrt{LC}}$$

The "time constant" of this circuit is $\sqrt{LC}$. The symbol ϕ represents a phase angle that determines how much of the initial energy at time $t = 0$ is electric and how much is magnetic; for the special case shown in Figure 1–5 (initially, all the energy is electric), $\phi = 0$.

Inductive-Resistive-Capacitive Circuits

A more realistic circuit is shown in Figure 1–6; it contains a resistor, a capacitor, and an inductor. It also contains a source of alternating current EMF (an AC generator). This is a device that generates an EMF $\varepsilon(t)$ that varies sinusoidally over time (which is exactly what the electric power company generators do) with a frequency of f (typically 60 Hz for line power in the United States). This circuit is governed by the following general equation:

$$-L\left(\frac{di}{dt}\right) - iR - \frac{q}{C} + \varepsilon(t) = 0$$

$$\varepsilon(t) = L\left(\frac{d^2q}{dt^2}\right) + R\left(\frac{dq}{dt}\right) + \left(\frac{1}{C}\right)q$$

If the AC generator is omitted, the circuit behaves as a damped oscillator. If the capacitor is initially charged up and then switched into the circuit, the current in the circuit varies sinusoidally over time, but its amplitude decays exponentially to zero because of power dissipation in the resistor.

If, on the other hand, the AC generator supplies energy continuously at a fixed frequency f, as expressed by the equation

$$\varepsilon(t) = \varepsilon_0 \sin(2\pi ft)$$

then the current flow in the circuit also varies sinusoidally with the frequency f of the driving EMF,

$$i(t) = i_0 \sin(2\pi ft + \phi)$$

although there generally is a phase shift ϕ (effectively, a time delay) between the current and the driving EMF. For such a circuit, it can be shown that the amplitude of the current flow i_0 is directly proportional to the amplitude of the driving EMF ε_0, that is, $i_0 = \varepsilon_0/Z$. This is analogous to the situation in a direct current (DC) circuit containing a battery and a resistor, in which current is proportional to the EMF of the battery ($i = \varepsilon/R$, Ohm's law). The quantity in the AC circuit that is analogous to resistance in a DC circuit is *impedance* (Z); like resistance, it is measured in ohms. Impedance is determined by all three circuit elements (L, C, and R); also, unlike DC resistance, impedance is a function of the frequency f of the AC generator.

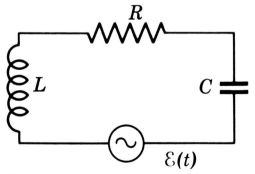

Figure 1–6. An AC circuit containing a generator, $\varepsilon(t)$, inductor, L, resistor, R, and capacitor, C. (From Halliday, D and Resnick, R: Physics, Part II, rev. ed 2. John Wiley & Sons, New York, 1962, p 952, with permission.)

Root-Mean-Square Potentials or Currents

Often, when measuring potentials and currents in AC circuits that vary sinusoidally with time, it is more convenient to deal with an average potential or current than with the amplitude (the peak positive or negative value) over a cycle. A simple average is not useful, however, because it is always zero; that is, the values are positive for half the cycle and negative for the other half. The most useful "average" quantity that can be used is the *root-mean-square* (rms) potential or current. This is defined as the square root of the average of the squares. Because the square of a quantity, whether negative or positive, is always positive, the rms value over a cycle is nonzero. It can be shown that the rms value of a sinusoidally varying quantity is equal to the amplitude divided by $\sqrt{2}$ (approximately 0.707 times the amplitude). When it is said that the "line voltage" for electric service in the United States is 120 V, an rms value is implied; the amplitude of the voltage variation is actually ± 170 V. Similarly, the rating of a fuse or circuit breaker is an rms current rather than a current amplitude.

Calculation of Reactance

In general, impedance is made up of three parts: the resistance (R), the reactance of the capacitor (X_C), and the reactance of the inductor (X_L). *Reactance*, which is measured in ohms, is the opposition that a capacitor or inductor offers to the flow of AC current; it is a function of frequency. The reactance of a capacitor is calculated as follows:

$$X_C = \frac{1}{2\pi f C}$$

Similarly, the reactance of an inductor is calculated as follows:

$$X_L = 2\pi f L$$

The reactance of a capacitor is least at high frequencies, becomes progressively greater at lower frequencies, and is infinite at zero frequency (DC), because an ideal capacitor uses a perfect insulator between the plates that is not capable of carrying any direct current. (The only reason a capacitor appears to conduct AC current is that AC current is constantly reversing direction; the capacitor in this case is merely being charged, discharged, and charged again in the opposite polarity [see Fig. 1–5].) The reactance of an inductor is zero at zero frequency (DC) and increases progressively with increasing frequency. This happens because the effect of an inductor is to oppose changes in current, and the more rapidly the current changes, the greater the induced EMF opposing that change will be.

Calculation of Impedance and the Phenomenon of Resonance

After the reactances and resistances have been calculated, impedance is calculated as follows:

$$Z = \sqrt{R^2 + (X_L - X_C)^2}$$

The right triangle shown in Figure 1–7 (with impedance Z being the hypotenuse) symbolizes in geometric form the relationship among impedance, resistance, and reactance. Note that the capacitive reactance X_C is *subtracted* from the inductive resistance X_L in calculating impedance. This, together with the frequency dependence of the reactances described previously, leads to an im-

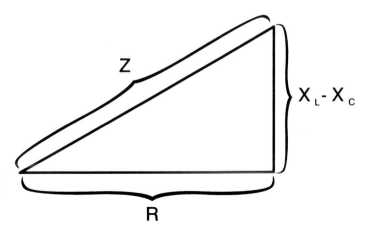

Figure 1–7. A right triangle symbolizing the relationship among resistance (R), inductive and capacitive reactance (X_L–X_C, and impedance (Z).

portant phenomenon in AC circuits: there is always one frequency at which the impedance of an *LRC* circuit is a minimum. This frequency may be calculated by setting $X_C = X_L$, because when this is true, impedance Z equals resistance R (the smallest possible value):

$$\frac{1}{2\pi fC} = 2\pi fL$$

$$(2\pi f)^2 = \frac{1}{LC}$$

$$f = \frac{1}{2\pi\sqrt{LC}}$$

Thus, the frequency at which impedance is a minimum is exactly equal to the frequency of oscillations of the *LC* or *LRC* circuit *without* an AC generator (see the frequency of the circuit shown in Figure 1–5). This can be restated as follows: when an *LRC* circuit is driven by an AC source of EMF, the largest current flow occurs when the frequency of the driving EMF is exactly equal to the natural, or resonant, frequency of the circuit. The current flow at driving frequen-

cies above or below the resonant frequency is less; that is, the impedance of the circuit is greater. This phenomenon is known as *resonance*; it is exploited in all tuner circuits to select a signal of one particular frequency (i.e., one broadcast station) and reject signals of all other frequencies. Similar circuits can be used as narrow-band-pass filters to eliminate all but a narrow range of frequencies from a signal or as notch filters to eliminate a narrow range of frequencies from a signal.

FILTER CIRCUITS

High-Pass Filters

Figure 1–8*A* shows a simple high-pass (low-frequency) filter circuit, which consists of a capacitor *C* in series with and a resistor *R* in parallel with the output circuit. The input potential V_{in} is applied between the input terminal and ground, and the output potential V_{out} is developed across the resistor *R*. This circuit may be analyzed in two ways. First, it may be treated as an *RC* circuit (see

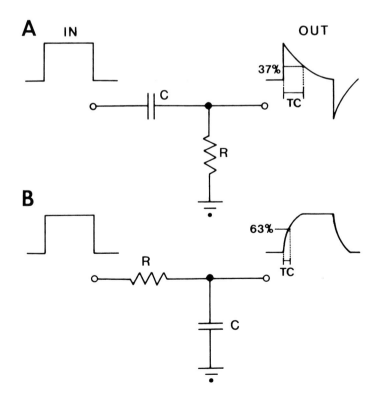

Figure 1–8. (*A*) A high-pass (low-frequency) filter circuit. (*B*) A low-pass (high-frequency) filter circuit. C = capacitor; R = resistor; TC = time constant. (From Fisch, BJ: Spehlmann's EEG Primer, rev. ed 2. Elsevier Science Publishers, Amsterdam, 1991, p 56, with permission.)

Fig. 1–3), with the input potential V_{in} taking the place of the battery EMF. The output potential V_{out} is proportional to the current i in the circuit. This decreases exponentially to zero with time constant (TC) equal to RC when the input potential is "turned on" and the capacitor charges, and it becomes negative and decreases exponentially when the input potential is "turned off" and the capacitor discharges (see Fig. 1–4). This accounts for the shape of the output in response to a square-wave calibration pulse.

Alternatively, one can imagine applying a sinusoidal AC potential to the input of this filter circuit. As demonstrated by the following equations, the current i is then equal to the input potential divided by the impedance Z of the circuit, which can be calculated from the resistance and capacitive reactance. With some algebraic manipulation, the ratio of the output to the input potentials can be calculated as a function of frequency:

$$i = \frac{\varepsilon}{Z} = \frac{V_{in}}{Z}$$

$$Z = \sqrt{R^2 + X_C^2} = \sqrt{R^2 + \left(\frac{1}{2\pi fC}\right)^2}$$

$$V_{out} = iR = \frac{V_{in}R}{Z} = V_{in}\frac{R}{\sqrt{R^2 + \left(\frac{1}{2\pi fC}\right)^2}}$$

$$\frac{V_{out}}{V_{in}} = \frac{2\pi fRC}{\sqrt{(2\pi fRC)^2 + 1}}$$

From this formula, it is seen that the output is strongly attenuated at low frequencies (when f is near zero) but is essentially equal to the input at high frequencies (when f is large). The "cut-off" frequency f of the high-pass filter is usually specified as the frequency at which the attenuation factor V_{out}/V_{in} is $1/\sqrt{2}$, or 0.707; this occurs when $f = 1/2\pi RC$. Equivalently, the time constant of the filter is given by $1/2\pi f$, where f is the filter cut-off frequency.

Low-Pass Filters

Figure 1–8B shows a simple low-pass (high-frequency) filter circuit, which consists of a resistor R in series and a capacitor C in parallel with the output circuit. This is also an RC circuit, but the output potential V_{out} in this case is developed across the capacitor and is proportional to the charge on the capacitor. Comparison with Figure 1–4 shows that the output potential increases exponentially to a maximum with time constant equal to RC when the input potential is "turned on" and the capacitor charges, and it decreases exponentially when the input potential is "turned off" and the capacitor discharges. This accounts for the shape of the output in response to a square-wave calibration pulse. The time constant of a high-frequency filter is discussed less often than that of a low-frequency filter in practice, because it is much smaller, for example, only 2 milliseconds for a 70-Hz filter, and cannot be measured on electroencephalographic (EEG) or electromyographic (EMG) tracings made at standard paper speeds by visual inspection of the calibration pulse.

This filter circuit can also be analyzed in the context of a sinusoidal AC input potential. The output potential V_{out} is equal to the current flow times the capacitive reactance X_C (the equivalent of Ohm's law for a capacitor, with reactance substituting for resistance in an AC circuit), and the current i is equal to the input potential V_{in} divided by the total impedance Z, which is the same as before. By using these facts and performing some algebraic manipulation, the ratio of the output to the input potential can be calculated as a function of frequency, as follows:

$$i = \frac{\varepsilon}{Z} = \frac{V_{in}}{Z}$$

$$Z = \sqrt{R^2 + X_C^2} = \sqrt{R^2 + \left(\frac{1}{2\pi fC}\right)^2}$$

$$V_{out} = iX_C = \frac{V_{in}}{2\pi fCZ}$$

$$= \frac{V_{in}}{2\pi fC\sqrt{R^2 + \left(\frac{1}{2\pi fC}\right)^2}}$$

$$\frac{V_{out}}{V_{in}} = \frac{1}{\sqrt{(2\pi fRC)^2 + 1}}$$

From this formula, it is seen that the output is attenuated at high frequencies but becomes nearly equal to the input at low frequencies. The "cut-off" frequency f of the low-pass filter is usually specified as the frequency at which the attenuation factor V_{out}/V_{in} is $1/\sqrt{2}$, or 0.707; this occurs when $f = 1/2\pi RC$, as for the high-pass filter.

Note that the only essential difference between a high-pass filter and a low-pass filter is the source of the output potential (to be fed to the amplifier); the high-pass filter develops its output potential across the resistor R, whereas the low-pass filter develops its output potential across the capacitor C.

TRANSISTORS AND AMPLIFIERS

Semiconductors and Doping

Transistors are constructed of materials called *semiconductors*, which have resistivities intermediate between those of good conductors, such as metals, and insulators. Silicon and germanium are the most frequently used substances. They are very poor conductors when in pure form, but when "doped" with trace quantities of elements capable of acting as electron donors or acceptors, they become semiconductors. The resistivity of the semiconductor can be altered by controlling the doping process.

Doping the tetravalent base material, silicon or germanium, with a pentavalent element such as arsenic provides extra "free" electrons that can conduct an electric current. Because these electrons carry negative charges, the semiconductor that results is referred to as an *n-type semiconductor*. Alternatively, the base material can be doped with a trivalent element such as gallium. An ab-

sence of sufficient electrons to fill all the orbitals is the result; the unfilled, or electron-deficient, areas are called *holes* and behave as positive charges that are free to move and, thus, conduct a current. The resulting semiconductor is referred to as a *p-type semiconductor*. (What actually happens is that electrons from a neighboring atom move to fill in the hole, resulting in a hole moving to a new position.) Thus, n-type semiconductors have electrons available for conducting current, whereas p-type semiconductors have holes–potential spaces for electrons–available for conducting current.

Diodes and Rectification

A useful electronic device can be made when two or more dissimilar semiconductors are adjacent. When an n-type semiconductor slab is fused along one face with a p-type semiconductor, electrons diffuse from the n region to the p region, filling some of the empty holes of the p region, up to the point at which the relative attraction of the holes for electrons is exactly counterbalanced by the effect of the electric field set up between the regions by the migration of electrons. This leaves the p region with a net negative charge and the n region with a net positive charge. If such a device is connected in a circuit to a source of EMF with the positive potential applied at the p region, a process called *forward biasing*, the electric field across the junction is reduced and further migration of electrons from n to p occurs, which constitutes a current flowing (by convention) from the p to the n terminal (Fig. 1–9). However, if the positive potential is applied at the n region, a process called *reverse bias-*

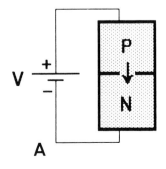

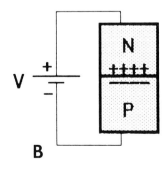

Figure 1–9. (*A*) A forward-biased pn semiconductor junction permits current flow. (*B*) A reverse-biased junction prevents current flow. V = Potential across the semiconductor junction. (From Misulis, KE: Basic electronics for clinical neurophysiology. J Clin Neurophysiol 6:41–74, 1989, with permission.)

ing, the electric field across the junction is actually increased and current flow is blocked, because in the region of the junction, the electrons that would carry current in the *n* region have been depleted and the holes that would carry current in the *p* region have been filled.

This type of device is called a *diode*; it allows current flow only in one direction. If a sinusoidal AC potential is applied to such a device, *rectification* results—the current flow is pulsatile but constrained to a single direction. This is the first step that occurs in power supply circuits that convert AC line voltage to DC voltages suitable for use in most electronic circuits.

Transistors and Amplification

A *transistor* is a device that controls the transfer of electrical charge across a resistor. The most common type are junction bipolar transistors, which can be made in two forms, called "npn" and "pnp." Both types are composed of a three-layer sandwich of semiconductors of different types. The names npn and pnp refer to the types of semiconductors and to their order in the sandwich; for example, npn transistors have a positively doped material in the middle layer and negatively doped materials in the outer layers. In an npn transistor, the primary movement of electrons is from one of the outer layers, the *emitter*, which is connected to the negative terminal of an external battery or power supply, to the other outer layer, the *collector*, which is connected to the positive terminal, through the middle layer, the *base*. Under resting conditions, little current flow occurs because although the emitter-base np junction is forward-biased and can conduct current, the base-collector pn junction is reverse-biased and blocks current (Fig. 1–10). However, if a small positive potential is applied to the base through an external connection, the junctional electric field at the base-collector pn junction is reduced, because some electrons entering the p-type base are allowed to leave through the external connection, preventing the filling of holes in the base. This allows a continuous movement of electrons from the emitter through the base into the collector. The middle, or base, layer is very thin, so that even a small positive potential applied to the base is sufficient to facilitate a large conductance between collector and emitter. Thus, a small controlling voltage across base and emitter governs the flow of a much larger current through the collector-to-emitter circuit. This, in effect, is the basis of an *electrical amplifier*.

The gain, or amplification, of a single transistor is limited; therefore, multiple stages of amplification are used in an EEG or EMG machine. For example, 6 stages of amplification, each with a gain of 10, give an overall amplification factor of 1,000,000. A complete EEG or EMG amplifier contains preamplification stages that typically boost the signal from the patient by a factor of 10 to 1000, followed by low-frequency, high-frequency, and 60-Hz notch filter circuits, fol-

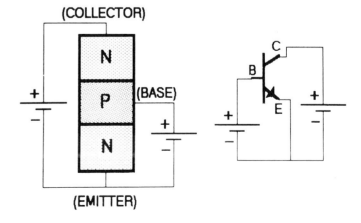

Figure 1–10. An npn junction bipolar transistor showing the potential applied between the emitter, *E*, and base, *B*, that controls the flow of current between the emitter and collector, *C*. The right-hand figure shows the circuit, using the conventional symbol for the transistor. (From Misulis, KE: Basic electronics for clinical neurophysiology. J Clin Neurophysiol 6:41–74, 1989, with permission.)

lowed by driver amplification stages that provide the remaining amplification and produce an output signal capable of driving the oscilloscope display or the pen motors of the paper display unit.

Differential Amplifiers

The type of amplifier used predominantly in clinical neurophysiology is the *differential amplifier*. This type of amplifier is constructed to amplify only the *difference* in the potential between its two inputs. This is one way of reducing contamination of the physiologic signal by electrical noise, for example, 60-Hz noise from line voltage devices, because this noise tends to be the same at all electrode positions and cancels out when a difference in potential is formed. On the other hand, physiologic signals are usually different at different electrode positions.

A differential amplifier is actually composed of two amplifier circuits (two transistors), one for the $G1$ input and one for the $G2$ input (the nomenclature $G1$ and $G2$ is still commonly used, even though it originated in the early days of EEG and EMG

when vacuum tube amplifiers were used; "G" was used to indicate a "grid" in the vacuum tubes). The $G2$ transistor is connected to a negative power supply voltage, and the $G1$ transistor is connected to a positive power supply voltage. The outputs of both transistors are connected (Fig. 1–11). In this fashion, increased current flow in the emitter-to-collector circuit of the $G1$ transistor produces a positive change in the output potential, whereas increased current flow in the emitter-to-collector circuit of the $G2$ transistor produces a negative change in the output potential.

Although an ideal differential amplifier would be sensitive only to the difference in potential between the two inputs, in practice, a large enough signal applied to both inputs simultaneously, called *common mode*, produces a small output signal. This occurs because the input impedance and the gain and frequency response of the two transistors in the differential amplifier are not perfectly matched. The common mode rejection ratio (CMRR) of a differential amplifier can be calculated as the applied common input potential divided by the output potential. For modern amplifiers, this ratio is

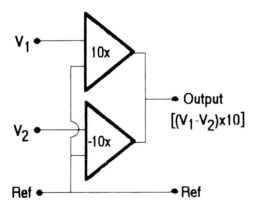

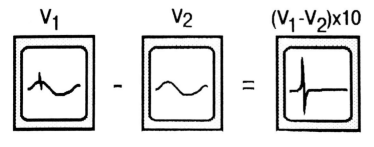

Figure 1–11. A differential amplifier constructed from two single-ended amplifiers, with input potentials V_1 and V_2 with respect to the reference (Ref), the second of which produces an inverted output; the net output is given by the product of the amplifier gain (10 in this case) and the difference in potential between the two inputs. (From Misulis, KE: Basic electronics for clinical neurophysiology. J Clin Neurophysiol 6:41–74, 1989, with permission.)

about 10,000. However, if the impedances of the electrodes placed on the patient are high or differ significantly between the $G1$ and $G2$ inputs, the effective signal perceived by the two transistors in the differential amplifier can differ significantly and the CMRR can be drastically reduced, thus allowing significant amounts of noise to contaminate the signals being recorded.

Even though the differential amplifier output depends on the difference in the potentials at its two inputs, each input potential as measured by the amplifier is relative to a common reference or ground potential. This ground potential is, in practice, equal to the potential at a single ground electrode that must be attached to the patient. For example, the ground electrode for EEG recording is typically placed on the mastoid process; occasionally, other locations, such as the frontal area, are used. As long as the electrode-to-patient connections are adequate (i.e., have low enough impedance), the location of the ground electrode does not matter. Because each input of the differential amplifier receives a potential that is relative to the same ground and because these potentials are subtracted in the output, the potential of the ground electrode cancels out. However, if there is a very poor electrode connection (i.e., one with high impedance), or in the extreme case, if one electrode is left unconnected, the differential amplifier input effectively becomes the ground electrode potential. In addition to introducing more 60-Hz and other noise into the recorded signal, artifacts and mislocalization of cerebral electrical activity can result by the unexpected introduction of a signal coming from an EEG ground electrode into one or more channels.

CHAPTER SUMMARY

This chapter reviews the principles of electrical and electronic circuits that are important to clinical neurophysiology. Knowledge of these principles and how to solve simple circuit problems is necessary for a complete understanding of the proper operation of equipment used in clinical neurophysiology and of the terminology and specifications given in equipment manuals.

CHAPTER 2

ELECTRICAL SAFETY IN THE LABORATORY AND HOSPITAL

Terrence D. Lagerlund, M.D., Ph.D.

ELECTRIC POWER DISTRIBUTION SYSTEMS

The electrical systems of buildings are designed to distribute electric energy from one central point of entrance to all the electrical appliances and receptacles. Power companies provide electric energy at high voltage (typically 4800 V for a hospital or medical clinic) to minimize transmission losses. A step-down transformer converts the high-voltage energy to safer, usable voltages (usually 120 V and 240 V). Figure 2–1 shows the wiring of a typical 120-V circuit. The secondary coil of the transformer has a center tap that acts as the return path ("neutral") for the circuit; it is connected to earth ground through a grounding stake at the transformer site. Each of the two outer ends of the 240-V secondary coil can be used to drive one 120-V circuit; this provides a "hot line" whose potential is 120 V (rms) from ground. This hot line incorporates a circuit breaker that limits current flow to a level (e.g., 20 A) that will not cause excessive heating in wiring in the building. For reasons explained later, each receptacle also includes a ground contact connected to earth ground through a conductor separate from the "neutral" conductor (see Fig. 2–1, wherein this separate conductor is the metal conduit that houses the wiring).

ELECTRIC SHOCK

Electric shock is the consequence of the flow of current through the body. The effect

18

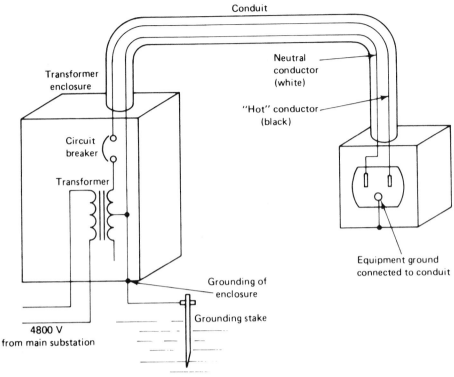

Figure 2–1. Scheme of a building's electric power distribution system showing a step-down transformer, circuit breaker, grounding stake, and equipment ground carried by metal conduit. (From Cromwell, L, et al: Biomedical Instrumentation and Measurements. Prentice-Hall, Englewood Cliffs, NJ, 1973, p 378, with permission.)

of electric shock depends both on the magnitude of the current flow and the path taken by the current in the body, which is determined by the points of entry and exit.

Requirements for Electric Current to Flow Through the Body

When an externally generated current flows through the body, it has a point of entrance and a point of exit. The current may be thought of as originating from some electrical apparatus, or source, flowing through a conducting material from the apparatus to the body, flowing through the body, and finally flowing through a second conducting material from the body to "ground." Thus, to have an electric shock, there must be at least two connections to the body: one to the current source and the other to ground. An apparatus can act as a source of current ei-

ther (1) because a point of connection between it and the body, such as an exposed metal part of the chassis or other metal contacts or terminals, is in direct continuity with the hot line through a very low resistance path caused by some fault such as a mechanical break in insulation or fluid spilled into the circuit or inadvertent direct connection of electrode lead wires to energized, detachable power-line-cord plugs,[1] or, more commonly, (2) because of a low-level leakage of current through a moderate resistance path, which may be inherent in the design of the apparatus. A further requirement for significant electric shock is that the entire pathway to, through, and out of the body must have a sufficiently low resistance.

An additional requirement for a *lethal* electric shock is that the current must take a path through the body that includes the heart (e.g., when current enters through one arm and exits through the other), because the mechanism of lethal shock is almost invariably the induction of ventricular fibrillation.

Physiologic Effect of Electric Current

For currents that enter and leave the body through the skin, the usual situation outside of hospitals, Figure 2–2 shows the approximate amounts of current associated with various physiologic effects, ranging from minimal perception (500 μA) to severe burns and physical injury (>10 A). Because current flow in a limb leads to involuntary muscle contraction through direct depolarization of muscle fibers, a victim of electric shock may not be able to let go of the source

of the current when it exceeds about 25 mA. The threshold for induction of ventricular fibrillation is approximately 100 mA. The externally applied current spreads out as it passes through the body, so that the fraction passing through the heart is small, less than 0.1%.

In hospitals, one of the two required contacts between an external source or ground and the body may be an intracardiac catheter. If current enters or leaves through this device, essentially the entire current flows through the myocardium. In this case, the threshold for inducing ventricular fibrillation is far less than for externally applied current. In humans, this threshold is estimated to be about 180 μA, but experiments in dogs have shown that as little as 20 μA is

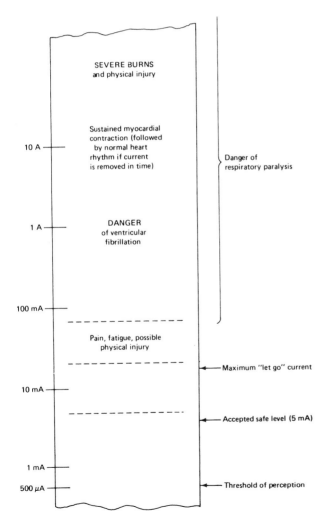

Figure 2–2. Effects of 60-Hz AC electrical current flow of various magnitudes produced by a 1-second external contact with the body. (From Cromwell, L, et al: Biomedical Instrumentation and Measurements. Prentice-Hall, Englewood Cliffs, NJ, 1973, p 374, with permission.)

sufficient.[4] The threshold may be significantly lower in persons with preexisting heart disease.

Factors Reducing Risk of Electric Shock

The risk of electric shock or electrocution from appliances is reduced by several factors, including the following:
1. Leakage currents that are available from most electrical appliances are relatively small.
2. People using appliances are often not connected to ground.
3. Contacts with the source of leakage current and with ground usually have high resistance, for example, dry, intact skin.
4. Healthy, alert people can withdraw from a source of current in most cases.
5. The hearts of healthy people require significant electric currents to induce ventricular fibrillation.

Factors Increasing Risk of Electric Shock in Hospitals

The risk of electric shock or electrocution from appliances in hospitalized patients is significantly greater because of the following factors:
1. Leakage currents that may be available from appliances are relatively large because patients may be attached to many instruments (thus providing multiple current sources), conducting fluids may get into instruments through spillage or leakage, and instruments may be used by many persons or used in many locations (or both), thus increasing the chance of fault due to misuse or wear. In the operating room, instruments such as electrosurgical units may present special risks to the patient if proper precautions for electrical safety are not followed.[2,3]
2. Through attached electrical instruments, patients are often grounded or they may easily contact grounded ob-

jects, for example, metal parts of beds, lamps, and instrument cases.
3. Contacts with the source of leakage current and with ground are often low resistance, because connections to monitoring devices purposely minimize skin resistance (e.g., electrodes applied with conducting paste) or bypass it altogether (e.g., indwelling catheters).

   Furthermore, patients with conductive intracardiac catheters, such as pacemaker leads and saline-filled catheters, have a direct low-resistance pathway to the heart. Because only tiny currents flowing in such a path may induce lethal ventricular fibrillation, such patients are called "electrically susceptible."
4. Weakened or comatose patients cannot withdraw from a source of current.
5. Patients' hearts may be more susceptible (through disease) to electric-current-induced ventricular fibrillation.

LEAKAGE CURRENT

Origin

Leakage current in an electrical apparatus may originate in several ways, including the following:
1. There is always a finite internal circuit resistance between the power line (hot wire) and the instrument chassis, known as "instrument ground"; this may be increased by faults in the wiring or by breakdown of insulation. A resistance as large as 5 MΩ still allows 24 μA to flow between the "hot" conductor and ground, which may be enough to induce ventricular fibrillation in an electrically susceptible patient.
2. The capacitance between the "hot" conductor and the chassis resulting from internal circuitry or external cabling may provide a relatively low-impedance pathway for alternating current. A capacitance as small as 440 picofarad (pF) still allows 20 μA to flow between the "hot" conductor and ground.

3. The inductive coupling between power-line circuits and other circuit loops, such as ground loops when there are multiple ground connections to the patient, can induce ground-path current flow as well.

In addition to the leakage currents available from equipment-to-patient ground connections, leakage currents may be introduced by similar mechanisms into other leads or connections to the patient.

Methods by Which Leakage Current Reaches Patients

Leakage currents may reach patients when contact is made either directly or through another person to exposed metal parts or to the chassis of electrical equipment. Leakage currents may also reach patients through a direct connection of the chassis ("equipment ground") to the patient; such a "patient ground" connection is necessary for noise-free recording of physiologic signals. Finally, leakage currents may reach patients through resistive or capacitive (or possibly inductive) coupling to leads other than the patient ground.

Methods to Reduce Leakage Current Reaching Patients

Many methods are used in modern hospital electrical distribution systems and in biomedical instruments to decrease the risk of electric shock by reducing the available leakage currents, including the following:

1. The chassis and all exposed metal parts of electrical appliances are grounded through a separate ground wire, through the round pin of electric plugs. Any leakage currents that would otherwise flow to a subject in contact with the chassis are instead shunted through this low-resistance pathway to ground. Because the leakage currents in properly functioning equipment are small, the ground wire in the building power distribution system usually carries very little current, unlike the neutral wire, which carries the full operating current. Hence, the potential drop between the equipment chassis and the earth ground connection located at the electrical distribution panel of the building is minimal, and the equipment ground potential remains very close to the earth ground potential (Fig. 2–3).

2. Hospital rooms that have exposed metal parts, for example, window frames, bathroom plumbing, shelving, and door frames, may also connect these to earth ground through the same grounding system used for electrical outlets. All such grounded points in one room should be connected to a single ground wire, an "equipotential grounding system." Also, all biomedical equipment connected to a patient should draw power from the same

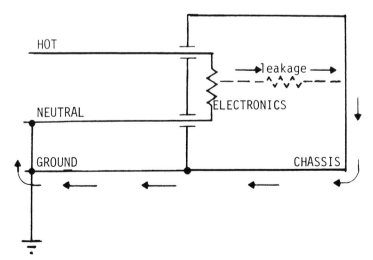

Figure 2–3. The power cord ground wire conducts leakage current from an electric apparatus chassis to earth ground. (From Seaba, P: Electrical safety. American Journal of EEG Technology 20:5, 1980, with permission.)

Figure 2–4. A faulty floor polisher produces excessive leakage current, which flows through the building ground wiring. A portion of this current also flows through the large ground loop involving the patient and the two electrical devices connected to the patient (i.e., an ECG monitor attached to one leg and an EEG machine). This ground loop was created by the use of two outlets physically distant from each other. (From Seaba, P: Electrical safety. American Journal of EEG Technology 20:10, 1980, with permission.)

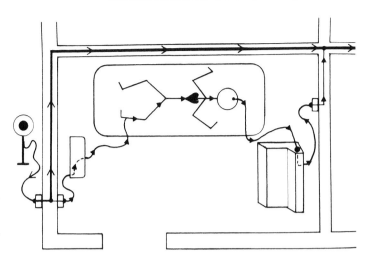

group of outlets to avoid large "ground loops" (Fig. 2–4).

3. When necessary, isolation transformers may be used to eliminate the neutral-to-ground connection entirely, thereby reducing the risk of shock when a patient connected to a biomedical instru-ment comes in contact with an earth ground (Fig. 2–5). However, isolation transformers do not entirely eliminate the risk of shock, because they themselves may have significant leakage currents; furthermore, they are expensive, and also generate heat, noise, and elec-

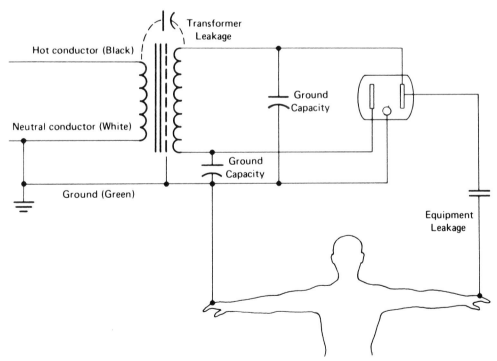

Figure 2–5. An isolation transformer used to reduce equipment leakage current. The equipment is connected to the secondary coil of the transformer, which is electrically isolated from the power line hot and neutral conductors. (From Cromwell, L, et al: Biomedical Instrumentation and Measurements. Prentice-Hall, Englewood Cliffs, NJ, 1973, p 387, with permission.)

trical interference. Therefore, their use is usually limited to special settings, such as operating rooms.

4. Appliances can be constructed with nonmetallic cases to minimize the chance of patients contacting the equipment chassis.

5. Appliances should have short line cords, and the use of extension cords should be avoided to minimize capacitive and resistive leakage currents between the hot and the ground wires. Note that each foot length of cord unavoidably introduces about 1 μA of leakage current into the ground connection.

6. If at all possible, direct connections of patients to ground should be avoided.

In particular, inadvertent electrical paths between ground and patients that bypass the normally high skin resistance (especially paths provided by intracardiac catheters) should be avoided by using nonconductive materials.

7. Current that may flow through all connections between a patient and the equipment, both signal and ground, should be limited to no more than 20 μA through the use of current limiters (Fig. 2–6) or inductive or optical coupling devices (Fig. 2–7). Alternatively, when practical, battery-powered equipment that has no direct connection to line voltage or to ground can be used.

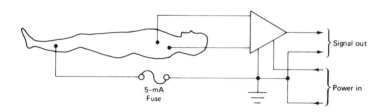

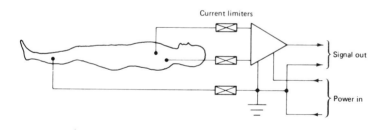

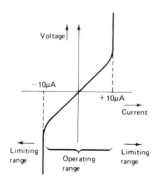

Figure 2–6. Current-limiting devices. (*Top*) Circuit used in older biomedical equipment includes a 5-mA fuse in the patient ground connection; current flow through this one connection is limited to 5 mA, which is insufficient protection for electrically susceptible patients. (*Middle*) Circuits in modern biomedical equipment use special current limiters. (*Bottom*) Electrical characteristics of current limiters. In the operating range, current is proportional to voltage (Ohm's law), but currents outside this cannot exceed 10μA. (From Cromwell, L, et al: Biomedical Instrumentation and Measurements. Prentice-Hall, Englewood Cliffs, NJ, 1973, p 384, with permission.)

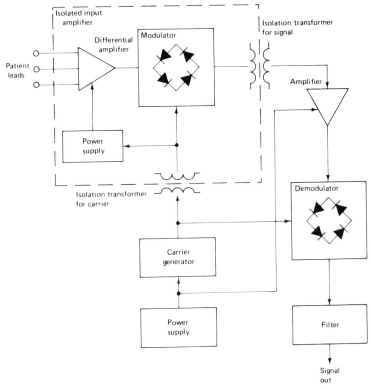

Figure 2–7. The input circuit of an inductively isolated biomedical instrument. Isolated patient leads are achieved by using a carrier wave amplifier with transformer coupling. (From Cromwell, L, et al: Biomedical Instrumentation and Measurements. Prentice-Hall, Englewood Cliffs, NJ, 1973, p 385, with permission.)

ELECTRICAL SAFETY PRINCIPLES AND IMPLEMENTATION

Equipment Grounding

Proper grounding of electrical equipment (i.e., providing a low-resistance pathway from the equipment chassis to an earth ground) is usually accomplished through the grounding wire in the line cord that connects to the round pin in the plug and thence to the building electrical grounding system. Failure of this ground connection may occur in several ways. There can be a failure of attachment of the grounding wire in the line cord to the equipment chassis, a break in continuity of the grounding wire within the cord, or a failure of connection of the grounding wire to the grounding pin. Also, the grounding pin may make poor contact with the wall receptacle because of a re-

duction in contact tension caused by mechanical wear. The grounding pin can also be deliberately bypassed using a so-called cheater (3-prong to 2-prong) adapter. Defects also can occur in building wiring, such as an improper or omitted connection of the wall receptacle's grounding terminal to a ground wire or an interruption of the ground connection somewhere in the building's wiring, especially if the metal conduit, which is subject to corrosion and loss of mechanical contact, provides the ground connection. Particularly in newly constructed or remodeled rooms and buildings, it is advisable to visually inspect and to electrically test the ground connection in all wall receptacles. Because the ground connection is only for electrical safety purposes, the lack of it in no way affects operation of the electrical equipment and, therefore, will remain undetected if not specifically checked.

Tests for Equipment Grounding and Leakage Current

Each hospital, laboratory, or clinic should establish an electrical safety program that includes selecting equipment that meets appropriate safety standards, testing new equipment after purchase to verify that standards are met, inspecting and retesting equipment periodically thereafter to ensure that damage through use and misuse does not compromise safety, educating all those who use the equipment (especially technicians) in electrical safety principles, and ensuring that certain basic minimal safety tests are performed each time a biomedical apparatus is plugged in, turned on, and connected to a patient.

Tests that should be performed on building wiring at the time of installation include the following:

1. Visually inspect the wiring of all wall receptacles to ensure that it is correct.

2. Measure the resistance between each wall receptacle ground (and other grounded objects in the room) and a ground known to be adequate, such as a cold water pipe or an independent grounding bus. This resistance should be less than 0.1 Ω.

3. Measure the contact tension provided by the wall receptacle, that is, the force required to withdraw the ground pin of a test plug from the receptacle. This should be at least 10 ounces.

These tests, except the first, should also be performed periodically, for example, every 6 to 12 months, and the receptacles whose contact tension has degraded below 10 ounces should be replaced.

Tests that should be performed on each biomedical instrument at the time of purchase and periodically thereafter, for example, every 6 to 12 months, include the following:

1. Visually inspect the line cord and plug for signs of damage, wear, or breakage.

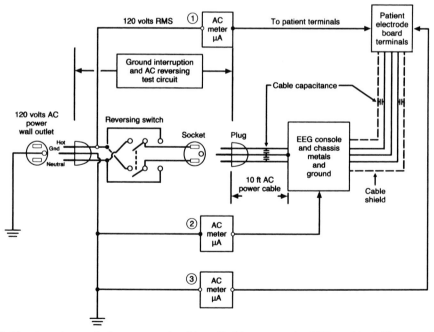

Figure 2–8. Circuit used to test leakage current in a biomedical instrument (an EEG machine). The reversing switch allows the test to be made with both standard and reverse polarity of line voltage. The three leakage currents measured are (1) from the power line hot wire to each patient terminal; (2) from the equipment chassis to earth ground; and (3) from each patient terminal to earth ground. (From Grass, ER: Electrical Safety Specifically Related to EEG (Bulletin X757C78). Grass Instrument Company, Quincy, MA, 1978, p 9, with permission.)

2. Measure the resistance between the ground pin of the plug and the instrument chassis. This should be less than 0.1 Ω.
3. Measure the chassis-to-earth ground leakage current using the circuit shown in Figure 2–8. This measurement should be made with the equipment's grounding pin disconnected (to ensure safety even if the building grounding system is faulty) and under four separate conditions, including both normal and reverse polarity of the hot and neutral wires (to ensure safety even if the wall receptacle is erroneously wired with opposite polarity) and with the equipment power switch "on" and "off." In all four conditions, this current should not exceed 100 μA.
4. Measure the leakage current from each terminal that connects to a patient, including the patient ground to earth ground, under the same four conditions. This is the maximal leakage current that the equipment can supply *to* a patient who is grounded through a second connection. For use with electrically susceptible patients, this current should not exceed 20 μA.
5. Measure the leakage current from the power-line hot wire to each terminal that connects to a patient, including the patient ground, under the same four conditions. This is the maximal current that can be absorbed by the equipment *from* a patient who accidentally comes in contact with a 120-V power line. For use with electrically susceptible patients, this current should not exceed 20 μA.

Rules for Electrical Safety

In addition to a program of periodic testing and inspection, electrical safety requires that all persons using electrical equipment in the laboratory or hospital be familiar with the following rules:

1. Do not ever directly ground patients or allow patients to come in contact with grounded objects while connected to a biomedical instrument. Patient-ground connections should be made to only one instrument at a time; thus, if a patient is already connected to a device such as an ECG monitor, a separate EEG or EMG ground should be omitted unless it is absolutely necessary for obtaining a recording free of artifact and noise.
2. Ensure that every electrical device or appliance, for example, lamps, electric beds, electric shavers, and radios that a patient might accidentally come in contact with is connected to an adequate earth ground, such as through use of an approved three-prong grounded plug.
3. Use only safe, properly designed, and pretested electrical equipment. All biomedical devices directly connected to patients must have isolation or current-limiting circuits if they are to be used with electrically susceptible patients. All line-powered equipment should have a three-prong grounded plug. In general, patients should not be allowed to bring their own electrical appliances from home for use in a hospital room.
4. Ensure that all electrical equipment in use has had a safety inspection recently (within 6 to 12 months), as indicated by a dated electrical safety inspection tag or sticker.
5. Connect all patient-connected equipment to outlets in the same area or cluster to avoid large ground loops.
6. Never use an extension cord on patient-connected equipment, because this adds leakage current through its internal capacitance and resistance, and thus provides another chance for ground connection failure.
7. Cover all electrical connections to intracardiac catheters with insulation, such as a surgical rubber glove, to eliminate electrical continuity between external devices or ground and the catheter whenever possible.
8. Have a defibrillator available at all building locations where patients have cardiac catheters in place.
9. Do not ignore the occurrence of any electric shocks, however minor; inves-

tigate their cause. Thoroughly test any equipment that may have been involved before putting it back in service. Also, do not ignore any abnormal 60-Hz interference or artifact in an electrophysiologic recording; this finding may indicate that some device is leaking current into the patient.

10. Follow certain safety procedures, including routine safety checks, each time an electrical device is to be connected to a patient.

Electrical Safety Procedures for Technicians

The following procedures should be followed by technicians while performing an electrophysiologic test requiring line-powered equipment on a patient, especially portable studies performed in a patient's room:

1. Check the physical condition of the equipment. Is there any evidence of liquid spills, cord wear, or damage? Is the plug bent or broken? Is the equipment labeled with a current electrical safety inspection sticker?
2. Inspect the patient area for any two-wire ungrounded appliances. Have them unplugged and removed.
3. Inquire about any other instruments attached to the patient. Are they labeled with a current electrical safety inspection sticker? If there is already a patient ground connection to one of these other instruments, try recording without any other patient ground connection.
4. Choose an outlet in the same area or cluster used by other patient-connected devices. Before plugging in the equipment, check the contact tension

of the chosen receptacle with a simple device that should be carried with all portable equipment.

5. Turn on the instrument and calibrate it before connecting it to the patient. Major electrical problems may show up during calibration; furthermore, electrical surges occur as the instrument is turned on and leakage currents may be higher while it is warming up.
6. Disconnect the patient from the instrument before turning it off.

CHAPTER SUMMARY

This chapter reviews the principles of electrical safety that are relevant to clinical neurophysiologic studies. Knowledge of these principles is necessary both for those involved in evaluating and purchasing test instruments and those involved in maintaining and using them. The legal responsibility for patient safety, including electrical safety, is shared by all those who order, perform, interpret, or supervise electrophysiologic testing.

REFERENCES

1. Risk of electric shock from patient monitoring cables and electrode lead wires. Health Devices 22:301–303, 1993.
2. Litt, L and Ehrenwerth, J: Electrical safety in the operating room: Important old wine, disguised new bottles (editorial). Anesth Analg 78:417–419, 1994.
3. McNulty, SE, Cooper, M, and Staudt, S: Transmitted radiofrequency current through a flow directed pulmonary artery catheter. Anesth Analg 78:587–589, 1994.
4. Starmer, CF, McIntosh, HD, and Whalen, RE: Electrical hazards and cardiovascular function. N Engl J Med 284:181–186, 1971.

CHAPTER 3

VOLUME CONDUCTION

Terrence D. Lagerlund, M.D., Ph.D.

PRINCIPLES

Electrophysiologic studies involve recording potential differences from electrodes in contact with the body; usually, these electrodes are placed on the skin surface. The bioelectric potentials are generated by sources inside the body, for example, brain, peripheral nerve, and muscle, that may be some distance from the recording electrodes. These sources may be either active or passive. Active sources are ionic channels that open or close in response to changes in transmembrane potential, neurotransmitter binding, intracellular calcium, other second messengers, and so forth, allowing small currents to flow into or out of the cell body, axon, or dendrite. Passive sources are areas of neuronal membrane that permit current flow into or out of the cell by passive leakage or capacitive effects. These current sources (active or passive) lead to widespread extracellular currents that flow in the entire conducting medium surrounding the neurons, which is called the volume conductor. Some of the extracellular currents in the volume conductor reach the skin surface, where, according to Ohm's law, the current causes a potential drop across the space between two electrodes. This drop in potential can be detected and amplified by a differential amplifier. The properties of the volume conductor determine the potentials recorded by a given array of electrodes in response to a given set of generators.

In general, a volume conductor such as the body can be characterized by its conductivity (or resistivity) and its dielectric constant (capacitive properties), which may vary among tissues. This may be modeled by dividing the entire volume conductor into multiple small regions, each of which is assumed to be homogeneous; that is, the conductivity and dielectric constant are the same throughout. For many purposes, at those frequencies of interest in making neurophysiologic recordings, the effects of capacitive properties of the volume conductor are relatively small and may be ignored; if this is done, only a conductivity value for each region, together with the geometry of the region, is needed to characterize the volume conductor. In such a noncapacitive (purely resistive) volume conductor, the recorded potentials are always in phase, or synchronous, with the current sources, and the conductive properties of the medium are independent of frequency. For example, the relative attenuation of high-frequency source potentials is the same as that of low-frequency source potentials as they are conducted to the recording electrode. However, in the more general case of a resistive-capacitive medium, volume conduction is frequency dependent, and the source currents and recorded potentials are out of phase.

CALCULATING POTENTIALS IN INFINITE HOMOGENEOUS MEDIA

The simplest form of a volume conductor is a homogeneous medium without boundaries in which generators and recording electrodes are embedded. In this situation, the recorded potential can easily be calculated from the configuration of source currents.

Monopole, Dipole, and Quadrupole Sources

A single source, or sink, of current is referred to as a *monopole*. The potential relative to a distant reference at distance r from a monopole source in an infinite homogeneous medium of conductivity σ (resistivity $\rho = 1/\sigma$) is given by

$$V = \frac{I}{4\pi\sigma r}$$

where I is the magnitude of the monopolar current source. Thus, the potential of a monopole falls off inversely with distance from the source, and equipotential lines for a monopole form circles around the current source or sink (Fig. 3–1A).

Two adjacent current monopoles of opposite polarity constitute a current dipole. This is a more realistic generator than an isolated monopole, because the current emanating from the source can flow through the medium to the sink, where it is absorbed. In respect to the potentials they produce at distant recording sites, many neuronal current generators may be described in terms of a current dipole. The potential relative to a distant reference measured at distance r from a dipole source in an infinite homogeneous medium of conductivity σ is given by

$$V = \frac{Id(\cos\,\theta)}{4\pi\sigma r^2}$$

where I is the magnitude of the dipole current source, d is the pole separation, and θ is the angle between the dipole axis and the line from the dipole to the measurement point (Fig. 3–2A); this formula is valid only for $r \gg d$. Thus, the potential of a dipole falls off inversely with the square of the distance from the source. The lines of current flow around a dipole form curved paths (Fig. 3–2B). The equipotential surfaces are perpendicular to the lines of current flow and have a figure eight configuration around the dipole. The zero potential surface is a plane halfway between the two poles of the dipole, because on this plane $\theta = 90$ degrees and $\cos\theta = 0$ (Fig. 3–2B).

Two adjacent current dipoles of opposite orientation placed end-to-end constitute a current quadrupole. The potential relative to a distant reference measured at distance r from a quadrupole source in an infinite homogeneous medium of conductivity σ is given by

$$V = \frac{Id^2(3\,\cos^2\,\theta - 1)}{4\pi\sigma r^3}$$

where I is the magnitude of the dipole current source, d is the separation between the dipoles (and the separation between individual poles in each dipole), and θ is the angle between the axis of the two dipoles and the line from the quadrupole to the measurement point. This formula is valid only for $r \gg d$. Thus, the potential of a quadrupole falls off inversely with the cube of the distance from the source, and the equipotential

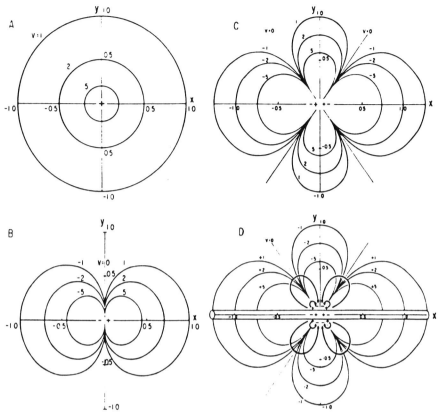

Figure 3–1. Equipotential lines in a volume conductor for various current source distributions. (*A*) A point source or monopole; (*B*) a dipole; (*C*) a quadrupole (two oppositely directed dipoles); and (*D*) an action potential source propagating along an axon. In (*D*) the arrows represent some lines of current flow. (From Stein,[4] p 81, with permission.)

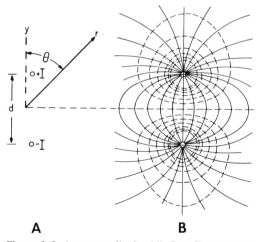

A   **B**

Figure 3–2. A current dipole. (*A*) Coordinate system showing definition of r and θ. (*B*) Lines of current flow (*solid*) and equipotential lines (*dashed*) in a volume conductor. (From Nunez,[2] p 118, with permission.)

surfaces around the quadrupole have a cloverleaf configuration (Fig. 3–1*C*). A quadrupole is a fair approximation of the potential generated by an action potential propagating along an axon, because the axonal membrane has a negative polarity outside and a positive polarity inside at the peak of the action potential. However, on either side of this peak, the membrane is positive outside and negative inside (Fig. 3–1*D*).

Spatial Distributions of Potentials From Various Sources

To understand the spatial distributions of potentials generated by various types of sources, the formulas already given can be used to calculate the potential recorded from successive points along a given line or axis in the volume conductor surrounding the source. For a monopole source, the potential recorded relative to a distant reference at points along a line at distance l from the source in an infinite medium of conductivity σ is

$$V = \frac{I}{4\pi\sigma(l^2 + x^2)^{1/2}}$$

where I is the source current magnitude and x is the distance along the line between the source and the recording point. The potential distribution is a single peak whose sharpness increases with decreasing distance l from the source (Fig. 3–3A).

For a dipole source, the potential re-

corded relative to a distant reference at points along a line *perpendicular* to the dipole axis in an infinite medium of conductivity σ is

$$V = \frac{Idl}{4\pi\sigma(l^2 + x^2)^{3/2}}$$

where I is the dipole magnitude, d is the pole separation, l is the distance from the dipole to the line, and x is the distance along the line between the dipole axis and the recording point (Fig. 3–4A). This situation, for example, applies to scalp potentials produced by a radially oriented cortical dipole generator. The potential distribution is a single peak whose sharpness increases with decreasing distance l from the source (Fig. 3–5A).

For the same dipole source, the potential difference recorded between two closely spaced electrodes, that is, bipolar recording,

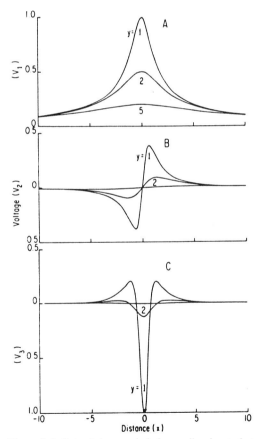

Figure 3–3. Potentials recorded along a line located at various distances from a current source (1, 2, and 5 cm) as a function of position along the line: for (*A*) a current monopole, (*B*) a dipole, and (*C*) an action potential source. (From Stein,[4] p 84, with permission.)

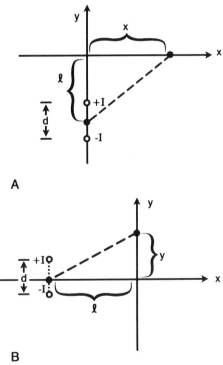

Figure 3–4. (*A*) Coordinate system for potentials recorded along a line *perpendicular* to the axis of a current dipole source. (*B*) Coordinate system for potentials recorded along a line *parallel* to the axis of a current dipole source.

at points along a line *perpendicular* to the dipole axis is

$$\Delta V = \left(\frac{dV}{dx}\right)\Delta x$$

where Δx is the electrode separation and dV/dx (the derivative of V with respect to x) is the slope of the V versus x curve. The potential distribution obtained with this recording method is biphasic (Fig. 3–5B).

For a dipole source, the potential recorded relative to a distant reference at points along a line *parallel* to the dipole axis in an infinite medium of conductivity σ is

$$V = \frac{Idy}{4\pi\sigma(l^2 + y^2)^{3/2}}$$

where I is the dipole magnitude, d is the pole separation, l is the distance from the dipole to the line, and y is the distance along the line between the dipole center and the recording point (Fig. 3–4B). This situation, for example, applies to scalp potentials produced by a tangentially oriented cortical dipole generator. The potential distribution is biphasic, and its sharpness increases with decreasing distance l from the source (Fig. 3–3B and 3–5C).

For the same dipole source, the potential difference recorded between two closely spaced electrodes, that is, bipolar recording,

at points along a line parallel to the dipole axis is

$$\Delta V = \left(\frac{dV}{dy}\right)\Delta y$$

where Δy is the electrode separation and dV/dy (the derivative of V with respect to y) represents the slope of the V versus y curve. The potential distribution obtained using this recording method is triphasic (Fig. 3–5D).

For a quadrupole source, the potential recorded relative to a distant reference at points along a line *parallel* to the quadrupole axis in an infinite medium of conductivity σ is

$$V = \frac{Id^2(2y^2 - l^2)}{4\pi\sigma(l^2 + y^2)^{5/2}}$$

where I is the quadrupole magnitude, d is the pole separation, l is the distance from the quadrupole to the line, and y is the distance along the line between the quadrupole center and the recording point. This situation approximates the potential expected from an action potential propagating along an axon parallel to the line of the recording electrodes.[1,4] The potential distribution is triphasic, and its sharpness increases with decreasing distance l from the source (Fig. 3–3C).

Figure 3–5. Potentials recorded along a line located at various distances from a current dipole (*solid curve*, 1 cm; *dashed curve*, 2 cm; *dotted curve*, 3 cm) as a function of position along the line. (*A*) Referential recording, with the line perpendicular to the dipole axis; (*B*) bipolar recording for line perpendicular to dipole axis; (*C*) referential recording, with the line parallel to the dipole axis; (*D*) bipolar recording for line parallel to dipole axis.

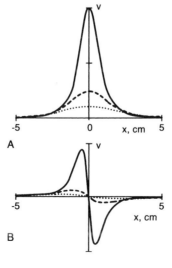

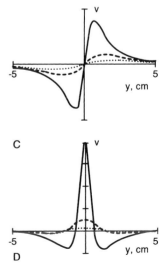

POTENTIALS IN NONHOMOGENEOUS MEDIA

Planar Interfaces and Hemi-Infinite Homogeneous Media

When a monopolar source is located in a volume conductor divided into two hemi-infinite regions of differing conductivity with a planar interface between them, configurations of the lines of current flow change at the interface. This occurs because the current density (current per unit area) flowing in the direction parallel to the interface is less in the region of higher conductivity. Consequently, if the source is located in the region of lower conductivity, the lines of current flow bend outward as they enter the region of higher conductivity (Fig. 3–6A). If the source is located in the region of higher conductivity, the lines of current flow bend inward as they enter the region of lower conductivity (Fig. 3–6B).

One of the major inhomogeneities in the body as a volume conductor is the interface between the body and the external environment. For sources located close to the skin surface, such as action potentials propagating in fairly superficial nerves or electroencephalographic (EEG) activity generated by cortical sources about 2 to 3 cm deep, this interface can be approximated as a plane, with a region of high conductivity on one side containing the embedded source and a region of essentially zero conductivity on the other side. In this situation, no current flow penetrates from the high conductivity region to the zero conductivity region; thus, the lines of current flow are completely reflected at the interface. Consequently, potentials measured at the interface, that is, surface-recorded potentials, due to underlying generators, are twice as large as they would be for the same generators and recording site in an infinite homogeneous medium (Fig. 3–7A).

Homogeneous Sphere Model

For sources located in the head, such as cortical and subcortical generators of EEG and evoked potentials, a spherical volume conductor model is a reasonable approxi-

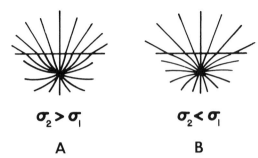

$\sigma_2 > \sigma_1$ $\qquad$ $\sigma_2 < \sigma_1$

A $\qquad\qquad$ **B**

Figure 3–6. Lines of current flow due to a monopolar source below a planar interface when (A) the conductivity of the upper region is larger than that of the lower region ($\sigma_2 > \sigma_1$), or when (B) the conductivity of the upper region is less than that of the lower region ($\sigma_2 < \sigma_1$). (From Nunez,[2] p 145, with permission.)

mation of the actual geometry. The simplest model assumes a uniform conductivity σ within a sphere of radius a, and zero conductivity outside the sphere. For a dipole source located at the *center* of the sphere, such as a brain stem generator of one of the short latency auditory evoked potential peaks, the potential at a point inside the sphere located at distance r from the dipole and at an angle θ from the dipole axis is

$$V = \frac{Id(\cos \theta)}{4\pi\sigma r^2} \left(1 + \frac{2r^3}{a^3} \right)$$

where I is the magnitude of the dipole and d is the pole separation. At points near the center of the sphere ($r \cong 0$), the potential is the same as that expected in an infinite homogeneous medium, but at the surface of the sphere, that is, for scalp surface recordings, it is three times as great (Fig. 3–7D). This is because the lines of current flow are confined to the spherical volume and the current density (and hence electric potential) is correspondingly greater.

Multiplanar and Multiple Sphere Models

A single air-body interface surface is only one of the inhomogeneities that affect volume conduction. For EEG and scalp-recorded evoked potentials, other important inhomogeneities are the differing conductivities of brain, cerebrospinal fluid (CSF), skull, and scalp. To investigate the effects of

Figure 3–7. The approximate effects on measured surface potentials due to a dipole source of various inhomogeneities in a volume conductor. ϕ_H is the potential that would have been recorded at the same point were the same source dipole immersed in an infinite homogeneous medium; ϕ is the potential actually recorded in the inhomogeneous medium. (*A–C*) are based on a planar interface and apply to superficial, or cortical, dipole sources: (*A*) body-air interface; (*B*) brain-CSF and CSF-skull interfaces; (*C*) brain-CSF, CSF-skull, skull-scalp, and scalp-air interfaces. (*D*) and (*E*) are based on spherical interfaces and apply to deep, or brain stem, dipole sources: (*D*) body-air interface; (*E*) brain-skull, skull-scalp, and scalp-air interfaces. CSF = cerebrospinal fluid. (From Nunez,[2] p 174, with permission.)

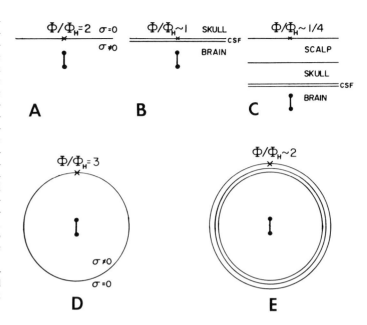

these regions, multiplanar and multiple sphere models have been used.

For dipole sources located in cortex and for subdural recording electrodes, a model using two planar interfaces (brain-CSF and CSF-skull) can be used. This model predicts that the measured potentials would be approximately equal to those that would be recorded in an infinite homogeneous medium (Fig. 3–7B). For cortical dipole sources with scalp surface recording electrodes, a model using five regions (brain, CSF, skull, scalp, and air) and four planar interfaces predicts that the measured potentials would be approximately equal to one quarter of those that would be recorded in an infinite homogeneous medium (Fig. 3–7C). This may be compared with the factor-of-2 augmentation of potentials predicted by the single planar interface model (see Fig. 3–7A); the predicted relative attenuation (by a factor of 8) is mainly due to the effects of the poorly conducting skull, whose conductivity is only about 1/80 that of the brain or scalp.

For deep dipole sources, for example, brain stem auditory evoked potential generators, a multiple sphere model with four regions (brain, skull, scalp, and air), three spherical interfaces, and a dipole in the center is appropriate. For scalp surface recording electrodes, this model predicts that the measured potentials would be approxi-

mately equal to twice those that would be recorded in an infinite homogeneous medium (Fig. 3–7E). This may be compared with the factor-of-3 augmentation of potentials predicted by the homogeneous sphere model; the predicted relative attenuation (by a factor of 2/3) due to the poorly conducting skull is not nearly as great for deep sources as for superficial, or cortical, generators.[2]

APPLICATIONS OF VOLUME CONDUCTION PRINCIPLES

"Stationary" Potentials Produced by Propagating Generators

When recording from various electrodes placed at different locations along the path of a complex propagating generator, such as an action potential source, the time at which the propagating potential is seen at each location is different, because of the finite velocity of propagation of the generator. (Note that volume conduction of electric potentials from the generator to the recording electrode is essentially instantaneous, because electric disturbances propagate at the speed of light in a conducting medium.) However, when a propagating generator passes through an interface between volume-conducting regions of different sizes or conductivities, a potential can be induced *simultaneously* at all recording electrodes during the time when the generator is crossing the boundaries between regions with differing properties. Such a potential, which does not appear at different times in different recording locations, has been referred to as a *stationary potential* (Fig. 3–8). This effect may be observed in recordings of somatosensory evoked potentials as a consequence of the change in geometry of the volume conductor as a propagating nerve impulse travels from a limb to the trunk. The same effect may also influence the morphology of a brain-stem auditory evoked potential recording, because of changes in volume conductor properties along the central auditory pathways caused by the complicated anatomy of the posterior fossa. Such a stationary potential can only be seen in derivations in which the first and second electrodes between which the potential is measured are on opposite sides of the boundary between regions with different sizes or conductivities; generally, this only occurs when recording with respect to a relatively distant reference electrode.[3]

EEG Applications

In recording scalp EEG activity, it is desirable to measure the potentials at each scalp electrode position with respect to a distant, totally "inactive" reference electrode. In fact, it is not possible to find an inactive ref-

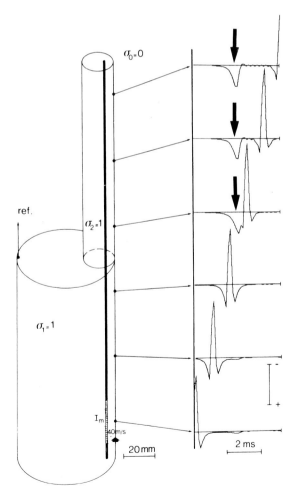

Figure 3–8. Action potential propagation from a larger to a smaller volume conductor region, modeled here as two joined cylinders of unequal diameter but equal conductivities. The expected potentials as a function of time at various recording positions at the surface of the cylinders are shown, recorded with respect to a reference electrode, *ref.*, on the larger cylinder, for an action potential propagating upward at 40 m/s. Recording locations on the surface of the smaller cylinder, above the junction, show a "stationary" potential (*arrows*) at the time corresponding to the propagating potential traversing the junction region. (From Stegeman, et al.[3] p 181, with permission.)

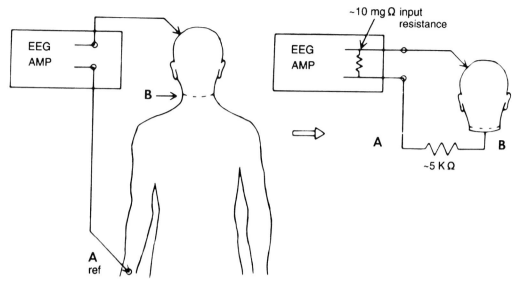

Figure 3–9. The futility of using a distant reference for scalp EEG recording. (*A*) The right arm reference is electrically equivalent to a (*B*) neck reference, except for a slight additional resistance in series. (From Nunez,[2] p 181, with permission.)

erence. Even if a physically distant reference position is chosen, such as on a limb, volume conduction between the head and the distant position will make the reference "active," that is, the reference electrode will be electrically equivalent to a reference electrode placed at the neck (still relatively close to intracranial generators), with a slight additional resistance that is negligible in its effects because of the large input resistance of the EEG amplifier (Fig. 3–9). In addition, such a distant reference would have unacceptable characteristics in that large electrical artifacts, for example, those produced by the electrocardiogram, movement, and muscle, will probably be seen in the recording.

Properties of the volume-conducting medium between intracranial generators and scalp electrodes can have a major effect on the recorded potentials. When the poorly conducting skull is breached by openings, for example, naturally occurring openings such as the orbits or external auditory meati or iatrogenic openings, such as craniotomy defects, long current paths through the openings may cause appreciable electrical potentials to be recorded in areas that are, in fact, far from the generators (Fig. 3–10). Amplitude asymmetries, that is, differences between homologous regions on the oppo-site side of the head, are the most commonly observed effects of skull defects, with higher amplitudes occurring on the side of the opening. A regional increase in the thickness of the conducting medium between

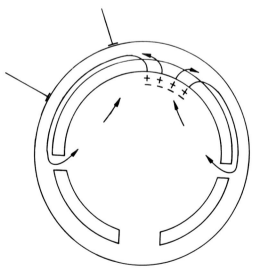

Figure 3–10. The effect of skull openings on scalp recorded potentials; the large skull resistivity—80 times that of scalp or brain—leads to long current paths through skull openings that may cause appreciable potentials to be recorded far from the generators. (From Nunez,[2] p 189, with permission.)

intracranial generators and overlying electrodes may lead to significant focal attenuation of electrical activity, as in the case of a subdural hematoma or fluid collection. For *extracranial* generators such as the eyes, which have a constant electric dipole moment that induces changing electrical potentials when they move, the effect of skull openings is reversed, that is, the amplitude of potentials due to these generators is usually attenuated in the region of a large skull defect.

Dipole Source Localization

In many clinical neurophysiologic studies, particularly EEG and evoked potential studies, conducted for clinical or research purposes, localization of the generators of a particular waveform or activity is of paramount importance. Volume conductor theory, as discussed in this chapter, provides a way to calculate the surface potential distribution that will result from a known configuration of intracranial generators, the *forward problem*. However, it is also desirable to have a way to determine the type, location, strength, and orientation of all generators of a given scalp surface-recorded waveform or activity, the *inverse problem*. Unfortunately, for any given potential distribution recorded from the scalp surface, the number of possible configurations of generators that could

equally well produce that distribution is infinite; in general, the inverse problem does not have a unique solution. If the problem can be constrained by independent anatomic or physiologic data, however, then a solution to the inverse problem may be possible.[5] For example, if there is reason to believe that a particular EEG waveform or evoked potential peak is generated by a single intracranial source that may be well represented as a single electric dipole, then a model of the volume conductor properties of the head, such as the three-sphere model discussed previously, can be coupled with an appropriate mathematical algorithm to find the location, orientation, and strength of the single dipole whose predicted scalp-potential distribution best fits the observed scalp potential.

The inherent uncertainties in the geometry and electrical properties of the volume conductor limit the accuracy with which dipole localization based on scalp-recorded potentials can be performed. A new technique for improving source localization derives from the magnetic field that any electric current generates. Thus, intracranial current sources generate magnetic fields that, with appropriate sensing devices, may be detected at various locations outside the head. These magnetic fields, unlike scalp-recorded potentials, are unaffected by the intervening medium, and calculations of source locations from magnetic field maps

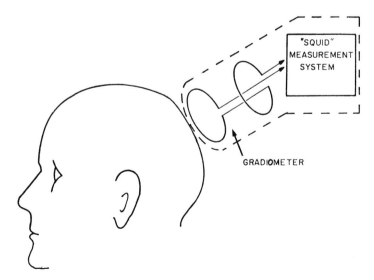

Figure 3–11. A magnetoencephalographic recording device consists of a magnetic field gradiometer (two coils with opposite polarities connected in series) and a superconducting quantum interference device (SQUID) amplifier. (From Nunez,[2] p 211, with permission.)

thus may be performed without a need for complex and possibly inaccurate volume conduction methods. With a sensitive magnetic detector, for example, a magnetometer or magnetic gradiometer, and a special type of high-gain, low-noise amplifier, specifically, a superconducting quantum interference device, or SQUID, it is possible to record the magnetic equivalent of the EEG or evoked potentials which is known as the *magnetoencephalogram* (MEG) or the *evoked magnetic fields* (EMF). Although this technique is still largely confined to research rather than clinical applications, it has allowed intracranial generators to be localized accurately and noninvasively (Fig. 3–11).

CHAPTER SUMMARY

This chapter reviews the principles of volume conduction as applied to the potentials recorded in clinical neurophysiologic studies. Knowledge of these principles is necessary for a proper interpretation of neurophysiologic recordings in order to extract information concerning the function and location of the neural structures that generate the recorded activity or waveforms.

REFERENCES

1. Dumitru, D and DeLisa, JA: AAEM minimonograph 10: Volume conduction. Muscle Nerve 14:605–624, 1991.
2. Nunez, PL: Electric Fields of the Brain: The Neurophysics of EEG. Oxford University Press, New York, 1981.
3. Stegeman, DF, Van Oosterom, A, and Colon, EJ: Farfield evoked potential components induced by a propagating generator: Computational evidence. Electroencephalogr Clin Neurophysiol 67:176–187, 1987.
4. Stein, RB: Nerve and Muscle: Membranes, Cells, and Systems. Plenum Press, New York, 1980, pp 65–86.
5. Van Oosterom, A: History and evolution of methods for solving the inverse problem. J Clin Neurophysiol 8:371–380, 1991.

CHAPTER 4

DIGITAL SIGNAL PROCESSING

Terrence D. Lagerlund, M.D., Ph.D.

DIGITAL COMPUTERS IN CLINICAL NEUROPHYSIOLOGY

Utility

Digital computers can perform types of signal processing that are not readily available with ordinary electrical circuits, which are analog devices. Because of their large storage capacities and rapid, random-access retrieval, digital computers can make the process of obtaining, storing, retrieving, and viewing clinical neurophysiologic data easier. Also, because of their sophisticated computational abilities, they may aid in extracting information from waveforms that is not readily obtainable using visual analysis alone. Furthermore, they are well suited for quantification of key features of waveforms. This may be useful in accurate clinical diagnosis of electroencephalographic (EEG), electromyographic (EMG), and evoked potential studies, and it also lends itself to serial comparisons between studies performed on the same subject at different times or between two subject groups in scientific investigations. They may also partially automate the interpretation of clinical neurophysiology studies. This chapter discusses uses of digital signal processing and storage that are common to many types of physiologic studies.

Construction of Digital Systems

A digital (computerized) system for acquisition, storage, and display of physiologic waveforms has the following key components: (1) electrodes, (2) amplifiers and filters, (3) analog-to-digital converters (ADCs), (4) solid state digital memory, (5) digital processor (central processing unit), (6) magnetic or optical disk (or tape) storage, and (7) screen or printer for waveform display.

The electrodes, amplifiers, and filters in a digital system are essentially identical to those in an all-analog system. The amplified signal for each channel is sent to an ADC, which converts it by the process of *digitization* to digital form and stores it in solid state memory. A digital processor is capable of moving digital data around in memory and of processing or manipulating them. It may also send data to a permanent magnetic, optical disk, or a tape storage medium, or it may generate displays of waveforms and related textual annotations on a monitor screen or a computer printer.

DIGITIZATION

Principles

Electrical signals derived from an electrode or from some other type of transducer may be used to represent electrical or nonelectrical physiologic quantities (such as potential in μV, current in mA, pressure in millimeters of mercury, or oxygen saturation in percentage) in one of two ways. An *analog* signal takes on any potential (voltage) within a specific range (e.g., -3 to $+3$ V); the potential is generally directly proportional to the physiologic quantity represented by the signal so that potential is an *analog* of the physiologic quantity. Analog signals are for the most part *continuous* in the sense that the

potential varies continuously as a function of time. In contrast, a single digital signal may take on only one of two possible potentials (e.g., 0 or 3 V); such a signal may represent one of two possible states ("on" or "off"; "yes" or "no") or one of two possible digits ("0" or "1") and is said to represent one *bit* (*bi*nary dig*it*) of information. Multiple digital signals may be used to represent a physiologic quantity as a binary number (a series of 0s and 1s forming a quantity in a base 2 number system; that is, the rightmost digit has a value of $2^0 = 1$, the second digit from the right has a value of $2^1 = 2$, the third digit has a value of $2^2 = 4$, and so on). Digital signals are *discrete* and *discontinuous*, that is, they have only two possible states, and the nearly instantaneous transition from one state to another is made only at specific times. This is the only format in which digital computers can store and process information, and it is most suited to performing complex and accurate arithmetic operations (e.g., adding, subtracting, multiplying, dividing) or logical operations (logical conjunction, disjunction, negation). Analog representations are more suited for human interpretation, for example, a waveform display generally uses vertical displacement as an analog to the physiologic quantity, such as the potential, being displayed and horizontal displacement as an analog to elapsed time.

Analog-to-Digital Conversion

Digitization, or analog-to-digital conversion, is the process by which analog signals are converted to digital signals. It is the transformation of *continuous* potential changes in an analog signal representing a physiologic quantity to a sequence of discrete digital numbers (binary integers). Digitization is performed by a complex circuit known as an *analog-to-digital converter* (ADC). There are two aspects to digitization, *quantization* and *sampling*.

Quantization

The term quantization describes the assignment of a digital number to the instantaneous value of the potential input to the ADC. A simple example is shown in Figure 4–1, which shows a 4-bit ADC, whose input is an analog signal in the range of 0 to 16 V and whose output is a 4-digit binary number that can take on the values 0, 1, 2, . . . , 15. In this example, any input potential between 12 and 13 V will result in the same output (12); thus, the resolution of the ADC (also known as the quantum size) is 1 V. The input range of the ADC is 0 to 16 V.

In general, parameters described by the following three terms characterize quantization:

1. *Quantum size* (ADC resolution)—this determines the minimum potential change that can be detected by the ADC and corresponds to a change of 1 in the least significant bit (2^0). A typical value might be 1 mV (for the *amplified* signal reaching the ADC).

2. *Number of bits in ADC* (n)—this determines the range of digitized (output) values. For an ADC that can accept positive or negative inputs, one bit is required for sign (+ or −), and the fractional resolution is then 1 part in 2^{n-1}. A typical value might be 9 to 12 bits (corresponding to ± 1 part in 256 to 1 part in 2048).

3. *Input range*—this determines the maximum and minimum input potentials. Input potentials above or below the maximum/minimum are called *overflow* or *underflow*, respectively. A typical value might be ± 2 V.

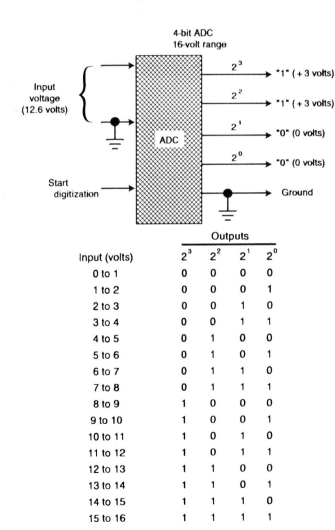

Input (volts)	2^3	2^2	2^1	2^0
0 to 1	0	0	0	0
1 to 2	0	0	0	1
2 to 3	0	0	1	0
3 to 4	0	0	1	1
4 to 5	0	1	0	0
5 to 6	0	1	0	1
6 to 7	0	1	1	0
7 to 8	0	1	1	1
8 to 9	1	0	0	0
9 to 10	1	0	0	1
10 to 11	1	0	1	0
11 to 12	1	0	1	1
12 to 13	1	1	0	0
13 to 14	1	1	0	1
14 to 15	1	1	1	0
15 to 16	1	1	1	1

Figure 4–1. Scheme of a four-bit analog-to-digital converter (ADC). Inputs consist of a continuous signal to be digitized and a start digitization pulse from a clock used to initiate digitization at appropriate times. Outputs consist of four digital signals (+3 or 0 V representing "1" and "0") that together can encode a four-bit integer (range 0 to 16).

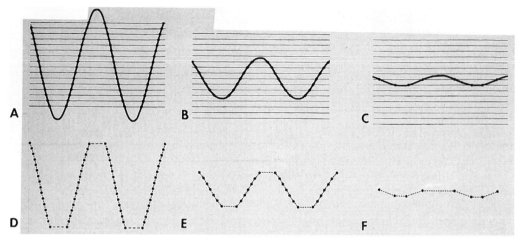

Figure 4–2. Effect of quantization parameters, analog-to-digital converter resolution, and input range on the fidelity with which an analog signal can be represented digitally. In (*A*) the signal exceeds the input range and its digital representation (*D*) is clipped. In (*B*) the signal uses more than 50% of the input range and is relatively well represented (*E*). In (*C*) the signal uses less than 15% of the input range, and because of the limited analog-to-digital converter resolution, it is poorly represented (*F*). (From Spehlmann,[4] p 46, with permission.)

The three quantization parameters are related by the formula

$$\text{Input range} = \pm \text{ADC resolution} \times (2^{n-1} - 1)$$

where n = number of bits. Note that the input range of the ADC should match the expected range of amplified potentials as closely as possible. If the range of a signal exceeds the range of the ADC, the ADC will either overflow or underflow, and the resulting signal will be distorted ("clipped"), but if the range of a signal is too small compared with the ADC range, much of the resolution of the ADC is wasted and the effective resolution may be insufficient, again significantly distorting the signal (Fig. 4–2).

Sampling

In digitization, the conversion of the continuous analog signal to digital form is usually performed at discrete equidistant time intervals. The following two terms characterize sampling:

1. *Sampling interval*—this determines the temporal resolution of the digitizer. A typical value may range from 0.01 millisecond (for brain-stem auditory-evoked potentials) to 5 milliseconds or more (for EEG).
2. *Sampling frequency*—this is the reciprocal of the sampling interval and is measured in hertz (s^{-1}).

In addition to determining the temporal resolution of the digitizer, the sampling frequency determines the maximum frequency in the signal to be digitized that can be adequately represented. The *sampling theorem*, or *Nyquist theorem*, states that if a signal contains component frequencies ranging from 0 to f_N, then the minimum sampling frequency that can be used for the digitized data to adequately represent the frequency content of the original signal is $2f_N$, where f_N is the Nyquist frequency. The Nyquist frequency can be calculated from the sampling interval as $f_N = 1/(2 \times \text{sampling interval})$. For example, if $f_N = 50$ Hz, then the sampling frequency must be at least 100 Hz (sampling interval of 0.01 seconds or less). This sampling frequency is the *minimum* necessary to avoid *gross* distortion of the input signal; a larger sampling frequency (by a factor of 3 to 5) may be necessary in many applications to achieve adequate resolution of fine details in the waveforms being digitized.

Aliasing

Sampling at a frequency lower than $2f_N$ produces *aliasing*. Aliasing is distortion of a signal due to "folding" of frequency components in the signal *higher* than f_N onto lower frequencies. For example, a sine wave of 75 Hz, if sampled at 100 Hz, will appear in the digitized data as a sine wave of frequency 25 Hz and not 75 Hz. *Aliasing must always be avoided* or else the digitized data will grossly misrepresent the true signal. In practice, aliasing is avoided by *filtering* the input signal before digitization to remove all frequencies above the Nyquist frequency (Fig. 4–3).

For example, if the sampling interval in use is 5 milliseconds, the Nyquist frequency is 100 Hz. A 70-Hz low-pass filter with 6 dB per octave slope would attenuate frequencies of 100 Hz to 0.57 times their original amplitude, which may not be enough. A 50-Hz low-pass filter with 12 dB per octave slope will attenuate frequencies of 100 Hz to 0.2 times their original amplitude, which may be sufficient to prevent significant contamination of the digitized signal by aliased frequency components, provided that the initial amplitude of the faster components in the original signal is relatively small.

COMMON USES OF DIGITAL PROCESSING

The most common use of digital signal processing in clinical neurophysiology is signal averaging, particularly in evoked-potential and sensory-nerve conduction studies. Averaging may also be applied to repetitive transient waveforms and event-related potentials (such as movement-associated potentials). A second major use of digital signal processing is in digital filtering. Less common, but still important, uses are found in time/frequency analysis, including interval and Fourier (spectral) analysis, autocorrelation analysis, statistical analysis, and automated pattern recognition. Other uses tend to be more specialized for particular types of clinical neurophysiologic studies; some of these are discussed elsewhere in this book.

AVERAGING

Evoked Potentials and Nerve Conduction Studies

Digital averaging devices for nerve conduction studies and evoked potentials are used routinely in clinical neurophysiology.

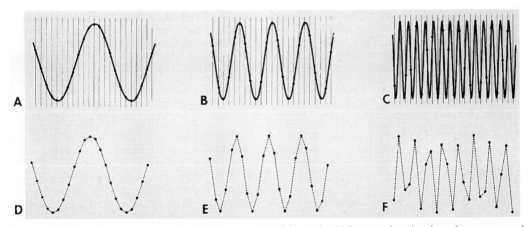

Figure 4–3. Effect of sampling interval and aliasing on the fidelity with which an analog signal can be represented digitally. In (*A*) the sampling frequency is 14 times that of the signal frequency, and the signal is well represented (*D*). In (*B*) the sampling frequency is only 6 times the signal frequency, and the representation is less accurate but still acceptable (*E*). In (*C*) the sampling frequency is only 1.5 times the signal frequency and therefore less than the Nyquist frequency; the consequent aliasing causes the digital representation (*F*), to be entirely misleading in that it appears to have a frequency that is about half the true frequency. (From Spehlmann,[4] p 44, with permission.)

Their function is similar regardless of the type of signal averaged, although for different types of studies, the epoch length for averaging differs significantly. Epoch lengths between 200 and 500 milliseconds are typical for visual and long-latency auditory-evoked potentials. Epoch lengths between 30 and 100 milliseconds are typical for middle-latency auditory evoked potentials and for nerve conduction studies. Epoch lengths between 10 and 20 milliseconds are typical of brain stem auditory evoked potentials and electrocochleograms.

Basic operation of an averager is shown in Figure 4–4. After each stimulus, the input signal is digitized at multiple discrete sampling times within a fixed-length epoch that begins with the stimulus. Digitized values of potential at each discrete sampling time— each characterized by its latency, or time after the stimulus—are averaged for many stimuli; the resulting averaged signal may be displayed on a screen or printed out. The stimulus-dependent portions of the signal, that is, the evoked potential or nerve action potential, are similar in amplitude and latency in each epoch averaged and appear in the averaged result, whereas the stimulus-independent, or random, portions of the signal (noise, background neuronal activity, among others) differ substantially from epoch to epoch and are suppressed by averaging. The suppression factor, which is often called the *signal-to-noise ratio,* for truly random signals is $\sqrt{n}$, where n is the number of epochs averaged.[4] For example, achieving a signal-to-noise ratio of 20 requires averaging 400 epochs. The required signal-to-noise ratio and, hence, the number of epochs depend on the type of signal being averaged and the amount of background activity, or noise. For example, typical brain-stem auditory-evoked potentials are about 0.5 μV in amplitude, whereas background EEG activity may be 50 μV or more, requiring a signal-to-noise ratio of 100 (10,000 epochs averaged). In contrast, sensory nerve action potentials are typically 10 μV or more in amplitude, with noise that is comparable, requiring a signal-to-noise ratio of only 2 to 3 (4 to 9 epochs averaged).

Repetitive Transient Waveforms

Repetitive transient waveforms that are not stimulus-related may also be averaged, such as epileptic spikes in an EEG or itera-

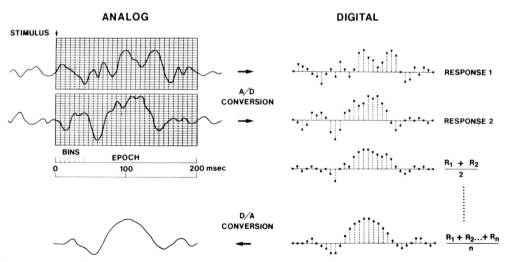

Figure 4–4. Operation of an averager. Analog signals recorded after each stimulus are digitized by an analog-to-digital converter during a fixed-length time window, or "epoch," after the stimulus. The resulting digital representations are totaled and divided by the number of epochs averaged. The digital result can be displayed by an analog device such as an oscilloscope after conversion from digital to analog form. A/D = analog-to-digital; D/A = digital-to-analog. (From Spehlmann,[4] p 37, with permission.)

tive EMG discharges. The epoch for averaging in this case is a time "window" around the waveform, for example, from a specified time before the peak of the waveform to a specified time after it. The waveforms to be averaged and the reference times defining the windows around them may be determined manually by positioning a cursor over the peak of each successive waveform to be averaged or automatically by using sophisticated transient detection programs capable of identifying all waveforms of interest and locating their peaks and onsets. After the epochs have been defined, averaging proceeds in the same way as for evoked potentials.

Movement-Associated Potentials

Movement-associated potentials—one class of event-related potentials—are a cerebral activity associated with, and generally preceding, voluntary or involuntary movement. They are obtained by simultaneously recording several EEG channels and one or more EMG channels, the latter to determine the time of occurrence and other characteristics of the movement. An EMG channel may act as a "trigger" for the averager, but the epoch for averaging usually begins *before* the onset of the muscle activity recorded by the EMG channel and may extend up to or beyond the time of the muscle activity. Thus, this type of averaging is often called *back-averaging*. It requires somewhat more sophisticated processing than ordinary *forward averaging*, because the signal being averaged must be digitized continuously and stored in memory so that when a "trigger" occurs, digitized data for the *previous* 0.5 to 1 seconds may be included in the average.

DIGITAL FILTERING

Types of Digital Filters

A *digital filter* is a computer program or algorithm that can remove unwanted frequency components from a signal.[3] As with analog filters, they may be classified as low-pass, high-pass, band-pass, or notch filters. Most digital filters function by forming a linear combination (weighted average) of signal amplitudes at the current time and various past times. The two types of commonly used digital filters are the *finite impulse response* (FIR) filter and the *infinite impulse response* (IIR) filter. The FIR filter output is a linear combination of only the input signal at the current time and past times. This type of filter has a property such that its output necessarily becomes zero within a finite amount of time after the input signal drops to zero. The IIR filter output is a linear combination of both the input signal at the current time and past times ("feedforward" data flow) and the output signal at past times ("feedback" data flow). This type of filter has a property such that its output may persist indefinitely in the absence of any further input, because the output signal itself is fed back into the filter. However, IIR filters can be unstable and also have the undesirable property of noise buildup, because noise terms created by arithmetic round-off errors are fed back into the filter and amplified. For these reasons, FIR filters are easier to design, although they often require less computation than FIR for comparable sharpness in their frequency responses and, hence, are often used for filtering of signals in "real time."

Characteristics of Digital Filters

Digital filters have several characteristics that distinguish them from analog filters. First, they can be constructed and modified easily, because they are software programs rather than hardware devices. Second, they can easily be designed to have relatively sharp frequency cut-offs if desired, for example, much sharper than the typical 6 dB/octave roll-off of an analog filter. Third, they need not introduce any time delay (phase shift) in the signal, as invariably happens with ordinary analog filters; thus, time relationships between different channels can be preserved even if different filters are used for each.[1]

Figure 4–5 shows an example of a segment of EEG contaminated by muscle artifact as it appears before and after application of a digital filter.

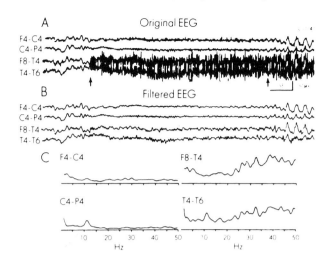

Figure 4–5. Example of digital signal filtering. An EEG contaminated by (*A*) scalp EMG artifact was filtered using (*B*) a low-pass digital filter, with (*C*) the frequency spectra of the unfiltered signals, showing the large high-frequency (> 20 Hz) muscle activity components in the last two channels before filtering. (From Gotman, et al.[1] p 633, with permission.)

TIME AND FREQUENCY DOMAIN ANALYSIS

Interval Analysis

Interval analysis is a method of determining the frequency or repetition rate of waveforms similar to that done by visual inspection. It is based on measuring the distribution of intervals between either zero or other level crossings or between maxima and minima of a signal. A zero crossing occurs when the potential in a channel changes from positive to negative or vice versa. A level crossing (less often used than a zero crossing) occurs when the potential in a channel changes from greater than to less than a given value (e.g., 50 μV) or vice versa. The number of zero crossings or other level crossings per unit of time is related to the dominant frequency of the signal (Figs. 4–6*d* and 4–6*e*). For example, a sinusoidal signal that crosses zero 120 times every second has a frequency of $\frac{1}{2}(120) = 60$ Hz.

Autocorrelation Analysis

Autocorrelation analysis may be used to recognize the dominant rhythmic activity in a signal and to determine its frequency. It is based on computing the degree of interdependence (correlation) between *successive* values of a signal. A signal that is truly random, such as white noise, will have no cor-

relation between successive values. In contrast, a signal with rhythmic components has an autocorrelation significantly different from zero. The *autocorrelation function* (ACF) is defined as the correlation between a signal and that same signal delayed by a time *t*, expressed as a function of *t*. The ACF at $t = 0$ is always 1, that is, 100% correlation. For a periodic signal (one with rhythmic components), it is an oscillating function of *t* with a frequency like that of the dominant rhythmic component in the original signal (see Fig. 4–6c).

Fourier (Spectral) Analysis

Fourier analysis is the representation of a periodic function as a Fourier series, a sum of trigonometric functions, that is, sines and cosines. A Fourier series may be used to approximate any periodic function. The greater the number of terms in the series, the greater the accuracy of the approximation will be. The Fourier transform is a mathematical method to analyze a periodic function into a sum of a large number of cosine and sine waves with frequencies of *f*, 2*f*, 3*f*, 4*f*, and so on, where *f* is the lowest, or *fundamental*, frequency in the function analyzed, and 2*f*, 3*f*, and so forth, are *harmonics*. The input of the Fourier transform is the periodic function, or signal, to be analyzed. The output is the amplitude of each sine and cosine wave in the series.

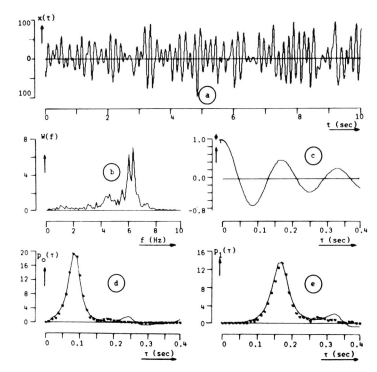

Figure 4–6. Examples of several types of signal analysis. (*a*) An EEG signal, (*b*) its power spectrum, (*c*) its autocorrelation function, and (*d*) the distribution density function of intervals between any two successive zero crossings. (*e*) The distribution density function of the intervals between successive zero crossings at which the signal changes in the same direction, that is, from positive to negative or vice versa. (From Lopes da Silva,[2] p 67, with permission.)

In dealing with *discrete* data, that is, a signal sampled at equal time intervals 0, T, $2T$, $3T, \ldots , (N-1)T$, a discrete Fourier transform is used. In this case, the fundamental frequency can be shown to be $f = 1/NT$. The input of the discrete Fourier transform is the N digitized value representing the signal at times 0, T, $2T, \ldots , (N-1)T$, where T is the sampling interval. The output is the amplitude of $N/2 - 1$ sine waves at frequencies f, $2f, \ldots , (N/2 - 1)f$ and of $N/2 + 1$ cosine waves at frequencies of 0, f, $2f, \ldots , (N/2)f$. Note that there is no data reduction: the number of input values (N) equals the number of output values.

By convention, both cosine and sine waves of a given frequency are often lumped together into one quantity describing the "amount" of that frequency present in the signal. This quantity is called the spectral *intensity* or *power*, and a plot of this as a function of frequency is the *power spectrum* (see Fig. 4–6b). The intensity, or power, is the square of the *amplitude*. The *phase* (phase angle) for any given frequency describes how much of that frequency is in the form of a cosine wave and how much is a sine wave.

The following formulas relate these quantities; here C is cosine wave amplitude and S is sine wave amplitude:

$$\text{Intensity } I = A^2 = C^2 + S^2$$
$$\text{Amplitude } A = \sqrt{I} = \sqrt{C^2 + S^2}$$
$$\text{Phase } \phi = \text{Arctan}\left(\frac{S}{C}\right)$$

Note: $\phi = 0$ for pure cosine wave,
$\phi = 90$ degrees for pure sine wave.

Statistical Analysis

Statistical analysis of a digitized signal may be a useful data-reduction technique. In effect, it treats digitized values at successive time points as independent values of a random variable. In this technique, one may plot the amplitude distribution—the number of digitized samples of the signal having a given amplitude value versus the amplitude itself—and visually inspect the shape of the distribution. Alternatively, one may calculate the *moments* of the *probability distribu-*

tion of signal amplitudes, including the following:[2]

First moment, mean voltage m_1 (*center* of distribution)

Second central moment, variance m_2, and standard deviation $\sigma = \sqrt{m_2}$ (*width* of distribution)

Third central moment m_3 and skewness $\beta_1 = m_3/(m_2)^{3/2}$ (*asymmetry* of distribution)

Fourth central moment m_4 and kurtosis excess $\beta_2 = m_4/(m_2)^2$ (*peakedness* or *flatness* of distribution).

Pattern Recognition

Pattern recognition algorithms are designed to detect a specific waveform in a signal that has characteristic features, such as a motor-unit potential in an EMG or a sharp wave in an EEG. The characteristic features may be defined in the time domain (e.g., durations, slopes, and curvature of waveforms), in the frequency domain (after filtering signal), or in both. One common approach in developing a pattern recognition algorithm is as follows: (1) Define a set of candidate features and how to calculate them. (2) Calculate the chosen features for a visually selected collection of waveforms of the type to be detected, the "learning set," and for a collection of similar waveforms determined *not* to be of the required type, the "controls." (3) Determine by statistical analysis whether it is possible to reliably separate the two groups of waveforms on the basis of the calculated features. For example, one could calculate the rising and falling slopes of candidate sharp waves, compute these slopes for true epileptic sharp waves and for other transients such as muscle artifacts and/or nonepileptic sharp transients in background activity, and determine whether a certain range of slopes characterizes the true sharp waves. These techniques have been used with some success to detect spikes and sharp waves, spike-and-wave bursts, sleep spindles and K complexes, and seizure discharge in EEG recordings and to detect motor unit potentials, fibrillation potentials, and other iterative discharges in EMG recordings.

CHAPTER SUMMARY

This chapter reviews the principles of digitization, design of digitally based instruments for clinical neurophysiology, and several common uses of digital processing, including averaging, digital filtering, and some types of time-domain and frequency-domain analysis. An understanding of these principles is necessary in order to appropriately select and use digitally based instruments and to understand their unique features.

REFERENCES

1. Gotman, J, Ives, JR, and Gloor, P: Frequency content of EEG and EMG at seizure onset: Possibility of removal of EMG artefact by digital filtering. Electroencephalogr Clin Neurophysiol 52:626–639, 1981.
2. Lopes da Silva, FH: Computerized EEG analysis: A tutorial overview. In Halliday, AM, Butler, SR, and Paul, R (eds): A Textbook of Clinical Neurophysiology. John Wiley & Sons, Chichester, England, 1987, pp 61–102.
3. Maccabee, PJ and Hassan, NF: AAEM minimonograph #39: Digital filtering: Basic concepts and application to evoked potentials. Muscle Nerve 15:865–875, 1992.
4. Spehlmann, R: Evoked Potential Primer: Visual, Auditory, and Somatosensory Evoked Potentials in Clinical Diagnosis. Butterworth, Boston, 1985, pp 35–52.

CHAPTER 5

ELECTROPHYSIOLOGIC GENERATORS IN CLINICAL NEUROPHYSIOLOGY

Terrence D. Lagerlund, M.D., Ph.D.

PHYSIOLOGIC GENERATORS
Resting Membrane Potential
Electrotonic Membrane Potentials
Action Potentials
Postsynaptic Potentials
Other Membrane Potentials
STRUCTURAL GENERATORS
Peripheral Nerves
Muscles
Sweat Glands
Spinal Cord
Brain Stem
Special Sensory Receptors
Optic and Auditory Pathways
Cerebral Cortex

PHYSIOLOGIC GENERATORS

The bioelectric potentials recorded in the performance of electrophysiologic studies are generated by sources inside the body which may be some distance from the recording electrodes. Sources may be classified according to mode of generation at the cellular level as electrotonic membrane potentials, action potentials, postsynaptic potentials, or other membrane potentials. These sources all derive from ionic current flow into or out of cells, which is made possible because all excitable cells, such as neurons and muscle fibers, have a resting membrane potential.

Resting Membrane Potential

The *resting membrane potential* of neurons and muscle fibers is defined as the electric potential inside the cell minus that outside the cell. It typically has a value of around -66 mV and is caused by ionic concentration gradients across the semipermeable cell membrane.

Ionic concentrations inside and outside a cell are characteristically different. For example, potassium is more concentrated intracellularly and sodium extracellularly. The value of the resting membrane potential of the cell depends on the concentration gradient for each ion species, as well as the relative permeability of the membrane to each ionic species. The Nernst potential for each ionic species is the resting membrane potential that would result if the membrane were permeable to that species alone. Nernst potentials for four ionic species are shown in Table 5–1.

Driving forces of two types are responsible

50

Table 5–1. **NERNST POTENTIAL FOR FOUR IONIC SPECIES**

Ion Species	Intracellular Concentration, mmol/L	Extracellular Concentration, mmol/L	Nernst Potential, mV
K^+	68	4	−76
Na^+	15	142	+60
Cl^-	9	105	−66
Ca^{2+}	0.0005	0.125	+74

for fluxes of ions across the cell membrane. The *chemical* driving force causes ions to move from a region of high concentration to one of lower concentration. The *electrical* driving force causes positively charged ions to move from a region of higher potential to one of lower potential, whereas negatively charged ions move in the opposite direction. The *net* driving force is the sum of the chemical and electrical driving forces. The net ionic flux across the membrane is the product of the net driving force and the membrane permeability to the ion. In the resting state, there is a net flux of sodium *into* the cell and a net flux of potassium *out of* the cell, but there is no net flux of chloride because the net driving force is zero. The steady-state concentration gradient of sodium and potassium ions across the cell membrane is maintained by an active, energy-consuming process that pumps potassium into the cell and sodium out of the cell.

An electrical equivalent circuit for the membrane can be constructed in which the chemical driving force for each ionic species is represented as an electrical battery (a seat of electromotive force given by the Nernst potential for that ion) and the membrane permeability for each ionic species is represented as a resistor. The flux of each ionic species represents a flow of electric current across the membrane.

Important membrane electrical properties include the electromotive force for each ion (the Nernst potential) determined by the ionic concentration, the conductance per unit area (g_{Na}, g_{Cl}, g_K) for each ion (the product of the number of ion channels per unit area and the conductance of each channel), and the capacitance per unit area (C_m) of the membrane (typically, 20 mF/m^2).

The equilibrium potential (resting potential) of the membrane can be calculated from the equivalent circuit for the membrane. Any change in membrane permeability to one or more ionic species or any externally injected current across the membrane alters the membrane potential from its equilibrium value.

Electrotonic Membrane Potentials

Electrotonic membrane potentials result from the passive electrical properties of the cell. These are the membrane properties mentioned above (total membrane conductance per unit area g_{tot} and membrane capacitance per unit area C_m) as well as the intracellular axial resistance per unit length r_a.

The membrane time constant determines the behavior of a localized portion of the membrane when an additional current is injected into the cell. For example, this current could be physiologic in origin (due to activation of ion channels) or could be introduced experimentally by a microelectrode impaling the membrane. Because of its resistive and capacitive properties, the membrane potential changes exponentially from its resting value to a new equilibrium value when the current is turned on and returns exponentially to its resting value when the current is turned off (Fig. 5–1). The time required for the potential to reach 63 percent of the new value is the time constant, given by $\tau = C_m/g_{tot} = RC$. Here, $R = 1/g_{tot}A$ is the total cell input resistance, and $C = C_mA$ is the total cell capacitance (A is the total cell membrane area).

Cable properties of axons and dendrites (or muscle fibers) determine the behavior of

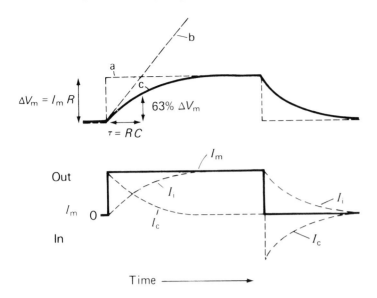

Figure 5–1. The actual shape (*c*) of the response of the membrane to a rectangular current pulse is intermediate between that of a pure resistive element (*a*) and a pure capacitive element (*b*). The product of the membrane's resistance (*R*) and the capacitance (*C*) is called the "membrane time constant" (*τ*). The total membrane current (I_m) is shown by the solid line in the lower half of the figure; dotted lines I_i and I_c show the time course of ionic and capacitive current, respectively. (From Koester, J: Passive membrane properties of the neuron. In Kandel, ER, Schwartz, JH, and Jessell, TM [eds]: Principles of Neural Science, ed 3. Appleton & Lange, Norwalk, CT, 1991, p 97, with permission.)

a long cylindrical cell process when an additional current is injected into one location. Because of the resistive and capacitive properties of the process, the membrane potential falls off exponentially with increased distance from the site of current injection (Fig. 5–2). The distance in which the potential change from the resting value reaches 37% of its maximum is the membrane space constant λ, given by

$$\lambda = \sqrt{\frac{1}{\pi a r_a g_{tot}}}$$

where *a* is the radius of the axon or dendrite cylinder.

Action Potentials

Action potentials are regenerative changes in membrane potential that propagate along a cell process (axon, dendrite, or muscle fiber). They have a depolarizing phase, in which the membrane potential reverses sign from negative to positive, followed by a repolarizing phase, in which the potential returns to its resting value.

Propagation velocity of an action potential in myelinated axons is faster than that seen in unmyelinated axons, because of the effect of the myelin on the cable properties of the axon. Specifically, myelin decreases the membrane capacitance and conductance, decreases the time constant, and increases the space constant of the segment of axon between the nodes of Ranvier. In unmyelinated axons, the action potential propagation velocity $v \propto a^{1/2}$, whereas in myelinated axons, $v \propto a$, where *a* is the axon radius.

The voltage clamp technique is used experimentally to measure the membrane conductance changes underlying an action potential. In this technique, an external apparatus connected to a microelectrode impaling the cell maintains the membrane potential clamped at a fixed value and measures the net current flow into or out of the cell (from which conductance can be calculated). To separate the contributions of various ionic channels to the net current flow, pharmacologic agents are used to block channels selectively; for example, tetrodotoxin blocks sodium channels and tetraethylammonium blocks potassium channels.

With these techniques, the basis of the ionic fluxes that occur during action potentials has been determined. Rapidly opening voltage-sensitive sodium channels are responsible for the depolarization phase of the action potential. Because of the factor of 400

A

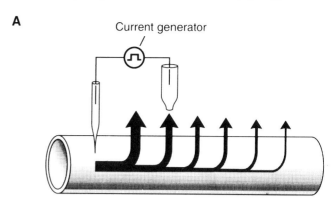

B

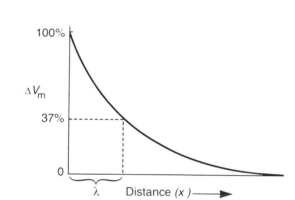

Figure 5–2. (*A*) Current injected into a neuronal process with a microelectrode follows the path of least resistance to the return electrode in the extracellular fluid. (*B*) The change in V_m produced by focal current injection decays exponentially with distance along the length of the process. (From Koester, J: Passive membrane properties of the neuron. In Kandel, ER, Schwartz, JH, and Jessell, TM [eds]: Principles of Neural Science, ed 3. Appleton & Lange, Norwalk, CT, 1991, p 99, with permission.)

increase in membrane sodium permeability at the peak of the action potential as compared with its resting value, the membrane potential comes close to the Nernst potential for sodium. More slowly opening voltage-sensitive potassium channels are partially responsible for the repolarization of the membrane (and may produce a transient hyperpolarization) after the action potential, but closing of the sodium channels due to an intrinsic inactivation time constant also contributes to the process and is probably the dominant mechanism in mammalian neurons.

Hodgkin and Huxley[2] formulated a mathematical model of the action potential in giant squid axons based on their experimental measurements of the time and voltage dependence of sodium and potassium conductances. This model was able to successfully reproduce the observed time course of the action potential (Fig. 5–3). Similar models have subsequently been developed for many types of vertebrate neurons.

Postsynaptic Potentials

Postsynaptic potentials (PSPs) are one type of nonregenerative and nonpropagating cell potential. They are membrane potentials caused by neurotransmitter-induced opening or closing of ion channels in a postsynaptic neuron or muscle fiber and underlie normal synaptic transmission.

Excitatory PSPs (EPSPs) are depolarizing potentials in the postsynaptic membrane caused by the effects of an excitatory neurotransmitter. Commonly, EPSPs are produced by the opening of channels to sodium and potassium and less commonly by the closing of potassium channels.

Inhibitory PSPs (IPSPs) are hyperpolarizing potentials in the postsynaptic membrane caused by the effects of an inhibitory neurotransmitter. Commonly, IPSPs are produced by the opening of potassium or chloride channels and less commonly by the closing of sodium or calcium channels. Note that postsynaptic inhibition may occur in the

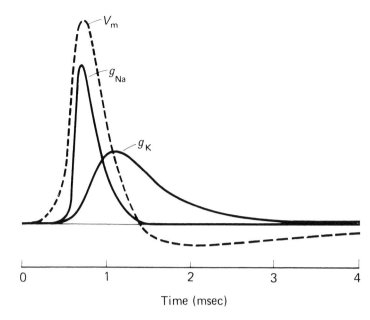

Time (msec)

Figure 5–3. The action potential can be reconstructed from the changes in g_{Na} and g_K that result from the opening and closing of Na$^+$ and K$^+$ voltage-gated channels. (From Koester, J: Voltage-gated ion channels and the generation of the action potential. In Kandel, ER, Schwartz, JH, and Jessell, TM [eds]: Principles of Neural Science, ed 3. Appleton & Lange, Norwalk, CT, 1991, p 110, as adapted from Hodgkin, AL: The Conduction of the Nervous Impulse. Liverpool University Press, Liverpool, 1964, p 63, with permission.)

absence of a detectable IPSP, because opening of potassium or chloride channels may merely decrease the cell input resistance (thereby making the cell harder to depolarize) without significantly hyperpolarizing the cell.

Other Membrane Potentials

Other types of membrane potentials include those caused by other ionic channel types, especially in neurons in the central nervous system. These include calcium channels as well as other types of sodium and potassium channels. These additional channel types are responsible for the rich variety of behaviors seen in central nervous system neurons and probably contribute to such phenomena as rhythmic oscillations and epileptic bursts in cortical neurons and adaptation and recruitment patterns in spinal motor neurons, although interneuronal connectivity is also important in these phenomena.

Slow calcium spikes are depolarizing potentials mediated by slow calcium channels. These channels are activated by membrane depolarization; they are inactivated by repolarization, by an intrinsic slow inactivation time constant, and by increasing intracellu-

lar calcium concentration. Slow calcium spikes are seen in cortical pyramidal cells and other neuron types. They are involved in intrinsic bursting behavior (in which multiple high-frequency fast sodium "spikes" or action potentials occur on the slow calcium spike) and the epileptic paroxysmal depolarization shift (Fig. 5–4).

Afterdepolarizations are prolonged or late depolarizing potentials that follow a sodium-mediated action potential. Early afterdepolarizations commonly are mediated by the slow calcium channels described previously. Several types of late afterdepolarizations have also been described, including rebound calcium spikes (which occur after hyperpolarizing events and are due to activation of low-threshold inactivating calcium channels) and persistent long-lasting depolarizations due to activation of slowly inactivating (persistent) sodium channels. Because late afterdepolarizations may lead to oscillations, they may be responsible for many types of rhythmic cortical activity.

Afterhyperpolarizations are prolonged hyperpolarizing potentials that follow a sodium- or calcium-mediated depolarizing potential. They commonly are mediated by slow or "calcium-dependent" potassium channels. These channels are activated by increasing intracellular calcium concentration

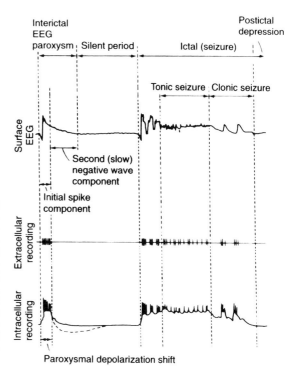

Figure 5–4. Relationship between surface-recorded electroencephalographic (*EEG*) discharges and intracellular and extracellular activity in a cortical epileptic focus in an experimental animal. (From Martin, JH: The collective electrical behavior of cortical neurons: The electroencephalogram and the mechanisms of epilepsy. In Kandel, ER, Schwartz, JH, and Jessell, TM [eds]: Principles of Neural Science, ed 3. Appleton & Lange, Norwalk, CT, 1991, p 787, as adapted from Ayala GF, et al: Genesis of epileptic interictal spikes. New knowledge of cortical feedback systems suggests a neurophysiological explanation of brief paroxysms. Brain Res 52:1–17, 1973, with permission.)

or by membrane depolarization and are inactivated by decreasing intracellular calcium concentrations, membrane repolarization, and a relatively long intrinsic inactivation time constant. Afterhyperpolarizations may last for hundreds of milliseconds in some neuron types (e.g., spinal motor neurons) and thus limit the maximum firing rate of the cell. Even more prolonged afterhyperpolarizations may occur because of activation of so-called early potassium channels, which are activated by membrane hyperpolarization and act to delay the return of membrane potential to equilibrium.

Dendritic spikes are small depolarizing potentials recorded from dendrites. They are mediated primarily by a type of slow calcium channel in certain regions of dendritic membrane ("active" or "booster" zones) and provide amplification and regeneration of postsynaptic potentials on dendritic branches far from the soma (i.e., more than 1 or 2 length constants distant) which would otherwise not be able to have a significant effect on the soma potential if only electrotonic (passive) conduction were involved.[3]

STRUCTURAL GENERATORS

The variety of clinical neurophysiologic studies corresponds to a variety of structural generators in the body, including muscles and sweat glands, peripheral nerves, and various components of the central nervous system. Each structural generator may have associated with it several different physiologic potential types. In some cases, the activity resulting from different physiologic potentials can be easily distinguished, but in many cases (particularly sensory pathways in the central nervous system assessed by evoked potential studies), complete knowledge of the physiologic generator underlying any recorded waveform is lacking.

Peripheral Nerves

Peripheral nerves consist of axons and supporting structures (i.e., myelin-producing Schwann cells, connective tissue, and blood vessels). They contain three types of axons. *Motor axons* originate from neurons in the spinal cord (spinal motor neurons) or brain stem (cranial nerve motor neurons); they synapse on muscle fibers. *Sensory axons* originate from neurons in spinal nerve or cranial nerve ganglia; these axons terminate in skin, muscle, or other organs in specialized sensory receptors (such as pacinian corpuscles or muscle spindles) or as "bare" nerve terminals. *Autonomic axons* originate either in neurons of the spinal cord or brain stem (preganglionic neurons) or in neurons in autonomic ganglia (postganglionic neurons located in the sympathetic trunk ganglia for many sympathetic neurons or in visceral ganglia near the end-organ innervated for all parasympathetic and some sympathetic neurons). The sum of the propagating action potentials of all stimulated sensory axons in a motor or mixed nerve can be recorded as a sensory nerve action potential during sensory nerve conduction studies. These primarily involve the large diameter (IA and IB) sensory axons in the nerve, because only they are stimulated by conventional electrical stimuli. Motor and autonomic axons are generally tested only indirectly by stimulating the nerve and observing the postsynaptic effects in muscle and sweat glands.

Muscles

Muscles consist of muscle fibers plus some connective tissue; both motor and sensory axons traverse muscles. Contraction of muscle fibers is initiated by neuromuscular synaptic transmission (with acetylcholine as the neurotransmitter), which leads to a propagated muscle action potential. This in turn causes calcium to enter the muscle fibers (calcium is the intracellular trigger for the contractile process).

Muscle end-plate potentials are excitatory postsynaptic potentials in muscle fibers originating at the end-plate (the neuromuscular junction). Both sodium and potassium ions flow through the channel opened by acetylcholine-receptor binding, so that the reversal potential (the membrane potential in a voltage clamp experiment at which the ionic current flow reverses from net inward to net outward) for this channel is near zero volts, intermediate between the Nernst potentials of sodium and potassium. Normal end-plate potentials are caused by the simultaneous release of hundreds of quanta (packets) of acetylcholine caused by an action potential reaching the nerve terminal. However, miniature end-plate potentials are much smaller and are caused by the spontaneous, random release of a single packet of acetylcholine from the nerve terminal. End-plate noise recorded from needle electrodes in the vicinity of the muscle end-plate is caused by miniature end-plate potentials. *End-plate spikes* are action potentials of muscle fibers caused by mechanical activation (by the electromyographic needle) of nerve terminals in the end-plate region and are mediated by normal neuromuscular synaptic transmission.

Muscle action potentials are similar to nerve action potentials but have a generally slower propagation velocity. They propagate outward (often in two opposite directions simultaneously) from the vicinity of the neuromuscular junction to the ends of the muscle fiber. Muscle action potentials may occur spontaneously in individual muscle fibers (fibrillation potentials), simultaneously in all muscle fibers that are part of the same motor unit (e.g., voluntary motor units or involuntary fasciculation potentials), or nearly simultaneously in all muscle fibers supplied by one motor or mixed nerve (leading to the surface-recorded compound muscle action potential in motor nerve conduction studies).

Sweat Glands

Sweat glands are end-organs innervated by sympathetic postganglionic neurons that have acetylcholine as their neurotransmitter. Many of these axons have multiple branches in the skin, so that a single axon may innervate many sweat glands. Release of acetylcholine by sympathetic nerve terminals in response to various stimuli (such as an electric stimulus applied to a contralateral limb) leads to a prolonged depolarization of sweat glands which can be recorded diffusely from the skin surface in certain areas (a peripheral autonomic surface potential). Func-

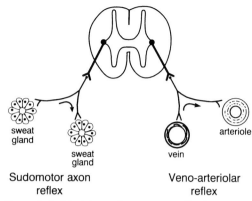

sweat gland

sweat gland

vein

arteriole

Sudomotor axon reflex

Veno-arteriolar reflex

Figure 5–5. Physiology of two postganglionic axon reflexes: the sudomotor axon reflex and the venoarteriolar reflex. (From Low, PA: Quantitation of autonomic function. In Dyck PJ, et al: [eds]: Peripheral Neuropathy, ed 3. WB Saunders, Philadelphia, 1993, p 731, with permission.)

tion of peripheral sympathetic axons and sweat glands is also assessed by use of the quantitative sudomotor axon reflex test. In this test, a controlled release of acetylcholine into one skin region leads to reflex depolarization of autonomic axons that propagate antidromically to a "Y" branch point and orthodromically from there to activate sweat glands synaptically in a nearby region of skin, thus producing sweat. Sweat gland activity (sweat production) is quantified by measuring water output of the glands[1] (Fig. 5–5).

Spinal Cord

The spinal cord contains cell bodies of motor, autonomic, and other neurons and axons of central and peripheral neurons arranged in various ascending and descending tracts. Many synapses occur on spinal cord neurons. Potentials originating from generators in the spinal cord therefore include EPSPs, IPSPs, and action potentials. Although the health and quantity of spinal motor neurons in the ventral horn can be assessed to some degree by standard electromyographic and nerve conduction studies, specific potentials produced by spinal motor neurons themselves are not routinely measured. During somatosensory evoked potential studies, recordings from cervical or cer-

vical-cranial derivations may demonstrate an N11 waveform that is thought to be generated primarily at the root entry zone in the dorsal horn at the level of spinal segments C-5 and C-6. Because it manifests a latency shift with recording at progressively higher levels, it may be due to a propagated action potential in the sensory axons that ascend in the dorsal columns.

Brain Stem

The brain stem somatosensory pathways include the dorsal column nuclei in the lower medulla and the medial lemniscus. Recordings during somatosensory evoked potential studies demonstrate an N13 waveform thought to be generated by the upper cervical cord dorsal columns (presynaptic action potential) or by the dorsal column nuclei (postsynaptic potential) or by both. A P14 waveform may also be seen; it is thought to be generated by the medial lemniscus (action potential).

Special Sensory Receptors

Sensory receptors in the visual and auditory systems generate characteristic potentials that can be recorded with appropriate evoked potential studies.

The retina contains receptors (rods and cones) and other types of neurons (bipolar, horizontal, amacrine, and ganglion cells). Retinal visual evoked potentials (the electroretinogram) can be recorded from electrodes placed near or on the eye and are thought to be due to summed postsynaptic potential activity in the retinal neurons.

Hair cells are sensory receptors in the cochlea. They release a neurotransmitter to activate the peripheral axons of the bipolar cells of the *spiral ganglion*, which in turn conduct action potentials to the central axons that make up the auditory nerve and synapse in the cochlear nucleus in the lower pons. Some components of the electrocochleogram (the cochlear microphonic and the summating potential) are thought to be produced by sensory receptor potentials in the hair cells. Wave I of the brain stem auditory evoked potential (BAEP) is due to the propagated action potentials in the auditory nerve.

Optic and Auditory Pathways

The optic and auditory pathways are generators of later components of the visual and auditory evoked potentials. Propagated action potentials in the optic tracts or optic radiations may contribute to some variable early components of diffuse light-flash visual evoked potentials, but these have little clinical usefulness. Wave II of the BAEP is thought to be generated either by propagated action potentials in the auditory nerve as it enters the brain stem or by postsynaptic potentials in the cochlear nucleus. Wave III may be generated by postsynaptic potentials in the superior olivary nucleus and waves IV and V by the lateral lemniscus (propagated action potentials) or the inferior colliculus (postsynaptic potentials). Later waves (VI and VII) may arise from the medial geniculate body and the auditory radiations.

Cerebral Cortex

The cerebral cortex is the generator of essentially all electroencephalographic (EEG) activity recorded without averaging, as well as of late components of evoked potentials (including the P100 component of the pattern-reversal visual evoked potential). Cortical neurons include both pyramidal cells (excitatory neurons that provide the major output of cortex) and stellate cells (excitatory or inhibitory interneurons). The long apical dendrites of pyramidal cells are perpendicular to the cortical surface. Each pyramidal cell also has an extensive dendritic tree on which 1000 to 100,000 synapses may occur.

The neocortex consists of six cellular layers (layer I, most superficial, to layer VI, deepest layer). Layer I contains mainly glial cells and axonal and dendritic processes. Layer IV is most developed in sensory areas of the cortex and receives much of the specific thalamocortical projections. Layer V is most developed in motor areas of the cortex, in which many of the pyramidal cells are exceptionally large and project particularly to distant sites, including the brain stem and spinal cord.

Brodmann divided the cortex into 52 cortical areas on the basis of cell size, neuron density, myelinated axon density, and number of layers. The primary somatosensory cortex (areas 3, 1, and 2) is the likely generator of the scalp components of the somatosensory evoked potential. Primary visual cortex (area 17) and visual association cortex (areas 18 and 19) are the likely generators of the P100 component of the visual evoked potential. Auditory cortex (areas 41 and 42) may be the generator of late components of the long-latency auditory evoked potential.

It is known that postsynaptic potentials—not action potentials—in cortical neurons are responsible for all scalp-recorded electrical activity. This is because postsynaptic potentials are of long duration (tens to hundreds of milliseconds), involve large areas of membrane surface, occur nearly simultaneously in thousands of cortical pyramidal cells, and occur especially in pyramidal cell dendrites that are uniformly directed perpendicular to the cortical surface. These properties all allow postsynaptic potentials to summate effectively to produce a detectable scalp potential. In contrast, action potentials are brief (1 millisecond), involve small surface areas of membrane (axons), occur at random widely spaced intervals in various neurons, and propagate along axons that are oriented in many directions, all of which make effective summation impossible.

CHAPTER SUMMARY

This chapter reviews the generators of electrophysiologic potentials in terms of basic cellular electrophysiology and the anatomic structures that generate electrophysiologic potentials of clinical interest. Knowledge of the generators of the potentials recorded in clinical neurophysiologic studies is helpful in understanding the characteristics and distribution of the recorded potentials and is the first step in correlating the alterations seen in disease states with the pathologic changes demonstrated in the underlying generators.

REFERENCES

1. Dyck PJ, Thomas PK, Griffin JW, et al: Peripheral Neuropathy, ed 3. WB Saunders, Philadelphia, 1993.

2. Hodgkin, AL and Huxley, AF: A quantitative description of membrane current and its application to conduction and excitation in nerve. J Physiol (Lond) 117:500–544, 1952.

3. Kandel, ER and Schwartz, JH: Principles of Neural Science, ed 2. Elsevier Science Publishing, New York, 1985.

CHAPTER 6

WAVEFORM PARAMETER CLASSIFICATIONS

Jasper R. Daube, M.D.

CONTINUOUS WAVEFORMS
EVENT RECORDING

The many structures of the central, peripheral, and autonomic nervous systems and muscle described in Chapter 5 generate various electrical signals. These signals are classified according to variables that are measured in order to identify and characterize the signals and to recognize the abnormalities that occur in disease. Each of the signals is a variation in the potential difference between two points over time. A potential difference between the two points being tested that does not change is recorded as a flat line with no signal, despite a difference in voltage. Variations in the difference in potential are broadly classified as continuous and intermittent changes.

CONTINUOUS WAVEFORMS

A continuously varying signal (Figure 6–1) is described by the rate of change of the signal (cycles per second), the size or amplitude of the signal (peak-to-peak), the character of the waveform, and the consistency of the signal over time. Continuous waveforms recorded from living tissue are usually sinusoidal, as shown in Figure 6–1. The

traces also show waveforms varying at five rates. The rate of fluctuation of a signal is known as its *frequency* and is measured in cycles per second. The term more commonly used to describe this rate of alteration is "hertz" (Hz). A 10-Hz signal is a waveform that has continuous variation 10 times per second. The continuous waveforms generated by physiologic generators can be described by their component frequencies. There are examples of 3-Hz, 3.5-Hz, 6-Hz, 10-Hz, and 20-Hz activity in Fig. 6–1. The change of a continuous waveform from one frequency to another may occur because a single generator changes its rate of activity or because one generator working at 3 Hz becomes inactive while a second one working at 6 Hz becomes active.

If the recording electrodes are located near two structures simultaneously generating signals of different frequencies, a more complex signal is recorded. In Figure 6–1, signals illustrating the combination of two frequencies are shown. Note that the combination of signals with closer frequencies results in less recognizable waveforms. Even as the physiologic waveforms recorded in clinical neurophysiology become more complex, they can still be described in terms of the component frequencies of the signal. Waveforms become even more complex when differently shaped waves of the same frequency summate (Fig. 6–2). All continu-

60

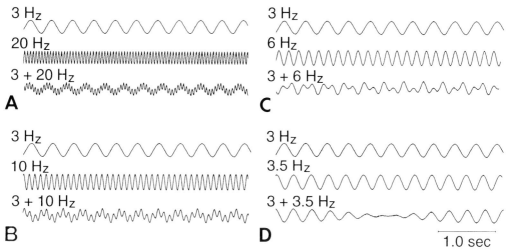

Figure 6–1. Simple, regular, sinusoidal waves of different frequencies combine to form complex waveforms. (*A*) A 3-Hz and a 20-Hz waveform summate into a waveform in which both are still recognizable. (*B*) A 3-Hz and a 10-Hz waveform summate to form a more complicated waveform in which the components are less recognizable. (*C*) Summation of a 3-Hz and a 6-Hz waveform results in an apparently regular, more complex waveform. (*D*) Summation of a 3-Hz and a 3.5-Hz waveform results in fluctuation of the waveform as the components go in and out of phase.

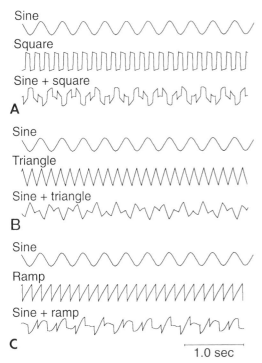

Figure 6–2. Summation of waveforms with different shapes but the same frequency (3 Hz) results in more complex waveforms. A sine wave is combined with (*A*) a square wave, with (*B*) a triangular wave, and with (*C*) a ramp wave.

ously varying signals in physiologic recording are described in terms of their frequency components (Fig. 6–3). Many of them, such as electroencephalographic waveforms, often have a predominant frequency that at any given time characterizes the signal at a pair of electrodes. The ability to dissect a waveform into its component frequencies does not mean that there are distinct structures generating each of the frequencies. For example, a neuron with synaptic input from multiple sources displays postsynaptic potentials that summate at the cell body and produce a complex, varying intracellular potential. Many frequencies would be identified by a frequency analysis, but there would be no single generators active at any of the component frequencies. Indeed, frequency analysis can be performed with automated electronic systems that can define the frequency components of any signal as a histogram, as shown in Figure 6–4. A frequency analysis of this sort can be made on signals of different time duration. A frequency analysis of the first two traces (see Fig. 6–1) shows two frequency components, whereas a frequency analysis of either the first or the second trace has only single frequency components.

Therefore, a continuously varying signal is described by its frequency components,

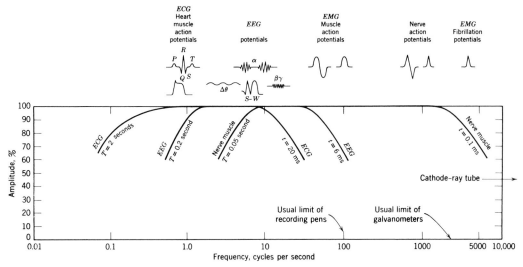

Figure 6–3. The frequency components of electrical activity recorded in clinical neurophysiology are shown on a logarithmic scale. *T* and *t* are the upper and lower time constants of each frequency cutoff. ECG = electrocardiogram; EEG = electroencephalogram; EMG = electromyogram. (From Geddes, LA and Baker, LE: Principles of Applied Biomedical Instrumentation. © John Wiley & Sons, New York, 1968, p 317, with permission.)

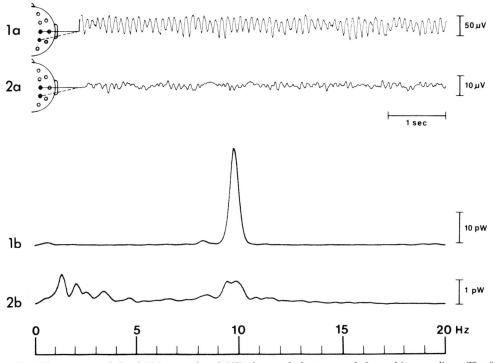

Figure 6–4. Frequency analysis of (*1a*) normal and (*1b*) abnormal electroencephalographic recordings. The frequency components of *1a* are shown in *1b* as a predominantly 10-Hz signal. Tracing *2a* is a combination of the 10-Hz activity, with the abnormal 1- to 4-Hz activity shown in *2b*. (From Fisch, BJ: Spehlmann's EEG Primer, ed 2. Elsevier Science Publishers, Amsterdam, 1991, p 132, with permission.)

their amplitude, and the intervals of time over which they occur. This approach to describing signals is most commonly used with electroencephalographic potentials. Also, a description of signals always includes the location at which the signal is recorded. For example, the frontal region of the head may have a 30-μV, 20-Hz signal occurring for 10 seconds of a 20-second recording, while the occipital region simultaneously has a 10-Hz signal throughout the 20-second recording epoch.

EVENT RECORDING

A second approach to the description of the waveforms seen in clinical neurophysiology is used primarily in recordings made up of a sequence and combination of well-defined events such as the one shown in Figure 6–5; in the top trace (see Fig. 6–5A), brief spikes occur during a recording that includes no other activity. Description and measurement of events use terminology similar to that used for continuous recording but with some differences in meaning. The spikes occurring in the first 5 seconds of the recording can be described as occurring at a frequency of three per second and those in the next 5 seconds as six per second. The entire recording of 37 events in 10 seconds could be described as spikes occurring 3.7 times per second. However, this is an average of their occurrences, as opposed to the actual recurrence of three and six times per second. Therefore, events are better described as the number per unit time, the regularity of their occurrence in time, and the pattern of recurrence. In the middle tracing (see Fig. 6–5B), groups of eight spikes recur at 2-second intervals. This is described as a burst pattern of 10-per-second spikes recurring regularly at 2-second intervals. A recording from a pair of electrodes often includes the activity of multiple generators, each of which may be generating events in different patterns. A single generator may generate regularly recurring events, events that occur in a changing but definable pattern, and events that recur in an unpredictable or random pattern. The activity of two independent generators (see Fig. 6–5C) gives a complex pattern of events that can be separated into the recurrence of distinct events coming from different generators. This distinction is made most accurately by analyzing the pattern of events, but recognizing the unique character of the individual events helps in distinguishing them. The latter becomes unreliable, however, when the generated events are similar or identical.

In summary, the characterization of the occurrence of events is (1) to identify recurrent patterns of individual events and classify those that are occurring in bursts into the rate of firing during a burst and the recurring rate of the burst and (2) to classify those that are nonbursting by the rate and regularity of the firing.

Each event is further characterized on the basis of its own variables. The event often

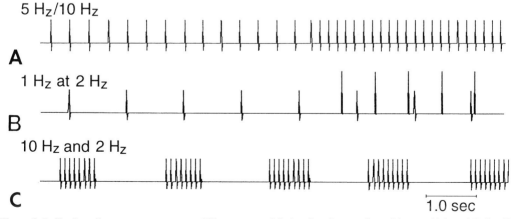

Figure 6–5. Single spikes occur as events at different rates. (*A*) A spike changes from 5/sec to 10/sec. (*B*) A spike occurring at 1/sec combines with a second, independent spike occurring at 2/sec. (*C*) A single spike occurs in bursts of 10/sec every 2 sec.

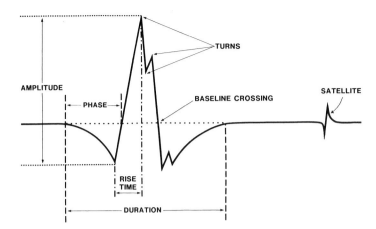

Figure 6–6. Each single event, such as the spikes in Figure 6–5, can be described quantitatively by the variables illustrated here. The same event could also be analyzed into its component frequencies, which could range from less than 1 Hz to more than 10,000 Hz.

called *discharge* has an amplitude that is measured either from the baseline to the peak or from peak-to-peak (Fig. 6–6). The discharge has a duration from onset to termination. Discharges have configurations that may be monophasic, biphasic, triphasic, or more complex with multiple phases. If there is more than one phase, each phase can be described according to its amplitude, duration, and configuration. Each component of a discharge has a *rise time* or a *rate of rise* from the positive peak to the negative peak. The rise time is a direct function of the distance of the recording electrodes from the generator and can be used to determine how close a generator is to the recording electrodes. Short duration or rapid rise times occur when the recording electrodes are close to the generator.

A typical discharge is triphasic when it is recorded from a nearby generator. If the waveform is moving, the initial positive portion of the discharge is recorded from the potential when the potential is distant from the recording electrode. A negative component is obtained when the potential is adjacent to the active electrode, and the late positive component when the potential leaves the electrode. The complexity of a discharge is a function of the number of generators that contribute to the discharge and the synchrony of their firing. Greater synchrony of firing produces a simpler waveform of larger size.

Individual events also can be characterized according to their frequency spectrum. The analysis of an event—regardless of its size, shape, or configuration—can break it down into a summation of the activity of different frequencies. Event recording is used in recording from single axons and single neurons and in clinical electromyography. In these settings, the measurements are generally those of the pattern of firing and the characteristics of the individual discharges. The frequency component of such potentials is not a useful measure.

Divison of electrical activity into continuous waveforms and events is somewhat arbitrary, and similar measurements can be made with either designation. The frequency of a continuous waveform is the inverse of the duration of a cycle. For example, a 10-Hz signal has a 100-millisecond interval between recurrences. The broad sine wave of the 10-Hz signal when repeated through one cycle could be considered an event and characterized as such. In contrast, the designation of events and their patterns of recurrence and complexity of appearance occasionally requires characterizing them in terms of frequency, amplitude, and rate of recurrence.

CHAPTER SUMMARY

The waveforms recorded in clinical neurophysiology are divided into continuous waveforms and discrete waveforms or events. Continuous waveforms are described by their frequency components, amplitude, and distribution. Discrete waveforms are described by their individual amplitude, duration, and configuration as well as by their pattern of occurrence and distribution.

CHAPTER 7

ALTERATION OF WAVEFORMS AND ARTIFACTS

Jasper R. Daube, M.D.

Abnormalities of the various waveforms generated by the central nervous system, nerves, or muscles can be assessed only in terms of a change in the waveform of a specific generator. No waveform itself can be defined as abnormal without reference to the generator. Normal waveforms arising from one generator would be abnormal if they arose from a different generator. For example, the spike activity normally generated by muscle has features that are similar to those of an epileptic spike generated by the cerebral cortex. Therefore, alterations in waveform must be considered in relation to the categories described in Chapter 6. In contrast, electrical artifacts often have distinct waveforms that do not arise from any physiologic generator. Artifacts are best defined as electrical activity of no clinical significance originating from nonphysiologic sources. Artifacts are considered separately in the last section of this chapter.

PHYSIOLOGIC ALTERATION OF WAVEFORMS

Changes in single potentials, such as a single well-defined electromyelographic (EMG) spike, are described by the characteristics of the single event. In contrast, changes in continuously varying signals, such as encephalographic (EEG) waves, are described by the characteristics of a series of waves. The distinction between a single event and continuous waves is not always clear, but they usually can be separated. The alterations of single potentials and of waves are considered separately below.

Single Potential

Electrical activity generated by neural or muscle tissue often appears as a single discrete event, a *single potential*, with no activity or only unrelated activity around it. Single potentials may be normal or abnormal. Changes in individual potentials are described by measuring the variables of the po-

65

tential to determine whether they are outside the normal range (Table 7–1). To describe single potentials, four sets of variables are measured. The first set is the size, which includes amplitude (peak-to-peak or baseline-to-peak), area, and duration of the potential. The second set is the waveform configuration, which includes the rate of change of the components of the potential, the number and timing of changes in the direction of the current flow, and the components of the potential. The components include the phases and turns of the potential. The third set is the pattern and frequency of occurrence of the potential. A spike might occur at a regular, low rate (e.g., a voluntary motor unit potential) or as high-frequency, short bursts (e.g., a myokymic discharge). The fourth set is the distribution or field of occurrence. For example, an epileptic spike might occur in the frontal lobe or in the temporal lobe. A waveform may have different variables in different parts of its field. A potential may be described by its relationship to other events, such as the latency of a response. Disease can alter these variables, initiate a new waveform, or eliminate a normal waveform.

If a single potential recurs over time, another set of variables are measured, including stability, rate, pattern, and the type of change that occurs with time. The alteration of waveforms with disease is defined by which variable is outside the normal range.

In disease, each variable should be considered for measurement. The methods for measuring these variables must be defined because the results can vary with the method of measurement.

Continuous Waves

Much of the electrical activity that is generated by neural and muscle tissues occurs as continuously varying potentials that may persist over long periods. These potentials usually have a sinusoidal configuration. Recurrent single events recorded at a considerable distance from the generator may appear as continuously varying waves. Continuous waves are characterized by variables similar to, but different from, those variables that characterize single events (Table 7–2).

Variables that are used to measure the size of continuous waves include amplitude (peak-to-peak or base-to-peak), root mean square (square root of mean amplitude over time), and power (square of the amplitude). In most situations, the major variable to measure is the frequency, or the number of cycles of the wave per second. Frequency can be simply measured as baseline or zero-crossings per second. More complex automated analyses of frequency spectra are the fast Fourier transforms and autoregressive modeling. Continuous waves may be simple, with a single frequency, or they may be complex, with more than one frequency contributing to the waveform. The addition of multiple frequencies changes the appearance of the

Table 7–1. **MEASURABLE VARIABLES OF SINGLE POTENTIALS**

Size	Amplitude, area, duration
Configuration	Rate of change, direction, number, and timing of reversal of direction
Recurrence	Rate, pattern, timing
Distribution	Field, area, location
Relationship to other waveforms or events	Time-locked, latency, order of interpotential interval
Stability	Pattern and type of change with recurrence over time

Table 7–2. **MEASURABLE VARIABLES OF CONTINUOUS WAVES**

Size	Peak-to-peak amplitude, root mean square, power
Frequency	Cycles per second, zero crossing
Appearance	Usually sinusoidal, frequency bands
Distribution	Field or area, symmetry
Relationship to other waves	Phase relation, synchrony

Table 7–3. **VOLTAGES, DISPLAY TIMES, AND FREQUENCY OF COMMON SIGNALS IN CLINICAL NEUROPHYSIOLOGY**

	Voltage, mV	Time, ms	Frequency, Hz
Electromyography	50–1000	20–1000	32–16,000
Nerve conduction studies	1–20,000	10–500	1–8000
Electroencephalography	1–2000	5000–200,000	0.1–1000
Brain stem auditory evoked potentials	0.1–2	5–20	
Somatosensory evoked potentials	0.1–20	50–200	20–3000
Visual evoked potentials	1.0–200	100–200	20–3000
Skin potentials	100–5000	1000–10,000	0.1–100
Electrocardiography	1000–5000	10,000–50,000	0.5–100
Respiratory movements	50–2000	5000–200,000	32–10,000
Electronystagmography	1000–5000	5000–100,000	1–2000
Electroretinography	1000–5000	500–2000	1–500
Vascular reflexes	1000–5000	1000–5000	0.1–100

wave from a simple sinusoidal pattern to a more complex, varying one. Continuous waves can be analyzed with regard to their frequency components and the power of each component. Polarity or direction is seldom described because the waves are continuous. Frequency analysis can provide a precise measurement of the waveform, but it requires defining the amount of each of the component frequencies. This is sometimes done with frequency bands (Table 7–3).

The distribution of continuous waves is another variable that can be measured. It is usually described as broad areas, and comparisons are made between homologous areas of the body for symmetry. The relationship of continuous waves to other waves in the same or other areas is another important variable that is measured to identify alterations produced by disease. Waveforms may occur in synchrony for defined periods or they may not be in synchrony but still have a definable time relationship, the *phase relation*. Waves may be in phase or out of phase.

Measurement of the timing, frequency, and the spatial distribution of the waves can provide valuable information about the presence and the stage of disease.

Signal Display

The single potentials and continuous waves generated by neural and muscle tissues can be recorded as analog or digital signals. Most modern equipment uses a digital format that allows the signals to be readily stored for subsequent review and analysis. This capability makes it possible to analyze signals without displaying the raw data, showing only the processed data. Although this can improve the recording efficiency, it has the risk of recording and analyzing unwanted signals, such as the artifacts described in the following section. Thus, it is preferable to display the raw, unprocessed signal for review before proceeding to analyze the information. The human eye and ear are better than automated systems for recognizing artifact. For example, the raw signal recorded during evoked potential testing should be displayed along with the averaged potential during data collection.

Unprocessed signals are best displayed as a horizontal trace in which the horizontal axis (sweep) is time and the vertical axis is voltage change. The sweep speed and amplification vary widely with the many different forms of signals (Table 7–3). Multiple signals from different areas are often recorded simultaneously as vertically separated lines.

There are many formats for displaying processed data. The commonest one is a line format, as used for averaged signals. Results may also be shown as histograms, bar graphs, numerical tables, topographic maps, or frequency plots (e.g., compressed spectral arrays). Statistical analysis of the data is used with many of these displays. The assumptions of any statistical analysis performed must be understood and appropriate for the problem to be solved.

ARTIFACTUAL WAVEFORMS

Artifacts are unwanted signals that are generated by sources other than those of interest. Artifacts are signals that are not of clinical value, which can be classified as signals from living tissue (physiologic artifact) or as signals from other sources (nonphysiologic artifact).

Physiologic Artifacts

Physiologic artifacts are unwanted noise which in other settings are the signals of in-

terest. These include: (1) the electrocardiogram—a relatively high-amplitude, widely distributed potential generated by heart muscle during any clinical neurophysiologic recording; (2) electromyographic signals that accompany muscle contractions during EEG and evoked potential recordings; (3) potentials that occur with the movement of electrically charged structures (e.g., tongue movement, eye movement or blink); and (4) autonomic nervous system potentials, such as those arising from changes in skin impedance with perspiration. Common artifacts are listed in Table 7–4. Each of these signals and other waveforms that may be recorded to study a particular structure in one setting may be an artifact that interferes with the recording of a different signal in another setting (Fig. 7–1). Although physiologic artifacts are phenomena that cannot be dissociated from normal function, they must somehow be circumvented. Decreasing the

Table 7–4. **COMMON FORMS OF ARTIFACT AND INTERFERENCE**

Source	Appearance
Movement of charged structures	
Eye movement	Slow positive, lateralized
Eye blink	V-shaped positive
Tongue movement	Slow positive
Eye flutter	Rapid, rhythmic, alternating
Normal activation	
Muscle potentials	Rapid, recurrent spikes
Perspiration	Very slow oscillation
Electrocardiogram	Sharp and slow, regular
Dental fillings touching	Short spikes
Transcutaneous stimulator	70- to 150-Hz spikes
Cardiac pacemaker	1-Hz spike
Paging and radio signals	Intermittent, recognizable sound
Recording system	
Electrode movement	Irregular, rapid spike
Wire movement	Irregular, slower waves
Poor electrode contact	Mixture of rapid spikes and 60 Hz
Rubbing materials, static	Sharp spikes
Display terminal	300 Hz, regular
White thermal noise	Random, high-frequency
Electromagnetic, external	
60 cycle, equipment	Regular, 60 Hz
Switch artifact	Rapid spikes
Diathermy	Complex, 120 Hz
Cautery	Dense, high-frequency spikes

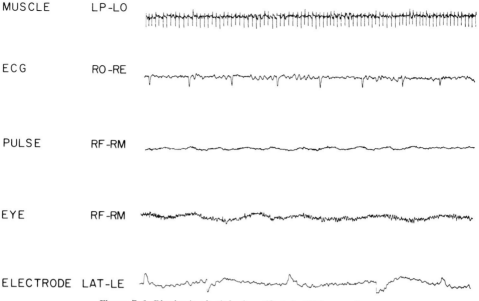

Figure 7–1. Rhythmic physiologic artifacts in EEG recordings.

level of muscle activity can help circumvent electromyographic artifact. Another method is to filter out unwanted frequencies. In some cases, physiologic artifacts need only to be recognized and then mentally discounted or subtracted electronically, for example, an eye movement artifact on EEG records.

Nonphysiologic Artifacts

Nonphysiologic artifacts are from technical sources, for example, the recording electrodes, the electrical amplification and display system, and the external electrical devices or wiring. The commonest source of such artifact is movement of the wires that connect the electrodes to the equipment or movement of the electrodes on the skin. Movement of the electrodes on the skin causes change in the electrical charge and capacitance that exist at the interface of the electrode and the skin. Alteration in static field or electromagnetic induction with wire movement can produce large artifacts. Artifacts can arise from the opening and closing of switches on the equipment; poor recording electrode connection, with high resistance of the electrodes; and the use of dissimilar metals. Spurious signals generated

within the recording apparatus are most commonly a 60-Hz or a 300-Hz signal.

Several external power sources generate specific artifacts; examples include the 60-cycle signal caused by electromagnetic radiation from power lines; the modified 60-cycle signal caused by fluorescent lights; the high-frequency, complex discharges from cautery and diathermic equipment; and the irregular waveforms from radio sources (Fig. 7–2).

Artifacts are sometimes referred to as *interference*, because they interfere with recording the activity of interest. By recognizing the nature and source of an artifact, clinical neurophysiologists can often reduce or eliminate it by changing the electrodes or by changing the location of the equipment or its relationship to the power source. At times, averaging can reduce activity if it is not time-locked to the stimulus. Differential amplification that is used in all modern recording equipment markedly reduces the external artifacts. Appropriate grounding can also help.

Continuously occurring artifacts are sometimes referred to as *noise* and compared with the signal as a *signal-to-noise ratio.* This ratio determines the likelihood of eliminating the artifact by averaging the signal.

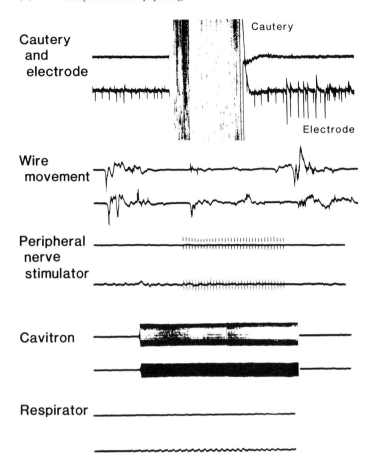

Figure 7–2. Nonphysiologic artifacts recorded during surgical monitoring of muscle activity.

CHAPTER SUMMARY

Any of the variables used to describe the continuous and discrete waveforms recorded in clinical neurophysiology can be altered. Changes in amplitude, frequency, and distribution of waveforms occur in continuous waveforms. Frequency change may include the addition of new, abnormal frequencies, the loss of normal frequencies, and either an increase or decrease in amplitude. Discrete events themselves may be abnormal. The configuration, distribution, size, and pattern of normally occurring discrete events may be changed by disease.

SECTION 2
Electrophysio-logic Assessment of Neural Function
PART A
Cortical Function

In clinical neurophysiology, neural function is assessed by measuring the electrical waveforms generated by neural tissue and the changes in these waveforms produced by disease. The characteristics of the waveforms and their alteration with disease are a function of the neural generators producing the waveform. A particular modality of recording in clinical neurophysiology reflects only the alteration in the area of the nervous system generating the activity. Thus, the topics in this book are arranged according to the generators of the electrical activity. For example, electroencephalography (EEG) records the waveforms arising from the cerebral cortex; the waveforms are recorded using electrodes placed on the scalp. These waveforms may change indirectly with disease elsewhere (e.g., the slowing that occurs in cerebral ischemia because of reduced cardiac output). The electroencephalographic (EEG) recordings described in Chapters 8 and 9 reflect disease processes that directly involve the cerebral cortex. Some diseases involving the cerebral cortex, such as Alzheimer's disease, do not cause a change in its electrical activity because the disease has no effect on the neural generators.

The EEG records both the ongoing spontaneous electrical activity of the cerebral cortex and the cortical response to external stimuli. These patterns of responses can provide important clues to the underlying disease process.

Variations of standard EEG recordings have been developed which provide other kinds of information in specific situations. Longer recordings are used to document an infrequent episodic event that is a symptom or a sign of a disease to better identify the nature, character, and spread of the event.

Each of these variables is important in determining the optimal evaluation and therapy for the disease.

Some abnormalities occur infrequently and may not be detected with a standard 30- to 60-minute EEG recording. For abnormal electrical activity that occurs only in an outpatient setting or under other specific circumstances, a fuller picture of the changes can be obtained with ambulatory recordings, described in Chapter 10, which are made while the patient is engaged in activities at home or at work. To help define the nature and origin of frequent seizures, a patient undergoes prolonged EEG recording in which the electrodes are left in place for several days while continuous recordings are made as described in Chapter 11. The EEG can also be recorded in other specialized situations, for example, in an intensive care unit, operating room, or with computerized quantitation as described in Chapter 12. Patients being considered for epilepsy surgery require highly specialized recordings, as described in Chapter 13. Cortical function can also be assessed with nonspontaneous potentials, such as those that occur before a planned movement or in response to external stimulation, described in Chapter 14.

CHAPTER 8

ELECTRO-ENCEPHALOGRAPHY: GENERAL PRINCIPLES AND ADULT ELECTRO-ENCEPHALOGRAMS

Donald W. Klass, M.D.
Barbara F. Westmoreland, M.D.

The Glossary of the International Federation of Clinical Neurophysiology defines *elec-troencephalography* (EEG) as "(1) The science relating to the electrical activity of the brain. (2) The technique of recording electroencephalograms."[14] These definitions encompass EEG both for use as a research tool and as a clinical diagnostic procedure. The purpose of this chapter, however, is to survey the clinical applications of the product (electroencephalograms) without consideration of the actual recording technique.[54,62,72] Two additional points should be mentioned. First, in practice it is often important to monitor other physiologic variables simultaneously with the EEG. Second, detection and interpretation of the EEG data derived from visual analysis involve matters of judgment and experience, which render clinical EEG an art as much as a science.

DISPLAY

Preference for the use of particular montages to display EEG activity varies considerably among different electroencephalographers. The American EEG Society has recommended that all EEG laboratories in-

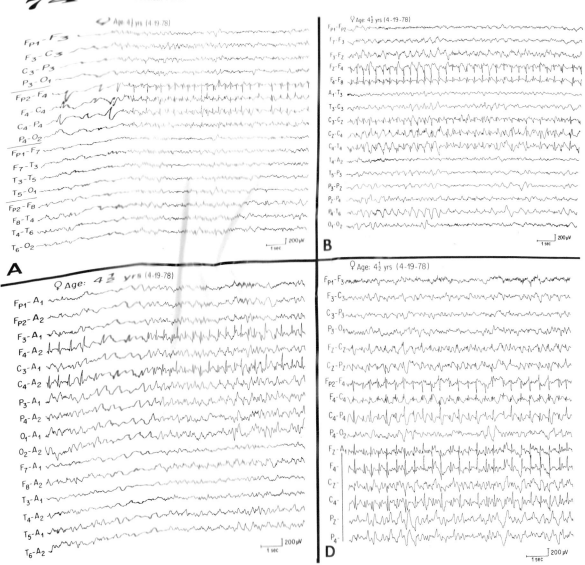

Figure 8–1. EEG from a 4½-year-old girl, showing the appearance of (*A*) focal right frontocentral (F₄,C₄) spikes in a longitudinal (anteroposterior) bipolar montage, (*B*) focal right frontocentral spikes in a transverse bipolar montage, (*C*) focal right frontocentral spikes in a referential montage, and (*D*) focal right frontocentral spikes in a montage combining bipolar and referential recording.

clude a minimum number of standard montages for basic recording to facilitate communication among different laboratories.[2] Included in the minimum recommendations are choices for longitudinal bipolar (Fig. 8–1*A*), transverse bipolar (Fig. 8–1*B*), and referential displays (Fig. 8–1*C*). All EEG laboratories are strongly encouraged to comply with these recommendations.

Because any one type of montage may be inadequate to solve a particular problem, most electroencephalographers consider it advisable to use a combination of display systems for each patient and to emphasize whatever array seems to be most useful in each case (Fig. 8–1*D*). One should be able to select the most advantageous arrays for particular types of activity to localize that activity adequately or to display it more prominently if it seems equivocal. Therefore, proper facility and flexibility in recording require, first, a clear understanding of the rationale for the different display systems and, second, considerable practice in visualization and recording of the actual EEGs.

ACTIVATION PROCEDURES

Activation procedures are commonly used in standard EEG practice to determine whether the response to provocation is normal or abnormal.[6,44,69] Commonly, these procedures are used to elicit abnormal activity, particularly paroxysmal activity, when the basic resting record has been uninformative. The simple act of eye opening is universally used in EEG practice to test for reactivity of normal or abnormal activity, but the act of eye closure can sometimes provoke paroxysmal discharges of the spike-wave variety. Eye closure can also provoke a nonspecific abnormality in the form of bisynchronous rhythmic posterior slow waves (posterior phi).

A second commonly used activation procedure is *hyperventilation*. In patients who can cooperate adequately, hyperventilation is performed for 3 to 5 minutes and recording is continued for at least 2 to 3 minutes afterward. In many normal persons, but particularly in children, hyperventilation induces bursts of generalized slow waves (Fig. 8–2).

This so-called hyperventilation buildup usually subsides promptly after cessation of the overbreathing. With hypoglycemia, the response may be exaggerated in magnitude and persistence. The most common abnormality elicited by hyperventilation is the diffuse and bilaterally synchronous 3-Hz spike-and-slow-wave discharge, which may sometimes be accompanied by the clinical manifestations of typical absence seizures. Less frequently, hyperventilation may activate focal spikes or increase the prominence of focal slow waves.

Another activation procedure used in most EEG laboratories is *stroboscopic intermittent photic stimulation*. Stimulation should include frequencies from 1 to 30 Hz and must include frequencies between 10 and 20 Hz, measuring first with the patient's eyes open and then closed. Most normal subjects show some evidence of flash-related responses over the posterior head regions (*photoentrainment* or *photic driving*), but absence of grossly detectable responses or transient asymmetries of the driving response are not considered abnormal.

Two other principal types of phenomena may be activated by photic stimulation. The *photoparoxysmal* response (previously termed *photoconvulsive*) usually takes the form of a generalized and bilaterally synchronous burst of spike-and-slow-wave complexes that may or may not outlast the photic stimulus.

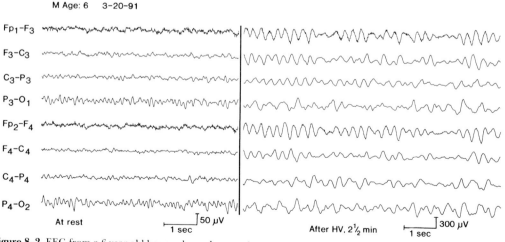

Figure 8–2. EEG from a 6-year-old boy awake and at rest (segment on left) and normal response to hyperventilation (*HV*) (segment on right).

These paroxysmal abnormalities have a high correlation with clinical seizures, particularly in patients who have a clinical history of light sensitivity in their natural environment.[44] The discharges may be associated with clinical manifestations of myoclonus or absence attacks. The technologist must terminate the photic stimulation when a photoparoxysmal response occurs; otherwise, prolonged stimulation may precipitate a generalized tonic-clonic seizure. Lesser degrees of photoparoxysmal responses may be confined to the posterior head regions; they are less likely to be associated with clinical manifestations.

Another type of response that needs to be differentiated from the photoparoxysmal response may resemble cerebral activity but is actually due to myogenic potentials that follow the flash frequency, particularly in the frontal regions. This latter type of response is known as the *photomyogenic* or *photomyologic* response (previously termed *photomyoclonic*).[44] Typically, it shows recruitment during the period of intermittent photic stimulation and never outlasts the flashes. Although this type of response is less common than the photoparoxysmal variety, it does not signify an abnormality unless the response is grossly exaggerated, widespread, and is associated with generalized myoclonus.

Rarely, photic stimulation can activate focal spikes in the occipital regions. Exaggerated responses to low flash frequencies recorded from the occipital regions may also be indicative of an encephalopathy, such as ceroid lipofuscinosis in children and Jakob-Creutzfeldt disease in adults. In some patients with sensitivity to light, activation of paroxysmal discharges may also occur when scanning geometric patterns, watching television, or playing video games.

Recording during sleep is an important means of activating focal or generalized paroxysmal abnormalities. Significant generalized spike-wave discharges and focal temporal sharp waves are usually activated best during deeper levels of nonrapid eye movement (NREM) (slow wave) sleep. Daytime sleep in the laboratory can usually be achieved after partial overnight sleep deprivation and by foregoing stimulating beverages or medications before the recording.

Total overnight sleep deprivation, however, may be an activator in itself. If the patient cannot fall asleep spontaneously, sedation may be required. The most commonly used sedative in EEG laboratories is chloral hydrate, because it does not induce beta (fast) activity that may complicate interpretation of the tracing, whereas barbiturates and most benzodiazepines typically induce beta activity and are generally avoided for that reason.

Other activation procedures may need to be used in special circumstances. Every effort should be made to attempt to reproduce the particular stimulus that may have triggered the patient's symptoms. In addition to the light sensitivity mentioned above, activation in the visual domain may result from reading, from stimuli in the auditory system (simple sounds or complex music), from somatosensory stimulation in a localized region of the body, or from mental activity such as arithmetic calculation, to mention only a few examples. Activation may involve highly individual stimulus-response characteristics.[44]

ARTIFACTS

Artifact in EEG refers to any electrical signal that is not generated directly by the brain. Artifacts are frequent contaminants of EEG recordings because of the high sensitivity of the instrumentation required to amplify the EEG, and artifacts can mimic almost every kind of EEG pattern.[66]

The technologist and electroencephalographer need to be constantly alert to the possibility of artifact.[72] Excessive artifact can render the EEG uninterpretable; even worse, confusion of artifact with brain wave activity can lead to serious misinterpretation.

NORMAL EEG ACTIVITY OF ADULTS

Awake State

The activity seen in the EEGs of awake adults consists of frequencies in the alpha and beta ranges, with the alpha rhythm constituting the predominant background activity.

ALPHA RHYTHM

Alpha activity refers to any activity in the range between 8 and 13 Hz, where *alpha rhythm* is a specific rhythm consisting of alpha activity occurring over the posterior head regions when the person is awake and relaxed and has the eyes closed (Fig. 8–3); it is attenuated by attention. The alpha frequency is very stable in a person, rarely varying by more than 0.5 Hz[39,46,52] over many years of adult life, and the frequency is identical on the two sides of the head.

The usual alpha amplitude in an adult is 15 to 50 μV. The maximal amplitude occurs over the occipital region, with variable spread to the parietal, temporal, and, at times, central leads. Often the alpha activity has a higher voltage and wider distribution over the right hemisphere.

The alpha rhythm should attenuate bilaterally and promptly with eye opening, alerting stimuli, or mental concentration. Some alpha rhythm may return when the eyes remain open for more than a few seconds. Failure of the alpha rhythm to attenuate on one side with either eye opening or mental alerting indicates an abnormality on the side that fails to attenuate.[3,75]

BETA ACTIVITY

Beta activity has a frequency greater than 13 Hz.[14] The average voltage is between 10 and 20 μV.[47] There are three main types of beta activity, based on distribution: (1) the precentral type occurs predominantly over the frontal and central regions, increases with drowsiness, and may attenuate with bodily movement; (2) posterior dominant beta activity can be seen in children up to 1 to 2 years of age; it also is enhanced by drowsiness; and (3) generalized beta activity maximal over the frontocentral regions is induced or enhanced by certain drugs, such as the benzodiazepines and barbiturates, and may attain an amplitude greater than 25 μV[39,47,52] (Fig. 8–4).

THETA ACTIVITY

Theta activity is only a minimal component of the EEG of a normal awake adult, but it is more prevalent during drowsiness. In young children, theta activity is a common component of the background activity. Adolescents retain somewhat more theta activity than middle-aged adults. In older patients, theta components can occur as single transients or as part of a mixed alpha-theta burst over the temporal regions.[39,46]

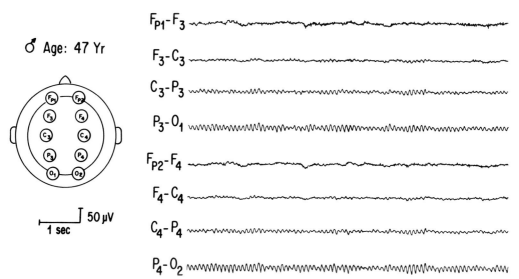

Figure 8–3. Normal EEG in a 47-year-old man, showing symmetric alpha rhythm predominantly in the occipital regions (O_1 and O_2).

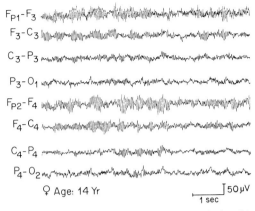

F_{P1}-F_3

F_3-C_3

C_3-P_3

P_3-O_1

F_{P2}-F_4

F_4-C_4

C_4-P_4

P_4-O_2

♀ Age: 14 Yr   50 µV
                              1 sec

Figure 8–4. EEG containing diffuse beta activity in a 14-year-old girl on diazepam therapy.

MU RHYTHM

Mu rhythm consists of arch-shaped waveforms with a frequency of 7 to 11 Hz. It typically occurs independently on the two sides of the head over the central regions.[24,42,47,63,64,67] Mu activity is functionally related to the sensorimotor cortex and is attenuated by active or passive movement of the extremities or by the thought of movement (Fig. 8–5). It can occur in an asymmetrical fashion or predominate over one hemisphere. Mu rhythm, which can be quite striking if there is an overlying skull defect, needs to be distinguished from pathologic spike activity.[17]

LAMBDA WAVES

The *lambda wave* has a configuration resembling the Greek letter "λ" and occurs over the occipital regions when the subject actively scans a picture.[22] Lambda waves appear to represent an evoked cerebral response to visual stimuli produced by movements of images across the retina with saccadic eye movements.[12] The waveforms are monophasic or diphasic, with the most prominent component usually surface-positive (Fig. 8–6). The amplitude is usually between 20 and 50 µV and the duration is from about 100 to 250 milliseconds. Lambda waves are bilateral and synchronous but may be asymmetrical.

EEG IN OLDER ADULTS

Previously, it was thought that a shift to slower background frequencies occurred in the older population. However, several recent studies have shown that the alpha frequency remains above 8 Hz in normal elderly subjects.[37,52] There is a tendency for the alpha rhythm to be of slightly lower voltage in older subjects and perhaps to show less reactivity.[5,52]

Benign temporal slow transients consist of sporadic slow-wave components in the theta and delta frequencies, occurring singly or in brief groups over the temporal region. Tem-

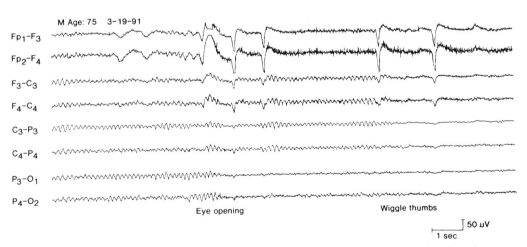

Figure 8–5. Normal EEG showing bilateral mu rhythm in the central regions; the rhythm persists when the eyes are opened and attenuates with movement of the thumbs. This is in contrast to the alpha rhythm (O_1 and O_2), which is attenuated by eye opening.

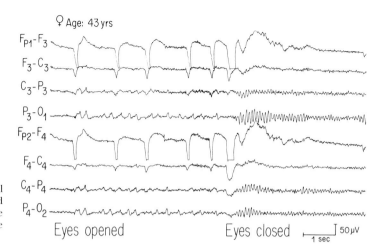

Figure 8–6. EEG showing normal lambda waves maximal in O_1 and O_2 when the patient's eyes are open and looking around the room.

poral transients can be seen in normal subjects after the age of 40. The transients have a left-sided preponderance, appear to be related to a normal aging process, and increase in prevalence after the sixth decade.[46] Drowsiness facilitates the appearance of temporal transients.

Drowsiness

In adults, drowsiness is typically associated with slowing of the background frequency, followed by disappearance of alpha activity and enhancement of theta activity. At times, bursts of generalized moderate-to-high-amplitude rhythmic 5- to 7-Hz theta activity can be present.[65] In older subjects, there may be an enhancement of theta and delta waves over the temporal regions. In addition, some admixed sharply contoured waveforms (*wicket spikes* or *wicket waves*) may also be present over the temporal regions.[61] This activity can occur in an asymmetrical fashion and may be maximal over the left temporal region.

The beta activity over the frontocentral regions is often increased in prominence during drowsiness. This usually has a frequency of 16 to 20 Hz, but occasional bursts of faster frequencies may occur.[46]

Mu activity may also be seen during drowsiness and may persist after the alpha rhythm disappears.

Sleep State

Sleep activity consists of slow waves, spindles, V waves, K complexes, and positive occipital sharp transients of sleep (POSTS).

In adults, the sleep spindles of stage 2 slow wave sleep (NREM sleep) usually have a frequency of 14 Hz and occur in a symmetrical and synchronous fashion over the two hemispheres in the central regions (Fig. 8–7). In a slightly deeper level of stage 2 sleep, the spindle frequency decreases to about 12 Hz and maximal amplitude is located more anteriorly. More continuous spindle activity may be seen in some patients who are receiving drug therapy, particularly benzodiazepines.

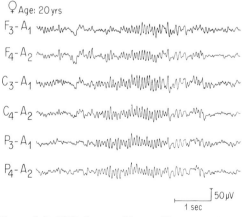

Figure 8–7. EEG from a 20-year-old woman during sleep, showing normal 14-Hz sleep spindles.

Changes during drowsiness and sleep occur in normal elderly subjects. During drowsiness, some subjects exhibit rhythmic trains of high-amplitude delta activity that need to be carefully distinguished from the pathologic intermittent rhythmic delta activity which they may closely resemble. During stage 2 of NREM sleep, the V waves and K complexes are lower in amplitude and less sharp in appearance than in young adults, and sleep spindles are less prominent.

VERTEX SHARP TRANSIENTS

Vertex sharp transients (V waves) are sharp-contoured transients that occur maximally over the central vertex region during sleep (Fig. 8–8). In children and young adults, V waves may have a sharp or spiky appearance and attain high voltages. They are typically symmetrical in the central leads but may show transient asymmetries at the time of sleep onset. F waves, or frontally dominant V waves, are often broader than the centrally dominant V waves and may extend asymmetrically into the lateral frontal regions.

The *K complex*[50] is a diphasic or polyphasic wave that is maximal at the vertex and is frequently associated with spindle activity. It represents a nonspecific response to afferent stimulation and is generally linked to the arousal mechanism.

POSTERIOR OCCIPITAL SHARP TRANSIENTS OF SLEEP

Posterior occipital sharp transients of sleep (POSTS) are sharp-contoured surface-positive transients that occur singly or in clusters over the occipital regions (Fig. 8–9). They are usually bilaterally synchronous but may be somewhat asymmetrical. POSTS are predominantly seen during light-to-moderate levels of sleep and should not be mistaken for abnormal sharp waves.

RAPID EYE MOVEMENT SLEEP

During rapid eye movement (REM) sleep, the EEG shows a low-voltage pattern that has some similarities to an awake pattern when the eyes are open, and it also shows intermittent groups of irregular eye movement artifacts. In addition, rhythmic groups of saw-toothed waves may occur intermittently

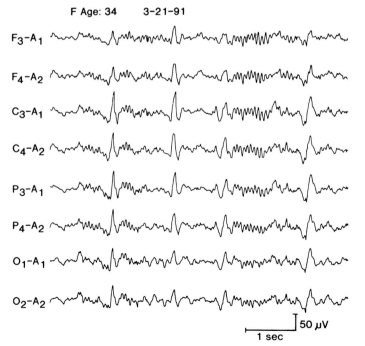

F Age: 34   3-21-91

F3-A1
F4-A2
C3-A1
C4-A2
P3-A1
P4-A2
O1-A1
O2-A2

50 µV
1 sec

Figure 8–8. EEG from a 34-year-old woman during sleep, showing normal V waves maximal in C_3 and C_4.

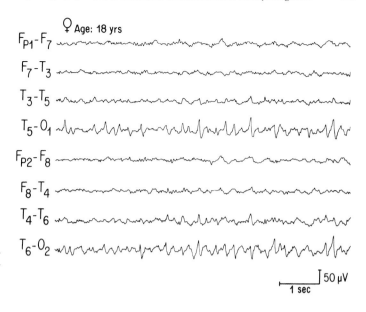

Figure 8–9. EEG during sleep showing prominent normal posterior occipital sharp transients of sleep (POSTS) maximal in O_1 and O_2.

over the frontal and central leads and may precede the rapid eye movements.[63]

BENIGN VARIANTS

Variants During Wakefulness

ALPHA VARIANTS

The alpha-variant patterns consist of activity over the posterior head regions, which has a harmonic relationship to the alpha rhythm and which shows a reactivity and a distribution similar to those of the alpha rhythm.[35] The *slow alpha variant* appears as dicrotic or notched waveforms that result from a subharmonic component of the alpha rhythm, usually in the range of 4 to 5 Hz. The *fast alpha variant* contains a frequency twice that of the resting alpha activity.

PHOTIC RESPONSES

Complex waveforms may be induced by photic stimulation when harmonics or subharmonic components are admixed with the fundamental frequency of the driving response. Occasionally, the resultant mixture of frequencies can produce waveforms that simulate epileptiform spikes or spike-wave complexes.

SUBCLINICAL RHYTHMIC ELECTROGRAPHIC DISCHARGE OF ADULTS

Subclinical rhythmic electrographic discharge of adults (SREDA) is an uncommon phenomenon that occurs mainly in elderly adults.[76] It closely resembles an epileptogenic seizure discharge but is never accompanied by any clinical symptoms and has no significance for the diagnosis of epileptic seizures (Fig. 8–10). The characteristics of SREDA are listed in Table 8–1.

BREACH RHYTHM

Various normal rhythms are enhanced in amplitude when recorded over a skull defect. The term *breach rhythm* has been used to refer to a focal increase in the amplitude of sharp-contoured EEG activity over or near the area of a skull defect.[17] Particularly when the mu and beta rhythms are involved, the activity can resemble epileptiform spikes (Fig. 8–11).[17]

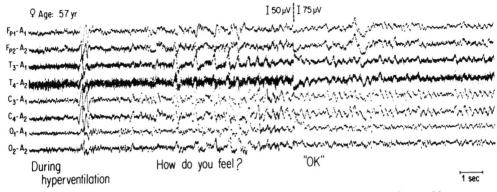

Figure 8–10. Onset of subclinical rhythmic electrographic discharge of adults in a 57-year-old woman.

Table 8–1. CHARACTERISTICS OF SUBCLINICAL RHYTHMIC ELECTROGRAPHIC DISCHARGE OF ADULTS

Feature	Characteristic
Onset	Segmented or abrupt
Repetition rate	Theta
Duration	Average, 1 minute
Distribution	Maximal parietal-posterior temporal
Laterality	Symmetrical or asymmetrical
State	Awake at rest or during hyperventilation
Background activity	Often visible during the discharge
Delta aftermath	None
Clinical accompaniment	None
Patient age	Mainly elderly

Variants During Drowsiness and Sleep

Several patterns that have distinctive characteristics but little or no clinical significance occur principally during drowsiness or light sleep.[45] Some of these have an appearance suggestive of epileptiform abnormality, but they have no importance for the diagnosis of seizures or cerebral lesions.

RHYTHMIC TEMPORAL THETA BURSTS OF DROWSINESS

The *rhythmic temporal theta bursts of drowsiness* (previously known as the *psychomotor-variant pattern*[31]) frequently assume a flat-topped or notched appearance because of the harmonics of the fundamental theta frequency. The bursts may occur bilaterally or independently over the two temporal regions, with a shifting emphasis from side to side (Fig. 8–12). This pattern differs from a true seizure discharge in that it does not evolve into other frequencies or waveforms. It is present predominantly in young adults.

14&6-Hz POSITIVE BURSTS

The *14&6-Hz positive bursts* (previously known as *14 and 6 per second positive spikes*[28,29]) are displayed best on long inter-electrode distance referential montages and are most prominent over the posterior temporal region during light sleep. As the name implies, the bursts occur at a rate of 14 Hz (Fig. 8–13A) or between 6 and 7 Hz (Fig. 8–13B) and are from 0.5 to 1 seconds in duration. They usually occur independently over the two hemispheres and vary from side to side in occurrence. They are most frequently seen in subjects between 12 and 20 years old.[49,80]

SMALL SHARP SPIKES

Small sharp spikes (SSS),[27] also known as *benign epileptiform transients of sleep (BETS)*[79] or

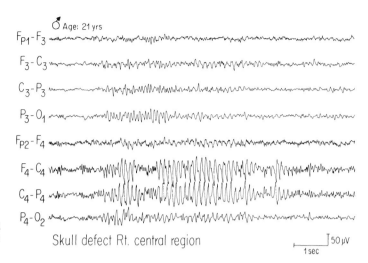

Figure 8–11. EEG from a 21-year-old man, showing breach rhythm (C₄) over the area of a skull defect.

Skull defect Rt. central region

benign sporadic sleep spikes (BSSS),[45] occur mainly in adults during drowsiness and light levels of sleep (Fig. 8–14). They are usually low-voltage, short-duration diphasic spikes with a steep descending limb. The SSS may have a single low-voltage aftercoming slow-wave component, but they do not distort the background and they are not associated with rhythmic slow-wave activity, as temporal sharp waves are. They are best seen with long interelectrode distance derivations, includ-ing the temporal and ear leads. Provided a long enough recording is obtained, they almost always have a bilateral representation, occurring either independently or synchronously over the two hemispheres. Their characteristics are summarized in Table 8–2. They need to be carefully distinguished from more important types of spikes,[59] because the SSS have no significance for the diagnosis of epileptic seizures.

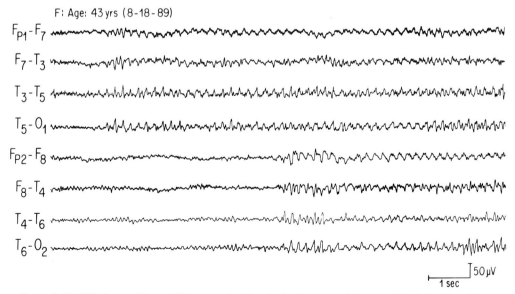

Figure 8–12. EEG from a 43-year-old woman, showing rhythmic temporal theta activity during drowsiness.

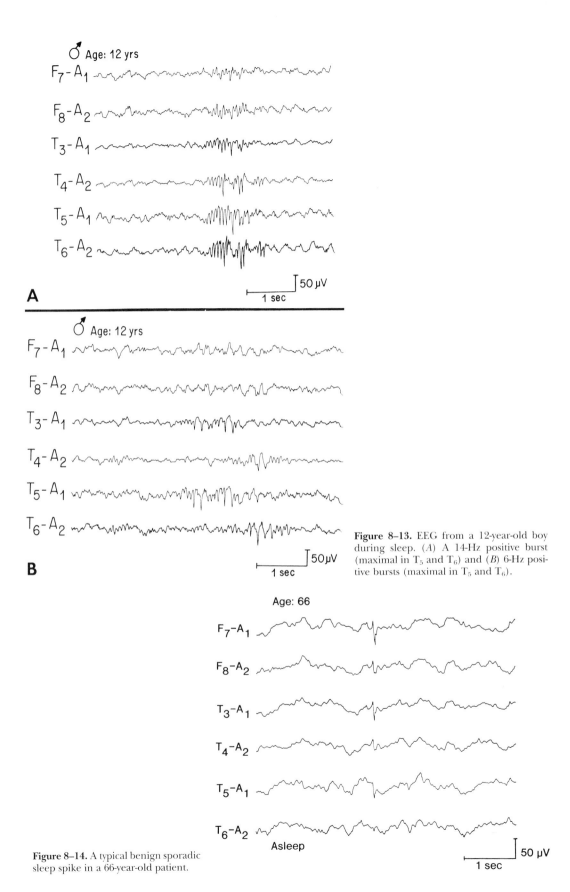

A

Age: 12 yrs

F_7-A_1

F_8-A_2

T_3-A_1

T_4-A_2

T_5-A_1

T_6-A_2

50 μV

1 sec

B

Age: 12 yrs

F_7-A_1

F_8-A_2

T_3-A_1

T_4-A_2

T_5-A_1

T_6-A_2

50 μV

1 sec

Figure 8–13. EEG from a 12-year-old boy during sleep. (*A*) A 14-Hz positive burst (maximal in T_5 and T_6) and (*B*) 6-Hz positive bursts (maximal in T_5 and T_6).

Age: 66

F_7-A_1

F_8-A_2

T_3-A_1

T_4-A_2

T_5-A_1

T_6-A_2

Asleep

50 μV

1 sec

Figure 8–14. A typical benign sporadic sleep spike in a 66-year-old patient.

Table 8–2. **CHARACTERISTICS OF BENIGN SPORADIC SLEEP SPIKES**

Feature	Characteristic
Amplitude	Low
Duration	Short
Morphology	Sharp, diphasic (steep descent)
Associated slow wave	None or minimal
Background activity	No disruption
Distribution	Widespread
Laterality	
Single	Maximal unilateral
Multiple	Bilateral
Occurrence	
Event	Sporadic
State	Drowsiness or light sleep
Patient age	Adult
Clinical accompaniment	None

6-Hz SPIKE-AND-WAVE

The *6-Hz spike-and-wave* pattern has also been called the *fast spike-and-wave*, because of its repetition rate, or the *phantom spike-and-wave*, because of its usual low-amplitude spike component. It occurs mainly during drowsiness and disappears during deeper levels of sleep (Fig. 8–15). It has no associated clinical manifestations and has no useful correlation with clinical seizures or other symptoms.[53,71] Its typical characteristics are listed in Table 8–3.

WICKET SPIKES

Wicket spikes[60] consist of single spike-like waveforms or trains of monophasic mu-like waveforms (Fig. 8–16). Wicket spikes, or wicket waves, have a frequency of 6 to 11 Hz and an amplitude ranging from 60 to 200 μV and are seen mainly in older adults. They occur during drowsiness and light sleep and become apparent when the alpha and other

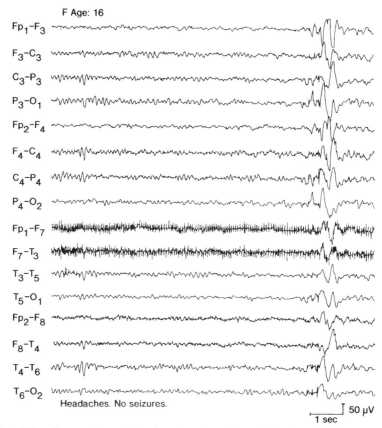

F Age: 16

Fp$_1$–F$_3$
F$_3$–C$_3$
C$_3$–P$_3$
P$_3$–O$_1$
Fp$_2$–F$_4$
F$_4$–C$_4$
C$_4$–P$_4$
P$_4$–O$_2$
Fp$_1$–F$_7$
F$_7$–T$_3$
T$_3$–T$_5$
T$_5$–O$_1$
Fp$_2$–F$_8$
F$_8$–T$_4$
T$_4$–T$_6$
T$_6$–O$_2$

Headaches. No seizures.

50 μV
1 sec

Figure 8–15. A 6-Hz spike-wave burst during drowsiness in a 16-year-old girl with headaches but no seizures.

Table 8–3. CHARACTERISTICS OF 6-Hz SPIKE-AND-WAVES

Feature	Characteristic
Frequency	6 ± 1 Hz
Repetition rate	Regular
Burst duration	Brief, <1–2 seconds
Spike duration	Brief
Amplitude	Low
Distribution	Diffuse, maximal anterior or posterior
Laterality	Bisynchronous, symmetrical
Clinical state	Drowsiness
Clinical accompaniment	None
Patient age	Young adult

awake patterns drop out. Wicket spikes are present over the temporal regions, occurring bilaterally or independently over the two temporal regions, and they may occur more frequently on one side, usually the left. When wicket spikes occur as a single waveform, they may be mistaken for a temporal-spike discharge; however, wicket spikes are not accompanied by aftercoming slow waves or a distortion or slowing of the background that occurs with a true epileptogenic temporal spike.

MITTEN PATTERNS

Mitten patterns are seen during sleep and consist of fast-wave and slow-wave compo- nents that resemble a mitten, with the thumb of the mitten formed by the last wave of a spindle and the hand portion by the slow component.[30] They may resemble, but should not be mistaken for, a spike-and-wave discharge.

PATHOLOGIC ACTIVITY

Nonspecific Types

Abnormal amounts of diffuse slow-wave activity are an indication of the severity of disturbed cerebral function but are considered nonspecific, because they give little clue as to the cause.[68] For example, similar changes may be caused by intrinsic cerebral degenerative disease, cerebral dysfunction from hypoxia, inflammatory disease involving the central nervous system, the effects of external toxins, or various systemic electrolyte or metabolic disorders. In general, with the more severe cerebral disturbance, the average frequency of abnormal activity is slower, spontaneous variability is lessened, abnormal patterns are less reactive to external stimuli, and the likelihood that normal physiologic activity is disrupted or lost is greater. Voltage increases to a point with increasing severity, but with the most severe impairment, voltage is decreased and, with cerebral death (see later section), is eventually lost entirely.

Even though nonspecific slow-wave abnormalities may provide no clue about the

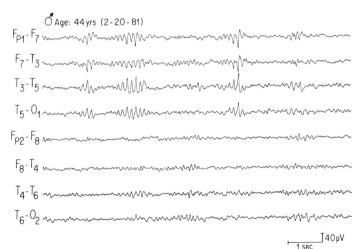

Figure 8–16. Wicket spikes in the left temporal region of a 44-year-old man.

cause, they can be helpful in indicating organic cerebral dysfunction in subjects in whom a distinction may need to be drawn between psychiatric and organic causes of mental symptoms.

Distinctive Epileptiform Patterns

Epileptiform discharges are paroxysmal waveforms with distinctive morphology that stand out from the ongoing background activity;[19,56,74] they are recorded predominantly from patients with epileptic seizures. *Interictal activity* refers to activity recorded between seizures. *Ictal discharges* are those occurring during a clinical seizure. The main types of epileptiform discharges are spikes, sharp waves, and spike-and-slow-wave complexes. *Spike discharges* are potentials with steep ascending and descending limbs, a pointed peak, and a duration <70 milliseconds.[14] *Sharp waves* are broader waveforms that have a duration of 70 to 200 milliseconds.[14] A *spike-and-slow-wave complex* consists of a spike followed by a slow wave and is often referred to as a *spike-wave pattern*. Epileptiform discharges may be focal or generalized in distribution.

INTERICTAL DISCHARGES

Focal

Anterior Temporal Spikes. The *anterior temporal spike*, or *sharp wave*, is the most frequent type of focal discharge in adolescents and adults (Fig. 8–17). Temporal spikes and sharp-wave discharges have a very high correlation with the presence of clinical partial seizures (90 to 95%).[27] Sleep markedly potentiates the presence of these discharges, especially moderately deep levels of NREM sleep.[27,40] About 30 to 50% of patients with temporal lobe epilepsy may show spikes or sharp-wave discharges during the awake recording. This increases to 90% when the patient goes to sleep. Therefore, the EEG evaluation for someone with suspected complex partial seizures should include a sleep recording.

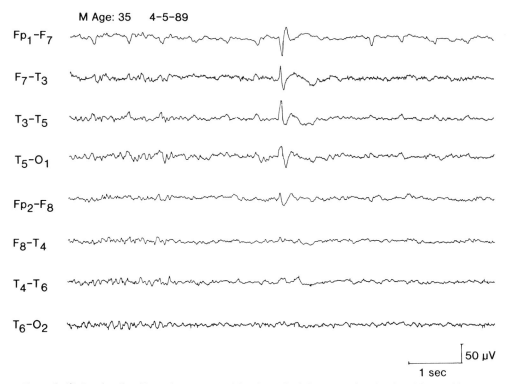

Figure 8–17. Focal epileptiform sharp waves arising from the left temporal region in a 35-year-old man.

Frontal Spike Discharge. Frontal spike discharges may occur at any age. This is another highly epileptogenic spike, and 70 to 80% of patients with a frontal spike discharge have seizures.[27,30,38] Often, some underlying pathologic condition can be demonstrated.

Centrotemporal (Rolandic or Sylvian) Spikes. The *centrotemporal (C-T) spike* is one of the more common types of focal spike discharges in children[4,33,48] (see Fig. 9–9). With standard electrode placement, the discharge is seen primarily in the central and midtemporal leads, but it is often maximal in leads halfway between these two positions (i.e., C_5 and C_6). The site of origin appears to be the lower rolandic area just above the sylvian fissure.[48] The spike discharges have a characteristic appearance: they often are high-amplitude, diphasic, blunt spikes, with an aftercoming slow wave. They may be frequent and may occur in brief clusters or trains, unilaterally, bilaterally, or shift from side to side. During sleep, C-T spikes are often activated or enhanced.[38,48] The EEG background activity is usually normal. Frequently, the surface distribution of the spike corresponds to a tangential dipole source with peak negativity in the tempororolandic region and positivity in the frontal leads.

The C-T spike discharge is seen primarily between the ages of 4 and 12 years. About 60 to 80% of children with the C-T spike discharge have seizures,[4,27,48] and the condition is termed *benign rolandic epilepsy of childhood (BREC)*. Seizures typical of this disorder, which have been termed *sylvian seizures*, consist of twitching of one side of the face or hand (or both); motor-speech arrest; excessive salivation or drooling because of difficulty swallowing; tingling of the side of the mouth, tongue, or cheek; progression to a generalized seizure; and more frequent occurrence during the night.[4,48] The seizures are easily controlled with anticonvulsant medications. The EEG pattern and seizures usually disappear spontaneously in the second decade of life, and no gross focal lesion can be demonstrated.

Occipital Spikes. Occipital spikes can be seen in young children, usually those younger than 3 to 5 years, and often resolve as the child gets older.[38] The spike discharges may be unilateral or bilateral. Occipital spikes in this age group are not highly epileptogenic;

only 30 to 50% of children with occipital spikes have seizures.[38,55] Instead, the presence of these spikes tends to correlate more with early-onset visual deprivation, and approximately 40% of the children with occipital spikes may have some associated visual problem. On the other hand, occipital spikes in older children and adults may be associated with a focal pathologic condition and associated seizures.

Diffuse. The main types of generalized epileptiform discharges are the 3-Hz spike-and-wave, slow spike-and-wave, atypical spike-and-wave discharges, and paroxysmal rhythmic fast activity.

3-Hz Spike-and-Wave. The *3-Hz spike-and-wave* pattern consists of stereotyped, generalized, bilaterally synchronous, symmetrical spike-and-slow-wave complexes that have an average repetition rate of 3 Hz (see Fig. 9–7). Usually, there is maximal voltage in the superior frontal ($F_{3,4}$) regions. These complexes are displayed best in ear reference montages. If the discharge lasts longer than 3 to 4 seconds, there is usually some type of clinical accompaniment such as staring, sursumvergence, clonic movements of the face, motor arrest, and unresponsiveness, all indicative of absence seizures.[34] The discharges and clinical seizures are enhanced by hyperventilation and hypoglycemia.[18] The interictal EEG record is usually normal. During sleep, the spike discharges occur in a more fragmented and less sustained fashion, often consisting of single spike-wave complexes or multispike-and-wave discharges of varying complexity that increase in number during deepening stages of NREM sleep.[18,19,56] The pattern is seen most often in children between the ages of 3 and 15 years.[18]

Slow Spike-and-Wave. The *slow spike-and-wave* pattern has also been referred to as *sharp-and-slow-wave complexes*, because the spike component has a duration that conforms more to the definition of a sharp wave.[9,14] The complexes occur rhythmically with a frequency of 1.5 to 2.5 Hz (see Fig. 9–8). The trains of slow spike-and-wave discharges often are not associated with any apparent clinical manifestation. However, if appropriate testing is performed, there may be some type of subtle impairment of psychomotor performance. The interictal background between the spike-and-wave bursts is often abnormal. Seizures in patients with the slow spike-and-wave pattern consist of various types and combinations of generalized seizures, such as tonic seizures, atypical absences, tonic-clonic seizures, akinetic seizures, or myoclonic seizures.[9,25,26] The slow spike-and-wave pattern is most often seen in children between 2 and 6 years of age, but it may persist through adolescence.[9,19,51]

The slow spike-and-wave pattern usually occurs in patients with some type of underlying organic pathologic condition and who also have clinical signs of cerebral damage. Many patients with this pattern of discharge have a severe convulsive disorder, with multiple types of seizures, poor response to anticonvulsants, and signs of mental and motor dysfunction. This constellation of clinical signs and EEG pattern has been referred to as the *Lennox-Gastaut syndrome.*[26,33]

Atypical Spike-and-Wave. The term *atypical spike-and-wave* refers to a generalized spike-and-wave discharge that lacks the regular repetition rate and stereotyped appearance of the 3-Hz spike-and-wave or slow spike-and-wave pattern. Atypical spike-and-wave complexes occur with varying frequencies, ranging between 2 and 5 Hz, and there may be admixed multiple spike components. When multiple spikes are a prominent feature, the discharges are known as *multispike-and-slow-wave complexes.* The paroxysms are usually brief, ranging from 1 to 3 seconds in duration. Sleep often activates these discharges.[56] The atypical spike-and-wave discharge may be seen in children or adults and occurs in patients with different types of generalized seizures, especially myoclonic and tonic-clonic seizures.

Paroxysmal Rhythmic Fast Activity. *Paroxysmal rhythmic fast activity,* or *paroxysmal tachyarrhythmia,* consists of rhythmic fast activity or repetitive spike discharges with a frequency of from 12 to 20 Hz, predominately visible in the parasagittal regions.[77] The paroxysms are usually synchronous but may be asymmetrical. They often culminate in one or more slow waves. This pattern often occurs with tonic-autonomic seizures and is seen most frequently during recordings made during sleep.

Hypsarrhythmia. The term *hypsarrhythmia* refers to a high-voltage continuous or nearly

continuous pattern consisting of a chaotic admixture of multifocal spikes or sharp waves and arrhythmic slow waves[27] (see Fig. 9–6). The hypsarrhythmic pattern is seen mainly in patients between the ages of 4 months and 4 years and is often associated with the clinical seizures known as *infantile spasms.*[36] The EEG accompaniment of infantile spasm usually consists of an initial high-amplitude diffuse spike and/or slow wave, followed by an abrupt generalized decrease of voltage with low-amplitude fast activity, called the *electrodecremental pattern.*[8] The patient may also have brief myoclonic seizures that are associated with a generalized high-amplitude spike or sharp-wave discharge in the EEG. The symptom complex of hypsarrhythmia and infantile spasms with arrest of psychomotor development, *West's syndrome,* is not a specific disease entity but reflects a response of the immature brain to a severe cerebral insult or dysfunction, usually occurring before 1 year of age. The cause is unknown in about half of the patients, but in the other half, this symptom complex may be the result of diverse prenatal, perinatal, or postnatal difficulties, such as encephalitis, congenital defects, or various biochemical or metabolic derangements.[33] One of the most frequent identifiable causes is tuberous sclerosis.

ICTAL DISCHARGES

Ictal EEG discharges consist of repetitive activity that has an abrupt onset and termination. Morphology of the activity can take any shape or form—spikes, sharp waves, spike-and-wave discharges, or rhythmic activity in the beta, alpha, theta, and delta frequency ranges,[20] or electrodecremental episodes, as mentioned previously. The ictal discharge pattern may consist of stereotyped waveforms throughout the discharge, as with the generalized 3-Hz spike-and-wave pattern during absence seizures, or the seizure pattern may evolve with changing frequencies, amplitudes, and waveforms and spreading distribution, as with most seizures of focal origin. In contrast to the 3-Hz spike-and-wave pattern of absence seizures, a generalized tonic-clonic seizure evolves in several stages.[21] The first phase consists of rhythmic fast activity (repetitive spikes) that is associated with the tonic phase of the seizure. The next phase consists of spike discharges and slow waves associated with the clonic phase of the seizure. These spike and slow-wave paroxysms gradually become separated by increasingly longer pauses and stop suddenly. Postictally, there is a temporary period of diffuse flattening followed by diffuse slow waves, which gradually subside over a variable period.

Ictal discharges of focal origin may begin in a localized fashion (Fig. 8–18), but frequently the onset is marked by an abrupt diffuse or asymmetrical decrease in voltage. Surface attenuation at onset is particularly common with discharges beginning in the temporal lobe.[40] Postictal slowing is typically focal unless the seizure has secondarily evolved into a generalized seizure. Ictal seizure discharges may last from a few seconds to 2 or 3 minutes. When a typical discharge occurs without clinical seizure manifestations, it is known as a *subclinical electrographic seizure discharge. Status epilepticus* consists of continuous or frequently repeated seizure discharges and only brief interruptions without clinical recovery.

EEG MANIFESTATIONS OF FOCAL INTRACRANIAL LESIONS

The practice of EEG has changed considerably since the advent of neuroimaging

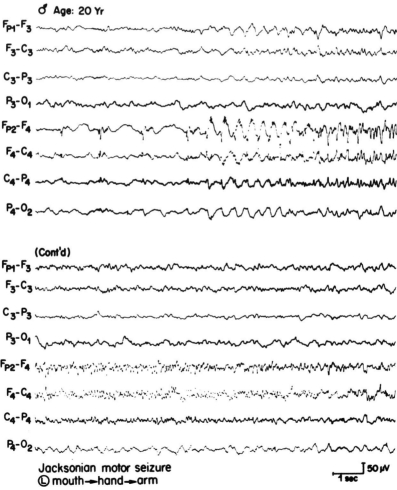

Figure 8–18. Focal seizure discharge arising from the right frontal region (F_4) and accompanied by the clinical manifestations of a jacksonian motor seizure.

techniques, and EEG is no longer the primary noninvasive screening device for focal intracranial lesions it once was. Nevertheless, electroencephalographers and referring physicians need to be aware of the electrophysiologic changes that can occur with focal lesions, because EEG offers a different measure of cerebral function than do CT and most MRI scans. The two types of procedures are complementary rather than mutually exclusive. The EEG is a measure of what the brain is capable of doing rather than a depiction of what damage has been done. Furthermore, when clinically indicated, frequent sequential EEG examinations are facilitated by virtue of being available at the bedside and of having negligible risk and modest cost.

Principles

Several different but interrelated factors influence the EEG expression of focal intracranial lesions. This concept may be summarized by the term *spatial-temporal biodynamics*. This term indicates that the EEG expression of a lesion is a dynamic physiologic process involving the dimension of time as well as of space.

Of the spatial factors, the first to consider is the extent of the lesion. A lesion, irrespective of type, may not produce any EEG abnormality until it attains sufficient size to be detectable in the usual scalp recording. A second spatial factor is density. This does not refer to the physical properties of the lesion but to the concentration of its effects on the adjacent brain. Because lesions such as neoplasms and cysts are electrically silent, the EEG abnormalities they produce are caused by the disturbance of nearby neuronal structures. Therefore, local pressure from a concentrated mass generally produces focal EEG abnormalities more readily than even a highly malignant neoplasm, which infiltrates among neurons more diffusely. A third spatial factor to consider is the location of the lesion. In general, the more superficial a lesion is in relation to the convexity of the cerebral hemisphere, the more likely it is to produce focal EEG abnormalities. Independent of pathologic type and size, lesions near the base of skull or in the posterior fossa may provoke no change or nonfocal abnormality in the scalp EEG.

Of the temporal factors, one needs to consider when the tracing is made with respect to the stage of evolution of the lesion. A single EEG recorded very early in the development of a neoplasm may be entirely normal and of no help diagnostically, but sequential recordings could be extremely helpful by contributing positive evidence for diagnosis if a focal abnormality emerges later. By way of contrast, sequential recordings made after infarction of a similar cerebral region would be expected to show maximal abnormality in the acute stage and diminishing abnormality during the course of resolution. The type and magnitude of the EEG abnormalities also depend on the age of the patient at the time the disease occurs. Some types of EEG abnormality are expressed only during a particular stage of maturation. The location and magnitude of some abnormalities also can differ between children and adults. Intermittent rhythmic delta activity produced by a deep-seated lesion, for example, is apt to be dominant over the occipital regions in children and over the frontal regions in adults. Cerebral insults often produce more prominent EEG abnormalities in children than might be expected to result from a process of similar severity in adults. Finally, the balance between destructive and reparative forces greatly affects the EEG findings. For example, a rapidly expanding neoplasm typically produces a predominance of slow polymorphic and highly persistent delta activity, in contrast to the slowly growing neoplasm whose only EEG manifestation may be a spike focus indistinguishable from the effects of scar formation.

Importance of Correct Identification

An accurate description is essential for correct identification of any type of EEG activity, whether normal, abnormal, or artifactual. Mistakes in identification can usually be avoided if the basic physical and physiologic elements are considered as completely as possible in formulating the description: (1) frequency, (2) voltage, (3) phase relationships, (4) quantity, (5) morphology, (6) topography, (7) reactivity, (8) manner of occurrence, and (9) polarity. These factors are important not only for identifying abnormalities but also for determining their degree of severity and significance.[41]

When attempting to determine the identity of any EEG phenomenon, another important consideration is its context. The matter of context includes other ongoing EEG activity and also the clinical state of the patient at the time the recording is made.[41]

EEG Manifestations

Focal intracranial lesions may produce almost any type of EEG abnormality or no EEG abnormality at all, depending on the spatial-temporal biodynamics of the individual case. Furthermore, the EEG manifestations of different pathologic types of focal lesions may be indistinguishable from one another.

LOCAL OR REGIONAL EFFECTS ON NORMAL ACTIVITY

Focal lesions located near the convexity of the cerebral hemispheres frequently produce localized attenuation of normal background activity. Depending on the location and size of the lesion, this may involve alpha activity, beta activity, sleep activity, or all background rhythms or one side of the head. Less frequently, focal lesions may result in an increase in amplitude of the background activity, a circumstance that may occur with indolent lesions or those involving subcortical nuclei. Before attributing an increase in amplitude to the lesion itself, one needs to exclude the effects of unequal interelectrode distances and underlying skull defects. One should also remember that some activity, such as the mu rhythm, may normally be asymmetrical.

Asymmetry of frequency is less common than asymmetry of amplitude. A consistent asymmetry of alpha activity of more than 1 Hz is abnormal and usually indicates a lateralized disturbance on the side of the slower frequency.

Lesions may cause defective reactivity of normal rhythms. A focal lesion may produce ipsilateral defective attenuation of alpha activity with eye opening or with mental activity when the eyes are closed.[3,75]

The significance of an asymmetry depends on its degree with respect to the limits of normal variation, its consistency during the recording, the concurrent asymmetry of other activity in the same region, and the presence of associated abnormal activity.

LOCAL EFFECTS—ABNORMAL ACTIVITY

Polymorphic focal delta activity is the hallmark of focal lesions affecting the cerebral hemispheres (Fig. 8–19). Although this type of activity does not arise directly from the lesion itself, it generally indicates proximity to the site of the maximal or active destructive process. Factors in judging the degree of severity include the following: morphology (irregular), frequency (slowest), persistence (high), amplitude (increased to a maximum and then decreased), reactivity (deficient), and associated destruction of underlying background activity. To some extent, these factors also represent acuteness.

Focal polymorphic delta activity is detected most clearly during alert wakefulness. Often, it is delineated less clearly during sleep, because of the presence of normal, more widespread delta activity. Persistence during sleep, however, has been related to superficial location and epileptogenicity of the lesion. In a patient with seizures and focal polymorphic delta activity during wakefulness, sleep may be a useful adjunct to determine whether focal distinctive epileptiform discharges arise from the same area or independently from a different area. Focal polymorphic delta activity does not help to distinguish the pathologic type of the lesion responsible for generating the abnormal activity. Although metastatic brain tumors may cause bilateral independent delta foci, that circumstance is uncommon.

Because of the important consequences of focal delta activity, artifacts that may mimic this type of abnormality need to be carefully excluded. Localized slow wave artifacts can be generated by faulty electrodes, eye or head movement, vascular pulsation, and electrodermal activity. Focal slow wave ab-

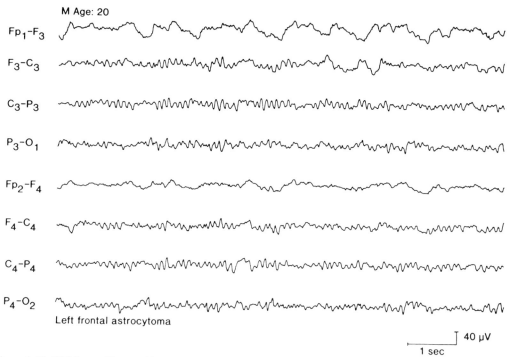

Figure 8–19. EEG from a 20-year-old man with a left frontal astrocytoma, showing persistent polymorphic delta activity predominantly in the left frontal region (F_{p1}).

normalities can also be masked by higher amplitude normal background rhythms, by diffuse abnormal rhythms, and by artifacts (Fig. 8–20). These masking effects can sometimes be diminished by maintaining the patient's alertness and by making ample recording with the patient's eyes open. Optimal techniques of recording focal slow waves include use of a long time constant, a high-frequency filter if faster activity is intrusive, a slow paper speed, and montages that display homologous derivations in adjacent channels. Various perceptual techniques have been used by individual electroencephalographers to aid in the detection of low-amplitude abnormal slow waves. According to the prevalent view, focal polymorphic delta activity is related most closely to lesions that involve the superficial white matter of the cerebral hemisphere and that interfere with cortical connections.

Focal spikes or sharp waves may accurately represent the focal origin of seizures, but they usually do not arise from the lesion itself. These distinctive epileptiform discharges may be slightly removed from the site of maximal destruction and may represent the pathophysiologic balance between destructive and reparative processes. Although these types of paroxysmal discharges usually signify a relatively chronic lesion, they may also be the first manifestation in the course of slowly progressive lesions. Well-defined spikes superimposed on normal background activity are more usually associated with chronic lesions, whereas ill-formed sharp waves superimposed on a disrupted background more often represent acute lesions. These paroxysmal abnormalities may appear only during drowsiness or sleep. Less frequently, they are activated by hyperventilation and only rarely by photic stimulation. Optimal detection requires the use of maximal high-frequency response of the recording instrument and appropriate adjustments of sensitivity. Artifacts that can simulate spikes or sharp waves include those resulting from faulty electrodes, electrostatic discharges, and chance superimposition of ECG artifact. Not all sporadic spikes have serious pathologic implications, however, as already discussed.

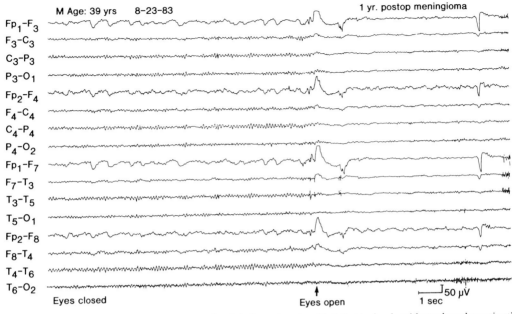

Figure 8–20. Focal polymorphic delta activity in the right frontal region (F_{p2}) that is clearly evident when the patient's eyes are open but masked by eye movement when the eyes are closed.

REGIONAL EFFECTS—ABNORMAL ACTIVITY

Some types of focal abnormal slow wave are likely to be somewhat more distant from the maximal site of the lesion, including focal delta activity that is more regular, more reactive, and less continuous than the persistent polymorphic delta activity already mentioned. Generally, this is also true of focal theta activity. Different types of slow wave abnormalities may coexist, and such contiguous rhythms need to be sorted out to determine their relative importance for localization. Artifacts such as those generated by abnormal electrocardiographic paroxysms or by rhythmic eye movements may simulate intermittent rhythmic delta activity or continuous focal theta activity. Some acute lesions are associated with periodic lateralized epileptiform discharges (PLEDs), which are helpful for demonstrating a unilateral hemispheric lesion (Fig. 8–21). In cases of ischemic cortical infarction, PLEDs and delta activity are typically present before computed tomography shows evidence of an anatomic lesion.

REMOTE EFFECTS—ABNORMAL ACTIVITY

Intermittent rhythmic delta activity is sometimes referred to as a *projected rhythm*, because when it results from a focal intracranial lesion it represents a distant effect of the lesion rather than a local superficial source. When intermittent rhythmic delta activity occurs in isolation, the lesion is generally situated deep within the cranium. However, this type of abnormality does not necessarily signify an intracranial lesion, because it is indistinguishable from abnormalities resulting from systemic metabolic or electrolyte disorders. This type of abnormality may also be a transient postictal effect. Characteristically, it is bilateral and diffusely distributed, bisynchronous, monorhythmic and monomorphic, and reactive to eye opening and alerting. It often is increased by hyperventilation, and it usually disappears during sleep. Sometimes the waveforms may be sawtoothed rather than sinusoidal. Artifacts that need to be distinguished from this type of abnormality include the electro-oculographic and glossokinetic potentials. When diffuse inter-

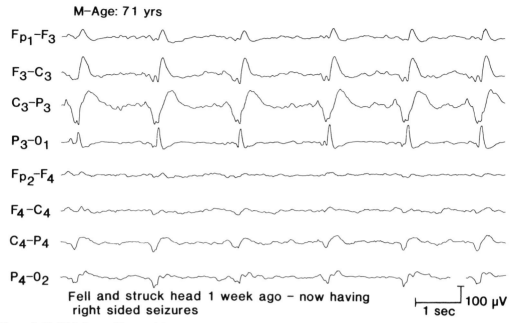

M–Age: 71 yrs

Fp_1–F_3

F_3–C_3

C_3–P_3

P_3–O_1

Fp_2–F_4

F_4–C_4

C_4–P_4

P_4–O_2

Fell and struck head 1 week ago – now having right sided seizures

100 µV

1 sec

Figure 8–21. EEG from a 71-year-old man with recent head trauma, showing left-sided periodic lateralized epileptiform discharges.

mittent rhythmic delta activity is secondary to a focal hemispheric lesion, it is often asymmetrical (more frequently of higher amplitude on the side of the lesion).

SEQUENTIAL ALTERATIONS

In a patient with seizures, development of focal delta activity despite seizure control may indicate the presence of a neoplasm. A progressive focal lesion should be suspected if focal polymorphic delta activity appears after previous EEGs have been normal or if previous EEGs have contained nonfocal, nonspecific abnormality. A progressive lesion in sequential recordings may show increased magnitude of focal polymorphic delta activity (factors of severity mentioned above), enlarged topographic distribution of the abnormality, ipsilateral attenuation of background activity, or the addition of diffuse intermittent rhythmic delta activity (Fig. 8–22). In contrast to the usual situation with neoplasms, after cortical infarction, the EEG typically shows maximal focal slow-wave abnormality acutely, and sequential recordings show decreasing focal abnormality. When

PLEDs occur with acute lesions (vascular or neoplastic), these discharges usually subside within a few days despite continued presence of the anatomic lesion.

For sequential follow-up of patients after intracranial surgical procedures, it is important to obtain a postoperative baseline EEG. Generally, this is best accomplished after about a week, when the scalp wound has healed and the acute effects of the surgical procedure have subsided. Sequential recordings may be helpful for detecting early or late complications and may also be useful for assessing the effects of radiation therapy or chemotherapy for intracranial lesions.

EEG MANIFESTATIONS OF DIFFUSE ENCEPHALOPATHIES

The EEG is helpful in the evaluation of diffuse disorders of cerebral function and serves as a measurement of the severity of the disturbance.[73]

Diffuse encephalopathies can be caused by various conditions, including metabolic,

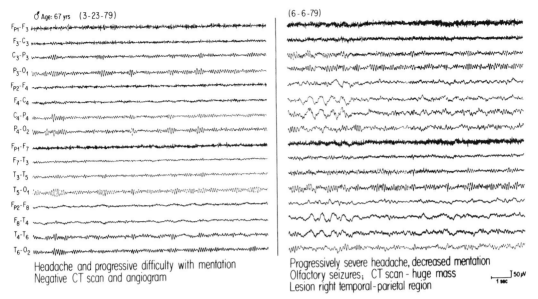

♂ Age: 67 yrs (3-23-79)　　　　　　　(6-6-79)

Headache and progressive difficulty with mentation
Negative CT scan and angiogram

Progressively severe headache, decreased mentation
Olfactory seizures; CT scan - huge mass
Lesion right temporal-parietal region

50 μV
1 sec

Figure 8–22. EEGs from a 67-year-old man, showing focal persistent polymorphic delta activity in the right temporal region (F_8 and T_4) (segment on left) and the addition 2½ months later (segment on right) of intermittent rhythmic delta activity caused by a rapidly progressive brain tumor.

toxic, inflammatory, posttraumatic, hypoxic, and degenerative disorders.

The type of diffuse disorder and whether it involves white or gray matter influence the EEG pattern.[32] Processes that predominantly affect white matter usually cause polymorphic delta slowing in the EEG, whereas processes that involve cortical and subcortical gray matter are more likely to cause intermittent bilaterally synchronous paroxysmal slow wave activity. Epileptiform abnormalities are seen more commonly in gray

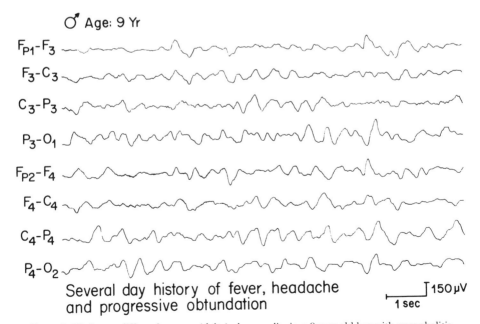

♂ Age: 9 Yr

Several day history of fever, headache
and progressive obtundation

150 μV
1 sec

Figure 8–23. Severe diffuse slow-wave (delta) abnormality in a 9-year-old boy with encephalitis.

matter disease than in white matter disease. Other factors that influence the degree and type of EEG abnormalities include the age and clinical state of the patient, the stage of the disease process, and other complicating factors such as infectious processes, metabolic derangements, or drug effects.

The most common type of EEG finding in diffuse disorders, or encephalopathies, consists of slowing of varying degrees[43,73] (Fig. 8–23). This may involve background activity, the theta frequency range, or generalized polymorphic delta range. Intermittent bursts of bilaterally synchronous rhythmic slow waves can occur in a generalized fashion or have a maximal expression over the anterior or posterior head regions. Usually the degree of slowing parallels the degree of disturbance of function or alteration in level of consciousness (or both). These findings can be caused by various diffuse disorders and, therefore, are considered nonspecific changes in that they are not diagnostic of any single condition, as mentioned previously.

At times, however, the EEG may show a more specific pattern, such as periodic patterns or the various distinctive coma patterns.

The periodic patterns include those associated with Jakob-Creutzfeldt disease, subacute sclerosing panencephalitis, and herpes simplex encephalitis.

Jakob-Creutzfeldt disease is a diffuse, subacute, and progressive disorder of the central nervous system that occurs in middle-age patients and is thought to be a transmissable prion disease. The disease is characterized

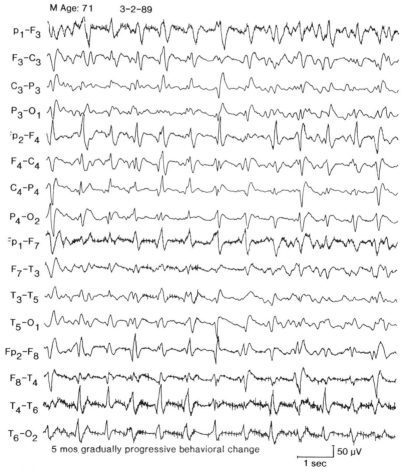

Figure 8–24. Diffuse periodic sharp waves in a 71-year-old man with Jakob-Creutzfeldt disease.

by dementia, motor dysfunction, myoclonus, and, when the disease is fully developed, a characteristic periodic EEG pattern consisting of generalized, bisynchronous, and periodic sharp waves recurring at intervals of 0.5 to 1 second, with a duration of 200 to 400 milliseconds, the prototype of the *periodic short-interval diffuse discharges (PSIDDs)*[10,11] (Fig. 8–24). Myoclonic jerks are often associated with the periodic sharp waves; however, there is not always a constant relationship between the two. Although other degenerative or toxic disorders may occasionally be associated with a quasiperiodic sharp wave pattern, the presence of periodic sharp waves, progressive dementia, and myoclonus is strongly suggestive of Jakob-Creutzfeldt disease.

Subacute sclerosing panencephalitis (SSPE) is a degenerative disorder that occurs in children and adolescents and is believed to be caused by the measles virus. This degenerative disorder is characterized by abnormal movements, intellectual deterioration, and a diagnostic, periodic EEG pattern. This consists of repetitive stereotyped high-voltage sharp-and-slow-wave complexes recurring every 4 to 15 seconds, the prototype of the *periodic long-interval diffuse discharges (PLIDDs)*[10,16] (Fig. 8–25). This pattern usually is present during the intermediate stages of the disease. The morphology of the complexes is stereotyped in a single recording from a single patient; however, the shape of the complexes can vary in different patients and change from time to time in the same patient at different stages of the disease. The complexes are usually generalized and bisynchronous, but at times they may be asymmetrical or more lateralized. Stereotyped motor jerks or spasms are often associated with the periodic complexes.

PLEDs are often seen with herpes simplex encephalitis.[57] They consist of periodic sharp waves that occur in a focal or lateralized manner over one hemisphere, particularly involving the temporal regions.

The coma patterns include the triphasic wave, alpha- and beta-frequency, spindle, and burst-suppression.

In hepatic coma, the EEG often shows a *triphasic wave pattern* consisting of medium- to high-voltage broad triphasic waves that occur rhythmically or in serial trains at a rate of 1 to 2 Hz in a bilaterally synchronous and symmetrical fashion over the two hemispheres and have a fronto-occipital or occipitofrontal time lag.[7] The triphasic waves usually have a frontal predominance and consist of a short-duration, low-voltage surface-negative component followed by a prominent positive sharp-contoured wave and then a longer duration surface-negative slow wave[23,58] (Fig. 8–26). Although triphasic waves are often associated with hepatic dysfunction, atypical triphasic waves can be seen in other conditions, including metabolic derangements, electrolyte disturbances, toxic states, and degenerative processes, or after a hypoxic episode.

The *alpha-frequency coma pattern* consists of diffusely distributed invariant alpha activity that shows little or no reactivity or variability. This type of pattern has been seen after cardiac arrest or hypoxic insult to the brain and with significant brain stem lesions.[78] When

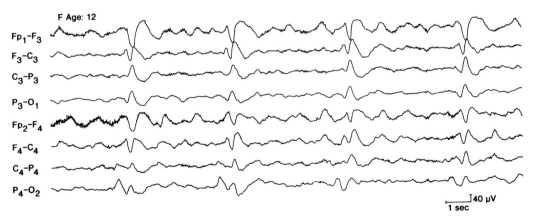

Figure 8–25. Diffuse periodic complexes in a 12-year-old girl with subacute sclerosing panencephalitis.

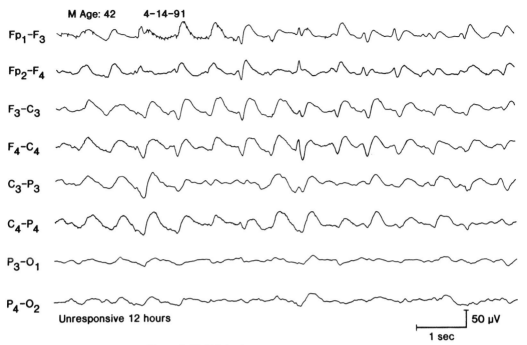

Figure 8–26. Triphasic waves in the frontal regions.

the alpha-frequency coma pattern is seen in the context of a hypoxic insult, it usually indicates a poor prognosis.

The *beta-frequency coma pattern* consists of generalized beta activity superimposed on underlying delta slowing. This pattern is usually associated with drug toxicity or anesthesia.

A *spindle coma pattern* resembles a sleep EEG and consists predominantly of spindle activity with some V waves but shows no reactivity.[15] This type of pattern can result from various causes, including head trauma,

hypoxic insults, or brain stem lesions. Depending on the type of underlying cause and severity of damage to the central nervous system, the pattern indicates that the potential for improvement exists. In many types of coma, spontaneous variability of EEG activity, including the sleep-like pattern, indicates a better prognosis than a prolonged invariant pattern.

The *burst-suppression pattern* consists of periodic or episodic bursts of activity, usually irregular mixtures of sharp waves or spikes, alternating with intervals of attenuation

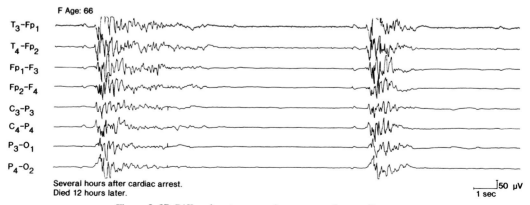

Figure 8–27. Diffuse burst-suppression pattern after cardiac arrest.

(Fig. 8–27). This pattern is often seen after a severe insult to the brain, such as a hypoxic or anoxic insult, in which case the pattern usually indicates a poor prognosis. However, the burst-suppression pattern can also be seen with potentially reversible conditions, such as anesthesia, drug intoxication, and hypothermia.

In summary, in patients with diffuse disorders, the EEG is useful in documenting a disturbance of cerebral function, in determining the degree of the disturbance, in monitoring changes and trends in the course of the disease process, and in helping to establish the diagnosis in certain conditions in which a characteristic EEG pattern is present. Also, the EEG sometimes helps to detect the presence of an additional, more focal cerebral process.

EVALUATION FOR SUSPECTED BRAIN DEATH

The EEG can provide confirmatory evidence of brain death, which is manifested by an absence of spontaneous or induced electrical activity of cerebral origin (Fig. 8–28).

Electrocerebral inactivity (ECI) is defined as "no EEG activity over 2 μV."[1] There are important minimal technical criteria for recording in patients with suspected cerebral death. These criteria include the following:[1]

1. A minimum of 8 scalp electrodes should be used.
2. Interelectrode impedances should be less than 10,000 Ω but more than 100 Ω.
3. The integrity of the entire recording system must be verified.
4. Interelectrode distances should be at least 10 cm.
5. The sensitivity should be at least 2 μV/mm for at least 30 minutes of recording.
6. Appropriate filter settings should be used.
7. Additional monitoring techniques should be used when necessary.
8. There should be no EEG reactivity to afferent stimulation.
9. The recording should be made by a qualified technologist.
10. A repeat EEG should be performed if there is doubt about the presence of electrocerebral silence.

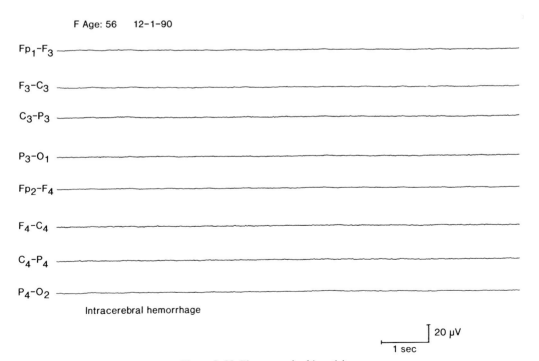

Figure 8–28. Electrocerebral inactivity.

Because temporary and reversible ECI can be caused by drug overdose and hypothermia, these conditions should be excluded before reaching a conclusion of brain death.[13] In young infants, because of uncertainties about the significance of ECI, one should exercise caution in the interpretation of this finding.[13] The guidelines of the Task Force for Determination of Brain Death in Children[70] recommend that for infants between 7 days and 2 months of age, two EEGs demonstrating ECI be performed at least 48 hours apart and that for infants between 2 months and 1 year of age, two records showing ECI be done at least 24 hours apart.

CHAPTER SUMMARY

This chapter provides an overview of the diverse patterns of activity seen in clinical electroencephalography. For accurate appraisal of the EEG, the many individual differences in normal activity need to be considered first, with appropriate attention to the limits of normal variability at different ages and at different states of wakefulness and sleep. Detection and correct identification of abnormal patterns of activity require considerable practice and experience. Finally, the usefulness of the EEG depends on an enlightened interpretation of the findings with regard to the specific clinical problem that the EEG is being used to solve.

REFERENCES

1. American Electroencephalographic Society: Guidelines in electroencephalography, evoked potentials, and polysomnography. J Clin Neurophysiol 11:1–147, 1994.
2. American Electroencephalographic Society: Guidelines in EEG and evoked potentials, 1986. J Clin Neurophysiol 3 Suppl 1:1–152, 1986.
3. Bancaud, J: Correlation of neuro-psycho-pathological and EEG findings in cases with cerebral tumours. Electroencephalogr Clin Neurophysiol Suppl 19:204–248, 1961.
4. Beaussart, M: Benign epilepsy of children with Rolandic (centro-temporal) paroxysmal foci: A clinical entity; study of 221 cases. Epilepsia 13:795–811, 1972.
5. Bennett, DR: Electroencephalographic and evoked potential changes with aging. Semin Neurol 1:47–51, 1981.
6. Bickford, RG: Activation procedures and special electrodes. In Klass, DW and Daly, DD (eds): Current Practice of Clinical Electroencephalography. Raven Press, New York, 1979, pp 269–305.
7. Bickford, RG and Butt, HR: Hepatic coma: The electroencephalographic pattern. J Clin Invest 34:790–799, 1955.
8. Bickford, RG and Klass, D: Scalp and depth electrographic studies of electro-decremental seizures (abst). Electroencephalogr Clin Neurophysiol 12:263, 1960.
9. Blume, WT, David, RB, and Gomez, MR: Generalized sharp and slow wave complexes: Associated clinical features and long-term follow-up. Brain 96:289–306, 1973.
10. Brenner, RP and Schaul, N: Periodic EEG patterns: Classification, clinical correlation, and pathophysiology. J Clin Neurophysiol 7:249–267, 1990.
11. Burger, LJ, Rowan, AJ, and Goldensohn, ES: Creutzfeldt-Jakob disease: An electroencephalographic study. Arch Neurol 26:428–433, 1972.
12. Chatrian, GE: The lambda waves. In Rémond A (ed): Handbook of Electroencephalography and Clinical Neurophysiology. Vol 6: The Normal EEG Throughout Life. Part A: The EEG of the Waking Adult. Elsevier Scientific Publishing, Amsterdam. 1976, pp 6A-123–6A-149.
13. Chatrian, G-E: Coma, other states of altered responsiveness, and brain death. In Daly, DD and Pedley, TA (eds): Current Practice of Clinical Electroencephalography, ed 2. Raven Press, New York, 1990, pp 425–487.
14. Chatrian, GE, Bergamini, L, Dondey, M, et al: A glossary of terms most commonly used by clinical electroencephalographers. Electroencephalogr Clin Neurophysiol 37:538–548, 1974.
15. Chatrian, GE, White, LE Jr, and Daly, D: Electroencephalographic patterns resembling those of sleep in certain comatose states after injuries to the head. Electroencephalogr Clin Neurophysiol 15:272–280, 1963.
16. Cobb, W: The periodic events of subacute sclerosing leucoencephalitis. Electroencephalogr Clin Neurophysiol 21:278–294, 1966.
17. Cobb, WA, Guiloff, RJ, and Cast, J: Breach rhythm: The EEG related to skull defects. Electroencephalogr Clin Neurophysiol 47:251–271, 1979.
18. Dalby, MA: Epilepsy and 3 per second spike and wave rhythms: A clinical, electroencephalographic and prognostic analysis of 346 patients. Acta Neurol Scand 45 Suppl 40:1–180, 1969.
19. Daly, DD: Use of the EEG for diagnosis and evaluation of epileptic seizures and nonepileptic episodic disorders. In Klass, DW and Daly, DD (eds): Current Practice of Clinical Electroencephalography. Raven Press, New York, 1979, pp 221–268.
20. Daly, DD: Epilepsy and syncope. In Daly, DD and Pedley, TA (eds): Current Practice of Clinical Electroencephalography, ed 2. Raven Press, New York, 1990, pp 269–334.
21. Drury, I and Henry, TR: Ictal patterns in generalized epilepsy. J Clin Neurophysiol 10:268–280, 1993.
22. Evans, CC: Spontaneous excitation of the visual cortex and association areas—lambda waves. Electroencephalogr Clin Neurophysiol 5:69–74, 1953.

23. Fisch, BJ and Klass, DW: The diagnostic specificity of triphasic wave patterns. Electroencephalogr Clin Neurophysiol 70:1–8, 1988.

24. Gastaut, H: Etude électrocorticographique de la réactivité des rhythmes rolandiques. Rev Neurol (Paris) 87:176–182, 1952.

25. Gastaut, H and Broughton, R: Epileptic Seizures: Clinical and Electrographic Features, Diagnosis and Treatment. Charles C Thomas, Publisher, Springfield, Illinois, 1972.

26. Gastaut, H, Roger, J, Soulayrol, R, et al: Childhood epileptic encephalopathy with diffuse slow spike-waves (otherwise known as "petit mal variant") or Lennox syndrome. Epilepsia 7:139–179, 1966.

27. Gibbs, FA and Gibbs, EL: Atlas of Electroencephalography, vol 2, ed 2. Addison-Wesley, Cambridge, Massachusetts, 1952.

28. Gibbs, FA and Gibbs, EL: Fourteen and six per second positive spikes. Electroencephalogr Clin Neurophysiol 15:553–558, 1963.

29. Gibbs, FA and Gibbs, EL: Atlas of Electroencephalography, vol 3, ed 2. Addison-Wesley Press, Cambridge, Massachusetts, 1964.

30. Gibbs, FA and Gibbs, EL: Medical Electroencephalography. Addison-Wesley Publishing, Reading, Massachusetts, 1967.

31. Gibbs, FA, Rich, CL, and Gibbs, EL: Psychomotor variant type of seizure discharge. Neurology (Minneap) 13:991–998, 1963.

32. Gloor, P, Kalabay, O, and Giard, N: The electroencephalogram in diffuse encephalopathies: Electroencephalographic correlates of grey and white matter lesions. Brain 91:779–802, 1968.

33. Gomez, MR and Klass, DW: Epilepsies of infancy and childhood. Ann Neurol 13:113–124, 1983.

34. Goode, DJ, Penry, JK, and Dreifuss, FE: Effects of paroxysmal spike-wave on continuous visual-motor performance. Epilepsia 11:241–254, 1970.

35. Goodwin, JE: The significance of alpha variants in the EEG, and their relationship to an epileptiform syndrome. Am J Psychiatry 104:369–379, 1947.

36. Jeavons, PM and Bower, BD: Infantile spasms. In Vinken, PJ and Bruyn, GW (eds): Handbook of Clinical Neurology. Vol 15: The Epilepsies. North-Holland Publishing, Amsterdam, 1974, pp 219–234.

37. Katz, RI and Harner, RN: Electroencephalography in aging. In Albert, ML (ed): Clinical Neurology of Aging. Oxford University Press, New York, 1984, pp 114–138.

38. Kellaway, P: The incidence, significance and natural history of spike foci in children. In Henry, CE (ed): Course in Clinical Electroencephalography: Current Clinical Neurophysiology; Update on EEG and Evoked Potentials. Elsevier/North-Holland Biomedical Press, Amsterdam, 1980, pp 151–175.

39. Kellaway, P: An orderly approach to visual analysis: Characteristics of the normal EEG of adults and children. In Daly, DD and Pedley TA (eds): Current Practice of Clinical Electroencephalography, ed 2. Raven Press, New York, 1990, pp 139–199.

40. Klass, DW: Electroencephalographic manifestations of complex partial seizures. Adv Neurol 11:113–140, 1975.

41. Klass, DW: Identifying the abnormal EEG. In Hallidav. AM, Butler, SR, and Paul, R (eds): A Textbook of Clinical Neurophysiology. John Wiley & Sons, Chichester, 1987, pp 189–199.

42. Klass, DW and Bickford, RG: Observations on the rolandic arceau rhythm (abstr). Electroencephalogr Clin Neurophysiol 9:570, 1957.

43. Klass, DW and Daly, DD: Current Practice of Clinical Electroencephalography. Raven Press, New York, 1979.

44. Klass, DW and Fischer-Williams M: Sensory stimulation, sleep and sleep deprivation. In Rémond A (ed): Handbook of Electroencephalography and Clinical Neurophysiology. Vol 3: Techniques and Methods of Data Acquisition of EEG and EMG. part D: Activation and Provocation Methods in Clinical Neurophysiology. Elsevier Scientific Publishing Company, Amsterdam, 1976, pp 3D-5–3D-73.

45. Klass, DW and Westmoreland, BF: Nonepileptogenic epileptiform electroencephalographic activity. Ann Neurol 18:627–635, 1985.

46. Kooi, KA, Tucker, RP, and Marshall, RE: Fundamentals of Electroencephalography, ed 2. Harper & Row, Publishers, Hagerstown, Maryland, 1978.

47. Kozelka, JW and Pedley, TA: Beta and mu rhythms. J Clin Neurophysiol 7:191–207, 1990.

48. Lombroso CT: Sylvian seizures and midtemporal spike foci in children. Arch Neurol 17:52–59, 1967.

49. Lombroso CT, Schwartz IH, Clark DM, et al: Ctenoids in healthy youths: Controlled study of 14- and 6-per-second positive spiking. Neurology (Minneap) 16:1152–1158, 1966.

50. Loomis, AL, Harvey, EN, and Hobart, GA, III: Distribution of disturbance-patterns in the human electroencephalogram, with special reference to sleep. J Neurophysiol 1:413–430, 1938.

51. Markand, ON: Slow spike-wave activity in EEG and associated clinical features: Often called 'Lennox' or 'Lennox-Gastaut' syndrome. Neurology (Minneap) 27:746–757, 1977.

52. Markand, ON: Alpha rhythms. J Clin Neurophysiol 7:163–189, 1990.

53. Maulsby RL: EEG patterns of uncertain diagnostic significance. In Klass, DW and Daly, DD (eds): Current Practice of Clinical Electroencephalography. Raven Press, New York, 1979, pp 411–419.

54. McGee, FE, Jr and White, RJ: EEG instrumentation. In Henry, CE (ed): Current Clinical Neurophysiology: Update on EEG and Evoked Potentials. Elsevier/North-Holland Biomedical Press, Amsterdam, 1980, pp 1–49.

55. Niedermeyer, E and Lopes da Silva, F: Electroencephalography: Basic Principles, Clinical Applications and Related Fields, ed 2. Urban & Schwarzenberg, Baltimore, 1987.

56. Pedley, TA: EEG patterns that mimic epileptiform discharges but have no association with seizures. In Henry, CE (ed): Current Clinical Neurophysiology: Update on EEG and Evoked Potentials. Elsevier/North-Holland Biomedical Press, Amsterdam, 1980, pp 307–336.

57. Rawls, WE, Dyck, PJ, Klass, DW, et al: Encephalitis associated with herpes simplex virus. Ann Intern Med 64:104–115, 1966.

58. Reiher, J: The electroencephalogram in the investigation of metabolic comas (abstr). Electroencephalogr Clin Neurophysiol 28:104, 1970.

59. Reiher, J and Klass, DW: Two common EEG patterns of doubtful clinical significance. Med Clin North Am 52:933–940, 1968.

60. Reiher, J and Lebel, M: Wicket spikes: Clinical correlates of a previously undescribed EEG pattern. Can J Neurol Sci 4:39–47, 1977.

61. Reiher, J, Lebel, M, and Klass, DW: Small sharp spikes (SSS): Reassessment of electroencephalographic characteristics and clinical significance (abstr). Electroencephalogr Clin Neurophysiol 43:775, 1977.

62. Reilly, EL: EEG recording and operation of the apparatus. In Niedermeyer, E and Lopes da Silva, F (eds): Electroencephalography: Basic Principles, Clinical Applications, and Related Fields, ed 3. Williams & Wilkins, Baltimore, 1993, pp 104–124.

63. Rémond, A: Handbook of Electroencephalography and Clinical Neurophysiology. Vol 6: The Normal EEG Throughout Life. Part B: The Evolution of the EEG From Birth to Adulthood. Elsevier Scientific Publishing, Amsterdam, 1975.

64. Rémond, A: Handbook of Electroencephalography and Clinical Neurophysiology. Vol 11: Clinical EEG, I. Part A: Semiology in Clinical EEG. Elsevier Scientific Publishing, Amsterdam, 1977.

65. Santamaria, J and Chiappa, KH: The EEG of Drowsiness. Demos Publications, New York, 1987.

66. Saunders, MG and Westmoreland, BF: The EEG in evaluation of disorders affecting the brain diffusely. In Klass, DW and Daly, DD (eds): Current Practice of Clinical Electroencephalography. Raven Press, New York, 1979, pp 343–379.

67. Schnell, RG and Klass, DW: Further observations on the Rolandic arceau rhythm (abstr). Electroencephalogr Clin Neurophysiol 20:95, 1966.

68. Sharbrough, FW: Nonspecific abnormal EEG patterns. In Niedermeyer, E and Lopes da Silva, F (eds): Electroencephalography: Basic Principles, Clinical Applications, and Related Fields, ed 3. Williams & Wilkins, Baltimore, 1993, pp 197–215.

69. Spehlmann, R: EEG Primer. Elsevier/North-Holland Biomedical Press, Amsterdam, 1981.

70. Task Force for the Determination of Brain Death in Children: Guidelines for the determination of brain death in children. Neurology 37:1077–1078, 1987.

71. Thomas, JE and Klass, DW: Six-per-second spike-and-wave pattern in the electroencephalogram: A reappraisal of its clinical significance. Neurology (Minneap) 18:587–593, 1968.

72. Tyner, FS, Knott, JR, and Mayer, WB, Jr: Fundamentals of EEG Technology. Vol I: Basic Concepts and Methods. Raven Press, New York, 1983.

73. Vas, GA and Cracco, JB: Diffuse encephalopathies. In Daly, DD and Pedley, TA (eds): Current Practice of Clinical Electroencephalography, ed 2. Raven Press, New York, 1990, pp 371–399.

74. Walter, WG: Epilepsy. In Hill, JDN and Parr, G (eds): Electroencephalography: A Symposium on Its Various Aspects. Macdonald & Company, London, 1950, pp 228–272.

75. Westmoreland, BF and Klass, DW: Asymmetrical attenuation of alpha activity with arithmetical attention (abstr). Electroencephalogr Clin Neurophysiol 31:634–635, 1971.

76. Westmoreland, BF and Klass, DW: A distinctive rhythmic EEG discharge of adults. Electroencephalogr Clin Neurophysiol 51:186–191, 1981.

77. Westmoreland, BF and Klass, DW: Unusual EEG patterns. J Clin Neurophysiol 7:209–228, 1990.

78. Westmoreland, BF, Klass, DW, Sharbrough, FW, and Reagan, TJ: Alpha-coma: Electroencephalographic, clinical, pathologic, and etiologic correlations. Arch Neurol 32:713–718, 1975.

79. White, JC, Langston, JW, and Pedley, TA: Benign epileptiform transients of sleep: Clarification of the small sharp spike controversy. Neurology (Minneap) 27:1061–1068, 1977.

80. Wiener, JM, Delano, JG, and Klass, DW: An EEG study of delinquent and nondelinquent adolescents. Arch Gen Psychiatry 15:144–150, 1966.

CHAPTER 9

ELECTRO-ENCEPHALOGRAPHY: ELECTRO-ENCEPHALOGRAMS OF NEONATES, INFANTS, AND CHILDREN

Barbara F. Westmoreland, M.D.
Donald W. Klass, M.D.

NEONATAL ELECTROENCEPHALOGRAPHIC PATTERNS

Normal

Recordings in newborn infants usually require monitoring of physiologic variables such as the electrocardiogram (ECG), respirations, eye movement, and chin myogram in addition to electroencephalographic (EEG) recording to help determine the state of the infant.[1,9,24]

In infants of less than 32 weeks' conceptional age, which is the age since the first day of the mother's last menses, the EEG consists of an intermittent or discontinuous pattern, with bursts of activity alternating with long quiescent periods. There is little distinction between the EEG of the awake and sleep states.[7,9,11,24]

Between 32 and 37 weeks' conceptional age, there is an EEG distinction between the awake state and the two types of sleep states, active and quiet sleep.[7,9,24] Active sleep, which is similar to rapid eye movement (REM) sleep in adults, manifests eye movements, body twitches, facial grimaces, reduction in muscle tone, and irregular respirations. During this state, the EEG shows a more continuous pattern, similar to that of the awake state. Quiet sleep has reduced eye and body movements, increased muscle tone, regular respirations, and a regular ECG. During quiet sleep, the EEG shows a discontinuous pattern, with bursts of mixed sharp and slow wave activity alternating with periods of flattening of the background. This is referred to as the *tracé alternant* pattern.

Other types of activity that are present at this age include: (1) occipital-dominant slow waves, which are broad, high-amplitude slow waves over the occipital head regions, shifting from side to side in prominence; (2) the spindle delta brush pattern, which consists of moderate-to-high-amplitude delta waves with superimposed 8- to 20-Hz activity; it is seen over the rolandic, temporal, and occipital head regions (Fig. 9–1); (3) multifocal sharp transients or random focal sharp waveforms occurring in multiple locations but most frequently over the frontal head regions (frontal sharp transients) (Fig. 9–2); (4) anterior slow waves, which consist of rhythmic delta slow waves over the anterior head regions; they are usually associated with frontal sharp transients; and (5) bursts or trains of sharp-contoured theta waves over the temporal and central vertex regions.[7,9,11,24] In the younger premature infant, these patterns are seen in both the awake and sleep states. As the child matures, these patterns are seen primarily during sleep and then disappear after about 44 weeks' conceptional age.[9,11,24]

Infants 38 to 42 weeks old show four basic patterns: (1) a low-voltage irregular pattern that is present during wakefulness and active sleep; (2) a high-voltage slow-wave pattern that is seen during quiet sleep; (3) a tracé alternant pattern that is also seen during quiet sleep; and (4) a mixed pattern of theta and delta waves seen during drowsiness and active sleep and as a transitional pattern between the various states.[9,11,24]

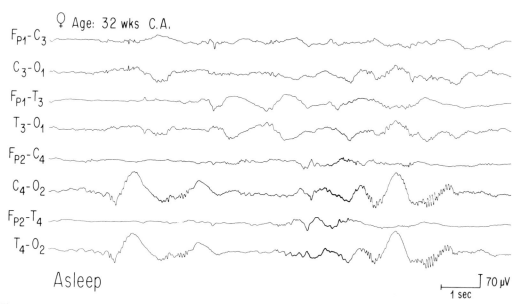

Figure 9–1. EEG from a normal premature infant at 32-weeks' conceptional age (C.A.) during sleep, showing the delta brush pattern.

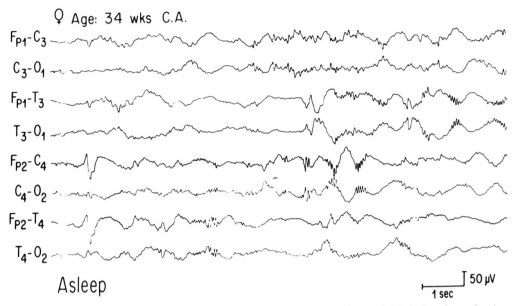

Figure 9–2. EEG from a normal premature infant at 34-weeks' conceptional age (C.A.) during sleep, showing a frontal sharp transient (F_{p2}). Less prominent sharp transients also occur in T_3 and T_4.

Abnormal

The EEG in premature newborn infants should be interpreted with care. The age of the infant is important because the EEG activity changes with maturation, and what might be normal at one age may be abnormal for a more mature infant. The background rhythms are the most significant EEG finding.[13] If these are appropriate for the infant's age, the infant usually has a fairly good prognosis; if abnormal, then the degree of abnormality usually reflects the degree of disturbance and the ultimate outcome of the infant.[12] Abnormalities may be divided into mild and significant types.

MILD ABNORMALITIES

Mild abnormalities, such as excessive multifocal sharp transients or immature and dysmature patterns, can be seen in stressed premature or term infants. These findings are nonspecific and rarely suggest a specific etiologic diagnosis.[11] They are often transient and usually disappear within a few days. Mild and transient focal abnormalities in the EEG are not usually associated with any obvious focal pathologic condition. On the other hand, persistent focal EEG abnormalities are often associated with structural lesions such as intracranial hemorrhage or congenital defects.[11,17]

SIGNIFICANT ABNORMALITIES

Significant abnormalities in the EEG are usually associated with an important disturbance of brain function and often indicate a poor prognosis or poor neurologic outcome. The more abnormal the pattern, the more severe the underlying encephalopathy or disturbance of brain function.[11,13,17,18] The following patterns are significantly abnormal:

Isoelectric EEG. An *isoelectric EEG* is a flat record that meets the criteria for electrocerebral inactivity. Infants with a single flat EEG may survive the neonatal period but usually suffer severe long-term neurologic sequelae.

Paroxysmal Burst-Suppression Pattern. *Paroxysmal burst-suppression pattern* consists of diffuse bursts of abnormal activity superimposed on an isoelectric or very low-amplitude background. This is an invariant pattern that does not change with state of sleep-wakefulness or in response to stimuli. It, too, is associated with severe encephalopathy and poor long-term prognosis.

Persistent Low Voltage. *Persistent low voltage* can occur in a generalized fashion and in association with a diffuse disturbance of function or, in a more focal fashion, in association with focal lesions such as porencephaly, subdural collection of fluids, or congenital abnormalities.

Epileptiform Activity. *Epileptiform activity* is one of the most frequent types of abnormalities seen in EEGs of neonates. It consists of focal or multifocal interictal and ictal discharges.[18] The interictal discharges usually take the form of spikes and sharp waves and broad slow waves. The ictal discharges consist of rhythmic activity that may take the form of spikes, sharp waves, rhythmic slow waves, or rhythmic activity in the alpha, beta, theta, or delta range and may evolve over relatively long periods. Ictal electrographic discharges may occur in association with clinical seizures or may be present without any clinical accompaniment (Fig. 9–3). If associated with seizures, the seizures usually take the form of clonic or tonic movements,[18] but there may be diverse and subtle manifestations that may not be easily recognizable as epileptic.

Positive Rolandic Sharp Waves. *Positive rolandic sharp waves* can be seen unilaterally or bilaterally and occur most commonly in the rolandic and midline areas. Initially, they were described in infants with intraventric-ular hemorrhage; however, they also occur in patients who have periventricular leukomalacia and deep white matter lesions.

Persistent Slowing. *Persistent slowing* may be diffuse or focal.

CONDITIONS GIVING RISE TO ABNORMAL EEG PATTERNS

Hypoxia-Ischemic Insult. This gives rise to severe EEG abnormalities and is currently thought to be the most common cause of neonatal seizures.[12]

Metabolic Causes. The most common metabolic conditions that produce abnormal EEG patterns are hypoglycemia and hypocalcemia, and in the past, they were one of the most frequent causes of neonatal seizures.[23] Metabolic derangements can be associated with focal or multifocal epileptiform abnormalities in the EEG; the EEG usually improves after the metabolic disorder is corrected.

Drugs and Drug Withdrawal. These are becoming more frequent causes of seizures and irritability in newborn infants. EEGs in these infants often show evidence of diffuse or multifocal cortical irritability, as manifested by focal or multifocal epileptiform activity and seizure discharges.

Infectious Diseases. Infectious diseases involving the central nervous system are fre-

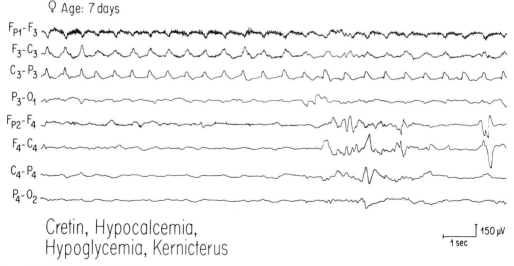

Figure 9–3. Focal subclinical EEG-seizure discharge arising from the left frontal region (F_3) in a 7-day-old infant with neonatal seizures.

quently associated with abnormalities in the EEG.[16] The most characteristic finding is seen in neonatal herpes simplex encephalitis and consists of *periodic lateralized epileptiform discharges (PLEDs)*.[19] Other findings associated with infectious processes include significant asymmetries, interhemispheric asynchronies, multifocal sharp waves, and seizure discharges.

Biochemical Disorders and Aminoacidurias. Biochemical disorders and aminoacidurias can be associated with abnormal EEG patterns.[20] Phenylketonuria was a common cause in the past, but other types of aminoacidurias have now been recognized and may present with neonatal seizures and epileptiform discharges. Pyridoxine deficiency and pyridoxine dependency can cause severe EEG abnormalities.

Dysgenetic Disorders or Neurocutaneous Disorders. Tuberous sclerosis presenting in infancy may be associated with a hypsarrhythmic pattern or focal or multifocal EEG abnormalities that may or may not be related to the location of the tubers.[10] Sturge-Weber syndrome is associated with an asymmetry of background activity and epileptiform activity on the side of the facial nevus.

Congenital Abnormalities. These may be associated with various types and combinations of abnormal patterns reflecting the abnormality.[16,20] Lissencephaly can be associated with diffuse fast activity.[25]

DEVELOPMENTAL CHANGES DURING INFANCY, CHILDHOOD, AND ADOLESCENCE

In the first 3 months of life, there is a transition from the neonatal to the infant EEG pattern.[11,24] The background consists of irregular low-amplitude delta activity. Rhythmic activity in the range of 5 to 6 Hz is present in the central regions and is probably a precursor of the mu rhythm. At 3 months of age, rhythmic occipital activity in the range of 3 to 4 Hz is present. This activity can be attenuated with eye opening[6,7] and represents a precursor of the alpha rhythm. Between 4 and 6 months of age, the central rhythm becomes better developed and shows a frequency of 5 to 8 Hz; a better de-

fined occipital rhythm in the range of 5 to 6 Hz is present when the eyes are closed. Between 6 months and 2 years of age, the central rhythm is well developed at a frequency of 5 to 8 Hz. After 6 months, the occipital rhythms become more prominent, and there is a gradual shift to higher amplitude and faster frequency activity, ranging from 6 to 8 Hz.[6,7] In patients between 2 and 5 years old, the central and occipital rhythms differentiate further. By 3 years of age, the occipital alpha rhythms range from 6 to 8 Hz, and the amplitude of this activity gradually increases.[14]

Between 6 and 16 years of age, there is a progressive increase in the alpha frequencies, and the typical adult frequency range of 9 to 10 Hz is usually reached by 10 to 12 years of age. Some interspersed theta activity may still be present, occurring predominantly over the anterior head regions.

During the first decade of life, there is a considerable amount of variability among children of the same age with regard to the amount of alpha, theta, and delta activity present.[3,4]

Waves of delta frequency are common over the posterior head regions in children and adolescents (posterior slow waves of youth). In normal teenagers, these occur sporadically rather than in consecutive trains, do not protrude much above the average amplitude of the alpha rhythm, and attenuate together with the alpha rhythm when the eyes are opened.

Hyperventilation

In children younger than 8 years old, the slowing produced by hyperventilation is often maximal over the posterior head regions (Fig. 9–4), whereas in those older than 8 years of age the buildup response usually occurs maximally over the anterior head regions.[2,14,22]

Photic Responses

Young children show responses in the occipital regions only at slower flash frequencies, and if they are younger than 9 years old, the driving response usually occurs up to a

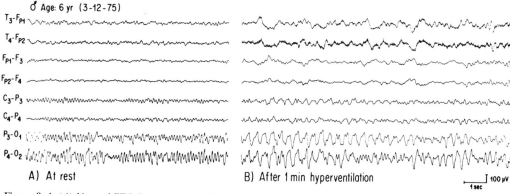

Figure 9–4. (*A*) Normal EEG from a 6-year-old boy during resting wakefulness and (*B*) during hyperventilation.

flash frequency of 10 Hz. After 10 years of age, driving can be seen up to 15 to 20 Hz.

Drowsiness

Drowsiness in young children is characterized by high-amplitude sinusoidal 4- to 5-Hz theta activity that is maximal over the frontal, central, and parietal regions. These slow waves initially occur in prolonged rhythmic trains and, in children between 1 and 9 years old, in bursts (Fig. 9–5). Sometimes the slow waves may have a notched or sharp appearance because of superimposed faster frequencies. During adolescence, characteristic

trains of monorhythmic sinusoidal theta activity occur over the frontal regions and may precede the disappearance of the alpha rhythm.

Sleep

During the first few months of life the tracé alternant pattern of newborns (see later in this chapter) is replaced by generalized slow wave activity during quiet sleep, and there is an increased percentage of time spent in nonrapid eye movement (NREM) sleep.[21] Spindles become apparent by 1 to 3 months old and are well developed and bi-

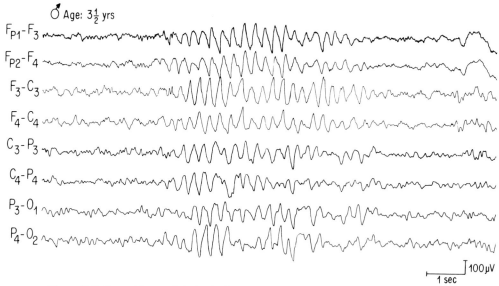

Figure 9–5. Normal burst of rhythmic slow waves during drowsiness in a 3½-year-old boy.

synchronous by 1 to 2 years old.[15] The spindles in the patient's first year of life may have a characteristic arciform or comb-like appearance and occur in prolonged and asynchronous trains. V waves are apparent by 3 to 5 months old. In children from 2 to 5 years old, V waves have a high amplitude and a sharp or spiky appearance and occur in groups. O waves are large, broad, biocciptal delta waves that are present over the occipital regions during drowsiness and sleep. Occasionally they may have a somewhat sharply contoured monophasic or diphasic waveform. Occipital slow waves are most prominent at 1 to 2 years old but may persist to at least 5 years old.

BENIGN VARIANTS IN CHILDREN

Several benign variants can be seen in the EEGs of children and adolescents. These include the slow lambdas of youth, O waves, and 14&6-Hz positive bursts.

Posterior Slow Waves Associated With Eye Blinks or Slow Lambdas of Youth

A phenomenon similar to that of the lambda waves is the posterior slow waves associated with eye blinks in some children (*slow lambdas of childhood*, or *shut-eye waves*). They are single, broad, and monophasic or diphasic waveforms that occur bilaterally over the occipital head regions after eye blinks or eye movements.[16,26] The amplitude is often 100 to 200 μV, and the duration is about 200 to 400 milliseconds. The predominant polarity is surface-negative. The waveform may have a sharp-contoured appearance, but it should not be misinterpreted as abnormal epileptiform activity. These waveforms can be seen in children who are between 6 months and 10 years old and are most prominent in those 2 to 3 years old.

O Waves, or Cone-Shaped Waves

In young children, high-voltage slow-wave transients, *O waves*, varying from a cone-

shaped appearance to diphasic slow-wave transients may be present over the occipital head regions and interspersed with occipital dominant slow delta waves of sleep.[14] This activity should not be mistaken for abnormal sharp-wave or slow-wave activity.

The 14&6-Hz positive bursts consist of bursts of positive waves occurring at a rate of 14 and 6 Hz and seen mainly over the posterior temporal regions. They can be seen in children but are most often seen in adolescents (see Fig. 8–15).

ABNORMALITIES

The EEGs of children show types of abnormalities similar to those seen in adult EEGs. The main types of abnormalities are epileptiform discharges and slowing.

Epileptiform

Almost any type of epileptiform abnormality can be seen in children. The types of epileptiform abnormalities most commonly seen include:

1. *Hypsarrhythmia*—a pattern seen in children ranging from 4 months to 4 years old. It consists of high-amplitude multifocal spikes, sharp waves, and slow waves. This type of epileptiform activity is often seen in association with infantile spasms (Fig. 9–6).[5,8]
2. The *3-Hz generalized spike-and-wave discharge*—a pattern usually seen in children between the ages of 3 and 15 years. It is associated with absence seizures (Fig. 9–7).[8]
3. The *generalized slow spike-and-wave pattern*—a pattern seen in young children with frequent seizures and mental retardation (the Lennox-Gastaut syndrome) (Fig. 9–8).[8]
4. *Focal occipital spike discharges*—a pattern seen in children who usually are younger than 5 years old.
5. *Central temporal spikes*—a pattern seen in children 4 to 12 years old. It occurs in children with the benign rolandic epilepsy of childhood (BREC) (Fig. 9–9). (These patterns are described more fully in Chapter 8.)

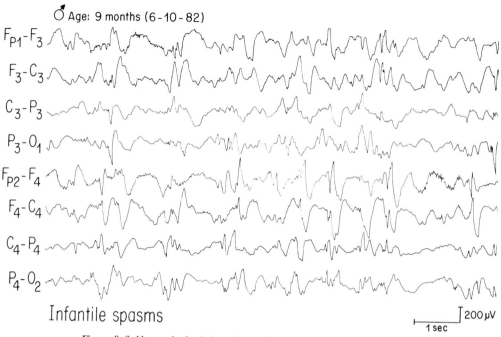

Figure 9–6. Hypsarrhythmia in a 9-month-old infant with infantile spasms.

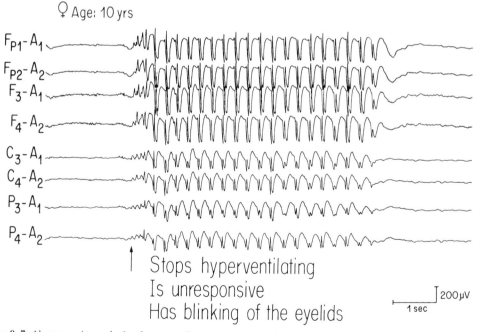

Figure 9–7. Absence seizure during hyperventilation accompanied by typical paroxysm of 3-Hz spike-and-slow-wave complexes.

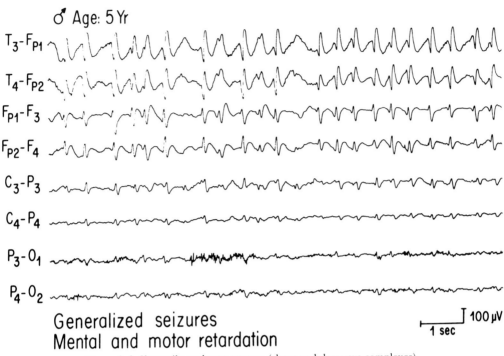

Figure 9–8. Slow spike-and-wave pattern (sharp-and-slow-wave complexes).

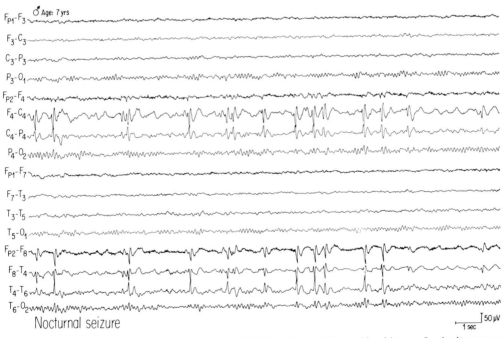

Figure 9–9. Central-temporal spikes (maximal in C_4 and T_4) in a 7-year-old boy with a history of a single nocturnal seizure.

Slowing

Slow wave abnormalities may be focal or diffuse and are often more prominent and take longer to resolve in children, as compared with adults. The slow wave abnormalities also tend to have a maximal expression over the posterior head region in children.

CHAPTER SUMMARY

This chapter discusses the normal and abnormal EEGs seen in neonates, infants, and children. In interpreting the EEGs of infants and children, one has to take into account age, maturational changes, and normal interindividual variation.

REFERENCES

1. American Electroencephalographic Society: Guideline two: Minimum technical standards for pediatric electroencephalography. J Clin Neurophysiol 11:6–9, 1994.
2. Bickford, RG: Activation procedures and special electrodes. In Klass, DW and Daly, DD (eds): Current Practice of Clinical Electroencephalography. Raven Press, New York, 1979, pp 269–305.
3. Blume, WT: Atlas of Pediatric Electroencephalography. Raven Press, New York, 1982.
4. Corbin, HPF and Bickford, RG: Studies of the electroencephalogram of normal children: Comparison of visual and automatic frequency analyses. Electroencephalogr Clin Neurophysiol 7:15–28, 1955.
5. Donat, JF: West syndrome: Seizure types and lesions. Am J EEG Technol 32:26–45, 1992.
6. Dreyfus-Brisac, C and Curzi-Dascalova, L: The EEG during the first year of life. In Rémond, A (ed): Handbook of Electroencephalography and Clinical Neurophysiology. Vol 6: The Normal EEG Throughout Life. Part B: The Evolution of the EEG From Birth to Adulthood. Elsevier Scientific Publishing, Amsterdam, 1975, pp 6B-24–6B-30.
7. Dreyfus-Brisac, C and Monod, N: The electroencephalogram of full-term newborns and premature infants. In Rémond A (ed): Handbook of Electroencephalography and Clinical Neurophysiology. Vol 6: The Normal EEG Throughout Life. Part B: The Evolution of the EEG From Birth to Adulthood. Elsevier Scientific Publishing, Amsterdam, 1975, pp 6B-6–6B-23.
8. Drury, I and Henry, TR: Ictal patterns in generalized epilepsy. J Clin Neurophysiol 10:268–280, 1993.
9. Ellingson, RJ: EEGs of premature and full-term newborns. In Klass, DW and Daly, DD (eds): Current Practice of Clinical Electroencephalography. Raven Press, New York, 1979, pp 149–177.
10. Gomez, MR: Tuberous Sclerosis, ed 2. Raven Press, New York, 1988.
11. Hahn, JS and Tharp, BR: Neonatal and pediatric electroencephalography. In Aminoff, MJ (ed): Electrodiagnosis in Clinical Neurology, ed 3. Churchill Livingstone, New York, 1992, pp 93–141.
12. Holmes, G, Rowe, J, Hafford, J, et al: Prognostic value of the electroencephalogram in neonatal asphyxia. Electroencephalogr Clin Neurophysiol 53:60–72, 1982.
13. Holmes, GL, and Lombroso, CT: Prognostic value of background patterns in the neonatal EEG. J Clin Neurophysiol 10:323–352, 1993.
14. Kellaway, P: An orderly approach to visual analysis: Characteristics of the normal EEG of adults and children. In Daly, DD and Pedley, TA (eds): Current Practice of Clinical Electroencephalography, ed 2. Raven Press, New York, 1990, pp 139–199.
15. Kellaway, P and Fox, BJ: Electroencephalographic diagnosis of cerebral pathology in infants during sleep. I. Rationale, technique, and the characteristics of normal sleep in infants. J Pediatr 41:262–287, 1952.
16. Kooi, KA, Tucker, RP, and Marshall RE: Fundamentals of Electroencephalography, ed 2. Harper & Row, Publishers, Hagerstown, Maryland, 1978.
17. Lombroso, CT: Neonatal EEG polygraphy in normal and abnormal newborns. In Niedermeyer, E and Lopes da Silva, F (eds): Electroencephalography: Basic Principles, Clinical Applications, and Related Fields, ed 3. Williams & Wilkins, Baltimore, 1993, pp 803–875.
18. Mizrahi, EM, and Kellaway, P: Characterization and classification of neonatal seizures. Neurology 37:1837–1844, 1987.
19. Mizrahi, EM and Tharp, BR: A characteristic EEG pattern in neonatal herpes simplex encephalitis. Neurology (NY) 32:1215–1220, 1982.
20. Niedermeyer, E and Lopes da Silva, F (eds): Electroencephalography: Basic Principles, Clinical Applications, and Related Fields, ed 3. Williams & Wilkins, Baltimore, 1993.
21. Petersén, I, Selldén, U, and Eeg-Olofsson, O: The evolution of the EEG in normal children and adolescents from 1 to 21 years. In Rémond, A (ed): Handbook of Electroencephalography and Clinical Neurophysiology. Vol 6: The Normal EEG Throughout Life. Part B: The Evolution of the EEG from Birth to Adulthood. Elsevier Scientific Publishing, Amsterdam, 1975, pp 6B-31–6B-68.
22. Rémond, A: Handbook of Electroencephalography and Clinical Neurophysiology. Vol 6: The Normal EEG Throughout Life. Part B: The Evolution of the EEG from Birth to Adulthood. Elsevier Scientific Publishing, Amsterdam, 1975.
23. Rose, AL and Lombroso, CT: Neonatal seizure states: A study of clinical, pathological, and electroencephalographic features in 137 full-term babies with a long-term follow-up. Pediatrics 45:404–425, 1970.
24. Stockard-Pope, JE, Werner, SS, and Bickford, RG: Atlas of Neonatal Electroencephalography, ed 2. Raven Press, New York, 1992.
25. Westmoreland, BF and Klass, DW: Unusual EEG patterns. J Clin Neurophysiol 7:209–228, 1990.
26. Westmoreland, BF and Sharbrough, FW: Posterior slow wave transients associated with eye blinks in children. Am J EEG Technol 15:14–19, 1975.

CHAPTER 10

AMBULATORY ELECTRO-ENCEPHALOGRAPHY

Kathryn A. Hirschorn, M.D.

HISTORY
TECHNIQUE
MONTAGES
ARTIFACTS
CLINICAL APPLICATION

One useful tool for the evaluation of patients with seizures and other paroxysmal events is ambulatory electroencephalography (AEEG). For many episodic behavioral events, the clinical history alone is inadequate to establish the diagnosis. Epileptiform abnormalities, when present on routine electroencephalography (EEG), can add confidence to the clinical diagnosis of seizures. However, many patients with seizures show no epileptiform abnormalities on routine recordings. An accurate diagnosis often requires capturing the clinical spells with simultaneous electrophysiologic and behavioral monitoring.

HISTORY

The Holter electrocardiographic (ECG) monitor, developed in 1947, allowed 24 hours of compressed ECG to be recorded in a single channel. In 1972, a multichannel tape was developed by Marson and McKinnon.[14] Ives and Woods[10,11] soon applied this technology to EEG, and 24-hour AEEG was born. In the early stages, an ink-jet EEG machine with a high-frequency response of 600 to 700 Hz was used to print the entire EEG in a highly compressed fashion. Ictal events could be identified and then played back at more conventional paper speed. Later development of an audio-video playback unit allowed the EEG to be reviewed at a maximum speed of 1 hour/1 minute. Quy,[15] in 1978, developed miniature preamplifiers that could be applied directly to the head. This development resulted in fewer artifacts and more stable recordings. Even with the earliest 3- to 4-channel recordings, it was obvious that continuous EEG recording was capable of capturing most generalized and focal ictal events. By the mid-1980s, an 8-channel machine was introduced, and AEEG gained wider acceptance. In the last few years, event-triggered systems, some of which have computer-detection programs, have been developed that use 16 and 24 channels.[9]

TECHNIQUE

Because AEEG electrodes are worn by patients in unsupervised settings, electrode placement requires extra care. AEEG is recorded on an audiocassette tape. The small, lightweight recorder is attached to a belt around the waist or to a shoulder strap. A clock and an event button are on the outside of the unit. Technicians should use stable low-impedance electrodes and apply them carefully with collodion. The electrodes should lie flat and, when possible, the patient's hair should be glued over them for added stability. Patients should be told to wash their hair the night before the electrodes are placed and to avoid using hair spray and other styling products. The AEEG recorder clock and the patient's watch should be synchronized for accurate timing of events. New batteries for the recorder are needed every 24 hours.

Each patient is given a diary to record daily activities. Specific activities, such as toothbrushing, riding in a car, or eating should be noted in the diary. Withdrawal of antiseizure medications in the outpatient setting is not recommended because of the risk of worsening seizures or of status epilepticus. Before the patient leaves the EEG laboratory, the technologist should check the integrity of the recorder and run a brief hard copy of EEG. While the EEG is being printed, the patient can be asked to induce typical artifacts that will be seen on the recording. This can be accomplished by having patients move their eyes, chew, scratch their heads, and turn their heads back and forth. Patients need to be told not to chew gum and to avoid getting the recorder wet. While a hard copy of the EEG is being made, 5 minutes of hyperventilation can be performed. This may help to identify abnormalities such as generalized spike-and-wave that may be present and, thus, anticipated on the cassette tape to be scanned later. A family member ought to be assigned to observe any specific behavioral events and to note them in the diary.

When the patient returns to the laboratory, the tape can be checked to be certain that it recorded properly. Spells can be quickly reviewed and the decision made whether to continue recording another day.

The playback units for the cassette tape can allow review with several options. Tape speed can be varied from slow to as fast as 60 times normal. In addition to being displayed visually,[17] EEG data can be heard as an audio playback of compressed EEG. Seizures and spikes have recognizable sounds; for example, a complex partial seizure has a roaring sound as the seizure evolves. A 3-Hz spike-and-wave discharge sounds like a "plop," whereas an atypical spike-and-wave or polyspike discharge sounds like a scratchy "plop." In early stages of drowsiness, when epileptiform activity is often greatest, one can review the tape at slow speed, whereas in daytime, when more artifacts are present, faster speeds can be used. Filters and sensitivity can be changed on the monitor. Selected epochs of EEG can be printed as a hard copy.

MONTAGES

The newer 16- to 24-channel recorders allow fairly complete head coverage similar to standard 16- to 21-channel EEG. For recorders with only 4 or 8 channels available to record EEG, it is vital to maximize the chances of recording abnormalities (Table 10–1). The work of Ebersole and Leroy[5,13] nicely demonstrated that the most important areas are the anterior temporal and frontal regions. Most abnormalities were recorded in the anterior temporal regions, fewer were found in the frontal region, and the fewest in the posterior head regions. The anterior temporal electrodes were able to record temporal as well as generalized abnormalities.

Eye movement artifacts are so prominent in the frontopolar (Fp) electrodes that they are rarely helpful.

ARTIFACTS

Numerous artifacts plague AEEG and make its interpretation difficult.[4] They are most problematic in wakefulness. The patient's diary can help explain these various artifacts. The most common ones are eye movements, chewing, brushing the teeth, use of electric appliances such as shavers or hair dryers, and scratching or rubbing the electrodes. Riding in a car can produce an unusual seizure-like discharge. One should use caution in interpreting questionable ac-

Table 10–1. SELECTED AMBULATORY ELECTROENCEPHALOGRAPHIC MONTAGES

8-CHANNEL AEEG						
T5-T3	F3-C3	F3-F7	T5-T3	T5-T3	Fp1-F3	01-T5
T3-F7	T5-T3	F7-T3	T3-F7	T3-F7	F3-C3	T5-T3
F7-F3	T3-F7	T3-T5	F7-F3	F7-F8	C3-P3	T3-F7
F3-C3	F7-F3	T5-01	F3-F4	F3-C3	P3-01	F7-F3
C4-F4	F4-C4	F4-F8	F4-F8	F4-C4	Fp2-F4	F4-F8
F4-F8	T6-T4	F8-T4	F8-T4	F8-T4	F4-C4	F8-T4
F8-T4	T4-F8	T4-T6	T4-T6	T4-T6	C4-P4	T4-T6
T4-T6	F8-F4	T6-02	T6-02	ECG	P4-02	T4-T6

4-CHANNEL AEEG			
F3-F7	F7-T5	F3-P3	F7-T5
T3-T5	F3-P3	F4-P4	F8-T6
F4-F8	F4-P4	T3-T4	T3-T4
T4-T6	F8-T6	ECG	ECG

ECG can substitute for one channel as needed.

AEEG = ambulatory electroencephalography; ECG = electrocardiography.

tivity in wakefulness because this activity may well represent artifact. Because AEEG is battery operated, 60-Hz artifact is not present.

CLINICAL APPLICATION

AEEG can be performed as an outpatient or an inpatient procedure. Spells that are too infrequent to be captured on routine EEG can often be captured on AEEG.[2,3] Patients may carry on their normal daily activities with little disruption. Because of the limited number of channels, it is often difficult to use AEEG for definitive ictal localization. Lateralization, however, is frequently possible with AEEG.[7,18] With careful screening of the cassette tape, interictal abnormalities, which can support the diagnosis of seizures, can often be seen.[1,5,6] If the AEEG shows no change during a spell, however, one cannot be certain that the spell is nonepileptic, because simple partial seizures commonly do not have an electrographic accompaniment at the scalp. Subtle complex partial seizures may also be missed when using only 4 or 8 channels of recording capability. Diagnosis of psychogenic seizures is difficult with AEEG. However, if the patient reportedly has a spell that resembles a gen-

eralized tonic-clonic seizure and all that is seen is diffuse muscle activity followed by an immediate return of alpha frequencies, the diagnosis of psychogenic seizure can be strongly suspected. If the AEEG shows a definite ictal discharge, confirmation of the epileptic nature of the spell can be made.

AEEG is particularly useful for seizure disorders with generalized spike-and-wave discharges; 3-Hz spike-and-wave discharges of absence seizures are easily identified on the AEEG. Muscle tone commonly is decreased during the generalized spike-and-wave discharge. The AEEG allows the number of 3-Hz discharges in a 24-hour period to be quantitated. This can be used as a guide to the effectiveness of medical treatment.

AEEG is also useful for episodic loss of consciousness associated with dizziness or other presyncopal symptoms. ECG can easily replace one of the EEG channels. Although arrhythmias are not commonly found on AEEG, when present they can be extremely helpful or even diagnostic of the cause of the syncope or presyncope.[12] An 8-channel recorder can also be used simultaneously with a Holter ECG monitor. In this situation, it is important to synchronize the two different recorder clocks.

AEEG is usually helpful in classifying the

seizure type. After the seizure type is known, effective therapy can often be established.

In children who are cooperative, the AEEG can often be helpful in separating childhood generalized and focal epilepsies from the other common spells of childhood, including syncope, breathholding, night terrors, aggressive outbursts, migraine, psychogenic seizures, and panic attacks. Also, AEEG can be modified for neonatal recording and polysomnography.[8,16]

The principal disadvantage of AEEG is lack of the clinical correlation obtained with simultaneous videotape recordings. Because AEEG recording is performed away from the EEG technician, problems with artifacts are more common than with other monitoring systems.

CHAPTER SUMMARY

AEEG allows outpatient ambulatory monitoring of most episodic events; it is especially good for recording situational spells. Because generalized spike-and-wave discharges are easily seen on AEEG, it can be useful in quantitating the number of discharges in a given period to assess response to medication. Lack of behavioral correlation is the principal problem with AEEG. A seizure discharge seen on EEG confirms the presence of a seizure disorder, but lack of such a discharge does not disprove a seizure disorder. Simple partial seizures and some mild complex partial seizures may have no definite scalp EEG change or the change may be missed because of the limited number of recorded channels and other technical factors.

REFERENCES

1. Bridgers, SL and Ebersole, JS: The clinical utility of ambulatory cassette EEG. Neurology 35:166–173, 1985.
2. Callaghan, N and McCarthy, N: Twenty-four-hour EEG monitoring in patients with normal, routine EEG findings. In Dam, M, Gram, L, and Penry, JK (eds): Advances in Epileptology: The 12th Epilepsy International Symposium. Raven Press, New York, 1981, pp 357–359.
3. Davidson, DLW, Fleming, AMM, and Kettles, A: Use

of ambulatory EEG monitoring in a neurological service. In Dam, M, Gram, L, and Penry, JK (eds): Advances in Epileptology: The 12th Epilepsy International Symposium. Raven Press, New York, 1981, pp 319–321.
4. Ebersole, JS, Bridgers, SL, and Silva, CG: Differentiation of epileptiform abnormalities from normal transients and artifacts on ambulatory cassette EEG. Am J EEG Technol 23:113–125, 1983.
5. Ebersole, JS and Leroy, RF: A direct comparison of ambulatory cassette and intensive inpatient EEG monitoring in detecting interictal abnormalities (abstr). Electroencephalogr Clin Neurophysiol 53:21P, 1982.
6. Ebersole, JS and Leroy, RF: An evaluation of ambulatory, cassette EEG monitoring: II. Detection of interictal abnormalities. Neurology (NY) 33:8–18, 1983.
7. Ebersole, JS and Leroy, RF: Evaluation of ambulatory cassette EEG monitoring: III. Diagnostic accuracy compared to intensive inpatient EEG monitoring. Neurology (Cleveland) 33:853–860, 1983.
8. Eyre, JA: Clinical utility of cassette EEG in neonatal seizure disorders. In Ebersole, JS (ed): Ambulatory EEG Monitoring. Raven Press, New York, 1989, pp 141–155.
9. Gotman, J: Automated analysis of ambulatory EEG recordings. In Ebersole, JS (ed): Ambulatory EEG Monitoring. Raven Press, New York, 1989, pp 97–110.
10. Ives, JR, Hausser, C, Woods, JF, and Andermann, F: Contribution of 4-channel cassette EEG monitoring to differential diagnosis of paroxysmal attacks. In Dam, M, Gram, L, and Penry, JK (eds): Advances in Epileptology: The 12th Epilepsy International Symposium. Raven Press, New York, 1981, pp 329–336.
11. Ives, JR and Woods, JF: 4-Channel 24 hour cassette recorder for long-term EEG monitoring of ambulatory patients. Electroencephalogr Clin Neurophysiol 39:88–92, 1975.
12. Lai, C-W and Ziegler, DK: Syncope problem solved by continuous ambulatory simultaneous EEG/ECG recording. Neurology (NY) 31:1152–1154, 1981.
13. Leroy, RF and Ebersole, JS: An evaluation of ambulatory, cassette EEG monitoring: I. Montage design. Neurology (NY) 33:1–7, 1983.
14. Marson, GB and McKinnon, JB: A miniature tape recorder for many applications. Control Instrumentation 4:46–47, 1972.
15. Quy, RJ: A miniature preamplifier for ambulatory monitoring of the electroencephalogram. J Physiol (Lond) 284:23P–24P, 1978.
16. Reitman, M: Techniques of cassette polysomnography. In Ebersole, JS (ed): Ambulatory EEG Monitoring. Raven Press, New York, 1989, pp 243–255.
17. Stores, G, Hennion, T, and Quy, RJ: EEG ambulatory monitoring system with visual playback display. In Wada, JA and Penry, JK (eds): Advances in Epileptology: The Xth Epilepsy International Symposium. Raven Press, New York, 1980, pp 89–93.
18. Tuunainen, A and Nousiainen, U: Ictal recordings of ambulatory cassette EEG with sphenoidal electrodes in temporal lobe epilepsy: Comparison with intensive videomonitoring. Acta Neurol Scand 88:21–25, 1993.

CHAPTER 11

PROLONGED VIDEO ELECTRO-ENCEPHALOGRAPHY

Kathryn A. Hirschorn, M.D.

Prolonged video electroencephalography (PVEEG) is an important tool in the evaluation of patients with seizures and other paroxysmal or episodic events.[8] Often, the clinical history alone is not adequate to make an accurate diagnosis. Even with activation procedures such as sleep, sleep deprivation,[17] hyperventilation,[18] and photic stimulation, epileptiform abnormalities may not be present on routine electroencephalographic (EEG) recordings. An accurate diagnosis often requires capturing the clinical spell with simultaneous electrophysiologic and behavioral monitoring. PVEEG allows behavioral correlation of the patient's clinical spells with the EEG. The most experienced examiners may miss subtle or even obvious clinical manifestations of seizures when allowed to view the event only once. Capturing the seizure on video tape permits repeated review

of the event and comparison between subsequent events. PVEEG allows physicians and allied health staff an extended period of observation that is often helpful in ferreting out psychosocial issues that may contribute to the patient's condition.

PVEEG can be used to diagnose epileptic versus nonepileptic events and to help classify seizure type.[23] Also, PVEEG is a vital part of localization of the epileptic zone in patients being considered for epilepsy surgery.[24,25] Modifications of this technique can allow evaluation of neonates, infants and children, patients in intensive care units, and patients with other episodic events such as sleep disorders. This chapter describes various techniques for PVEEG and discusses clinical applications for this procedure.

HISTORY

Hunter and Jasper,[12] in 1949, then Schwab and associates[29] and Stewart, Jasper, and Hodge[32] described systems for cinematographic and EEG recordings. These early split-screen systems used mirrors to record simultaneously the patient's behavioral manifestations and the EEG. These systems were bulky and usually could record only 1 hour of cinema before the tape had to be

118

changed.[29] The development of the video-tape player and more compact camera units led to greater acceptance of the procedure.[13,14,31] Even today, a simple prolonged EEG can be obtained using a video camcorder and a standard EEG. An obvious disadvantage of this technique is that paper must be printed for the entire monitoring session. In the last decade, video EEG units with digitally acquired EEG stored on a computer system or the videotape itself have become commercially available. Many medical centers now use telemetered EEG,[2,3] with either cable or radio telemetry.

EQUIPMENT

At present, there is great variability in the PVEEG systems used in laboratories throughout the world. Most large medical centers have several types of systems that can be tailored to individual patient needs. Also, most large medical centers can perform inpatient and outpatient PVEEG. Outpatient PVEEG can be used for 4- to 8-hour recordings for patients with frequent events. Most patients who undergo PVEEG are admitted to an inpatient epilepsy unit for 24-hour recordings. Twenty-four-hour monitoring capability is desirable, because a large number of seizures occur at night or in the early morning hours.

The simplest type of long-term monitoring can be performed with prolonged conventional EEG. This requires an EEG machine, a cooperative patient, and a technician. A videotape recorder or camcorder can be used to record ictal behavior. This system is labor intensive, and patient mobility is restricted. Large volumes of paper are used while waiting to capture an event. The event then can be displayed only in the montage that was selected for the conventional EEG, preventing further EEG analysis after the seizure.

Currently, several systems are commercially available for PVEEG. Most of these systems use either VHS or beta tape with either multiplexed EEG recorded on the audio channel or a system that records EEG digitally on a computer. The patient may be connected to the recording equipment with a specialized EEG cable or the EEG data may be sent through a radio signal.

Analog PVEEG systems allow good correlation of the EEG with clinical behavioral changes.[1] Portable analog systems exist that can be brought into an intensive care unit or other rooms in the hospital. Infrared cameras can be used at night to achieve reasonable quality recording in a darkened room. The analog systems are fraught by the appearance of more artifacts than digital systems, and off-time analysis is not possible.

Digital PVEEG systems acquire EEG and video data with cable or radiotelemetry and a sophisticated system. Digital PVEEG allows on-line processing of EEG activity, with automatic detection of seizures and interictal epileptiform activity and off-line seizure analysis. Data reduction is possible with digital PVEEG systems. EEG data can be stored on optical disc or other computer storage devices. With cable telemetry, patients have limited mobility. However, the technical quality of cable or "hardwired" systems is usually superior to that of radiotelemetered systems. With the radiotelemetered system, patients have greater mobility, which may or may not be an advantage given the circumstances of a patient's seizures, seizure frequency, and severity. Also, radiotelemetered systems usually have more artifacts than "hardwired" or cable systems. The cost of digital PVEEG systems can be high, but because of their superior recording capabilities, their use is gaining wider acceptance.

Currently, many centers record EEG data on computer systems, allowing off-line analysis of ictal and interictal activity.[10,11] Montage reformatting, filtering, frequency analysis, timing analysis, and correlation studies can be performed with these systems. This ability to carefully analyze a single seizure in many ways and to view it with several different montages has decreased the number of actual seizures that need to be recorded. Because of the risks of recording seizures and of tapering seizure medications and the expense of this procedure, the ability to obtain more information from as few seizures as possible is important.

Portable PVEEG systems can also be used in intensive care units, neonatal units, and in psychiatric facilities. These portable systems can be helpful in capturing events that are too infrequent to be seen on routine EEG. The systems available can include a video camcorder, a videocassette recorder, and an amplifier with or without a standard EEG machine. If EEG is acquired digitally, it can later be analyzed off-line with available computer software.

CLINICAL APPLICATION

Epileptic Versus Nonepileptic Events

PVEEG can often help answer the most fundamental question of whether the spell in question is epileptic in nature.[15,16,19,23,33] Many different episodic events can mimic seizures clinically.[35] Separating epileptic and nonepileptic events is critical for determining effective therapy. Obviously antiseizure medications are rarely of benefit for conditions other than seizures.

Syncope can result from various nonepileptic causes, including aortic stenosis, cardiac arrhythmias, vasovagal depression, orthostatic hypotension, or hyperventilation. Using PVEEG with simultaneous electrocardiography (ECG) and blood pressure monitoring may help prove the nonepileptic nature of these events.

Vertigo and nonspecific dizzy spells may be difficult to differentiate from seizures. Various movement disorders can also be confused with seizures. Simultaneous surface electromyographic (EMG) recording of agonist and antagonist muscles with continuous EEG recording can often be helpful in the diagnosis of movement disorders.

Sleep disorders and seizures may also be confused with one another. PVEEG allows behavioral correlation with EEG and other features, such as respiration, eye movement, surface EMG, ECG, and oxygen saturation. PVEEG may also help separate transient ischemic attacks, hypoglycemic attacks, and migraine attacks from seizures.

PVEEG can be especially helpful in differentiating types of childhood spells.[4] Daydreaming, breathholding, migraines, night terrors, and other parasomnias can be confused with seizures. Often, outpatient monitoring is adequate for classification of these types of spells. Repetitive movements are common in the intensive care unit setting. Some of these movements may be ictal and require antiseizure medications, although many are nonictal. Again, PVEEG with a portable system can be helpful in diagnosing these movements.

One of the obvious advantages of PVEEG is that it can easily be tailored to fit individual diagnostic situations. The video and EEG portions can be coupled with recordings from eye movements, surface EMG, respirations, ECG, a thermistor, continuous blood pressure monitoring, and oxygen saturation. Under most circumstances, the differentiation of epilepsy from nonepileptic physiologic events is possible with intensive EEG monitoring.

Psychogenic Seizures or Nonepileptic Behavioral Events

Psychogenic seizures are a difficult diagnostic problem.[27,30] An accurate diagnosis of psychogenic seizures can lead to appropriate psychiatric treatment and discontinuation of unnecessary and potentially dangerous medical therapy. In most instances, psychogenic seizures are not under conscious control. Gates, Ramani, Whalen, and Loewenson[9] compared the clinical manifestations of psychogenic generalized tonic-clonic events and true generalized tonic-clonic seizures. They found that out-of-phase clonic movements of the extremities, pelvic thrusting, lack of eye manifestations, side-to-side head movements, and early vocalizations were more common in the psychogenic group. Psychogenic seizures also tend to last longer than 2 minutes. Patients with psychogenic seizures often have long attacks, with frequent rest periods during attacks. Also, these patients often respond in some degree to verbal or noxious stimulation during the generalized movements. Although a spectrum of psychiatric diagnoses are represented in patients with psychogenic seizures, a large number have a history of sexual abuse, rape, or torture.

In patients with psychogenic seizures,

PVEEG allows more careful analysis of behavioral and EEG manifestations. Muscle and movement artifacts are usually prominent, but there is no ictal EEG change. After the movements stop, there is usually an immediate return of normal background EEG. Postictal slowing and suppression of the EEG, which are typically present after a true generalized tonic-clonic seizure, are not present after a psychogenic seizure.

Classification of Seizure Type

PVEEG may lead to a reclassification of seizure type in many patients with uncontrolled seizures, especially if the diagnosis is in question. This also can lead to improved medical management.[22,28]

Distinguishing primary generalized from the secondary generalized seizures is often difficult based on clinical history and routine EEG. Some patients have rapid secondary spread and may have an interictal EEG that shows generalized spike-and-wave discharges. However, ictal recording may show the focal nature of the seizure onset.[34] Other patients with true absence or true generalized tonic-clonic seizures may mistakenly be thought to have focal epilepsy. By clinical history alone, absence seizures may be indistinguishable from complex partial seizures.[20] Untreated, virtually all patients with absence seizures will activate with 3-Hz spike-and-wave discharges with hyperventilation. Patients with infrequent spells or those on medication may be more difficult to diagnose with certainty and may benefit from PVEEG.[26] Medication taper can be done in the hospital while the patient is carefully observed.

Monitoring can usually differentiate temporal from extratemporal seizures. A large number of patients with simple partial seizures have no EEG accompaniment. However, the majority of patients with complex partial seizures show an ictal EEG change. Temporal lobe seizures often begin with an attenuation of scalp activity followed by a rhythmic discharge, usually in the theta range, which increases in amplitude and becomes more widespread. Postically, there is

often focal slowing over the temporal region where the seizure began. Tachycardia is observed in most patients with complex partial seizures of temporal lobe origin. Lateralized ictal paresis has also been observed in many of our patients with temporal lobe seizures.

Frontal lobe seizures tend to be shorter and to cause less postictal confusion than temporal lobe seizures. Frontal lobe seizures are associated with frequent falls, because of the rapid bilateral spread. Focal tonic or "fencing" postures may be seen. Certain frontal lobe seizures may mimic absence seizures and at times have been referred to as "pseudoabsences." Frontal lobe seizures often begin with low-amplitude fast activity but can be associated with some frontal sharp waves or spikes. PVEEG, especially when made with computer-assisted recordings, can often help identify these frontal lobe seizures.

PVEEG and Surgical Evaluation

Surgical treatment can frequently eliminate or decrease the frequency and severity of seizures in many focal epilepsies. Seizures arising from the temporal lobe are especially amenable to surgical treatment. A comprehensive evaluation is performed in those patients who are being considered for epilepsy surgery. Probably the most important part of the presurgical evaluation is the recording of the patient's typical seizures on continuous EEG with video monitoring.

PVEEG allows careful analysis of clinical and EEG changes and can be used with scalp, sphenoidal, foramen ovale, or implanted depth or surface cortical grid recordings.[5–7,21] With the computer-assisted recordings, montage reformatting, and off-time analysis of seizures, there is less need for invasive EEG recordings. For surgical monitoring, digital video systems with 24 to 124 channels are commonly used. For focal cortical resection, typical seizures must be shown to arise from a single region of the brain. An adequate number of seizures must be recorded so that the seizure can be localized confidently. More information about surgical evaluation is found in Chapter 13.

CHAPTER SUMMARY

PVEEG is an important tool for diagnosis of seizures and other episodic events. Various monitoring systems exist. Careful monitoring allows appropriate diagnosis of most seizure types and is a necessary part of the evaluation of patients being considered for surgical treatment of their seizures. Finally, this type of monitoring allows extended observation of patients and can lead to a much better understanding of their medical problems and any superimposed psychosocial issues.

REFERENCES

1. Aarts, JHP, Binnie, CD, Smit, A, and Wilkins, AJ: Performance testing during epileptiform EEG activity. In Burr, SH (ed): Mobile Long-Term EEG Monitoring (Proceedings of the MLE Symposium). Gustav Fischer, Stuttgart, 1982, pp 161–168.
2. Binnie, CD: Telemetric EEG monitoring in epilepsy. Recent Adv Epilepsy 1:155–178, 1983.
3. Binnie, CD, Rowan, AJ, Overweg, J, et al: Telemetric EEG and video monitoring in epilepsy. Neurology 31:298–303, 1981.
4. Connolly, MB, Wong, PK, Karim, Y, et al: Outpatient video-EEG monitoring in children. Epilepsia 35:477–481, 1994.
5. Devinsky, O, Sato, S, Kufta, CV, et al: Electroencephalographic studies of simple partial seizures with subdural electrode recordings. Neurology 39:527–533, 1989.
6. Engel, J Jr: Surgical Treatment of the Epilepsies. Raven Press, New York, 1987, pp 553–571.
7. Engel, J Jr: Seizures and Epilepsy. FA Davis, Philadelphia, 1989, pp 443–474.
8. Engel, J Jr, Burchfiel, J, Ebersole, J, et al: Long-term monitoring for epilepsy. Report of an IFCN committee. Electroencephalogr Clin Neurophysiol 87:437–458, 1993.
9. Gates, JR, Ramani, V, Whalen, SM, and Loewenson, R: Ictal characteristics of pseudoseizures. Arch Neurol 42:1183–1187, 1985.
10. Gotman, J: Automatic recognition of epileptic seizures in the EEG. Electroencephalogr Clin Neurophysiol 54:530–540, 1982.
11. Gotman, J: Seizure recognition and analysis. Electrophysiol Clin Neurophysiol Suppl 37:133–145, 1985.
12. Hunter, J and Jasper, HH: A method of analysis of seizure pattern and electroencephalogram: A cinematographic technique. Electroencephalogr Clin Neurophysiol 1:113–114, 1949.
13. Ives JR: Video recording during long-term EEG monitoring of epileptic patients. Adv Neurol 46:1–11, 1987.
14. Ives, JR, Thompson, CJ, and Gloor, P: Seizure monitoring: A new tool in electroencephalography. Electroencephalogr Clin Neurophysiol 41:422–427, 1976.
15. Kaplan, PW and Lesser, RP: Long-term monitoring. In Daly, DD and Pedley, TA (eds): Current Practice of Clinical Electroencephalography, ed 2. Raven Press, New York, 1990, pp 513–534.
16. Mattson, RH: Value of intensive monitoring. In Wada, JA and Penry, JK (eds): Advances in Epileptology: The 10th Epilepsy International Symposium. Raven Press, New York, 1980, pp 43–51.
17. Mattson, RH, Pratt, KL, and Calverley, JR: Electroencephalograms of epileptics following sleep deprivation. Arch Neurol 13:310–315, 1965.
18. Miley, CE and Forster, FM: Activation of partial complex seizures by hyperventilation. Arch Neurol 34:371–373, 1977.
19. Penry, JK and Porter, RJ: Intensive monitoring of patients with intractable seizures. In Penry, JK (ed): Epilepsy: The 8th International Symposium. Raven Press, New York, 1977, pp 95–101.
20. Penry, JK, Porter, RJ, and Dreifuss, FE: Simultaneous recording of absence seizures with video tape and electroencephalography: A study of 374 seizures in 48 patients. Brain 98:427–440, 1975.
21. Perry, TR, Gumnit, RJ, Gates, JR, and Leppik, IE: Routine EEG vs. intensive monitoring in the evaluation of intractable epilepsy. Public Health Rep 98:384–389, 1983.
22. Porter, RJ, Penry, JK, and Lacy, JR: Diagnostic and therapeutic reevaluation of patients with intractable epilepsy. Neurology 27:1006–1111, 1977.
23. Porter, RJ, Sato, S, and Long, RL: Video recording. Electrophysiol Clin Neurophysiol Suppl 37:73–82, 1985.
24. Porter, RJ, Theodore, WH, and Schulman, EA: Intensive monitoring of intractable epilepsy: A two year follow-up. In Dam, M, Gram, L, and Penry, JK (eds): Advances in Epileptology: The 12th Epilepsy International Symposium. Raven Press, New York, 1981, pp 265–268.
25. Quesney, LF and Gloor, P: Localization of epileptic foci. Electroencephalogr Clin Neurophysiol Suppl 37:165–200, 1985.
26. Quy, RJ, Fitch, P, Willison, RG, and Gilliatt, RW: Electroencephalographic monitoring in patients with absence seizures. In Wada, JA, and Penry, JK (eds): Advances in Epileptology: The 10th Epilepsy International Symposium. Raven Press, New York, 1980, pp 69–72.
27. Ramani, SV, Quesney, LF, Olson, D, and Gumnit, RJ: Diagnosis of hysterical seizures in epileptic patients. Am J Psychiatry 137:705–709, 1980.
28. Riley, TL, Porter, RJ, White, BG, and Penry, JK: The hospital experience and seizure control. Neurology 31:912–915, 1981.
29. Schwab, RS, Schwab, MW, Withee, D, and Chock YC: Synchronized moving pictures of patient and EEG. Electroencephalogr Clin Neurophysiol 6:684–686, 1954.
30. Scott, DF: Recognition and diagnostic aspects of nonepileptic seizures. In Riley, TL, and Roy, A (eds): Pseudoseizures. Williams & Wilkins, Baltimore, 1982, pp 21–33.

31. Stalberg, E: Experiences with long-term telemetry in routine diagnostic work. In Kellaway, P and Petersen, I (eds): Quantitative Analytic Studies in Epilepsy. Raven Press, New York, 1976, pp 269–277.
32. Stewart, LF, Jasper, HH, and Hodge, C: Another simple method for simultaneous cinematographic recording of the patient and his electroencephalogram during seizures. Electroencephalogr Clin Neurophysiol 8:688–691, 1956.
33. Sutula, TP, Sackellares, JC, Miller, JQ, and Dreifuss, FE: Intensive monitoring in refractory epilepsy. Neurology 31:243–247, 1981.
34. Theodore, WH, Porter, RJ, Albert, P, et al: The secondarily generalized tonic-clonic seizure: A videotape analysis. Neurology 44:1403–1407, 1994.
35. Wada, JA: Differential diagnosis of epilepsy. Electroencephalogr Clin Neurophysiol Suppl 37:285–311, 1985.

CHAPTER 12

ELECTRO-ENCEPHALOGRAPHIC SPECIAL STUDIES

Terrence D. Lagerlund, M.D., Ph.D.

QUANTITATIVE METHODS OF ELECTROENCEPHALOGRAPHIC ANALYSIS

Digital computers can aid in extracting information from electroencephalographic (EEG) waveforms that is not readily obtainable with visual analysis alone. This type of computer may also be used for quantification of key features of waveforms, which may be useful in accurate EEG interpretation and also in making serial comparisons between EEGs performed on the same subject at different times or between two subject groups in scientific investigations. Digital computers may also partially automate the interpretation of EEGs, particularly in the setting of prolonged monitoring for epilepsy. This chapter discusses some of the uses of computers in EEG.

Fourier (Spectral) Analysis

The technique of spectral analysis was described in Chapter 4. EEG spectral analysis has been used to assess the dominant background frequencies in normal and disease states, including dementia, cerebral infarction, cerebral neoplasms, and various toxic and metabolic disorders. Most abnormal conditions increase the amount of slow-wave activity in an EEG and may decrease the amount of faster frequency activity. Quantitative measures that may be used include: (1) the percentage power in the delta, theta, or delta + theta bands; (2) the percentage power in the alpha, beta, or alpha + beta bands; (3) various other power ratios (e.g., alpha to delta); (4) the mean frequency in the entire spectrum or in a portion of the spectrum, such as the alpha band; and (5) the "spectral edge" frequency, often taken to be the frequency below which 95% of the power in the entire spectrum occurs. Spectral analysis has also been used in intraoperative monitoring to assess changes in depth of anesthesia or cerebral ischemia. Although spectral analysis per se can be applied to only one channel at a time, useful comparisons can be made between the spectra of recordings done in different regions of the head. As expected, focal brain lesions tend to produce focal changes in power spectra, whereas diffuse processes produce generalized changes. Quantitative measures of asymmetry can be calculated, such as the left-to-right ratio of power in the entire spectrum or in individual frequency bands, alpha, delta, and so forth. Some studies of power spectra during transient cerebral events, such as epileptic seizures, have also been conducted. The power spectra of a partial complex seizure of temporal lobe origin, for example, tends to be complex, with several frequency components (often a fundamental with one or more harmonics, i.e., integral multiples of the fundamental frequency) whose amplitudes in various regions of the head may differ. The spectrum also evolves over time during the course of the seizure.

Spike, Sharp-Wave, and Seizure Detection

Detection of interictal and ictal epileptic discharges in an EEG by computer algorithms has received much attention recently because of the increasing use of prolonged EEG monitoring for epilepsy. In prolonged EEG monitoring, multiple EEG channels (21 or more in most cases) are recorded over many hours, days, or weeks, thus generating huge amounts of data that must be reviewed to assess the frequency, nature, and localization of epileptic discharges. Computer algorithms may be used to locate candidate interictal or ictal discharges for further study by an electroencephalographer and, thus, the need to scan manually many pages of EEG tracings may be avoided. Techniques of pattern recognition are discussed in Chapter 4. Spike and sharp-wave detection generally relies on "sharpness" criteria applied to individual waveforms on a channel-by-channel

basis, although more advanced algorithms may use context information and multichannel correlation to improve specificity. Seizure detection algorithms are more complex, because of the great variability in types of ictal discharges, and generally are based on detecting a sudden change in rhythmicity and amplitude of background activity simultaneously in multiple channels.

Current state-of-the-art commercially available software has a reasonably good overall sensitivity, although for the seizures of certain patients the sensitivity may be inadequate. Consequently, prolonged monitoring systems usually do not rely exclusively on computer detection of seizures but also make use of trained observers and/or patient and family members to recognize and log seizure occurrence. The specificity of currently used algorithms is also fairly good; however, in many prolonged monitoring situations, a large variety of artifacts arising from patient activities and various waveforms of cerebral origin, for example, V waves in sleep, cause false detections that may outnumber true epileptic discharges by an order of magnitude in some patients. Thus, all discharges detected by the computer must be reviewed by a physician or technician, but the ability to markedly decrease the amount of data to be reviewed still makes automated detection algorithms of great practical value.

Montage Reformatting

Montage reformatting allows EEG montages to be selected when the data are reviewed, independent of the montage used to acquire and to store the data. The same EEG segment can be viewed with a variety of different montages. To generate a derivation such as $X_1 - X_2$, which does not exist in the EEG as recorded, the computer looks for two existing channels, one of which records X_1 against a reference electrode and another which records X_2 against the same reference, and then subtracts the two. That is, $X_1 - X_2 = (X_1 - R) - (X_2 - R)$, where R is the reference. This allows new referential and bipolar montages to be formed.[4] For example, suppose that recorded EEG data include the following channels:

Channel	Derivation
1	F_{p1}-C_Z
2	F_{p2}-C_Z
3	F_3-C_Z
4	F_4-C_Z
5	C_3-C_Z
6	C_4-C_Z
7	P_3-C_Z
8	P_4-C_Z
9	O_1-C_Z
10	O_2-C_Z
11	A_1-C_Z
12	A_2-C_Z

Channel	Derivation
1-11	F_{p1}-A_1
2-12	F_{p2}-A_2
3-11	F_3-A_1
4-12	F_4-A_2
5-11	C_3-A_1
6-12	C_4-A_2
7-11	P_3-A_1
8-12	P_4-A_2
9-11	O_1-A_1
10-12	O_2-A_2

Then, a new referential montage with ipsilateral ear reference can be created by subtracting pairs of channels as follows:

New reference electrodes can also be created by averaging data from two or more existing electrodes. For example, suppose an EEG is recorded with a C_Z reference. A new derivation such as O_1-$A_{1/2}$, where $A_{1/2}$ represents the average of the ear electrodes, can

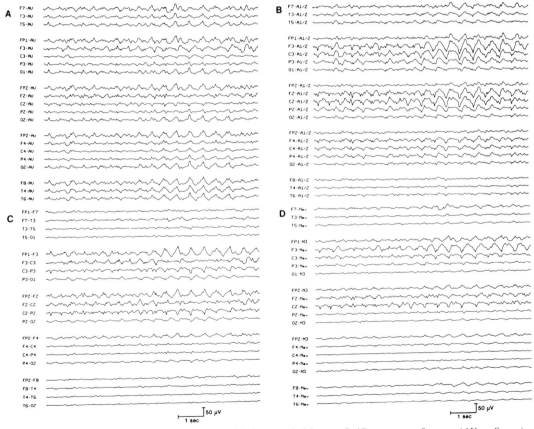

Figure 12–1. An EEG recorded during a seizure. (*A*) As recorded from a C_3/C_4 average reference (*AV* on figure), which is active. (*B*) Reformatted to average ear reference (*A1/2* on figure). (*C*) Reformatted to longitudinal bipolar montage. (*D*) Reformatted to longitudinal laplacian montage (Nav = 4-neighbor average; N3 = 3-neighbor average). (From Lagerlund,[4] p 318, with permission.)

be calculated as follows: $O_1\text{-}A_{1/2} = (O_1\text{-}C_Z) - 0.5(A_1\text{-}C_Z) - 0.5(A_2\text{-}C_Z)$. Other types of montages, such as a common average reference montage (in which the reference for each electrode is the average potential of all recorded electrodes) or a laplacian (source) montage[3,12] (in which the reference for each electrode is the average of the four nearest neighbors to that electrode), may also be easily generated by the computer. Figure 12–1 shows the same ictal discharge viewed with various montages (one viewed as recorded and three after reformatting).

Cross-Correlation Analysis

Cross-correlation analysis quantifies the relationship between EEG signals recorded from *different* derivations. The *cross-correlation function* is the correlation between EEG signal no. 1 and EEG signal no. 2 delayed by a time t, expressed as a function of t. If EEG signal no. 2 is identical to no. 1, the cross-correlation function is the same as the autocorrelation function described in Chapter 4. If signal no. 2 is similar to no. 1 but delayed by the time T, then the cross-correlation function will be a maximum at $t = T$, and periodically thereafter if both signals are periodic. Thus, examination of cross-correlation functions may be used as a means to assess small interchannel time differences (Fig. 12–2); one possible use of this is in determining whether a bilateral epileptic discharge in fact represents secondary bilateral synchrony or is generalized from onset.

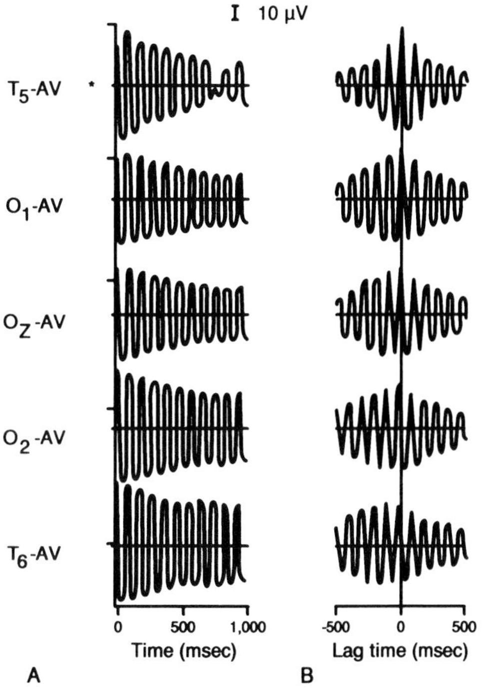
I 10 µV
T5-AV
O1-AV
OZ-AV
O2-AV
T6-AV
0
500
1,000
Time (msec)
-500
0
500
Lag time (msec)
A
B

Figure 12–2. (*A*) Autocorrelation functions and (*B*) cross-correlation functions for a 1-s epoch of alpha activity recorded using a C_3/C_4 average reference (*AV* in figure). The autocorrelations demonstrate the rhythmicity of the signal but cannot be used to assess interchannel time differences. The cross-correlations demonstrate a systematically increasing time delay between channels as one goes from right, T_6, to left, T_5, across the head.

Cross-Spectral Analysis

The *cross-power spectrum* between EEG signal no. 1 and no. 2 is the Fourier transform of the cross-correlation function. From the cross-power spectrum, one can calculate two additional functions:

1. *Coherence function*—a normalized function that takes on values between 0 and 1. It is defined as the absolute square of the cross-power spectrum of signals no. 1 and 2, divided by the product of the power spectrum of signal no. 1 and that of signal no. 2. It is a convenient measure of the degree of correlation between similar frequency components in the two signals (0 to 100%), expressed as a function of the frequency.
2. *Phase spectrum*—the relative phase angle (0 to 360 degrees) between corresponding frequency components of signal no. 1 and those of signal no. 2; it is expressed as a function of the frequency.

The phase angle, in turn, is related to timing differences between the two signals. Because the phase angle between two sine or cosine waves delayed by a fixed time T equals 360 degrees multiplied by the product of the time delay T and the frequency f, the time delay between the two signals can be calculated from the phase spectrum ϕ by the formula,[2]

$$T = \frac{\phi}{360° \cdot f}$$

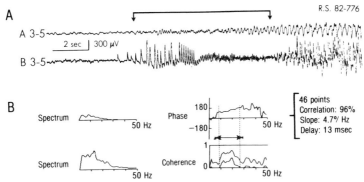

Figure 12–3. (*A*) Onset of a seizure discharge recorded with a bipolar montage from intracerebral electrodes in a patient with epilepsy; *A 3-5* are in amygdala; *B 3-5* are in hippocampus. (*B*) Power spectra of each channel (*left*) and phase angle and coherence spectra (*right*); the upper coherence graph is the coherence itself; the lower graph is the lower boundary of the 99% confidence interval of the coherence. For the range of frequencies indicated by the *arrows*, the phase is linearly related to frequency with slope 4.7 degrees/Hz, corresponding to a delay of 13 milliseconds between channels. Positive slope indicates that the first channel is leading. (From Gotman,[2] p 502, with permission.)

Thus, if a plot of phase angle versus frequency is linear over some range of frequencies, the time difference between the signals can be derived by taking the slope of the plot divided by 360° (Fig. 12–3).

Interpolation Techniques

Interpolation techniques are used to estimate the electric potential at intermediate scalp positions from its known value at each electrode position. This procedure is necessary when constructing maps of scalp potential, and it may form a preliminary step in the application of various techniques of spatial, or topographic, analysis of EEG, such as calculating accurate estimates of the laplacian of the scalp potential, multivariate statistical analyses, cortical projection techniques, and source dipole localization. Some interpolation methods that are commonly used include the following:

1. Linear interpolation from the potentials at three or four nearest neighbor electrodes.
2. Two-dimensional surface splines, which conceptually minimize the "bending energy" of an infinite, elastic plate constrained to pass through known points.[9]
3. Spatial Fourier analysis, which models the distribution of potentials over the head as a two-dimensional Fourier series in *x* and *y* coordinates.
4. Spherical splines, splines that are applied to a spherical surface instead of a rectangular one.[8]
5. Three-dimensional splines, which interpolate potentials in three dimensions, after which results may be projected as needed onto spherical or ellipsoidal surfaces.[5]
6. Spherical harmonic expansion, the equivalent of a Fourier series used in spherical coordinates instead of rectangular ones.[7]
7. Source distribution and volume conduction models, which localize source dipoles and then predict the scalp potential that these sources would generate.

Note that methods 1 to 3 are based on a rectangular planar model of the scalp surface, whereas methods 4 to 7 assume a spherical head and should be more accurate. Only method 1, the simplest and least accurate, however, has been widely used in commercial EEG systems.

Topographic Displays (Mapping)

EEG topographic, or spatial, maps are a way of displaying EEG data that differs significantly from the conventional multichannel amplitude-versus-time plots (montages).

However, topographic mapping is *not* in itself a method of EEG analysis, because it only displays the "raw" data in a different way. In its simplest form, a topographic map of scalp potential is a "snapshot" of the EEG at one instant in time; it shows the distribution of potentials on the head surface at that time and may facilitate localization of EEG abnormalities (while making it more difficult to appreciate their variation in time). Because potentials are actually measured at only a few points on the head, that is, the electrode locations, an interpolation technique must be used to estimate potentials at all other scalp points;[1] thus, the information conveyed by a topographic display is only as

accurate as the interpolation technique used (Fig. 12–4). Various methods of topographic display include three-dimensional plots (potential on z axis versus x and y coordinates), contour plots (connecting all points on the head with the same value of potential), gray scale intensity plots (degree of darkness at each point on a map of the head corresponds to the potential at that point), and color plots (color at each point on a map of the head corresponds to the potential at that point). As a general rule, topographic maps should be used only as an adjunct to ordinary time series EEG displays and not as a replacement for them, because 75% or more of the information derived from an EEG is

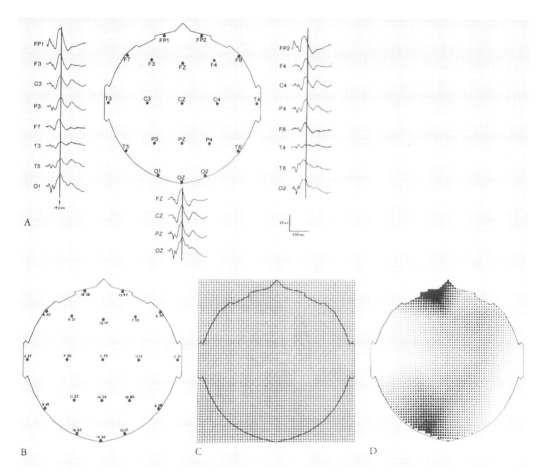

Figure 12–4. Example of topographic map construction for visual evoked-potential (EP) signals recorded referentially from 20 scalp electrodes. Each EP is divided into 128 intervals of 4 milliseconds. (*A*) Individual EPs for the indicated electrode locations. (*B*) Mean voltage values at each electrode location for the 4-millisecond interval beginning 192 milliseconds after the stimulus (corresponding to the vertical line in *A*). (*C*) The grid of interpolation points (64 × 64) used; each of the 4096 points is assigned a voltage value by linear interpolation from the three nearest known points. (*D*) The topographic map of this EP, using an equal interval intensity scale to represent voltage at each location. (From Duffy, et al,[1] p 311, with permission.)

related to temporal rather than spatial characteristics of waveforms.

In addition to maps of unprocessed EEG, topographic mapping has been used to display the spatial distribution of the results of various techniques of multichannel EEG analysis. For example, maps of power spectra—power within a specific frequency band at various scalp locations—can be generated. Similarly, maps of the laplacian of the scalp potential, of principal components of the scalp potential (from multivariate statistical analysis), or of estimated cortical potentials (from a cortical projection technique) can also be produced.

Multivariate Statistical Methods of Topographic Analysis

The purpose of multivariate statistical methods of EEG topographic analysis is to achieve data reduction from many simultaneously recorded EEG signals. This is possible because of the redundancy of multichannel EEG data—many activities or waveforms appear simultaneously in many different channels. Mathematical techniques such as factor analysis, principal component analysis, or eigenvector analysis are used to reduce the observed EEG signals in multiple channels to a minimum number of independent, or "orthogonal," component signals. Individual components may be displayed spatially as topographic maps or temporally as "derived" EEG channels. Although these methods may provide data reduction, their major drawback is that the resultant independent signals are not always recognizable as traditional "pure" EEG activities such as alpha or mu, and comparison between analyses made on the same EEG recording at different times or on different subjects is difficult. Also, these methods generally do not give information on the nature and location of physiologic generators of EEG.

Cortical Projection Techniques

Cortical projection techniques such as spatial deconvolution[6] are designed to reverse the "smearing" effect of the skull on scalp EEG by using a model of volume conduction in the head (such as a three-sphere model of brain, skull, and scalp). The electric potential at selected points on the brain surface is calculated from the electric potential at the scalp surface, thus noninvasively providing a distribution of electrical activity at the cortical surface. Cortical potentials may be displayed as a time series (montage format) or as a topographic display (map format). This technique has proved capable of resolving two adjacent dipole sources in cortex that could not be resolved by inspection of the scalp EEG signals alone and has been used successfully to localize median nerve somatosensory evoked potentials and interictal epileptic discharges on the cortical surface, with confirmation by recordings from subdural electrode grids.

Source Dipole Localization

The ultimate goal of localization of abnormal activity, such as epileptic discharges, from EEG is to find the intracranial sources generating a given distribution of scalp potentials. This is sometimes called the *inverse problem* (the *forward problem* refers to finding the distribution of scalp potentials resulting from a known distribution of intracranial sources). Although the forward problem has a unique solution, the inverse problem does not, that is, there are an infinite number of different sets of intracranial generators that could produce any given distribution of scalp activity. To constrain the problem, certain physiologically based assumptions must be made about the number and approximate location of generators. Most approaches to solving the inverse problem have concentrated on finding the *location, orientation,* and *strength* of a single dipole generator whose potential field best matches actual data. This is done with a least-squares minimization algorithm, which varies the dipole coordinates and direction to minimize the sum of squares of the differences between the predicted and actual potentials at each electrode location on the head.[10,11] The assumption of a single dipole generator is most useful for small generators, such as the generators of certain evoked potential peaks or of some epileptic spikes (Fig. 12–5).

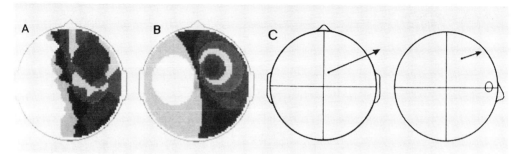

Figure 12–5. Dipole modeling of a spike discharge recorded with a sternoclavicular reference from a patient with epilepsy. (*A*) Map of measured spike discharge distribution. (*B*) Map of distribution of potential based on fitted dipole. (*C*) Fitted dipole located in right frontocentral region (*arrow tail*), with *arrowhead* indicating orientation of dipole negativity. (From Thickbroom, et al,[11] p 275, with permission.)

CHAPTER SUMMARY

This chapter reviews several quantitative analysis techniques that may be applied to digitized EEG data. Many of these techniques are, at present, used primarily as research tools, but as they become more widely available, they probably will have increasing impact on EEG interpretation and diagnosis.

REFERENCES

1. Duffy, FH, Burchfiel, JL, and Lombroso, CT: Brain electrical activity mapping (BEAM): A method for extending the clinical utility of EEG and evoked potential data. Ann Neurol 5:309–321, 1979.
2. Gotman, J: Measurement of small time differences between EEG channels: Method and application to epileptic seizure propagation. Electroencephalogr Clin Neurophysiol 56:501–514, 1983.
3. Hjorth, B: An on-line transformation of EEG scalp potentials into orthogonal source derivations. Electroencephalogr Clin Neurophysiol 39:526–530, 1975.
4. Lagerlund, TD: Montage reformatting and digital filtering. In Luders, H (ed): Epilepsy Surgery. Raven Press, New York, 1991, pp 318–322.
5. Law, SK, Nunez, PL, and Wijesinghe, RS: High-resolution EEG using spline generated surface Laplacians on spherical and ellipsoidal surfaces. IEEE Trans Biomed Eng 40:145–153, 1993.
6. Nunez, PL: Methods to estimate spatial properties of dynamic cortical source activity. In Pfurtscheller, G, and Lopes da Silva, FH (eds): Functional Brain Imaging. Hans Huber Publishers, Toronto, 1988, pp 3–9.
7. Pascual-Marqui, RD, Gonzalez-Andino, SL, Valdes-Sosa, PA, and Biscay-Lirio, R: Current source density estimation and interpolation based on the spherical harmonic Fourier expansion. Int J Neurosci 43:237–249, 1988.
8. Perrin, F, Pernier, J, Bertrand, O, and Echallier, JF: Spherical splines for scalp potential and current density mapping. Electroencephalogr Clin Neurophysiol 72:184–187, 1989.
9. Perrin F, Pernier, J, Bertrand O, et al: Mapping of scalp potentials by surface spline interpolation. Electroencephalogr Clin Neurophysiol 66:75–81, 1987.
10. Salu, Y, Cohen, LG, Rose, D, et al: An improved method for localizing electric brain dipoles. IEEE Trans Biomed Eng 37:699–705, 1990.
11. Thickbroom, GW, Davies, HD, Carroll, WM, and Mastaglia, FL: Averaging, spatio-temporal mapping and dipole modelling of focal epileptic spikes. Electroencephalogr Clin Neurophysiol 64:274–277, 1986.
12. Wallin, G and Stålberg, E: Source derivation in clinical routine EEG. Electroencephalogr Clin Neurophysiol 50:282–292, 1980.

CHAPTER 13

ELECTRO-ENCEPHALOGRAPHIC RECORDINGS FOR EPILEPSY SURGERY

Gregory D. Cascino, M.D.

CANDIDATES FOR EPILEPSY SURGERY
EXTRACRANIAL EEG IN PARTIAL
   EPILEPSY
CHRONIC INTRACRANIAL MONITORING
ELECTROCORTICOGRAPHY IN EPILEPSY
   SURGERY

Partial (focal or localization-related) epilepsy is the most common seizure disorder encountered.[8,21,34] Approximately 800,000 people in the United States have partial epilepsy, and nearly 45% of these patients are refractory to antiepileptic drug (AED) medication.[8,21,34,45] Surgical therapy for intractable partial epilepsy is an effective alternative treatment to AED medication.[11] An estimated 75,000 patients in the United States are potential candidates for epilepsy surgery.[8] Approximately 80% of all partial seizures emanate from the anterior temporal lobe, and the majority of extratemporal seizures are of frontal lobe origin.[47] Successful surgical therapy may free patients from seizures and produce improvement in cognitive, social, and behavioral functions.[41,42] The most common surgical procedure performed to alleviate partial epilepsy is a focal

cortical resection, or corticectomy, of epileptic brain tissue in the temporal lobe.[11] Long-term outcome studies have revealed that approximately 60% of patients are rendered seizure-free after anterior temporal lobe surgery.[43] The site of seizure onset, that is, the epileptogenic zone, is localized by preoperative electroencephalography (EEG) and electrocorticography (ECoG).[13] This chapter describes the usefulness of chronic intracranial EEG recordings and ECoG in patients with intractable partial epilepsy.

CANDIDATES FOR EPILEPSY SURGERY

Patients with medically refractory partial seizures who have a lowered quality of life because of seizures and AED therapy (or both) may be potential candidates for epilepsy surgery.[8,14] The type of partial seizure and the localization of the epileptogenic zone need to be considered in patient selection.[11] A comprehensive presurgical evaluation is performed to select appropriate surgical candidates.[11] Preoperative testing may include neuropsychologic studies, speech/language evaluation, visual perimetry, magnetic resonance imaging (MRI), and positron emission tomography (PET).[10,11,14] Patients may be considered appropriate candidates for a presurgical evaluation after 1 to 2 years of a medically refractory seizure disorder.[10,14] Contraindications to epilepsy surgery include active psychiatric disease, significant medical problems other than epilepsy, inappropriate or insufficient trials of AED medication, and an inability to cooperate with presurgical studies.[10,14] Patients ought to be selected for surgical treatment only when it is likely that a reduction of the number of seizures will be associated with an improvement in the quality of life.[10,14]

EXTRACRANIAL EEG IN PARTIAL EPILEPSY

EEG is the most frequently performed and single most important neurodiagnostic study in the evaluation of patients with intractable partial epilepsy.[5,25,46] EEG studies are usually performed between seizure episodes, that is, interictally, because of the episodic nature of the disorder.[5,25,27,46]

Interictal EEG recordings in patients with epilepsy should be performed according to the methodology established by the American EEG Society.[5,25] Standard activation procedures, for example, hyperventilation and photic stimulation, should be included.[3,5,25,46] Sleep deprivation and the recording of nonrapid eye movement sleep may be used to increase the sensitivity of the EEG to demonstrate interictal epileptiform alterations.[3,5,9,19,25] The EEG technologist should obtain information regarding the type(s) of seizures, the timing of the last seizure, and the identification of seizure-precipitating events, for example, reading epilepsy. Current AED medication and recent drug levels should also be noted.

There are several *limitations* of extracranial EEG that must be recognized for appropriate interpretation of these studies in potential surgical candidates.[27] *Interictal* scalp-recorded EEG studies in patients with partial epilepsy may fail to demonstrate specific epileptiform activity (e.g., spike, sharp wave, or spike-and-wave discharges) during brief recording periods, despite serial EEG studies. Epileptiform activity generated in cortical areas remote from the scalp electrodes, for example, amygdala and hippocampus, may not be associated with interictal extracranial EEG alterations.[27,30] Attenuation of spike activity by the dura mater, bone, and scalp limits the sensitivity of extracranial EEG recordings.[5] Approximately 20 to 70% of cortical spikes will be recorded on the scalp EEG.[1] Patients with intractable partial epilepsy may have repetitively normal interictal EEG studies.[2,5,27]

Most surgical epilepsy centers perform ictal EEG recordings, that is, during seizure activity, before considering a focal corticectomy, because of the limitations of the interictal EEG.[13] *Ictal extracranial EEG*, however, may also provide inaccurate information about the localization of the epileptogenic zone.[13] Scalp-recorded EEG studies may fail to detect specific alterations localized to the site of seizure onset, for example, amygdala, only to reveal distant, more widespread cortical excitability, such as the frontotemporal cortex.[27,30] The sensitivity and specificity of ictal extracranial EEG in partial epilepsy depend on the seizure type(s) and the localization of the epileptogenic zone.[30] The majority of patients have a focal or generalized scalp-recorded EEG alteration during complex partial seizures.[32] However, simple partial seizures may not be associated with an EEG change.[7] Ictal extracranial EEG recordings are also more sensitive and specific in patients with seizures of temporal lobe origin.[32] The scalp-recorded EEG reveals a localized epileptiform abnormality in only a minority of patients with frontal lobe epilepsy.[30]

The sensitivity and specificity of extracranial EEG studies may increase with the use of *supplementary electrodes*, that is, sphenoidal and inferior lateral temporal scalp (T_1, T_2, F_9, F_{10}) electrodes and closely placed scalp electrodes may be useful to delineate the topography of the interictal and ictal activity.[30–32,35,36,40] Sphenoidal electrodes may record epileptiform activity emanating from the mesiobasal limbic region and assist in localizing the epileptogenic zone before anterior temporal lobectomy is undertaken.[32] Results are conflicting about the sensitivity of sphenoidal electrodes compared with scalp electrodes in patients with temporal lobe epilepsy.[31] Nasopharyngeal electrodes are artifact-prone, poorly tolerated by patients (and, thus, may interfere with the sleep recording), and have not been demonstrated to be more specific or more sensitive than the inferior lateral scalp electrodes.[40] Computer-assisted EEG monitoring may be used for automatic seizure recognition and off-line seizure analysis.[20]

CHRONIC INTRACRANIAL MONITORING

"Chronic intracranial EEG monitoring" (CIM) is a generic term for invasive EEG recordings that are used preoperatively in the assessment of patients with intractable partial epilepsy. These studies are performed after it has been determined that scalp-recorded EEG studies alone are inadequate to localize the epileptogenic zone.[13,26] Most epilepsy centers perform these studies only in select patients before focal surgical therapy.[13,26] Prolonged recordings with implanted electrodes can be used with EEG telemetry and video monitoring over several days to several weeks to evaluate spontaneous EEG activity. Patients should be shown to have a medically refractory seizure disorder and to be appropriate surgical candidates before being considered for intracranial recordings.[13,18] Identifying an epileptogenic lesion with neuroimaging may obviate CIM in some patients. Intracranial electrodes must be inserted with the patient under general anesthesia. There is, moreover, approximately a 4% risk of hemorrhage or infection associated with CIM.[29]

There are several different methodologies available for the performance of CIM.[13,18] The most frequently used techniques are depth electrode and multielectrode grid recordings.[13] The results of the presurgical evaluation determine the appropriate intracranial technique and the anatomic location of the electrodes. Common *indications* for performing CIM include:

1. Difficulty with lateralization (right or left temporal lobe)
2. Difficulty with localization but appropriate knowledge of lateralization (temporal lobe or extratemporal lobe epilepsy)
3. Need to delineate functional anatomy
4. Need to determine the extent of the epileptogenic zone when localization is determined[12]

Depth electrode recordings were the first form of CIM used in the evaluation of patients with intractable partial epilepsy.[12] Early studies confirmed the increased sensitivity and specificity of these recordings compared with extracranial EEG monitoring (Fig. 13–1).[12] Depth electrodes are used most commonly to determine the lateralization of seizure onset in patients with scalp-recorded bitemporal seizures.[37] Depth electrodes can now be implanted stereotaxically with MRI and PET guidance. Common artifacts identified with extracranial recordings, such as muscle activity during seizures, do not occur with depth electrode studies. Depth electrode investigations in patients with temporal lobe epilepsy are performed with electrodes implanted in the mesiobasal limbic region. Montages are used that contain surface derivations, for example, scalp or multielectrode grid electrodes and recordings from the limbic region.[12]

Interpretation of the depth electrode studies involves examination of interictal and ictal epileptiform discharges and assessment of background EEG activity. Depth-recorded interictal epileptiform activity may not be reliable for determining the region of seizure onset.[12] Interictal spiking with depth electrodes may be more prominent contralateral to the epileptogenic zone in 10 to 20% of patients.[15] Bitemporal interictal epileptiform discharges do not necessarily preclude a successful surgical outcome after an anterior temporal lobectomy.[37,38] Ictal EEG is the most important parameter used to determine the site of seizure onset with depth electrodes.[12] The pattern of seizure onset recorded with depth electrodes may have an important predictive value in determining the response to focal corticectomy.[12] A regional pattern of seizure onset indicates a less favorable seizure outcome.[12] Focal seizure onset with an electrographic alteration at a single electrode contact implies more precise localization of the site of seizure onset.[12] Initial ictal EEG changes may include

GAIN 100%   LR (C3/4) -32.1 (DEPTH)

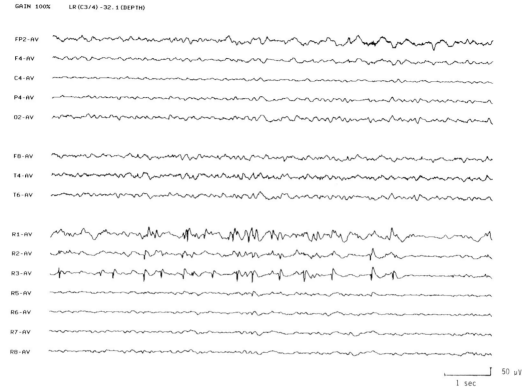

FP2-AV

F4-AV

C4-AV

P4-AV

O2-AV

F8-AV

T4-AV

T6-AV

R1-AV

R2-AV

R3-AV

R5-AV

R6-AV

R7-AV

R8-AV

50 µV

1 sec

Figure 13–1. Depth electrode recording shows right anterior temporal lobe spike discharges (R_1 to R_8). Scalp-recorded EEG is unremarkable. AV, C_3–C_4 reference.

attenuation of background EEG activity with the development of a low-amplitude, high-frequency rhythmical discharge. Spread to other electrode contacts may then occur with the development of clinical symptomatology. In most patients, several (5 to 10) spontaneous seizures are recorded with depth electrodes before surgical determination.[18] The propagation of seizure activity to the opposite cerebral hemisphere may also have important prognostic value in determining the response of focal corticectomy.[23] A shorter interhemispheric conduction time may be associated with a less favorable surgical outcome after anterior temporal lobectomy.[38] Background activity is not as valuable as ictal EEG in determining the localization of the epileptiform zone. Common nonepileptiform alterations in the region of the epileptogenic zone include attenuation of EEG activity and focal slowing.[12]

The *diagnostic yield* of depth electrode

studies has been assessed in patients with intractable partial epilepsy. The reported agreement between scalp-recorded EEG and depth localization has varied from 50 to 100%.[6,13,28] Studies that have combined the diagnostic accuracy of extracranial EEG with tests of focal functional deficit (e.g., PET, neuropsychologic studies, or both), showed the depth and scalp studies to be discordant in only 13% of patients.[15] Patients with favorable surgical outcomes may be more likely to have an agreement between scalp and depth localization.[39] The predictive value of depth electrode recordings has also been compared with extracranial EEG. A statistically significant difference has not been found in surgical outcome between patients monitored with the two techniques.[13,28] In select patients, extracranial EEG, performed in the context of a comprehensive presurgical evaluation, may be adequate without using CIM to localize the epileptogenic zone.[13] Depth studies, however, may demonstrate patients to be appropriate surgical candidates when the scalp recordings provide nonlocalizing information or reveal bitemporal epileptiform abnormalities.[37]

Multielectrode grid recordings are a later development in CIM.[12] Multielectrode grids are used to evaluate patients with neocortical epilepsy of temporal or extratemporal origin.[24] Subdural or epidural electrodes are placed over the cortical surface and used for localization of the epileptogenic zone and to delineate functional cortical anatomy.[24] These studies are most useful in patients in whom the lateralization of seizure onset is already known through presurgical evaluation.[13] Multielectrode grid recordings may reveal the extent of neocortical involvement and provide information about the precise localization of seizure onset. The grids contain multiple electrode contacts that are evenly spaced in a plastic insulating material.[12] The electrodes can be placed either under the temporal lobe to record from the inferior temporal region or in the interhemispheric fissure to record electrographic activity of mesiofrontal origin. Multielectrode grids are not as sensitive as depth electrodes for detecting epileptiform abnormalities emanating from the mesiobasal temporal lobe.[12]

Extraoperative *cortical stimulation* studies may be performed with multielectrode grids to determine the localization of functional cortex.[22,24] These studies may be very useful in patients who cannot cooperate with intraoperative cortical stimulation.[24] Intraoperative studies are limited in duration and may be difficult to perform in some patient groups, such as children. Extraoperative cortical stimulation during intracranial monitoring may be used to delineate the limits of primary cortical areas that may directly affect the extent of cortical resections.[22,24]

ELECTROCORTICOGRAPHY IN EPILEPSY SURGERY

Intraoperative intracranial EEG recordings were used initially at several epilepsy centers to guide the surgical resection of epileptic brain tissue.[13] The authoritative work of Wilder Penfield and Herbert Jasper in Montreal and A. Earl Walker in Baltimore demonstrated the importance of ECoG during focal cortical resective surgery for patients with intractable partial epilepsy.[28,44] Penfield and Jasper[28] relied on predominantly interictal extracranial EEG and ECoG to determine the region of cortical resection, because long-term EEG monitoring was still in its infancy. The methodology for ECoG established by the investigators in Montreal and Baltimore continues to be used today in the surgical treatment of epilepsy.[28,44]

The need for ECoG depends on the results of the presurgical evaluation, the localization of the epileptogenic zone, and the type(s) of extraoperative EEG monitoring.[13] If the epileptogenic zone has been localized by CIM, then the ECoG may be less important in determining the region of cortical resection.[13] Patients in whom depth electrode studies reveal a mesiobasal limbic origin of seizures may subsequently undergo *en bloc* resection of the temporal lobe without ECoG. Traditional focal corticectomies have used a "tailored resection," with ECoG localizing the site of seizure onset.[28]

Several techniques for ECoG are available that include the use of subdural strip electrodes or a rigid electrode holder with graphite tip or cotton wick electrodes.[48] Subdural strip electrodes are particularly useful in recordings obtained from the inferior temporal lobe and the mesiofrontal region.[48] The pre-excision ECoG at the Mayo Clinic uses three subdural strips (each with eight electrode contacts) placed on the superior and inferior temporal lobe gyri and in the suprasylvian region (Fig. 13–2). Three depth electrodes are also placed by hand in the region of the amygdala and hippocampus. A baseline extracranial EEG recording is performed before the surgical procedure with the patient awake, and a contralateral ear reference is used for the intracranial recordings. Samples of EEG activity are obtained before and after cortical resection. ECoG may be performed with the patient under light local or general anesthesia. The traditional procedure of the Montreal group was to obtain the ECoG recordings in awake patients.[28] Electrical stimulation was then used intraoperatively to assist in localizing the epileptogenic zone and functional anat-

PRE-EXCISION ECoG

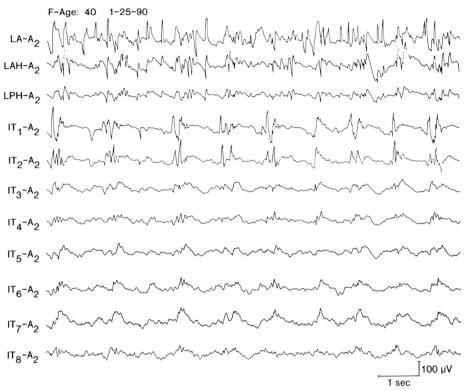

Figure 13–2. Electrocorticography (ECoG) performed at the time of left anterior temporal lobectomy. The three upper channels represent recording from the mesiotemporal region with depth electrodes. Prominent spiking is noted in the mesiotemporal region and in the lateral temporal cortex.

omy.[28] ECoG is performed at the Mayo Clinic with the patient under general anesthesia. The anesthetic agents used include nitrous oxide, fentanyl, or a low-volume percentage of isoflurane. A 16- or 32-channel EEG recording is obtained.

The strategy for the intraoperative EEG monitoring is discussed with the surgical team, the anesthesiologist, and the EEG technician. The electroencephalographer is present in the operating room for immediate interpretation of EEG data. Methohexital given intravenously may be used to enhance or to activate epileptiform activity in the pre-excision recordings. The extent of the focal corticectomy is based on the ECoG and the presurgical evaluation. The relationship between the localization of the epileptogenic zone and the eloquent cortex is considered at the time of the operation. A post-excision recording is obtained with a subdural strip electrode placed posterior to

the margin of the surgical resection (Fig. 13–3).

There are several important *limitations* of ECoG that must be recognized for appropriate interpretation of these studies.[13] Information about the localization and lateralization of the epileptogenic zone must be determined before the ECoG study. Other potential disadvantages of ECoG include the sampling of predominant interictal EEG activity, the restricted spatial distribution of the EEG recording, and the usually brief duration of the monitoring. ECoG has been used mainly to assess neocortical epileptiform activity and may be restricted in sampling from the orbitofrontal, mesiofrontal, and mesiotemporal regions. ECoG predominantly records epileptiform activity from the lateral surface of the temporal lobe, and this may not correlate with the region of seizure onset, for example, the mesiobasal temporal lobe.[13] The neurosurgical team must

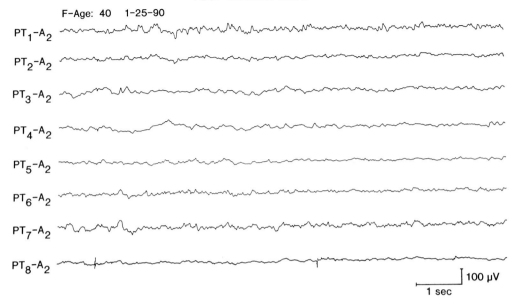

POST-EXCISION ECoG

F–Age: 40   1-25-90

PT$_1$–A$_2$
PT$_2$–A$_2$
PT$_3$–A$_2$
PT$_4$–A$_2$
PT$_5$–A$_2$
PT$_6$–A$_2$
PT$_7$–A$_2$
PT$_8$–A$_2$

100 μV
1 sec

Figure 13–3. Postexcision electrocorticography (ECoG) is performed with a subdural strip placed posterior to the margin of the resection. No definite residual spiking is noted.

delay the cortical resection until ECoG is obtained and the studies are interpreted. General anesthesia used during the surgical procedure may suppress epileptiform activity. High concentrations of certain anesthetic drugs, such as isoflurane, may make it difficult to document the neocortical extent of the epileptogenic zone. Enflurane may activate or increase interictal epileptiform activity; however, this may be nonspecific and may not correlate with the site of seizure onset.[13] The false positivity of methohexital-activated ECoG has also been observed.[13] Finally, certain ECoG-recorded post-excision spike discharges are not prognostically important.[13] Spikes recorded from the insula may not indicate the need for further cortical resection.[33]

There has been conflicting evidence on the *prognostic* importance of ECoG in determining long-term seizure control.[2,4,16,17] The location, morphology, and distribution of spiking in the post-excision recording may not affect seizure outcome. The presence of residual spiking may be associated statistically with an unfavorable seizure outcome.[4] The presence of residual spiking, however, may not preclude a successful surgical outcome after a focal corticectomy.[17]

CHAPTER SUMMARY

Extraoperative and intraoperative EEG studies are only one part of a comprehensive presurgical evaluation that is performed in patients with intractable partial epilepsy. The level of experience of individual EEG laboratories is important in selecting the methodology used during the presurgical evaluation. Ultimately, the preferred method for localizing the region of seizure onset is not a diagnostic technique, but is rather the demonstration of a long-term seizure-free outcome after focal corticectomy.

REFERENCES

1. Abraham, K and Ajmone Marsan, C: Patterns of cortical discharges and their relation to routine scalp electroencephalography. Electroencephalogr Clin Neurophysiol 10:447–461, 1958.
2. Ajmone-Marsan, C and Baldwin, M: Electrocorticography. In Baldwin, M and Bailey, P (eds): Temporal Lobe Epilepsy. Charles C Thomas, Publisher, Springfield, Illinois, 1958, pp 368–395.
3. Ajmone Marsan, C and Zivin, LS: Factors related to the occurrence of typical paroxysmal abnormalities in the EEG records of epileptic patients. Epilepsia 11:361–381, 1970.
4. Bengzon, ARA, Rasmussen, T, Gloor, P, et al: Prog-

nostic factors in the surgical treatment of temporal lobe epileptics. Neurology 18:717–731, 1968.

5. Daly, DD: Epilepsy and syncope. In Daly, DD and Pedley, TA (eds): Current Practice of Clinical Electroencephalography, ed 2. Raven Press, New York, 1990, pp 269–334.

6. Delgado-Escueta, AV and Walsh, GO: The selection process for surgery of intractable complex partial seizures: Surface EEG and depth electrography. Res Publ Assoc Res Nerv Ment Dis 61:295–326, 1983.

7. Devinsky, O, Sato, S, Kufta, CV, et al: Electroencephalographic studies of simple partial seizures with subdural electrode recordings. Neurology 39:527–533, 1989.

8. Dreifuss, FE: Goals of surgery for epilepsy. In Engel, J Jr (ed): Surgical Treatment of the Epilepsies. Raven Press, New York, 1987, pp 31–49.

9. Ellingson, RJ, Wilken, K, and Bennett, DR: Efficacy of sleep deprivation as an activation procedure in epilepsy patients. J Clin Neurophysiol 1:83–101, 1984.

10. Engel, J Jr: Approaches to localization of the epileptogenic lesion. In Engel, J Jr (ed): Surgical Treatment of the Epilepsies. Raven Press, New York, 1987, pp 75–97.

11. Engel, J Jr: Alternative therapy. In Engel, J Jr (ed): Seizures and Epilepsy. FA Davis, Philadelphia, 1989, pp 443–474.

12. Engel, J Jr and Crandall, PH: Intensive neurodiagnostic monitoring with intracranial electrodes. Adv Neurol 46:85–106, 1986.

13. Engel, J Jr and Ojemann, GA: The next step. In Engel, J Jr (ed): Surgical Treatment of the Epilepsies, ed 2. Raven Press, New York, 1993, pp 319–329.

14. Engel, J Jr and Shewmon, DA: Overview: Who should be considered a surgical candidate? In Engel, J Jr (ed): Surgical Treatment of the Epilepsies, ed 2. Raven Press, New York, 1993, pp 23–34.

15. Engel, J Jr, Sutherling, WW, Cahan, L, et al: The role of positron emission tomography in the surgical therapy of epilepsy. In Porter, RJ, Mattson, RH, Ward, AA Jr, and Dam, M (eds): Advances in Epileptology: The 15th Epilepsy International Symposium. Raven Press, New York, 1984, pp 427–432.

16. Falconer, MA: Discussion. In Baldwin, M and Bailey, P (eds): Temporal Lobe Epilepsy. Charles C Thomas, Publisher, Springfield, Illinois, 1958, pp 483–484.

17. Fiol, ME, Gates, JR, Torres, F, and Maxwell, RE: The prognostic value of residual spikes in the postexcision electrocorticogram after temporal lobectomy. Neurology 41:512–516, 1991.

18. Gates, JR: Epilepsy presurgical evaluation in the area of intensive neurodiagnostic monitoring. Adv Neurol 46:227–247, 1986.

19. Gibbs, EL and Gibbs, FA: Diagnostic and localizing value of electroencephalographic studies in sleep. Res Publ Assoc Res Nerv Ment Dis 26:366–376, 1947.

20. Gotman, J and Gloor, P: Automatic recognition and quantification of interictal epileptic activity in the human scalp EEG. Electroencephalogr Clin Neurophysiol 41:513–529, 1976.

21. Hauser, WA and Kurland, LT: The epidemiology of epilepsy in Rochester, Minnesota, 1935 through 1967. Epilepsia 16:1–66, 1975.

22. Lesser, RP, Lüders, H, Morris, HH, et al: Electrical

stimulation of Wernicke's area interferes with comprehension. Neurology 36:658–663, 1986.

23. Lieb, JP, Babb, TL, Engel, J Jr, et al: Interhemispheric propagation time of hippocampal seizures: Cell density and surgical outcome correlates (abstr). Electroencephalogr Clin Neurophysiol 58:38P, 1984.

24. Lüders, H, Lesser, RP, Dinner, DS, et al: Commentary: Chronic intracranial recording and stimulation with subdural electrodes. In Engel, J Jr (ed): Surgical Treatment of the Epilepsies. Raven Press, New York, 1987, pp 297–321.

25. Niedermeyer, E and Lopes da Silva F: Electroencephalography: Basic Principles, Clinical Applications and Related Fields, ed 2. Urban & Schwarzenberg, Baltimore, 1987.

26. Olivier, A, Gloor, P, Quesney, LF, and Andermann, F: The indications for and the role of depth electrode recording in epilepsy. Appl Neurophysiol 46:33–36, 1983.

27. Pedley, TA: Interictal epileptiform discharges: Discriminating characteristics and clinical correlations. Am J EEG Technol 20:101–119, 1980.

28. Penfield, W and Jasper H: Epilepsy and the Functional Anatomy of the Human Brain. Little, Brown & Company, Boston, 1954.

29. Pilcher, WH, Roberts, DW, Flanigin, HF, et al: Complications of epilepsy surgery. In Engel, J Jr (ed): Surgical Treatment of the Epilepsies, ed 2. Raven Press, New York, 1993, pp 565–581.

30. Quesney, LF, Constain, M, Fish, DR, and Rasmussen, T: Frontal lobe epilepsy: A field of recent emphasis. Am J EEG Technol 30:177–193, 1990.

31. Quesney, LF and Gloor, P: Localization of epileptic foci. Electroencephalogr Clin Neurophysiol 37 Suppl:165–200, 1985.

32. Quesney, LF, Risinger, MW, and Shewmon, DA: Extracranial EEG evaluation. In Engel, J Jr (ed): Surgical Treatment of the Epilepsies, ed 2. Raven Press, New York, 1993, pp 173–195.

33. Rasmussen, T: Surgical treatment of patients with complex partial seizures. Adv Neurol 11:415–449, 1975.

34. Rayport, M: Role of neurosurgery and management of medication-resistant epilepsy. In: Plan for Nationwide Action in Epilepsy. Vol 2. DHEW Publications, Washington, DC, 1977, pp 314–324.

35. Sharbrough, FW: Commentary: Extracranial EEG evaluation. In Engel, J Jr (ed): Surgical Treatment of the Epilepsies. Raven Press, New York, 1987, pp 167–171.

36. Sharbrough, FW: Electrical fields and recording techniques. In Daly, DD and Pedley, TA (eds): Current Practice of Clinical Electroencephalography, ed 2. Raven Press, New York, 1990, pp 29–49.

37. So, N, Gloor, P, Quesney, LF, et al: Depth electrode investigations in patients with bitemporal epileptiform abnormalities. Ann Neurol 25:423–431, 1989.

38. So, N, Olivier, A, Andermann, F, et al: Results of surgical treatment in patients with bitemporal epileptiform abnormalities. Ann Neurol 25:432–439, 1989.

39. Spencer, SS, Spencer, DD, Williamson, PD, and Mattson, RH: The localizing value of depth electroencephalography in 32 patients with refractory epilepsy. Ann Neurol 12:248–253, 1982.

40. Sperling, MR and Engel, J Jr: Electroencephalo-

graphic recordings from the temporal lobes: A comparison of ear, anterior temporal, and nasopharyngeal electrodes. Ann Neurol 17:510–513, 1985.

41. Taylor, DC: Mental state and temporal lobe epilepsy: A correlative account of 100 patients treated surgically. Epilepsia 13:727–765, 1972.

42. Taylor, DC: Epileptic experience, schizophrenia, and the temporal lobe. McLean Hospital Journal, Special Volume, 1977, pp 22–39.

43. Walczak, TS, Radtke, RA, McNamara, JO, et al: Anterior temporal lobectomy for complex partial seizures: Evaluation, results, and long-term follow-up in 100 cases. Neurology 40:413–418, 1990.

44. Walker, AE: Temporal lobectomy. J Neurosurg 26:642–649, 1967.

45. Ward, AA Jr: Perspectives for surgical treatment of epilepsy. Res Publ Assoc Res Nerv Ment Dis 61:371–390, 1983.

46. Westmoreland, BF: The electroencephalogram in patients with epilepsy. Neurol Clin 3:599–613, Aug 1985.

47. Williamson, PD, Wieser, HG, and Delgado-Escueta, AV: Clinical characteristics of partial seizures. In Engel, J Jr (ed): Surgical Treatment of the Epilepsies. Raven Press, New York, 1987, pp 101–120.

48. Wyler, AR, Ojemann, GA, Lettich, E, and Ward, AA Jr: Subdural strip electrodes for localizing epileptogenic foci. J Neurosurg 60:1195–1200, 1984.

CHAPTER 14

MOVEMENT-RELATED POTENTIALS AND EVENT-RELATED POTENTIALS

Joseph Y. Matsumoto, M.D.

Special electroencephalographic (EEG) studies have been designed to explore the cortical processes underlying movement, attention, and cognition. As a first step, these studies must define an "event," that is, either the cause or the effect of the higher cortical process under investigation. This event serves as the temporal reference point for computerized EEG averaging. Computers with the proper memory buffer can average the cortical activity that precedes the event (back averaging) and that which follows it.

MOVEMENT-RELATED CORTICAL POTENTIALS

Kornhuber and Deecke[11] defined the cortical activity that surrounds a self-paced, voluntary movement. Moreover, they found that the premotor cortical activity begins 1.0 to 1.5 seconds before the onset of movement. This activity is a gradually rising negativity that they termed the "Bereitschaftspotential." Studies in animals indicate that this activity reflects the feedforward processing of motor commands that are projected from the cerebellum through the thalamus to the motor and premotor areas of the cortex.[15]

Technique

One of the most challenging tasks in clinical neurophysiology is obtaining movement-related cortical potentials. The minimum scalp electrode montage should include the F_Z, C_Z, C_3, and C_4 positions referenced to linked ears. It is critical that the electro-oculographic activity also be recorded. By applying additional electrodes in grids over the sensorimotor cortex, the potentials can be mapped topographically. A low-frequency filter with a cut-off in the range of 0.05 Hz must be used to record the slow premovement negative wave.

A brisk voluntary movement, usually of a digit, acts as the timing event. The subject is instructed to stare straight ahead and to make self-paced, repetitive movements 3 to

10 seconds apart. The initial rise of rectified electromyographic (EMG) activity from the movement triggers data collection. The computer buffer is configured to collect the data for 2 seconds before and 1 second after the trigger. Data from 100 to 200 such movements are collected and stored for later analysis.

Averaging must be performed off-line to ensure proper artifact rejection. Each trace is reevaluated to detect any contamination by eye movement artifact. The tracing also must display a clear EMG take-off point. If acceptable, the tracing is aligned at the EMG onset and computed into the ongoing averaging process.[2]

Normal Waveforms

Movement-related potentials consist of four major waves. The earliest is the *Bereitschaftspotential*, a slowly rising negativity with a maximal amplitude at the vertex, beginning 1.0 to 1.5 seconds before the onset of movement.[11] Approximately 500 milliseconds before movement onset, the slope of the negativity turns more sharply upward. This period, termed *NS'*, localizes more focally to the contralateral central region.[16] Closely surrounding the EMG activity, the motor potential appears as a peaked wave in a scalp distribution over the contralateral motor cortex. The final wave, called the *reafferente potential* by Kornhuber and Deecke,[11] appears 90 milliseconds after movement. In fact, this is a series of positive and negative waves predominantly over the frontal and parietal areas which reflect sensory feedback from the movement.

The precise generators of each of the movement-related potentials is a subject of controversy. In particular, there is much speculation about the activity of the supplementary motor area (SMA) because of its central role in motor planning and control. The vertex predominance of the Bereitschaftspotential led to the hypothesis that it reflected planning activity in the SMA.[4,21] However, scalp topographic mapping, subdural grid recording, and magnetoencephalography indicate that multiple generators contribute to this potential.[3,12,20] Topographic mapping studies further suggest that

the predominant SMA activity occurs after, and not before, the movement.[20] To reconcile the conflicting data, one may theorize that multiple brain regions are active at any given time surrounding the onset of a voluntary movement and that the movement-related potentials are the sum of the activity in this distributed network.

Abnormalities in Disease

Early studies reported that the movement-related potentials in Parkinson's disease were normal, but this likely reflected triggering difficulties in patients who make slow, small movements. With the manual averaging technique, it was found that an early stage of negativity (650 milliseconds before movement) is depressed in parkinsonian patients, perhaps reflecting poor activation in the SMA.[6] Similarly, the Bereitschaftspotential associated with gait showed decreased activation in Parkinson's disease.[22] Levodopa augments the activity in this abnormal portion of the Bereitschaftspotential.[5] In tardive dyskinesia, the Bereitschaftspotential is increased in amplitude.[1] In cerebellar disease or after ventral thalamotomy, the Bereitschaftspotential is either reduced in amplitude or absent, indicating loss of the feedforward message in the cerebellothalamocortical loop.[18]

Jerk-Locked Averaging

The technique described above for movement-related potentials can be used to evaluate involuntary muscle jerks.[17] The examiner first performs a surface EMG study to define the muscle that leads the jerk. This muscle may then be used as the EMG trigger event. For this technique to be successful, the jerks must occur frequently enough to collect 50 to 100 tracings.

The technique is most helpful in uncovering the cortical event preceding a myoclonic jerk when the standard EEG is unrevealing (see Chapter 30). In cortical myoclonus, a focal or generalized transient potential is recorded which precedes the myoclonic jerk by 20 to 60 milliseconds. In reticular reflex myoclonus, jerk-locked averaging actually obscures the projected cortical transient potential because of the irregular interval between the jerk and the EEG event.[8] In nonepileptic myoclonus, jerk-locked averaging fails to reveal time-locked cortical activity.

Jerk-locked averaging has a limited role in evaluating other involuntary movements. In Tourette's syndrome, no cortical activity of any type precedes involuntary tics.[13] However, when the patient voluntarily mimics a tic, the Bereitschaftspotential appears. For this reason, hysteria is suspected if a slow premovement negativity is found preceding an apparently involuntary movement.

Contingent Negative Variation

In the contingent negative variation testing paradigm, a stimulus such as a click warns the subject to prepare to move. After 1 to 2 seconds, a second stimulus, such as a flash of light, signals the patient to begin moving. Contingent negative variation is a slow negative potential that appears in the interval between the warning stimulus and the second stimulus.[23] The distribution of this wave is predominantly bilateral and frontal, but it may shift with variations in the testing procedure.[9] The neural generators of contingent negative variation appear to be different from those of the Bereitschaftspotential.[10]

EVENT-RELATED POTENTIALS

Whereas standard somatosensory or visual evoked potentials map the cortical response to a simple sensory stimulus, event-related potentials record the cortical activity evoked by a stimulus charged with cognitive significance. As such, event-related potentials are more sensitive to the "endogenous" reaction to a stimulus than to the physical nature of the stimulus. Sutton and coworkers[19] were among the first to note a large late cortical positivity in reaction to stimuli to which a subject attached importance. Since that time, numerous techniques have been devised to study this phenomenon. A comprehensive review of this topic—spanning the

disciplines of physiology, psychology, psychiatry, and neurology—is beyond the scope of this book.

The P300

The P300 is the most commonly recorded event-related potential.[14] Generally, what is called the *oddball technique* of auditory stimulation is used, in which a *standard*, or *frequent*, *stimulus* is replaced at infrequent intervals by a stimulus of different tone, termed the *rare*, or *oddball*, *stimulus*. The subject is instructed to attend to or count the oddball stimuli. Only trials triggered by this rare event are averaged.

At times, the P300 is visible on a single raw tracing. Averaging clearly defines a wave with a peak latency of approximately 300 milliseconds and an amplitude of about 10 μV. The amplitude of the wave is increased by many factors, including the subject's attentiveness and the unpredictability of the oddball stimulus. The P300 has a bilateral, mid-parietal distribution. However, a single generator for the potential cannot be defined; the wave likely reflects activity in multiple areas of the brain. The role of the wave in cognition is also debated. It may be the electrophysiologic correlate of selected attention.

The P300 is abnormal in many diseases in which cortical processing is impaired. Amplitude is decreased and latency prolonged in all types of dementia.[7] Abnormalities have also been reported in mild metabolic encephalopathies, drug intoxications, multiple sclerosis, autism, and schizophrenia.[14]

CHAPTER SUMMARY

Special EEG averaging techniques may be used to study the cortical processes underlying movement and cognition. Movement-related potentials and contingent negative variation are observed prior to a voluntary movement. Jerk-locked averaging may detect cortical activity associated with involuntary movements. The P300 and other event-related potentials provide electrophysiologic correlates of perception and cognition.

REFERENCES

1. Adler, LE, Pecevich, M, and Nagamoto, H: Bereitschaftspotential in tardive dyskinesia. Mov Disord 4:105–112, 1989.
2. Barrett, G, Shibasaki, H, and Neshige, R: A computer-assisted method for averaging movement-related cortical potentials with respect to EMG onset. Electroencephalogr Clin Neurophysiol 60:276–281, 1985.
3. Cheyne, D and Weinberg, H: Neuromagnetic fields accompanying unilateral finger movement: Premovement and movement-evoked fields. Exp Brain Res 78:604–612, 1989.
4. Deecke, L and Kornhuber, HH: An electrical sign of participation of the mesial 'supplementary' motor cortex in human voluntary finger movement. Brain Res 159:473–476, 1978.
5. Dick, JPR, Cantello, R, Buruma, O, et al: The Bereitschaftspotential, L-DOPA and Parkinson's disease. Electroencephalogr Clin Neurophysiol 66:263–274, 1987.
6. Dick, JPR, Rothwell, JC, Day, BL, et al: The Bereitschaftspotential is abnormal in Parkinson's disease. Brain 112:233–244, 1989.
7. Goodin, DS, Squires, KC, and Starr, A: Long latency event-related components of the auditory evoked potential in dementia. Brain 101:635–648, 1978.
8. Hallett, M, Chadwick, D, Adam, J, and Marsden, CD: Reticular reflex myoclonus: A physiological type of human post-hypoxic myoclonus. J Neurol Neurosurg Psychiatry 40:253–264, 1977.
9. Hillyard, SA: The CNV and human behavior: A review. Electroencephalogr Clin Neurophysiol 33 Suppl:161–171, 1973.
10. Ikeda, A, Shibasaki, H, Nagamine, T, et al: Dissociation between contingent negative variation and Bereitschaftspotential in a patient with cerebellar efferent lesion. Electroencephalogr Clin Neurophysiol 90:359–364, 1994.
11. Kornhuber, HH and Deecke, L: Hirnpotentialänderungen bei Willkürbewegungen und passiven Bewegungen des Menschen: Bereitschaftspotential und reafferente Potentiale. Pflugers Arch 284:1–17, 1965.
12. Neshige, R, Lüders, H, and Shibasaki, H: Recording of movement-related potentials from scalp and cortex in man. Brain 111:719–736, 1988.
13. Obeso, JA, Rothwell, JC, and Marsden, CD: Simple tics in Gilles de la Tourette's syndrome are not prefaced by a normal premovement EEG potential. J Neurol Neurosurg Psychiatry 44:735–738, 1981.
14. Picton, TW: The P300 wave of the human event-related potential. J Clin Neurophysiol 9:456–479, 1992.
15. Sasaki, K, Gemba, H, Hashimoto, S, and Mizuno, N: Influences of cerebellar hemispherectomy on slow potentials in the motor cortex preceding self-paced hand movements in the monkey. Neurosci Lett 15:23–28, 1979.
16. Shibasaki, H, Barrett, G, Halliday, E, and Halliday, AM: Components of movement-related potentials and their scalp topography. Electroencephalogr Clin Neurophysiol 49:213–226, 1980.
17. Shibasaki, H and Kuroiwa, Y: Electroencephalographic correlates of myoclonus. Electroencephalogr Clin Neurophysiol 39:455–463, 1975.
18. Shibasaki, H, Shima, F, and Kuroiwa, Y: Clinical studies of the movement-related cortical potential (MP) and the relationship between the dentatorubrothalamic pathway and readiness potential (RP). J Neurol 219:15–25, 1978.
19. Sutton, S, Braren, M, Zubin, J, and John, ER: Evoked-potential correlates of stimulus uncertainty. Science 150:1187–1188, 1965.
20. Tarkka, I and Hallett, M: Topography of scalp-recorded motor potentials in human finger movements. J Clin Neurophysiol 8:331–334, 1991.
21. Toro, C, Matsumoto, J, Deuschl, G, et al: Source analysis of scalp-recorded movement-related electrical potentials. Electroencephalogr Clin Neurophysiol 86:167–175, 1993.
22. Vidailhet, M, Stocchi, F, Rothwell, JC, et al: The Bereitschaftspotential preceding simple foot movement and initiation of gait in Parkinson's disease. Neurology 43:1784–1788, 1993.
23. Walter, WG, Cooper, R, Aldridge, VJ, et al: Contingent negative variation: An electric sign of sensorimotor association and expectancy in the human brain. Nature 203:380–384, 1964.

SECTION 2
Electro-physiologic Assessment of Neural Function
PART B
Sensory Pathways

The sensory axons that conduct information from the periphery to the central nervous system generate electricity. However, directly recording the spontaneous electrical activity in nerves requires microneurography, a tedious procedure that is used only in special situations. In microneurography, fine electrodes are inserted directly into the nerve while recording the activity from a single channel. The recording represents the summation of the changes occurring in all the axons in the region of the electrode tip. Using this technique, the spontaneous activity of single axons or a group of axons and their response to external stimuli can be monitored.

In contrast, most electrophysiologic recordings from humans undergoing testing for possible neurologic or neuromuscular disease are summated responses made from specific generators in response to controlled external stimulation.

Somatic sensory and somatic motor axons can be tested by stimulation along the length of a nerve while recording the sensory or motor responses from peripheral nerve or muscle. Isolation of sensory axons for testing can be obtained by selective stimulation of sensory structures or by selective recording from generators that are purely sensory. The potentials recorded from sensory structures in response to specific stimulation are called *evoked potentials*. Evoked potentials are classified as nerve conduction studies (Chapter 15), somatosensory evoked potentials (Chapter 16), brain stem auditory evoked potentials (Chapters 17 and 19), and visual evoked potentials (Chapter 20). Movement-related potentials and event-related potentials (see Chapter 14 in Part A, Cortical Function) are also sometimes referred to as evoked potentials. Sensory evoked potentials may be recorded directly or averaged if the signal is small. Evoked potentials can be obtained from the somatosensory, somatic motor, auditory, and visual systems. Each has unique and specific waveforms that are altered in characteristic ways by disease.

CHAPTER 15

NERVE ACTION POTENTIALS

Bruce A. Evans, M.D.

In nerve action potential (NAP) studies, axons are evaluated by stimulating a nerve and recording a potential over the nerve at a different site. This technique is often used to study sensory axons selectively by stimulating a mixed nerve and recording distally over a cutaneous branch, by stimulating distal cutaneous nerve fibers and recording over the proximal nerve trunk, or by stimulating and recording over a pure sensory nerve or branch.

As recorded, NAPs represent the summation of the action potentials of individual fibers.[1]

METHODS

Nerve Stimulation

ELECTRODES

Surface Stimulation. Surface stimulation is convenient and adequate for routine use when stimulating commonly tested nerves at points where they are not deeply embedded in the tissue. Because distance measurements are taken from the skin, accuracy is not improved with the use of near-nerve stimulation needle electrodes and thus patient discomfort is comparable. Methods for magnetic and other forms of nonelectrical stimulation are being developed but are not yet reliable.[6,8,13]

Near-Nerve Needle Electrodes. These may be necessary to stimulate a nerve deep in the tissue. For example, unilateral absence of a surface-recorded superficial peroneal nerve action potential should be confirmed with needle stimulation of the nerve, because the nerve takes a course deep to the fascial plane in the lower leg in some patients and is not accessible to surface stimulation.

STIMULATING CUTANEOUS NERVES

Sensory axons may be stimulated either proximally or distally and either antidromic or orthodromic sensory NAPs may be recorded. Each method has limitations.

The orthodromic response is usually obtained by stimulating distal cutaneous sensory fibers and may have small amplitude; hence, it often cannot be recorded with surface electrodes at more proximal locations for conduction velocity determinations without averaging techniques. This orthodromic method is best suited for recording over a mixed nerve, with stimulation of distal cutaneous branches. When stimulating a ramifying cutaneous nerve, as in a digit in orthodromic sensory studies, circumferential ring electrodes are used to stimulate the maximum number of fibers, thus maximizing the amplitude of the recorded potential. Because NAP amplitude is directly related to distance between stimulating and recording distance, it is best measured at a standard distance.

STIMULATING MIXED NERVES

The antidromic response is obtained by stimulating a mixed nerve and recording over cutaneous branches distally. It usually has a larger amplitude than the orthodromic response, because of the lesser nerve-to-recording-electrode distance.[1] Also, the antidromic response can be elicited easily with stimulation at more proximal sites. However, the response is often contaminated with volume-conducted muscle responses resulting from mixed nerve stimulation. For this reason, the antidromic method is most suited for stimulating and recording from pure cutaneous nerves, for example, the sural and superficial radial nerves. NAPs may also be evoked by stimulation of mixed nerves at proximal locations with more proximal recordings, e.g., elbow to supraclavicular. Such potentials are mixed orthodromic and antidromic responses.

Stimulating electrodes used over a mixed or discrete nerve should have a small surface area and a cathode-to-anode separation of 1.5 to 3 cm. This provides a clear point for measurement of distances (from the position of the cathode) and minimizes current spread to other nerves.

Recording the Potential

ELECTRODES: SURFACE AND NEEDLE

Use of a near-nerve needle recording electrode produces significantly larger amplitude sensory or mixed nerve potentials than the more convenient surface electrodes and significantly increases signal-to-noise ratio.[10] The strong influence of needle-electrode-to-nerve distance on the amplitude of the recorded potential and the difficulty in reproducibly inserting the needle a standard distance from the nerve produce a wide range of amplitudes in normal nerves. However, the scatter of normal amplitudes is not greater than with surface electrode recording, because of the variations in nerve depth, skin resistance, and other factors associated with the latter technique.[1]

However, when observing a patient with sequential studies, the variability is less with surface recordings, because the nerve depth and, to some extent, the conductive properties of the skin and tissues are constant. Because of this reproducibility and the convenience of use, surface recording electrodes are most often used in routine sensory conduction studies.

When maximum information is to be derived from the NAP waveform, near-nerve recording needle electrodes are most appropriate. For example, an orthodromic sural sensory potential obtained with near-nerve recording in the calf and stimulation at the ankle with averaging of a large number of sweeps reveals individual components that reflect the activity in subpopulations of fibers conducting at velocities less than 15 m/s.[10] These small high-frequency components would be lost in surface recordings.

ELECTRODES: GEOMETRY

Surface recording electrodes may be arranged with an active electrode over the nerve and a reference electrode off the nerve, usually about 4 cm from the active electrode at right angles to the course of the nerve. This arrangement generally yields a triphasic response with a well-defined initial positive peak. When recording from the digits, use of different fingers for the active and reference electrodes approximates this arrangement as long as the cutaneous innervation of the fingers has no nerve in common (Fig. 15–1), although the response is usually biphasic, not containing a final positive peak. A volume-conducted response "seen" at the reference electrode may decrease the amplitude of the recorded potential by partially cancelling it, although this effect is minimal at a 4-cm distance. However, the reference electrode in this arrangement may record far field potentials generated by a nerve volley as it crosses the border of a region of geometric change in the volume conductor, which is to say, the tissue.[7] Areas of such geometric change include the transition from the shoulder to the upper arm and from the wrist to the hand. The far field, or junctional, potential recorded at the laterally placed reference electrode may precede the recording of the nerve action potential at the active electrode and cause distortion of that potential.

Surface electrodes may also be arranged with the two recording electrodes along the course of the nerve. This arrangement generally produces a larger amplitude potential than the previous method (Fig. 15–2). If the distance between the recording electrodes is such that the arrival of the negative peak of the potential at the first electrode (active, or G1) coincides with the arrival of the initial positive peak at the second electrode (reference, or G2), the amplitude is further enhanced. A longer interelectrode distance increases the noise of the recording and a shorter one decreases the amplitude of the response through common mode rejection. Because varying the interelectrode distance

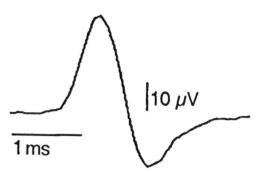

Figure 15–2. Potential recorded with an active electrode over the proximal index finger and a reference electrode placed 3 cm more distally on the index finger. Note the absence of an initial positive peak and a larger amplitude than the potential shown in Figure 15–1. The shoulder obscuring the initial negative takeoff of the nerve action potential represents a small far-field junctional potential.

changes the recorded amplitude, a constant standard electrode separation distance must be used in all studies.[14] This arrangement usually produces a triphasic potential. However, when recording from a digit, an initial positive peak generally is not seen before the main negative deflection. Junctional potentials may be seen with this recording configuration as well. A small junctional potential is shown in Figure 15–2 just before the onset of the sensory nerve action potential.

RECORDING FROM NERVE TRUNKS

Recording electrodes placed over a mixed nerve or a discrete sensory nerve should approximate the size of the nerve. Overlarge electrodes may reduce the size of the recorded potential due to shunting currents between the active portions and the inactive portions of a large electrode surface. One standard electrode for this use is a 5-mm tin disc.

RECORDING FROM CUTANEOUS NERVES

When recording over a ramifying cutaneous nerve with proximal stimulation of the mixed nerve, a larger electrode surface may increase the size of the recorded potential.

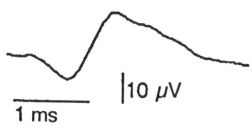

Figure 15–1. Triphasic median nerve potential recorded with an active surface electrode on the index finger and a reference electrode on the fifth digit.

Examples are the use of ring electrodes around a digit in median nerve antidromic studies and the use of 1-cm × 2-cm lead strip electrodes when attempting to record from the lateral femoral cutaneous nerve on the thigh.

SIGNAL AVERAGING

In most clinical situations, use of surface recording electrodes and appropriate techniques in routinely studied nerves yields easily recognizable sensory potentials without the use of averaging techniques. If one technique is unsuccessful, another often will yield a potential, as when a median sensory NAP response recorded over a digit cannot be obtained without averaging, but a response may be recorded over the larger mixed nerve at the wrist with stimulation of the nerve in the palm. In some situations, averaging may be necessary to obtain potentials for sensory conduction velocity studies in abnormal nerves, in which the potential with distal stimulation is already quite small or when using some of the more difficult and less reliable techniques, such as recording from the lateral femoral cutaneous nerve.

When averaging—especially if averaging more than a small number of responses— use of a hand-held stimulator may cause distortion or loss of amplitude, owing to movement of the stimulating electrode. It is best to substitute either a taped-on or strapped-on surface electrode pair or a small subcutaneous needle cathode taped to the skin to maintain a stable stimulation site. The number of responses to be averaged varies according to the need, for example, 8 to 16 responses are sufficient for improving resolution of a response that is visible but too noisy to be measured accurately, whereas as many as 1000 responses may be necessary to resolve responses less than 1 μV in size from the background noise. The degree of enhancement of the signal-to-noise ratio is roughly proportional to the square root of the number of trials, but it may not be adequate if noise levels are too high.[9] Time and patient discomfort become limiting factors in such a situation.

Measuring the Potential

AMPLITUDE

Amplitude of a NAP is measured from the initial positive peak (if present) to the negative peak. If no initial positive peak is present, the measurement is taken from the baseline to the negative peak (Fig. 15–3). Amplitude of the recorded potential is a measure of the number of axons able to conduct an action potential from the stimulation site to the recording site, assuming supramaximal stimulation has been achieved. Other factors, such as skin resistance and the distance from the recording electrode to the nerve, are also reflected in the recorded amplitude.

LATENCY

Latency is the lapsed time between the stimulation of nerve fibers and the arrival of the resulting potential at the recording electrode. It is measured from the artifact that represents the stimulus current to a point on the recorded potential (Fig. 15–4). The latency of the initial positive peak closely cor-

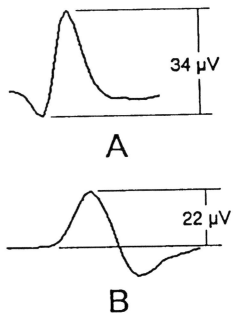

Figure 15–3. Amplitude measurement of a potential (A) with and (B) without an initial positive peak.

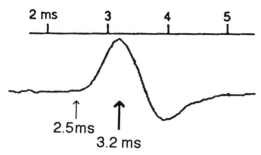

Figure 15–4. Distal latency measurement to the negative peak (*thick arrow*) and to the initial negative deflection (*thin arrow.*)

responds to the time of arrival of the action potentials in the fastest conducting fibers of the nerve. The potential recorded from a digit or recorded with a noisy background may not have a well-defined initial positive peak. Latency may be conveniently measured to the negative peak of the potential. Because this peak may be precisely located, it is conventionally used for latency measurements in the most distal segment of the nerve, that is, the "distal latency."

Because there is no delay in the sensory NAP appearance at the recording electrode attributable to something other than nerve

conduction velocity, for example, as in neuromuscular junction in motor studies, the conduction between the stimulating site and the recording site can be expressed as either latency (milliseconds) or conduction velocity (latency divided by distance, m/s). If distal latencies are performed at a standard distance, the distal latency is a preferable measure because it is a direct measurement and has less theoretical error than if it were divided into another value determined by measurement, namely, the distal distance. If, however, the distal site of stimulation is determined by landmarks, and therefore often at different distances, normal values probably should be normalized for distance by expressing them as conduction velocities.

CONDUCTION VELOCITY

Forearm or other more proximal conduction—always accompanied by wide variability in distance values among subjects—should be expressed as conduction velocity. Because distal conduction is already characterized by the distal latency or distal conduction velocity (or both) and because this distal segment is affected by normal distal

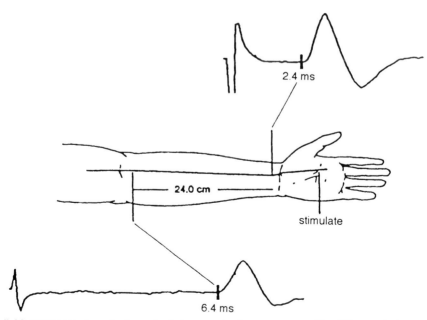

Figure 15–5. Measurement of conduction velocity in a proximal nerve segment. The difference in latencies between the proximal and distal stimulation sites is divided by the distance between the sites. To measure maximal conduction velocity, latencies are measured to the initial negative deflection.

axonal narrowing as well as sites of compression, the proximal conduction velocity is best expressed in terms of conduction between the proximal and distal stimulation sites rather than simply conduction between the proximal stimulation and recording sites. The potential disperses over distance; thus, subsets of fibers represented by the negative peak of the potential are different at different points along the nerve, and a negative peak latency difference used to derive a point-to-point velocity does not correspond to conduction in a single group of axons. Therefore, the latency measurements to the negative takeoff from the baseline or the initial positive peak, if present, are used to calculate conduction velocity between two points along the nerve. This accurately represents conduction in the fastest axons (Fig. 15–5).

Technical Precautions

Sensory NAPs are recorded at high amplification. This causes several difficulties in obtaining good recordings. Background noise caused by muscle activity can obscure the potential. This is reduced by allowing the patient to hear the amplified activity to aid relaxation. Shock artifacts can obscure short-latency responses. Skin abrasion, skin cleansing and drying, appropriate size and placement of ground electrodes, trial-and-error changes in the orientation of the anode and cathode with respect to the recording electrodes, and use of a standard stimulation isolation transformer help to eliminate unacceptable artifact.

Sensory nerves are often located superficially, and particular attention must be paid to skin temperature. The skin may need to be warmed to prevent amplitudes, latencies, and conduction velocities from deviating from normal values predicated on the basis of normal limb temperatures. Skin temperatures may be measured with a thermistor placed over the first dorsal interosseous muscle in the hand and 2 cm above the lateral malleolus in the leg. If hand temperature is less than 32°C or ankle temperature is less than 30°C, warm the limb with a radiant heat lamp. If hand temperature is less than

30.5°C or leg temperature less than 28.5°C, warm the limb in a 40°C water bath for 10 minutes.

When evaluating results against the given normal ranges for appropriate ages, take care that standard distances for stimulating-to-recording-electrode distance and standard separation distances of the two recording electrodes are those used when defining the normal values. Unavoidable deviations of distance, such as a large hand, must be corrected for by adjusting the normal values.

Selection of Nerves

Reproducible results with clearly defined normal ranges have been obtained with the sensory fibers in the median, ulnar, radial, sural, and plantar nerves (Table 15–1). Therefore, these are the nerves used in clinical studies in Mayo Clinic laboratories.

When consideration of a specific clinical problem suggests that additional studies would be useful, studies of the musculocutaneous nerve, medial and lateral antebrachial nerves of the forearm, superficial peroneal nerve, saphenous nerve, dorsal cutaneous branch of the ulnar nerve, or other proximal nerves can be performed. Because of reliability problems, anatomic variability, or undefined normal ranges, comparison of the two sides is essential.

Sensory recordings from the lateral femoral cutaneous nerve are possible in some subjects but are not sufficiently reliable for routine clinical use.

PATTERNS OF ABNORMALITY

Latency and Conduction Velocity

When compared with normal values obtained in the same segment of the same nerve with the same technique, latencies may be prolonged and nerve conduction velocities reduced in certain pathologic conditions. Conduction velocity of NAPs may be reduced in a specific nerve or segment because of generalized demyelination of the nerve or because of a focal demyelinating lesion located between the stimulation and

Table 15–1. **SENSORY NERVE CONDUCTION STUDIES—NORMAL VALUES FOR AMPLITUDE, DISTAL LATENCY, AND CONDUCTION VELOCITY IN FOREARM AND LEG SEGMENTS**

Nerve	Stimulate	Record	Amplitude, μV	Conduction Velocity, m/s	Distal Latency, ms
Median	Palm	Wrist	>50		<2.3
	Elbow			>55	
Median	Wrist	Digit 2	>14		<3.6
	Elbow			>56	
Ulnar	Palm	Wrist	>15		<2.3
	Elbow			>56	
Ulnar	Wrist	Digit 5	>10		<3.1
	Elbow			>56	
Superficial radial	Forearm	Dorsum of hand	>20		<2.8
	Elbow			>50	
Sural	Calf	Lateral ankle	>6		<4.5
	Proximal calf			>40	
Medial plantar	Plantar foot	Medial ankle	>7		<4.0

recording electrodes. A focal lesion is more likely to produce measurably abnormal conduction velocity or latency if the studied segment of the nerve is short, because the region of abnormally slow conduction is less likely to be offset by a long region of normal conduction velocity.

Slowed conduction velocities and prolonged latencies may also result from axonal narrowing, for example, as in regenerating nerves. Demyelination and axonal narrowing cannot be differentiated definitively on the basis of nerve conduction studies.

Amplitude

When compared with normal values obtained in the same segment of the same nerve with the same technique, the amplitude of NAPs may be reduced for several reasons. Conduction velocities of sensory nerve or mixed nerve fibers have a greater range than those of motor nerve fibers. For this reason, the normal recorded response shows greater dispersion and a greater relative reduction in amplitude with more distant points of stimulation than with motor studies. Consequently, the normal amplitude values are usually based on the more distal stimulation site with a short conduction distance.

Loss of axons in the nerve is reflected by a reduction in the amplitude of the NAP or by an absence of the potential if the axon loss is severe. An area of conduction block between the stimulating and recording electrodes due to a focal neuropractic lesion without anatomic abnormality or due to focal or generalized demyelination also produces decreased amplitude or absence of the recorded potential. Amplitude may also be reduced by nonuniform slowing of axonal conduction velocities without any conduction block, leading to an increased range of conduction velocities among the fibers. The resulting dispersion of the single-fiber NAPs leads to a lower amplitude-summated nerve action potential, especially when the distance between the recording and stimulating electrodes is great.

The greater range of conduction velocities among axons in sensory and mixed nerves than among motor nerve axons and the shorter duration of sensory NAPs than of compound muscle action potentials produces a greater cancellation in contribution to the compound potential, because of superimposition of positive and negative phases. For this reason, the area of the nerve action potential is a less useful parameter than the area of a compound muscle action potential as a reflection of the number of active axons except over very short distances.

Duration

In addition to reducing NAP amplitude, demyelination or axonal narrowing that nonuniformly affects axons in a nerve increases the duration of the potential, measured from the initial baseline reflection to the final return to the baseline. Generally, normal values for potential duration have not been systematically determined, and this parameter has not been significant clinically. Small late components of NAPs recorded with a near-nerve needle electrode and averaging techniques represent conduction in demyelinated, remyelinated, or regenerating fibers. These abnormal components may be an early or isolated finding in focal nerve injuries.[4,10]

CLINICAL CORRELATIONS

Focal Neuropathy

Focal neuropathies are characterized by abnormalities limited to a single nerve and a specific segment of that nerve. The abnormalities may include prolonged latency or reduced nerve conduction velocity over that segment of nerve and reduced amplitude nerve action potential. To detect and localize focal nerve lesions, nerve conduction techniques that examine short segments of nerve, including the likely site of the nerve lesion, and that have the least distance between stimulating or recording electrodes are the most sensitive. The most distal segments tested of several nerves include common sites of compression. The obtained distal latency may be prolonged because of nerve compression at those sites (Table 15–2). If the lesion is distal, a more proximal segment of the nerve usually shows normal conduction velocity or latency. Some nerves tested include common sites of compression in their more proximal segments. Reduced conduction velocity over those segments may result from nerve compression at those sites (Table 15–3). In either case, more severe injury may result in axonal destruction and reduction of the amplitude or loss of the recorded NAP.

Sensory studies are often the most sensitive for detecting mononeuropathy, as in carpal tunnel syndrome.[3,12] The appearance of additional small late components when recording a nerve potential from a mixed nerve with a needle electrode may give early evidence of a lesion localized to the segment evaluated, such as the ulnar nerve at the elbow.[11] Sensory studies may also aid in the localization of focal neurogenic lesions through the demonstration of a conduction block or focal slowing as an adjunct to motor studies, although the greater normal variability of amplitude due to dispersion mentioned earlier makes NAP studies less suitable for demonstration of conduction block than motor studies. Finally, sensory NAP studies may be the only way to document cutaneous nerve mononeuropathy.

Generalized Neuropathy

Sensory potentials may be selectively involved in sensory peripheral neuropathy and

Table 15–2. **NERVES USED FOR EVALUATION OF COMPRESSION AT COMMON DISTAL ENTRAPMENT SITES**

Nerve	Segment	Clinical Entrapment
Median	Digit 1, 2, 3, or 4 to wrist	Carpal tunnel syndrome
	Palm to wrist	Carpal tunnel syndrome
Ulnar	Digit 4 or 5 to wrist	Ulnar neuropathy at wrist
Plantar	Plantar foot to medial malleolus	Tarsal tunnel syndrome

Table 15–3. **NERVES USED FOR EVALUATION OF COMPRESSION AT COMMON PROXIMAL ENTRAPMENT SITES**

Nerve	Segment	Clinical Entrapment
Median	Wrist to elbow	Pronator syndrome
Ulnar	Above elbow to wrist	Cubital tunnel syndrome
	Above elbow to below elbow	
Superficial radial	Dorsum of hand to elbow	Superficial radial neuropathy

usually show abnormalities earlier than motor studies in a generalized sensory-motor neuropathy.[2] When detecting mild or diffuse nerve conduction slowing, nerve conduction techniques that examine long segments of nerve and that have the maximum distance between stimulating or recording electrodes are the most sensitive. In generalized neuropathy, multiple nerves show abnormalities. The abnormalities may be limited to distal segments and nerves with prolonged distal latencies; reduced NAP amplitudes or reduced nerve conduction velocities may be seen in proximal segments and nerves as well as in a more generalized disorder. Relatively preserved amplitudes in distal segments in conjunction with prolonged latencies and reduced conduction velocities suggest primarily a demyelinating cause. When NAPs are not obtainable, such a judgment cannot be made without reference to the motor potentials, which are usually preserved longer in both demyelinating and axonal neuropathies.

Intraspinal Lesions

Sensory studies aid in the localization of lesions to predorsal or postdorsal root ganglion regions (i.e., intraspinal or extraspinal).[5] Lesions that affect the dorsal root ganglia or more distal structures that contribute to peripheral nerves often lead to loss of peripheral sensory axons and reduction of the appropriate NAP amplitude. An intraspinal lesion that affects the same segment does not cause such axon loss peripherally, because only the central process of the sensory neuron is affected. With significant clinical sensory loss in the distribution of a sensory nerve, the presence of a normal NAP amplitude recorded from that nerve strongly suggests an intraspinal lesion, assuming the lesion was more than a few days old, to allow for completion of wallerian degeneration.

CHAPTER SUMMARY

Recording of NAPs is an integral part of clinical nerve conduction studies. The methods of stimulation and recording, including the choice of electrode and electrode geometry, affect the characteristics of the nerve action potential. These characteristics, including latency, amplitude, duration, and conduction velocity, can be measured with several alternative methods. Different methods may be most appropriate in different circumstances. Information gathered from the study of nerve action potentials is useful in evaluating focal and generalized neuropathies and in localizing intraspinal lesions that affect sensory function.

REFERENCES

1. Buchthal, F and Rosenfalck, A: Evoked action potentials and conduction velocity in human sensory nerves. Brain Res 3:1–122, 1966.
2. Buchthal, F and Rosenfalck, A: Sensory potentials in polyneuropathy. Brain 94:241–262, 1971.
3. Evans, BA and Daube, JR: A comparison of three electrodiagnostic methods of diagnosing carpal tunnel syndrome (abstr). Muscle Nerve 7:565, 1984.
4. Gilliatt, RW: Sensory conduction studies in the early recognition of nerve disorders. Muscle Nerve 1:352–359, 1978.
5. Gilliatt, RW, Le Quesne, PM, Logue, V, and Sumner, AJ: Wasting of the hand associated with a cervical rib or band. J Neurol Neurosurg Psychiatry 33:615–624, 1970.
6. Hashimoto, I, Gatayama, T, Yoshikawa, K, et al:

Compound activity in sensory nerve fibers is related to intensity of sensation evoked by air-puff stimulation of the index finger in man. Electroencephalogr Clin Neurophysiol 81:176–185, 1991.

7. Kimura, J, Mitsudome, A, Yamada, T, and Dickins, QS: Stationary peaks from a moving source in far-field recording. Electroencephalogr Clin Neurophysiol 58:351–361, 1984.

8. Maccabee, PJ, Eberle, L, Amassian, VE, et al: Spatial distribution of the electric field induced in volume by round and figure '8' magnetic coils: Relevance to activation of sensory nerve fibers. Electroencephalogr Clin Neurophysiol 76:131–141, 1990.

9. Normand, MM and Daube, JR: Interaction of random electromyographic activity with averaged sensory evoked potentials. Neurology 42:1605–1608, 1992.

10. Rosenfalck, A: Early recognition of nerve disorders by near-nerve recording of sensory action potentials. Muscle Nerve 1:360–367, 1978.

11. Tackmann, W, Vogel, P, Kaeser, HE, and Ettlin, T: Sensitivity and localizing significance of motor and sensory electroneurographic parameters in the diagnosis of ulnar nerve lesions at the elbow: A reappraisal. J Neurol 231:204–211, 1984.

12. Thomas, JE, Lambert, EH, and Cseuz, KA: Electrodiagnostic aspects of the carpal tunnel syndrome. Arch Neurol 16:635–641, 1967.

13. Verdugo, R and Ochoa, JL: Quantitative somatosensory thermotest. A key method for functional evaluation of small calibre afferent channels. Brain 115:893–913, 1992.

14. Wee, AS and Ashley, RA: Effect of interelectrode recording distance on morphology of the antidromic sensory nerve action potentials at the finger. Electromyogr Clin Neurophysiol 30:93–96, 1990.

CHAPTER 16

SOMATOSENSORY EVOKED POTENTIALS

C. Michel Harper, Jr., M.D.

Somatosensory evoked potentials (SEPs) are an electrical manifestation of the response of the nervous system to a specific stimulus. They are analogous to standard sensory nerve conduction studies but are typically lower in amplitude and are recorded over multiple levels of the central and peripheral nervous systems. Standard techniques involve electrical stimulation of large mixed or cutaneous nerves in the extremities with recording over peripheral nerve, plexus, spine, and scalp. Well-defined reproducible potentials of 1 to 50 μV can be recorded by averaging 500 to 1000 stimuli.

ORIGIN OF WAVEFORMS

Anatomy of Peripheral and Central Pathways

Although SEPs can be recorded after stimulation of almost any sensory nerve, large mixed nerves of the upper extremity (median, ulnar, or radial nerve) or lower extremity (tibial or peroneal nerve) produce potentials of the highest amplitude. The potentials recorded represent activity in the proprioceptive system conducted peripherally by large-diamter, myelinated, fast-conducting cutaneous and muscle afferents and centrally by the dorsal column–medial lemniscus pathway and spinocerebellar pathways, with multiple collaterals to gray matter at all levels.[13,16] These pathways are located in the posterior and lateral columns of the spinal cord.

Lesions affecting modalities carried by small-diameter sensory fibers or by central pathways in the ventral half of the spinal cord frequently do not produce somatosensory evoked potential abnormalities. Recently, somatosensory evoked potentials have been recorded by stimulation of small pain fibers with a carbon-dioxide laser, but the clinical usefulness of this technique has not been demonstrated.[22]

Stimulation of cutaneous nerves or dermatomes activates large cutaneous afferents

157

but produces SEPs of lower amplitude than stimulation of mixed nerves. Peripheral and spinal potentials are often absent in normal subjects, thus preventing disease in the central nervous system from being distinguished from that in the peripheral nervous system.[2]

Nomenclature

The morphology, polarity, and amplitude of SEP waveforms depend on the recording montage used. The most widely accepted method for identification of potentials uses "N" for negative and "P" for positive polarity, followed by a number representing the average latency of the peak. Thus, with a bipolar montage, the negative potential recorded approximately 9 milliseconds after stimulation of the median nerve at the wrist is termed "N9" Fig. 16–1).

Because a SEP is recorded with a differential amplifier, the polarity of a potential may not correlate with the polarity of its generator. For example, after stimulation of the median or ulnar nerve at the wrist, a broad

negativity can be recorded with a latency of 13 milliseconds over the posterior aspect of the cervical spine. This negativity arises from a horizontal dipole in the dorsal horn of the cervical spinal cord.[19] An electrode placed anteriorly over the thyroid cartilage or in the esophagus records the same activity as a broad positivity at 13 milliseconds[7] (see Fig. 16–1). Because the polarity of this potential depends on the montage used, it is frequently referred to as "N/P13."

Structural Origin of Potentials

Routine recordings of SEPs are made with surface electrodes placed over the extremities, spine, and scalp. The recorded activity represents the summated effects of nerve action potentials and synaptic potentials on the extracellular fluid of the body in its role as a volume conductor. Nerve action potentials that travel along nerves or fiber tracts are called "traveling waves." Potentials that remain localized in areas of nuclei or synapses are called "stationary waves." Wave-

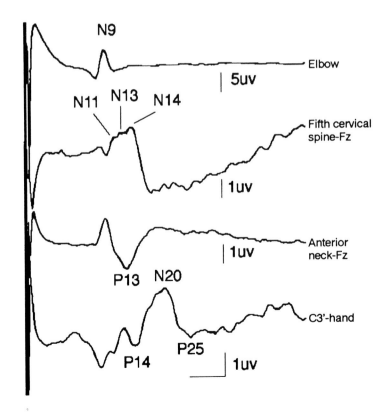

Figure 16–1. Normal median SEP illustrating electrical activity generated at the level of the brachial plexus (N9), cervical spinal cord (N11, N/P13), brain stem (N/P14), and cerebral cortex (N20, P25).

forms of SEPs recorded over peripheral nerves represent compound nerve action potentials traveling beneath the recording electrode. Waveforms recorded over the central nervous system are more complex and usually result from the combined activity of several different generators, including traveling waves in myelinated tracts and stationary synaptic potentials in nuclei. Because the size and shape of the volume conductor change in three dimensions, there may be changes in charge density and, therefore, voltage differences created by a traveling dipole moving through the volume conductor. These stationary far-field potentials may be positive or negative and frequently "contaminate" the SEP when certain recording montages are used (see discussion later in this chapter).[25]

Near-Field and Far-Field Potentials

In a volume conductor, the amplitude of the potential is inversely related to the square of the distance between the generator and the recording point. When the recording electrode is relatively close to the generator source, near-field potentials are recorded. Near-field potentials can be recorded with either a bipolar or referential montage, so long as the recording electrode is close to the generator. The particular montage used represents a compromise between the noise introduced by large interelectrode distances and the cancellation effect of more closely spaced electrodes. Bipolar montages are used to record peripheral potentials, whereas referential montages are used to record potentials over the spine and scalp. Traveling waves, stationary waves, or both may be recorded as near-field potentials.

An important principle in recording SEPs is that electrodes placed anywhere on the body are relatively active. Far-field potentials are recorded when both the recording and the reference electrodes are distant from the generator site.[9] Far-field potentials are invariably stationary potentials and can be generated by distant synaptic potentials or by nerve action potentials traversing an area where configuration of the volume conductor has changed.[25] Referential montages

with a recording electrode on the scalp and a noncephalic or linked-ears reference electrode preferentially record far-field potentials (see Fig. 16–1). The potential advantage of using far-field recordings is that information from many different levels of the nervous system can be obtained from a single recording montage. The disadvantages are excess noise introduced by long interelectrode distances and the presence of potentials created by changes in the configuration of the volume conductor which are not generated by fiber tracts or synapses in the nervous system. The final number, site, and type of recording montage used must be individualized to the needs of the patient and the resources of the laboratory. Recordings should be isolated from individual generators, both to record from as many levels of the nervous system as possible with a minimum number of channels, and to maximize signal-to-noise ratio.

Upper Extremity SEPs

After stimulation of the median or ulnar nerve at the wrist, activity can be recorded at the elbow, Erb's point, cervical spine, and scalp. The N5 potential recorded with a bipolar electrode at the elbow records the propagating nerve action potential in the median or ulnar nerve and provides a monitor of adequate stimulation and an estimation of peripheral conduction velocity.

The N9 potential recorded with an electrode at Erb's point (2 cm superior to the midpoint of the clavicle) referred to an electrode in the same location contralaterally represents ascending activity of sensory and antidromic motor fibers passing through the brachial plexus. The P9 potential recorded as a far-field potential in a scalp-noncephalic montage likely represents a change in current flow created by a change in the configuration of the volume conductor as the action potential travels from the arm to the trunk.[12]

An electrode over the spine of C-5 or C-7 referred to Fz is the most common montage for recording activity arising from the cervical spine and brain stem (see Fig. 16–1). This montage records three negative potentials, N11, N13, and N14. The N11 potential is a presynaptic traveling wave that arises from activity in the dorsal root entry zone or dorsal column of the cervical cord.[11] With median nerve stimulation, this activity enters the spinal cord from the C-6 to T-1 segments, but with ulnar nerve stimulation, it is confined to the C-8 to T-1 segments. The N13 potential is a standing horizontal dipole created by synaptic activation of neurons in the dorsal horn by collateral branches of the primary afferent fiber as it enters the dorsal column.[19] The negative end of the dipole faces posteriorly, whereas an electrode anterior to the spine records a positive potential. The N13 potential can be reduced or eliminated by a pathologic condition at the cervical level with enough sparing of conduction in the dorsal column to produce normal brain stem and cortical potentials[30] (Fig. 16–2). The N14 potential represents activity in the nucleus cuneatus or medial lemniscus (or both) at the cervicomedullary junction.[6] Separation of the N13 and N14 potentials is facilitated by recording from the anterior neck, which causes the N13 potential to reverse phase, or by using a scalp-noncephalic montage in which P13 and P14 are seen as well-defined far-field potentials (see Fig. 16–1).

The N20/P25 complex reflects synchronous postsynaptic potentials generated by neurons of the primary somatosensory cortex in response to the afferent thalamocortical volley.[8,24] With the more commonly used C3' (C4')-Fz montage, the N20/P25 complex probably is an average of independent generators in the parietal and frontal lobes, each of which can be eliminated selectively by pathologic conditions. This may go unnoticed in the bipolar scalp montage, because potentials will be recorded from either parietal or frontal electrodes. Therefore, when a cortical lesion is suspected, the recordings should be made from both parietal (C3' or C4') and frontal (F3 or F4) sites, using a noncephalic reference for both.[6] The activity recorded over the frontal lobe has a positive peak, P22, followed by a large negative peak, N30.

Lower Extremity SEPs

In the tibial SEP, the N8 potential is the peripheral nerve action potential recorded at the popliteal fossa (Fig. 16–3). An electrode over the L-1 vertebra referred to the iliac crest records a negative, sometimes bifid, potential. The initial, less regularly recorded N18 peak is a traveling wave that reflects conduction through the cauda equina and the dorsal column volley in lumbosacral spinal cord.[27] The second most prominent peak, N22, represents synaptic activity generated in the dorsal horn of the spinal cord and is analogous to the stationary N13 horizontal dipole in the cervical cord[19] (see Fig. 16–3).

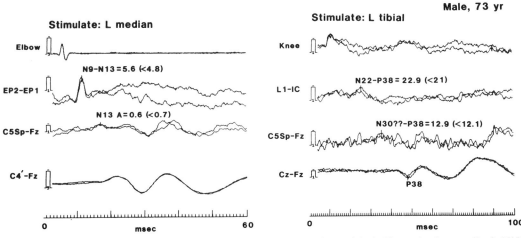

Figure 16–2. Upper and lower extremity SEPs in a patient with cervical spondylosis, illustrating low-amplitude N13 and prolonged interpeak latencies (N9 to N13, N22 to P38, and N30 to P38).

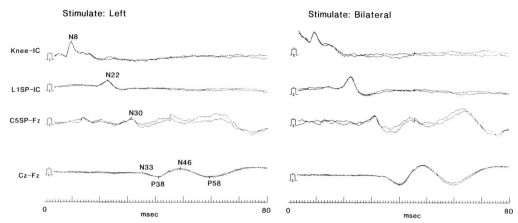

Figure 16–3. Normal tibial SEP with unilateral and bilateral stimulation, illustrating electrical activity generated at the level of the tibial nerve at the knee (N8), lumbar spinal cord (N22), cervical spinal cord (N30), and cerebral cortex (N33, P38, N46, P58).

The N30 potential recorded over the cervical spine represents activity in the fasciculus gracilis and, possibly, in the spinocerebellar pathways.[17] It is difficult to record in some subjects because of muscle artifact and progressive dispersion of the ascending volley in the dorsal columns.

The cortical generator for the lower extremity SEP resides in primary sensory cortex, deep in the interhemispheric fissure.[6] With a midline central montage (Cz–Fz), this activity is recorded as a small initial negativity, N33, followed by a prominent positivity, P38.

Stimulation of the peroneal nerve at the knee produces analogous potentials of shorter latency over the lumbar cord (N11), cervical cord (N19), and scalp (P27). A well-defined scalp potential can be recorded with cutaneous nerve (sural, superficial peroneal, femoral, pudendal nerves) or dermatomal stimulation, but spinal potentials are often absent in normal subjects.

Trigeminal SEPs

SEPs in response to electrical or mechanical stimulation of the trigeminal nerve have been reported.[4] These recordings are difficult to obtain because of shock artifact and contamination of the response with muscle artifact.

METHODOLOGY

Factors Affecting Signal-to-Noise Ratio

AVERAGING

Amplitude of the SEP signal is usually near the level of background noise created by brain and muscle activity, movement, or electromagnetic activity in the environment. Averaging summates activity that is time-locked to the stimulus trigger while gradually subtracting random background noise. Generally, 500 averages produce adequate SEPs. In the presence of excessive noise, increasing the number of stimuli averaged does not help extract a better signal. If noise becomes time-locked to the signal, it will be enhanced and may be mistaken for a physiologic signal. This can be avoided by (1) using a stimulation rate that is *not* a factor of 60 Hz (e.g., 2.1 Hz), (2) monitoring closely for movement and muscle artifact, and (3) recording two trials of averages to ensure reproducibility of the desired signal.

MONTAGE

Increasing the distance between the recording electrode and reference electrode enhances the amplitude of the signal but can also amplify unwanted noise. The bipolar montage has the lowest level of noise; therefore, it is used to record signals of high amplitude. A referential montage is used to record signals of low amplitude. Choosing the proper location of the reference electrode can minimize the amplification of noise.

MUSCLE

Electrical activity from nearby muscle frequently contaminates the SEP. This usually is not a problem at peripheral recording sites because of the high amplitude of the signal; however, activity in paraspinal and scalp muscles can easily obscure the spinal and cortical potentials of the SEP. Monitoring the audio signal from spine and scalp recording sites while recording the somatosensory evoked potential is essential in detecting excess muscle activity. This activity can be controlled effectively by relaxing the patient. Relaxation can be produced by keeping stimulus intensity as low as possible, making sure the patient is in a comfortable position, and providing a sedative.

ELECTRICAL NOISE

The two main sources of electrical artifact in recordings of SEPs are stimulus artifact and 60-Hz alternating current transmitted to the amplifier by the machine used to record the evoked potential or through electromagnetic radiation. Stimulus artifact can be decreased by using a stimulus-isolation device and a fast-recovery amplifier, by maintaining proper orientation and contact of the stimulating electrodes, and by avoiding higher than necessary stimulus intensity. Maintaining recording electrode impedance under 5000 Ω by cleaning the skin and proper grounding eliminates most 60-Hz noise from the SEP. Use of a 60-Hz "notch" filter is not recommended, because the SEP contains important physiologic information in this range. If different types of electrodes (e.g., surface and needle) are used at recording and reference sites, an impedance mismatch is created, thus amplifying 60-Hz interference.

FILTERING

The goal of filtering is to eliminate noise without reducing waveforms of interest.

With analog filters, a low-frequency setting of 30 Hz and a high-frequency setting of 3 kHz are usually satisfactory. Restricting low frequencies further with a filter setting of 150 Hz allows better visualization of some potentials (e.g., N11) and eliminates most 60-Hz interference. Unfortunately, this filter setting also reduces the amplitude of other important peaks (e.g., N13) and shortens the latency of all peaks. Changes in latency can be prevented with digital filters, but important information is still eliminated if filters are restricted much beyond the 30-Hz to 3-kHz range.

STIMULATION PARAMETERS

To a certain point, the relationship between stimulus intensity and amplitude of all waveforms of the SEP is direct. The stimulus intensity in a mixed nerve should be high enough to produce a visible twitch of the muscle, and in cutaneous nerves, it should be two to three times sensory threshold. Ideally, the nerve action potential recorded proximal in the limb should be maximal. Higher intensities are painful, thus making relaxation difficult, and do not increase the amplitude of the SEP. In the presence of peripheral neuropathy, higher stimulus intensities may be needed and are usually well tolerated. Stimulation with a needle near the nerve requires less current and may be tried when edema of the extremity is present or when surface stimulation is painful.

Stimulus repetition rates of 2 to 5 Hz are well tolerated by most patients. Lower rates may be required in sensitive or spastic patients to avoid flexor withdrawal reflex. Stimulation at 10 Hz or greater increases latency and decreases amplitude of some components.

CHOICE OF NERVE

The highest amplitude SEPs are obtained with stimulation of the median and tibial nerves. Reliable potentials at peripheral, subcortical, and cortical sites are also seen in ulnar, radial, and peroneal SEPs. The lowest amplitude SEPs are obtained with digital, superficial radial, sural, plantar, saphenous, and pudendal nerve and dermatomal stimulation. Peripheral and subcortical potentials are frequently absent, and latencies are longer than with mixed nerve stimulation.

UNILATERAL VERSUS BILATERAL STIMULATION

All SEPs are higher in amplitude when bilateral simultaneous stimulation is used. Unfortunately, this may mask an abnormality when the lesion is strictly unilateral. Bilateral stimulation should be reserved for instances when the tibial somatosensory evoked potential is completely absent or the scalp potential is abnormal, with absent spine potentials. In this situation, bilateral stimulation may produce adequate subcortical potentials and allow estimation of conduction time in central pathways.

Standard Methods

Table 16–1 outlines the method used in our laboratory to study SEPs. Surface electrodes are used for stimulation and for recording. Scalp electrodes are fixed with collodion, with spinal and Erb's point electrodes taped in place. Stimulating and peripheral recording electrodes are fixed with an elastic strap. Bipolar electrodes have a fixed interelectrode distance of 35 mm. Electrolyte gel is applied to all electrodes, and impedance is maintained under 5000 Ω. Despite differences between laboratories in the number of channels and recording montages used, the major recognized potentials are consistent. Two recording trials should be made to ensure waveform reproducibil-

TABLE 16–1. STANDARD METHODS FOR RECORDING SOMATOSENSORY EVOKED POTENTIALS

	Median or Ulnar Nerve	Tibial Nerve
Stimulation	Bipolar at wrist, cathode proximal	Bipolar at ankle, cathode proximal
Standard recording, montage:potential	Bipolar at elbow: N5 EPi-EPc:N9 C5s-Fz:N11, N13, N14 C3′(C4′)-Fz:N20/P25 Ground:lead plate arm	Bipolar at knee: N8 L1-ICc:N22 C5s-Fz:N30 Cz-Fz:N33/P38 Ground:lead plate leg

Optional recording	Anterior neck-C5s:P13
	C3′(C4′)-noncephalic:
	  P9, P13, P14 far-field
	  N20/P25 parietal
	F3′(F4′)-noncephalic:
	  N22/P30 frontal
		C3′-C4′:N33/P38

Machine settings	Stimulus rate, 1–5 Hz
	Stimulus intensity, slight
	  muscle twitch
	Amplifier sensitivity
	  10 μV spine and scalp
	  20 μV peripheral
	Analysis time, 50 ms
	Filter, 30–3000 Hz
	Number averaged, 500
		Stimulus rate, 0.5–2 Hz
		Stimulus intensity, slight
		  muscle twitch
		Amplifier sensitivity
		  10 μV spine and scalp
		  20 μV peripheral
		Analysis time, 80 ms
		Filter, 30–3000 Hz
		Number averaged, 500

Measurement	Height and F-distance
	Limb temperature
	Waveform amplitude
	  N5, N9, N13:onset-peak
	  N20:N20–P25
	  P25:P25–N35
		Height and F-distance
		Limb temperature
		Waveform amplitude
		  N8, N22, N30:onset-peak
		  N33:N33–P38
		  P38:P38–N46

Abbreviations: EPi-EPc = Erb's point ipsilateral to contralateral; C5s = spine of C-5 vertebra; L1-ICc = spine of L-1 vertebra to contralateral iliac crest.

ity. We routinely perform a motor conduction study in each extremity tested to help detect slowing of peripheral conduction.

Waveform Analysis

Abnormalities in the SEP include loss of amplitude and prolongation of latency. Amplitudes do not follow a normal distribution, so the range of control values is typically used for interpretation. Side-to-side differences are less reliable, but absence of a waveform easily recorded on the contralateral side is clearly abnormal. Subcortical or peripheral potentials may be low in amplitude or absent in isolation, because of central amplification and the presence of multiple parallel central pathways. Excess muscle activity, obesity, and inadequate stimulation are common technical causes for loss of amplitude. Increased amplitude of the cortical waveforms is seen in patients with cortical myoclonus.

Values for absolute and interpeak latencies are normally distributed in control populations. Normal values can be defined as the mean ± 2.5 standard deviations. The effects of height, limb length, and temperature can be eliminated by the use of interpeak latencies. Absolute latencies are used when peripheral or spinal potentials are absent, when it is necessary to use a regression curve of the absolute latency versus height, limb length, or F-distance for proper interpretation. Dispersion and greatly prolonged

latencies are more common in demyelinating disease, but it is difficult to predict pathologic conditions in individual patients.

CLINICAL APPLICATION

General Utility

SEPs can help detect lesions of the central nervous system when the results of clinical examination are normal or equivocal. This is particularly helpful in the search for a clinically silent "second lesion" in multiple sclerosis.[5b] In many cases, the location of one or more lesions can also be determined. In the upper limb, prolongation of the N9 to N13 interpeak latency indicates a lesion in the nerve roots or cervical spinal cord, whereas a prolonged N13 to N20 interpeak latency suggests a lesion between the cervical cord and cerebral cortex. A low-amplitude or absent N13 with preservation of N9 and N20 localizes the lesion to the cervical cord (see Fig. 16–2). In the lower limb, absence of the lumbar potential with a normal N8 suggests a lesion in the cauda equina or lumbar spinal cord. Prolongation of the N22 to P38 or the N22 to N30 interpeak latency indicates an abnormality rostral to the lumbar cord or in the thoracolumbar cord, respectively (Fig. 16–4). When reporting abnormalities of the SEP, the presence and location of lesions should be emphasized. Statements about pathologic conditions should be avoided because disease-specific changes are not ob-

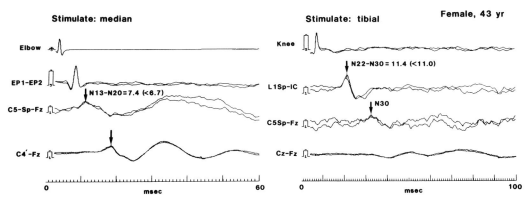

Figure 16–4. Upper and lower extremity SEPs in a patient with multiple sclerosis, illustrating electrophysiologic evidence of slowed conduction in two different regions of the central nervous system. The prolonged N13 to N20 interpeak latency indicates a lesion between the upper cervical spinal cord and cerebral cortex. The prolonged N22 to N30 interpeak latency indicates a lesion between the lumbar and cervical spinal cord.

served. SEPs are an objective and quantifiable measure of function; thus, they are particularly suitable for measuring changes in the physiology of proprioceptive pathways over time. Such is the case during surgical procedures, wherein SEPs help monitor and prevent injury to peripheral and central nervous system structures.[18]

Disorders of the Peripheral Nervous System

PERIPHERAL NEUROPATHY

Absolute latencies are prolonged with normal interpeak latencies. Amplitudes are low at both central and peripheral recording sites (Fig. 16–5). Because of central amplification, central potentials may be spared when peripheral potentials are absent.[10] In generalized neuropathies, SEPs can estimate peripheral sensory conduction velocity when peripheral sensory nerve action potentials are absent and can detect spinocerebellar degeneration, leukodystrophies, some cases of inflammatory neuropathy, and combined central and peripheral nervous system involvement in vitamin B_{12} deficiency. They can provide evidence for slowing in proximal segments of the peripheral nerve in inflammatory neuropathy and some types of hereditary neuropathy.[33] In addition, SEPs can often detect early axonal regeneration across focal traumatic lesions of the peripheral nerve before a sensory nerve action potential can be measured.[2]

BRACHIAL PLEXOPATHY

SEPs together with standard nerve conduction studies and electromyography may help to localize lesions and to determine prognosis in patients with traumatic plexopathy. Abnormalities reported include a low-amplitude or absent N9 potential, attenuation of all responses from the elbow rostrally, and a prolonged N9 to N13 interpeak latency.[33] The presence of a scalp SEP

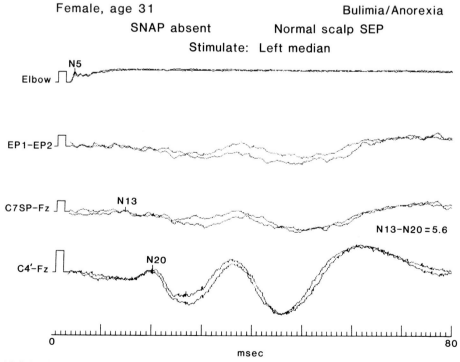

Figure 16–5. Median SEP in a patient with peripheral neuropathy due to vitamin B_6 abuse. Amplitudes are reduced at all recording locations, with prolonged absolute latencies and normal interpeak latencies. SNAP = sensory nerve action potential.

in combination with abnormal sensory nerve action potentials indicates at least partial continuity between peripheral and central structures. Conversely, the presence of a normal Erb's-point potential with absence of cervical and scalp potentials and normal sensory nerve action potentials suggests root avulsion. Unfortunately, localization often is not possible in the presence of severe lesions that involve both preganglionic and postganglionic elements of the plexus. With involvement of only one or two roots, results of stimulation of the median or ulnar nerve may be normal because afferents of mixed nerves enter the cord through multiple roots. Although this problem can be solved by stimulating individual segments, difficulty with recording the N9 and N13 potentials in normal subjects and the need for comparison with the contralateral side make cutaneous stimulation a time-consuming and often inadequate alternative. SEPs recorded intraoperatively over the spine and scalp with direct stimulation of the roots or other elements of the plexus can be very useful in detecting root avulsion and axonal continuity in postganglionic lesions.[26]

Abnormalities of the SEP have been reported in patients with thoracic outlet syndrome, even in the presence of normal findings on clinical examination, nerve conduction studies, and electromyography.[33] Several patterns have been described, including a low-amplitude N9 with a prolonged N9 to N13 interpeak latency and a low-amplitude N13, with or without attenuation of the N9 potential. The ulnar SEP is abnormal more often than the median.[32] Many studies are flawed because of vague definition of thoracic outlet syndrome and inadequate criteria for abnormalities of SEPs. SEPs are usually normal in patients with vascular or symptomatic thoracic outlet syndrome; moreover, they add little information to clinical examination and electromyography in neurogenic thoracic outlet syndrome.[2,32]

RADICULOPATHY

Median, ulnar, and radial SEPs are less sensitive than electromyography in the diagnosis of cervical radiculopathy.[33] Digital

SEPs have a higher sensitivity but lower specificity.[2] Tibial and ulnar SEPs are abnormal in 80 to 90% of patients who have combined cervical radiculopathy and myelopathy caused by spondylosis.[33] Abnormalities of the tibial SEP consist of a low-amplitude scalp potential with a prolonged N22 to P38 interpeak latency. The ulnar SEP usually shows a loss of N13 with preservation of N20 and prolongation of the N9 to N13 and the N9 to N20 interpeak latencies.

Although tibial, peroneal, and dermatomal SEPs have been reported to be more sensitive than electromyography in the diagnosis of lumbosacral radiculopathy,[28] this has not been the experience of others.[3] A potential problem is that mixed nerves carry the SEP signal over several different nerve roots, whereas dermatomal SEPs produce spine and scalp potentials of low amplitude and variable latency. Other studies reporting on the usefulness of SEPs in radiculopathy used questionable criteria for identifying abnormalities of SEPs or an inadequate standard for the diagnosis of radiculopathy.[23]

Disorders of the Central Nervous System

Abnormalities of SEPs have been described in many diseases of the central nervous system (Table 16–2).[5a] Diseases of myelin tend to produce prominent changes in latency, whereas axonal lesions preferentially affect the amplitude of central potentials. The overlap is so marked that pathologic conditions cannot be predicted by changes in the SEP. Despite limitations, these potentials are important in monitoring spinal, brain stem, and cortical function during various surgical procedures.[18] The following discussion is limited to those disorders in which SEPs appear to aid in diagnosis or management.

DEMYELINATING DISEASE

SEPs can be used to detect clinically silent lesions in patients with possible or probable multiple sclerosis. Median SEPs are abnormal in about two thirds of all patients with multiple sclerosis and in about half of those

TABLE 16–2. CENTRAL NERVOUS SYSTEM DISEASES ASSOCIATED WITH ABNORMAL SOMATOSENSORY EVOKED POTENTIALS

Demyelinating disease
Compressive lesions
   Spondylosis
   Extra-axial tumors
   Arnold-Chiari malformation
   Trauma
Intrinsic structural lesions
   Intra-axial tumors
   Arteriovenous malformation
   Syringomyelia
Spinocerebellar degenerations
Stroke
Dementia
Myoclonus
Motor neuron disease
Coma

edly prolonged central conduction times are seen in patients with leukodystrophies.[5a]

COMPRESSIVE LESIONS

Ulnar and tibial SEPs are more sensitive than median SEPs in cervical myelopathy due to spondylosis.[30] SEPs can be abnormal, with little or no objective evidence or myelopathy on clinical examination. The N13 potential is frequently attenuated or absent. This is in contrast to lesions at the foramen magnum (e.g., Arnold-Chiari malformation or tumor) in which the N13 potential typically is preserved and the N13 to N20 interpeak latency is prolonged (Fig. 16–6). In spinal trauma, preservation or return of SEPs early after onset is associated with a good prognosis.[33] However, damage confined to the corticospinal tract or anterior horn cells may produce severe neurologic disability with little or no abnormality of the SEP.

INTRINSIC STRUCTURAL LESIONS

SEPs are often normal in slow-growing tumors like gliomas that infiltrate but do not destroy sensory pathways.[5a] SEPs can help detect important feeder and draining vessels during selective embolization and surgical resection of arteriovenous malformations.[18] In syringomyelia, tibial SEPs are abnormal in most patients. Ulnar and median SEPs are usually normal in the presence of dissociated

without symptoms or signs of sensory involvement.[5b] Lower extremity SEPs are even more sensitive, probably because the white matter pathways are longer. Although the overall sensitivity of SEPs is lower than that of magnetic resonance imaging in multiple sclerosis, they may be better at detecting spinal cord lesions that are below the resolution of magnetic resonance imaging.[15,29] Mark-

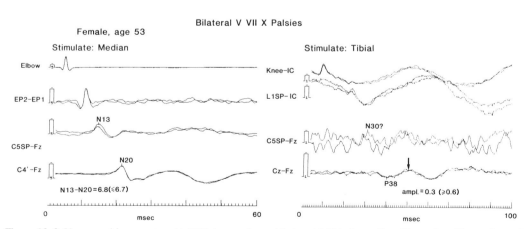

Figure 16–6. Upper and lower extremity SEPs in a patient with Arnold-Chiari type II malformation. The prolonged N13 to N20 interpeak latency is consistent with a lesion of the high cervical spinal cord or brain stem.

sensory loss and abnormal when all sensory modalities are impaired.[31]

SPINOCEREBELLAR DEGENERATION

The finding of slowed conduction in central and peripheral segments of the proprioceptive pathway may aid in differentiating spinocerebellar degeneration from peripheral neuropathy. In vitamin B_{12} deficiency, changes in central conduction usually appear before peripheral slowing, suggesting that the central process of a dorsal root ganglion cell is affected earlier than the distal process.[20] Early treatment lessens the abnormalities of the SEP.

DEMENTIA

Mild abnormalities of SEPs have been described in Huntington's disease and other illnesses associated with dementia.[5a] SEPs are frequently abnormal in multi-infarct dementia and may help distinguish it from Alzheimer's disease, which invariably is associated with normal SEPs.[1]

MYOCLONUS

The amplitude of the cortical SEP may be increased up to 10 times normal in patients with cortical reflex myoclonus[21] (Fig. 16–7). This is thought to reflect hyperexcitability of neurons receiving input from primary somatosensory cortex. Exaggeration of the cortical SEP is not seen in patients with myoclonus of brain stem or spinal cord origin.

MOTOR NEURON DISEASE

Patients with amyotrophic lateral sclerosis may show minor abnormalities of central conduction and amplitude reduction.[5a] Severe abnormalities in patients with suspected motor neuron disease should raise suspicion of other conditions that may mimic amyotrophic lateral sclerosis, for example, cervical spondylosis.

COMA

Several studies have shown that SEPs may help predict eventual outcome in patients with coma due to various causes.[5a] However, it is unclear whether these potentials provide any prognostic information in addition to that obtained by clinical examination and electroencephalography.[14] In brain-dead patients, all SEPs rostral to N13 are absent.[5a]

CHAPTER SUMMARY

SEPs recorded with surface electrodes represent volume-conducted activity arising from myelinated peripheral and central axons, synapses in central gray matter, and changes in the size and shape of the volume conductor. SEPs provide an objective measure of function in large-diameter myelin-

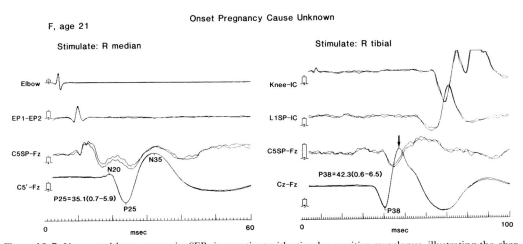

Figure 16–7. Upper and lower extremity SEPs in a patient with stimulus-sensitive myoclonus, illustrating the characteristic pattern of very-high-amplitude cortical potentials recorded from the scalp electrode derivations.

ated sensory afferents peripherally and in proprioceptive pathways centrally. Changes in amplitude and latency can be used to localize lesions in the nervous system, to identify objectively abnormalities in patients with few sensory manifestations or none at all, and to monitor function over time.

ACKNOWLEDGMENT

Illustrations courtesy of J. Clarke Stevens, M.D., Mayo Clinic Scottsdale.

REFERENCES

1. Abbruzzese, G, Reni, L, Cocito, L, et al: Short-latency somatosensory evoked potentials in degenerative and vascular dementia. J Neurol Neurosurg Psychiatry 47:1034–1037, 1984.
2. Aminoff, MJ: Use of somatosensory evoked potentials to evaluate the peripheral nervous system. J Clin Neurophysiol 4:135–144, 1987.
3. Aminoff MJ, Goodin DS, Parry GJ, et al: Electrophysiologic evaluation of lumbosacral radiculopathies: Electromyography, late responses, and somatosensory evoked potentials. Neurology 35:1514–1518, 1985.
4. Bennett, MH and Jannetta, PJ: Evoked potentials in trigeminal neuralgia. Neurosurgery 13:242–247, 1983.
5. Chiappa, KH: Short-latency somatosensory evoked potentials: Interpretation. In Chiappa, KH (ed): Evoked Potentials in Clinical Medicine, ed 2. Raven Press, New York, 1990, (a) pp 399–437; (b) p 400.
6. Desmedt, JE: Generator sources of early somatosensory potentials evoked by upper or lower limb stimulation somatosensory evoked potentials in man. In Morocutti, C and Rizzo, PA (eds): Evoked Potentials: Neurophysiological and Clinical Aspects. Elsevier Science Publishers, Amsterdam, 1985, pp 15–26.
7. Desmedt, JE and Cheron, G: Prevertebral (oesophageal) recording of subcortical somatosensory evoked potentials in man: The spinal P_{13} component and the dual nature of the spinal generators. Electroencephalogr Clin Neurophysiol 52:257–275, 1981.
8. Dinner, DS, Lüders, H, Lesser, RP, and Morris, HH: Cortical generators of somatosensory evoked potentials to median nerve stimulation. Neurology 37:1141–1145, 1987.
9. Dumitru, D and Jewett, DL: Far-field potentials. Muscle Nerve 16:237–254, 1993.
10. Eisen, A, Purves, S, and Hoirch, M: Central nervous system amplification: Its potential in the diagnosis of early multiple sclerosis. Neurology 32:359–364, 1982.
11. Emerson, RG, Seyal, M, and Pedley, TA: Somatosensory evoked potentials following median nerve stimulation. I. The cervical components. Brain 107:169–182, 1984.
12. Frith, RW, Benstead, TJ, and Daube, JR: Stationary waves recorded at the shoulder after median nerve stimulation. Neurology 36:1458–1464, 1986.
13. Giblin, DR: Somatosensory evoked potentials in healthy subjects and in patients with lesions of the nervous system. Ann N Y Acad Sci 112:93–142, 1964.
14. Goodwin, SR, Friedman, WA, and Bellefleur, M: Is it time to use evoked potentials to predict outcome in comatose children and adults? Crit Care Med 19:518–524, 1991.
15. Guérit, JM, and Argiles, AM: The sensitivity of multimodal evoked potentials in multiple sclerosis: A comparison with magnetic resonance imaging and cerebrospinal fluid analysis. Electroencephalogr Clin Neurophysiol 70:230–238, 1988.
16. Halliday, AM and Wakefield, GS: Cerebral evoked potentials in patients with dissociated sensory loss. J Neurol Neurosurg Psychiatry 26:211–219, 1963.
17. Halonen, J-P, Jones, SJ, Edgar, MA, and Ransford, AO: Conduction properties of epidurally recorded spinal cord potentials following lower limb stimulation in man. Electroencephalogr Clin Neurophysiol 74:161–174, 1989.
18. Harper, CM and Daube, JR: Surgical monitoring with evoked potentials: The Mayo Clinic experience. In Desmedt, JE (ed): Neuromonitoring in Surgery. Elsevier Science Publishers, Amsterdam, 1989, pp 275–301.
19. Jeanmonod, D, Sindou, M, and Mauguière, F: Three transverse dipolar generators in the human cervical and lumbo-sacral dorsal horn: Evidence from direct intraoperative recordings on the spinal cord surface. Electroencephalogr Clin Neurophysiol 74:236–240, 1989.
20. Jones, SJ, Yu, YL, Rudge, P, et al: Central and peripheral SEP defects in neurologically symptomatic and asymptomatic subjects with low vitamin B_{12} levels. J Neurol Sci 82:55–65, 1987.
21. Kakigi, R and Shibasaki, H: Generator mechanisms of giant somatosensory evoked potentials in cortical reflex myoclonus. Brain 110:1359–1373, 1987.
22. Kakigi, R, Shibasaki, H, Tanaka, K, et al: CO_2 laser-induced pain-related somatosensory evoked potentials in peripheral neuropathies: Correlation between electrophysiological and histopathological findings. Muscle Nerve 14:441–450, 1991.
23. Katifi, HA and Sedgwick, EM: Evaluation of the dermatomal somatosensory evoked potential in the diagnosis of lumbo-sacral root compression. J Neurol Neurosurg Psychiatry 50:1204–1210, 1987.
24. Kelly, DL Jr, Goldring, S, and O'Leary, JL: Averaged evoked somatosensory responses from exposed cortex of man. Arch Neurol 13:1–9, 1965.
25. Kimura, J, Mitsudome, A, Yamada, T, and Dickins, QS: Stationary peaks from a moving source in far-field recording. Electroencephalogr Clin Neurophysiol 58:351–361, 1984.
26. Landi, A, Copeland, SA, Wynn Parry, CB, and Jones, SJ: The role of somatosensory evoked potentials and nerve conduction studies in the surgical management of brachial plexus injuries. J Bone Joint Surg [Br] 62:492–496, 1980.
27. Phillips, LH II, and Daube, JR: Lumbosacral spinal

evoked potentials in humans. Neurology (NY) 30:1175–1183, 1980.

28. Sedgwick, EM, Katifi, HA, Docherty, TB, and Nicpon, K: Dermatomal somatosensory evoked potentials in lumbar disc disease. In Morocutti, C and Rizzo, PA (eds): Evoked Potentials: Neurophysiological and Clinical Aspects. Elsevier Science Publishers, Amsterdam, 1985, pp 77–88.

29. Turano G, Jones SJ, Miller DH, et al: Correlation of SEP abnormalities with brain and cervical cord MRI in multiple sclerosis. Brain 114:663–681, 1991.

30. Veilleux, M and Daube, JR: The value of ulnar somatosensory evoked potentials (SEPs) in cervical myelopathy. Electroencephalogr Clin Neurophysiol 68:415–423, 1987.

31. Veilleux, M and Stevens, JC: Syringomyelia: Electrophysiologic aspects. Muscle Nerve 10:449–458, 1987.

32. Veilleux, M, Stevens, JC, and Campbell, JK: Somatosensory evoked potentials: Lack of value for diagnosis of thoracic outlet syndrome. Muscle Nerve 11:571–575, 1988.

33. Yiannikas, C: Short-latency somatosensory evoked potentials in peripheral nerve lesions, plexopathies, radiculopathies, and spinal cord trauma. In Chiappa, KH (ed): Evoked Potentials in Clinical Medicine, ed 2. Raven Press, New York, 1990, pp 439–468.

CHAPTER 17

BRAIN STEM AUDITORY EVOKED POTENTIALS IN CENTRAL DISORDERS

Gregory D. Cascino, M.D.

Brain stem auditory evoked potentials (BAEPs) are electrophysiologic studies that usually have abnormal results in patients with lesions involving the auditory portion of cranial nerve VIII (CN VIII) or the auditory pathways in the brain stem or both. BAEPs may be performed in sedated patients or in those under general anesthesia. The rationale for these studies in patients with neurologic disease is the close correlation between specific auditory waveforms and structures in the brain stem. Alterations in auditory conduction may help in localizing lesions.

BAEPs are usually performed with a click stimulus that is delivered to each ear and that activates the peripheral and central auditory pathways. Auditory stimuli cause sequential activity in CN VIII, cochlear nucleus, superior olivary nucleus, lateral lemniscus, and inferior colliculus. Five prominent vertex positive waveforms, numbered I through V, are invariably present in normal subjects after peripheral auditory stimulation. The common BAEP alterations in patients with brain stem lesions include increased I to V interpeak latency, a low-amplitude or absent wave V, and an absence of all waveforms.

BAEPs may be useful in evaluating patients with suspected acoustic neuromas, multiple sclerosis, and brain stem gliomas. The studies also provide prognostically important information about comatose patients. For BAEPs to assess peripheral auditory function, the methodology must be altered.

This chapter reviews the methodology, interpretation, and clinical applicability of BAEPs in patients with neurologic disease.

AUDITORY ANATOMY AND PHYSIOLOGY

Knowledge of the auditory system is essential to understand the structures that are sequentially activated during BAEPs. The auditory system begins with the peripheral auditory apparatus. The cochlea and spiral ganglion must be activated (after monaural

171

stimulation) before the central auditory pathways can be assessed.[4,6,8,11,19] The initial central structure activated is the auditory portion of CN VIII, which enters the brain stem at the pontomedullary junction.[4,6,8,19] Sequential activation involves the cochlear nucleus in caudal pons, the superior olivary complex in caudal to mid-pons, the lateral lemniscus in mid-pons, and the inferior colliculus in caudal midbrain. Hearing in normal subjects is associated with bilateral auditory pathway activation, maximal contralateral to the ear stimulated. BAEPs almost exclusively activate the central auditory pathways ipsilateral to the ear stimulated. BAEPs may be associated with activation of the brain stem pathways involved with sound localization rather than with hearing.

AUDITORY EVOKED POTENTIALS IN NORMAL SUBJECTS

Stimulation of the peripheral auditory apparatus in normal subjects may produce seven vertex positive waveforms, labeled I to VII (note the use of Roman numerals)[4,5,11,15,19] (Fig. 17–1). Waves VI and VII are variably present and so are not useful clinically. Conventional audiometric ear-

phones are used to deliver a click, that is, an electrical square wave. The stimulus that optimally activates the central auditory system is maximal to the click threshold for each ear. Monaural stimulation is used, with the contralateral ear masked by white noise. Binaural stimulation may fail to reveal abnormality in a patient with a unilateral auditory lesion and so its use should be avoided. Usually, the preferred stimulus for waveform recognition is 65 to 70 dB above the click hearing threshold, that is, the click sensation level. The optimal stimulus repetition rate for waveform identification is approximately 10 per second. The appropriate stimulus intensity for clinical studies may also be determined by using wave V, which is recognized at stimulus intensities inadequate to generate the other waveforms. Electrodes are placed on each earlobe (A1 and A2) and at the vertex (CZ) to record the auditory waveforms (see Fig. 17–1). Mastoid electrodes are not routinely used, because of increased muscle artifact.

The specific generators of all the waveforms have not been clarified; however, the anatomic regions activated are known.* Importantly, proposed sites of waveform generation are often based on limited hu-

---

*1,2,4,5,8,11,14,16,19,21

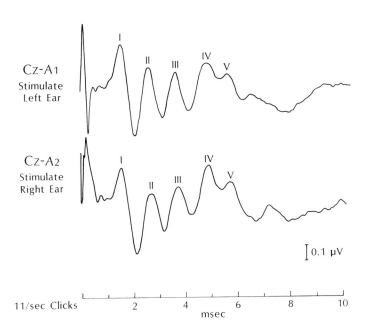

Cz–A1
Stimulate
Left Ear

Cz–A2
Stimulate
Right Ear

⌶ 0.1 μV

11/sec Clicks   2   4   6   8   10
msec

Figure 17–1. Normal brain stem auditory evoked potentials in a 26-year-old man. I to V = characteristic vertex positive waveforms. (From Daube, JR, et al: Medical Neurosciences, ed 2. Little, Brown & Company, Boston, 1986, p 361, with permission.)

man data from patients with brain stem lesions.[2,4–6,10,16,19,21] It is not known whether BAEPs are generated by nuclei or tracts or a combination of the two. *Wave I* represents the distal action potential of CN VIII and appears as a negative potential at the ipsilateral ear electrode. The absence of wave I does not allow an appropriate assessment of central auditory conduction. Supplementary electrodes, for example, a needle electrode in the external auditory canal, may help register this wave if it cannot be recorded with a conventional earlobe electrode. *Wave II* may be generated by either the ipsilateral proximal CN VIII or the cochlear nucleus. *Wave III* is likely related to activation of the ipsilateral superior olivary nucleus. *Wave IV* is produced by activation of the nucleus or axons of the lateral lemniscus. *Wave V* appears to result from activation of the inferior colliculus. *Waves VI* and *VII* are presumed to be generated by the medial geniculate body and the thalamocortical pathways, respectively.

Physiologic variables in normal subjects that may alter BAEPs include age, gender, and auditory acuity.[4,5,15,19,20] Stimulus repetition rate, intensity, and polarity may also affect BAEPs. The age of the patient may affect waveform morphology and latency. BAEPs can be recorded even in premature infants, but the absolute and interpeak latencies are more prolonged than those of older patients[4,8,11,21] (Fig. 17–2). By 2 years of age, the latencies approach the normal values of adult subjects. Persons older than 60 years have a statistically significant increase in BAEP latencies compared with those of younger subjects.[15,20] The BAEP interpeak latencies are significantly shorter in women than in men.

Patients with hearing loss require a higher stimulus intensity to activate the central auditory pathways. Significant peripheral auditory dysfunction may not allow brain stem auditory conduction to be assessed. The absence of waves II to V in patients with a normal wave I and an increased wave I to V interpeak latency cannot be explained on the basis of hearing loss alone. Common BAEP findings in patients with hearing loss are a delay in the absolute latencies of all waveforms, absence of only wave I with delayed waves II to V, or absence of all waveforms.[4]

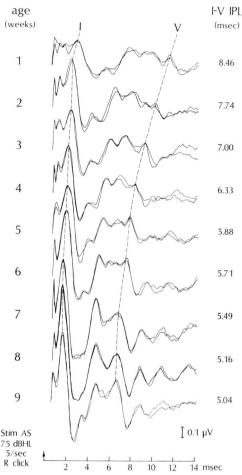

Figure 17–2. Effect of age (weeks) on wave I and wave V absolute latencies in an infant of 28-weeks' gestation. IPL = interpeak latency; Stim AS = stimulation of left ear. (From Stockard, JE and Westmoreland, BF: Technical considerations in the recording and interpretation of the brain stem auditory-evoked potential for neonatal neurologic diagnosis. American Journal of EEG Technology 21:31–54, 1981, p 44, with permission.)

The stimulus variables used may significantly affect the interpretation of the studies. Stimulation rates greater than 10 per second may be associated with a significant increase in absolute and interpeak latencies and a decrease in waveform amplitude. Reducing stimulus intensity affects the BAEPs much like hearing loss does. Stimulus intensities less than 65 to 70 dB above the sensation level increase absolute and interpeak latencies, decrease waveform amplitude, and alter waveform morphology. The polarity of the click produces movement of the earphone diaphragm away from the tympanic membrane (*rarefaction*) or toward the tympanic membrane (*condensation*). The former polarity may be preferred because of an increase in wave I amplitude. A mixture of the two polarities (*alternating*) is not used routinely because of alterations in waveform morphology and interpeak latencies.[4]

METHODS

The performance of BAEPs in patients with suspected neurologic disease begins with assessing peripheral auditory function.[4,5,21] A bedside test of auditory acuity should be conducted, for example, by using a wristwatch at a fixed distance from the patient's ears. The rationale for the study and the proposed methodology should be explained to the patient. The time required for performing BAEPs depends on several clinical factors, but the study can usually be completed in 30 to 45 minutes. The optimal filter bandpass is 100 to 3000 Hz. Averaging is performed on 10 milliseconds of data after auditory stimulation (see Fig. 17–1). The channel derivations include ipsilateral ear to vertex and contralateral ear to vertex. Wave I is identified in the ipsilateral ear derivation as a negative peak at the ear and as a positive peak at the vertex (see Fig. 17–1). The contralateral ear channel may be useful for distinguishing wave IV from wave V. At least two averages of 2000 to 4000 responses are obtained from each ear. Additional trials may be necessary for waveform recognition. If wave I is not identified, the following maneuvers may help: (1) increase stimulus intensity, (2) change the clock polarity to condensation, (3) slow the stimulus rate, or (4) use an external canal supplementary electrode. Sedation with chloral hydrate or diazepam given orally may be used in patients who are unable to relax or when excessive muscle artifact is present.

INTERPRETATION OF AUDITORY EVOKED POTENTIALS

The BAEP variables evaluated in patients with suspected neurologic disease include

measurement of absolute waveform latencies and interpeak latencies (I to III, III to V, and I to V) and determination of wave V/I amplitude ratio.[1,2,4,21] Normative data obtained by similar methods should be available to determine latency and amplitude criteria for abnormal BAEPs. Right and left ear evoked potential studies should only be compared by using identical stimulus variables.

BAEP abnormalities have been observed in several neurologic disorders, including cerebellopontine angle tumors, demyelinating disease, brain stem tumors, brain stem infarcts and hemorrhages, coma, and leukodystrophies.* The sensitivity and specificity of BAEPs in patients with these neurologic diseases have been determined. The commonest neurologic indication for BAEPs is evaluation for suspected acoustic neuroma or multiple sclerosis.[4,6,10,12] The rationale for BAEPs in patients with these suspected disorders is to demonstrate an electrophysiologic alteration indicative of a central nervous system abnormality and to provide information about the anatomic localization of the lesion. BAEPs may be particularly useful as a neurodiagnostic technique performed to "screen" for neurologic disease (e.g., acoustic neuroma in a patient with hearing loss).[10] BAEPs may be useful as a prognostic indicator in patients with coma.[4,7] BAEPs have been used to monitor the response to therapy.[9]

The proposed advantages of BAEPs compared with those of neuroimaging studies are lower cost, less discomfort for patients, shorter waiting period, and an easier ability to obtain sequential studies.[4] One disadvantage of BAEPs is a lack of specificity and sensitivity in patients with brain stem lesions not involving the central auditory pathways.[4] Clinical studies have indicated that BAEPs are complementary to magnetic resonance imaging (MRI) in patients with certain central auditory abnormalities, for example, acoustic neuromas.[4]

The commonest BAEP abnormality in patients with central nervous system disease is a prolonged I to V interpeak latency[1–6,10,16] (Fig. 17–3). Other alterations include the absence of all waveforms, a decreased V/I amplitude ratio, and preservation of wave I with poorly formed waves II to V (Fig. 17–4

---

*1–3,6,7,9,10,12–14,16,18,19

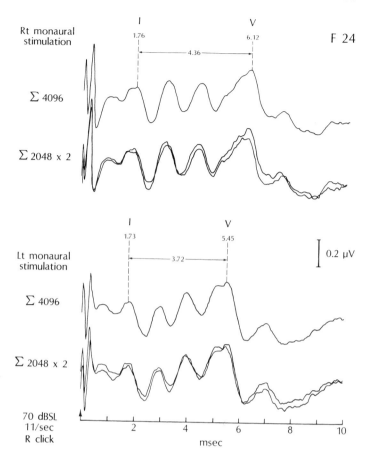

Rt monaural
stimulation

I
1.76

V
6.12

4.36

F 24

Σ 4096

Σ 2048 x 2

I
1.73

V
5.45

3.72

0.2 μV

Lt monaural
stimulation

Σ 4096

Figure 17–3. Patient with multiple sclerosis. Brain stem auditory evoked potentials are abnormal because of prolonged interside I to V interpeak latency. There were no symptoms or signs of brain stem disease in this patient, and the findings on neurologic examinations were unremarkable.

Σ 2048 x 2

70 dBSL
11/sec
R click

2   4   6   8   10

msec

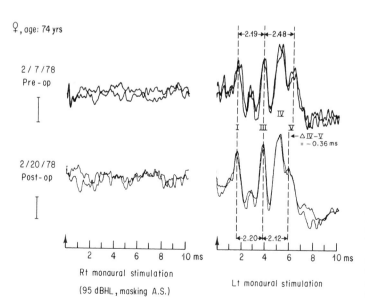

♀, age: 74 yrs

2/7/78
Pre-op

2/20/78
Post-op

2   4   6   8   10 ms

Rt monaural stimulation

(95 dBHL, masking A.S.)

←2.19→←2.48→

IV

I   III   V

←ΔIV–V
= –0.36 ms

←2.20→←2.12→

2   4   6   8   10 ms

Lt monaural stimulation

Figure 17–4. Brain stem auditory evoked potentials preoperatively and postoperatively in a 74-year-old woman with a right acoustic neuroma associated with brain stem compression. The tumor was resected. No response is observed after stimulation of right (Rt) ear either preoperatively or postoperatively. Postoperatively, stimulation of left (Lt) ear shows significant shortening of III to V interpeak latency related to resection of lesion. (From Stockard and Sharbrough,[18] p 245, with permission.)

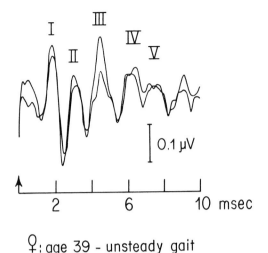

♀: age 39 - unsteady gait

Figure 17–5. 39-year old woman with multiple sclerosis. Brain stem auditory evoked potentials showing combined abnormalities of prolonged I to V interpeak latency and decreased V/I amplitude ratio.

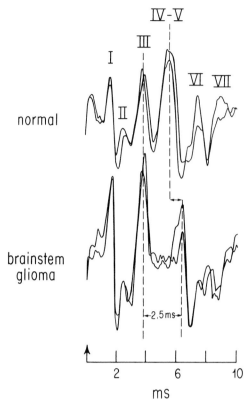

Figure 17–7. Brain stem auditory evoked potentials from a normal subject and from a patient with a brain stem glioma. The latter study shows prolonged III to V interpeak latency indicating a central auditory conduction defect between caudal pons and caudal midbrain.

through 17–6). The I to III and III to V interpeak latencies may be useful in determining the anatomic localization of auditory dysfunction. In patients with a prolonged I to III interpeak latency and a normal III to V interpeak latency, the auditory dysfunction is assumed to be between the distal part of CN VIII (near the cochlea) and the superior olivary nucleus ipsilateral to the ear stimulated. In patients with a prolonged III to V interpeak latency and a normal I to III interpeak latency, the auditory conduction defect is likely located between the superior olivary nucleus and the inferior colliculus ipsilateral to the ear stimulated (Fig. 17–7).

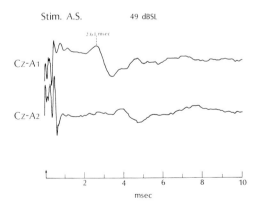

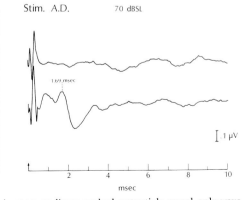

Figure 17–6. 30-year-old woman with multiple sclerosis. Brain stem auditory evoked potentials reveal only wave I bilaterally. Waves II through V are absent. Stim. A.S. = stimulation of left ear; Stim. A.D. = stimulation of right ear.

Acoustic Neuroma

BAEPs are a reliable indicator of the presence of cerebellopontine angle tumors affecting CN VIII (see Fig. 17–4).[4,10] Auditory conduction abnormalities are almost invariably found in patients with an acoustic neuroma, even in asymptomatic persons. BAEPs may be abnormal when the findings of other audiometric studies and even neuroimaging procedures fail to disclose an alteration.[4,10] Chiappa[4] has stated that "BAEPs are the most sensitive screening test when an acoustic neuroma is suspected." The characteristic BAEP changes are prolonged I to V and I to III interpeak latencies ipsilateral to the tumor.[10] The absolute or asymmetrical prolongation of the I to III interpeak latency may be the most sensitive BAEP variable.[4,10] Patients with autosomal dominant neurofibromatosis in whom bilateral acoustic neuromas develop may have normal BAEPs if the tumors are asymptomatic, small, and confined to the intracanalicular region.[4,10] Normal BAEP studies in a patient with symptoms suggestive of an acoustic neuroma, such as dizziness and hearing loss, argue strongly against the diagnosis.[4] MRI has a very low diagnostic yield in patients with normal BAEPs and suspected acoustic neuroma.[4] Chiappa[4] has suggested that neuroimaging procedures are unnecessary in most patients with suspected acoustic neuromas and normal BAEPs (if the studies are done appropriately). Other cerebellopontine angle tumors, for example, meningiomas, may not produce BAEP abnormalities until the lesion is large and involves CN VIII.[10] Thus, the diagnostic yield of BAEPs in patients with these tumors early in the course of the disease is low and neuroimaging is a more important neurodiagnostic technique.[10]

Demyelinating Disease

The frequency of BAEP abnormalities in patients with suspected multiple sclerosis is directly related to the likelihood the person has the disorder and the presence of clinical evidence for brain stem disease.[3,4,6] Abnormal BAEP studies are common in patients with clinically definite multiple sclerosis (see Figs. 17–3, 17–5, and 17–6).[4,6] In one study, 34 of 60 patients (57%) with clinically definite multiple sclerosis and symptoms or signs of brain stem lesions had abnormal BAEPs,[4,6] and 7 of 33 patients (21%) with brain stem disease and possible multiple sclerosis had central auditory conduction defects.[4,6] Importantly, BAEPs may be abnormal in patients suspected of having multiple sclerosis who do not have evidence of brain stem lesions.[4,6] Of patients without symptoms or signs of brain stem disease related to multiple sclerosis, 20 to 50% may have abnormal BAEPs.[6] BAEP results in patients with demyelinating disease include a unilateral or bilateral prolonged I to V interpeak latency and a decreased V/I amplitude ratio (see Figs. 17–3, 17–5, and 17–6). Approximately one third of patients with multiple sclerosis have both an increased interpeak latency and a wave V amplitude abnormality.[6] Characteristically, patients with bilateral central auditory conduction defects do not have auditory symptoms or abnormal click thresholds.[4] The usefulness of BAEPs in such cases includes identifying unsuspected brain stem lesions in patients with an anatomically unrelated disorder such as optic neuritis.

BAEPs may also provide confirmatory evidence of a central nervous system alteration when neurologic evaluation does not suggest the diagnosis of demyelinating disease. BAEPs may implicate brain stem disease in patients with vague, ill-defined, nonspecific symptoms. Potentially, BAEPs may be used in patients with multiple sclerosis to monitor the response to treatment. Ultimately, multiple sclerosis is a clinical diagnosis and the BAEP studies must be carefully interpreted along with the rest of the neurologic evaluation.

Intrinsic Brain Stem Lesions

BAEPs are abnormal in most patients with brain stem tumors, for example, pontine gliomas (Fig. 17–7 and Fig. 17–8).[1,2,4,14,16,18] The findings are similar to those in patients with other central auditory conduction defects (see earlier discussion). Brain stem tumors are often associated with bilateral

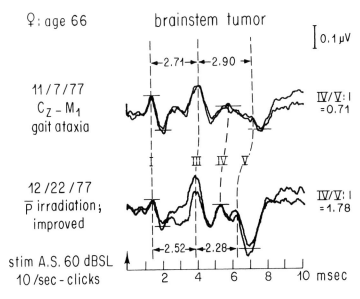

Figure 17–8. Brain stem auditory evoked potentials in a 66-year-old woman before and after ($\bar{p}$) radiation therapy for a brain stem tumor. Note the significant shortening of III to V interpeak latency after treatment. (From Stockard, et al,[21] p 402, with permission.)

BAEP abnormalities (maximal ipsilateral to the lesion).[1] Extra-axial tumors may not be associated with BAEP abnormalities unless there is direct compression and disruption of the brain stem.[10] Ependymomas of the fourth ventricle and cerebellar tumors may also be associated with central auditory conduction defects.[4,19]

The results of BAEP studies are variable in patients with brain stem strokes, that is, infarcts and hemorrhages (Fig. 17–9).[1,2,4,16,19] Abnormal BAEPs have been reported in these patients.[2,4,16,19] BAEPs are normal in patients with strokes that spare the brain stem auditory pathways, for example, infarction of the posterior inferior cerebellar ar-

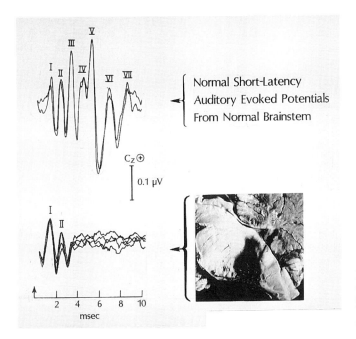

Figure 17–9. (*Top*) Brain stem auditory evoked potentials in a normal subject and (*Bottom*) in a patient who died of brain stem hemorrhage. Note absence of waves III through V in the latter study.

tery (lateral medullary syndrome).[1,4] Brain stem transient ischemic attacks are usually associated with normal BAEPs.[4]

Coma and Brain Death

BAEPs may be normal or abnormal in a comatose patient, depending on the underlying cause and the presence of a brain stem lesion.[4,7,14] BAEPs may provide prognostically important information about outcome. Patients with absence of BAEPs (if peripheral auditory dysfunction can be excluded) are unlikely to survive.[4] Lesions confined to the cerebral hemispheres are not associated with abnormal BAEPs unless the upper brain stem is functionally disrupted. BAEPs may be useful in assessing the integrity of brain stem structures when the findings on neurologic examination are unreliable, for example, after treatment with a high dose of barbiturates or with the patient under general anesthesia.[17] The brain-dead person invariably has abnormal BAEPs, with the characteristic findings of bilateral absence of all waveforms or presence of wave I and absence of waves II to V bilaterally.[4,7,13]

CHAPTER SUMMARY

Brain stem auditory evoked potentials are primarily performed in patients with suspected neurologic disorders to determine if there is evidence of a central nervous system lesion. These sensory evoked potential studies are highly sensitive to auditory conduction defects, but the findings are not pathologically specific. BAEPs provide data that are highly reproducible, objective, and lend themselves to sequential studies for comparison. BAEPs are noninvasive and can be performed not only in the clinical neurophysiology laboratory but also in a hospital room or in the intensive care unit. Patient cooperation is not critical because sedation is permitted. Important factors that need to be considered for accurate interpretation of BAEPs include patient age, gender, and auditory acuity. The diagnostic yield of BAEPs has been confirmed in patients with acoustic neuromas or intra-axial brain stem lesions involving auditory pathways. Finally, this sen-

sory evoked potential modality appears to be complementary to structural neuroimaging studies, for example, magnetic resonance imaging.

REFERENCES

1. Brown, RH Jr, Chiappa, KH, and Brooks, E: Brain stem auditory evoked responses in 22 patients with intrinsic brain stem lesions: Implications for clinical interpretations (abstr). Electroencephalogr Clin Neurophysiol 51:38P, 1981
2. Cascino, GD and Adams, RD: Brainstem auditory hallucinosis. Neurology 36:1042–1047, 1986.
3. Chiappa, KH: Pattern shift visual, brainstem auditory, and short-latency somatosensory evoked potentials in multiple sclerosis. Neurology (7, part 2) 30:110–123, 1980.
4. Chiappa, KH: Evoked Potentials in Clinical Medicine, ed 2. Raven Press, New York, 1990, pp 173–305.
5. Chiappa, KH, Gladstone, KJ, and Young, RR: Brain stem auditory evoked responses: Studies of waveform variations in 50 normal human subjects. Arch Neurol 36:81–87, 1979.
6. Chiappa, KH, Harrison, JL, Brooks, EB, and Young, RR: Brainstem auditory evoked responses in 200 patients with multiple sclerosis. Ann Neurol 7:135–143, 1980.
7. Goldie, WD, Chiappa, KH, Young, RR, and Brooks, EB: Brainstem auditory and short-latency somatosensory evoked responses in brain death. Neurology 31:248–256, 1981.
8. Hecox, K and Galambos, R: Brain stem auditory evoked responses in human infants and adults. Arch Otolaryngol 99:30–33, 1974.
9. Nuwer, MR, Packwood, JW, Myers, LW, and Ellison, GW: Evoked potentials predict the clinical changes in a multiple sclerosis drug study. Neurology 37:1754–1761, 1987.
10. Parker, SW, Chiappa, KH, and Brooks, EB: Brainstem auditory evoked responses (BAERs) in patients with acoustic neuromas and cerebellar-pontine angle (CPA) meningiomas (abst). Neurology 30:413–414, 1980.
11. Picton, TW, Taylor, MJ, and Durieux-Smith, A: Brainstem auditory evoked potentials in pediatrics. In Aminoff, MJ (ed): Electrodiagnosis in Clinical Neurology, ed 3. Churchill Livingstone, New York, 1992, pp 537–569.
12. Purves, SJ, Low, MD, Galloway, J, and Reeves, B: A comparison of visual, brainstem auditory, and somatosensory evoked potentials in multiple sclerosis. Can J Neurol Sci 8:15–19, 1981.
13. Starr, A: Auditory brain-stem responses in brain death. Brain 99:543–554, 1976.
14. Starr, A and Achor, LJ: Auditory brain stem responses in neurological disease. Arch Neurol 32:761–768, 1975.
15. Stockard, JE, Stockard, JJ, Westmoreland, BF, and Corfits, JL: Brainstem auditory-evoked responses: Normal variation as a function of stimulus and subject characteristics. Arch Neurol 36:823–831, 1979.

16. Stockard, JJ and Rossiter, VS: Clinical and pathologic correlates of brain stem auditory evoked response abnormalities. Neurology 27:316–325, 1977.

17. Stockard, JJ, Rossiter, VS, Jones, TA, and Sharbrough, FW: Effects of centrally acting drugs on brainstem auditory responses (abstr). Electroencephalogr Clin Neurophysiol 43:550–551, 1977.

18. Stockard, JJ and Sharbrough, FW: Unique contributions of short-latency auditory and somatosensory evoked potentials to neurologic diagnosis. Progress in Clinical Neurophysiology 7:231–263, 1980.

19. Stockard, JJ, Stockard, JE, and Sharbrough, FW: Detection and localization of occult lesions with brainstem auditory responses. Mayo Clin Proc 52:761–769, 1977.

20. Stockard, JJ, Stockard, JE, and Sharbrough, FW: Nonpathologic factors influencing brainstem auditory evoked potentials. American Journal of EEG Technology 18:177–209, 1978.

21. Stockard, JJ, Pope-Stockard, JE, and Sharbrough, FW: Brainstem auditory evoked potentials in neurology: Methodology, interpretation, and clinical application. In Aminoff, MJ (ed): Electrodiagnosis in Clinical Neurology, ed 3. Churchill Livingstone, New York, 1992, pp 503–536.

CHAPTER 18

AUDIOGRAM AND ACOUSTIC REFLEXES

Wayne O. Olsen, Ph.D.

Hearing tests with pure tone and speech stimuli are used to assess hearing function in patients for whom concerns exist about hearing or balance problems or who complain of difficulties with hearing or communication. These tests can help determine whether there are related balance and hearing problems, can document the basis of hearing complaints, and can help establish diagnosis. On the basis of the patterns of results of these tests, hearing loss can be categorized as follows:

Conductive—abnormality of the ear canal, tympanic membrane, and/or middle ear ossicles

Sensorineural—disorder of the cochlea or CN VIII

Mixed—a combination of conductive and sensorineural disorders

Acoustic reflexes are used primarily to help differentiate sensory (cochlear) from neural (CN VIII) lesions and to evaluate a portion of CN VII.[2,3]

AUDIOGRAM

Basic audiologic tests use pure tones delivered by standard earphones and bone vibrators to assess thresholds of hearing (just barely audible) for air-conducted and bone-conducted stimuli. Such tests are akin to Schwabach's and Rinne's tuning fork tests. However, presentation of electronically generated signals through standard transducers allows testing over a larger frequency and intensity range and with greater precision than is possible using tuning forks. The responses of the patient are plotted on a standardized chart called an "audiogram" and compared with internationally established reference levels for normal hearing.

Figure 18–1 shows a conventional audiogram format with intensity in decibels (dB) of hearing level (HL) (the American National Standards Institute) on the ordinate and frequency (125 to 8000 Hz) on the abscissa. The O-line represents the average hearing sensitivity for young people with normal hearing. The normal range indicated on the right side of the audiogram extends to 25 dB HL because persons with hearing thresholds in this range, at least for frequencies of 500 to 4000 Hz, generally do not report difficulty hearing and understanding conversational speech in quiet surroundings. People whose hearing thresholds are

181

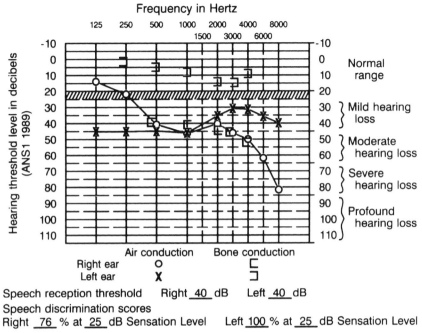

Figure 18–1. Audiogram showing pure tone and speech test results for sensorineural hearing loss (right ear) and conductive hearing loss (left ear). Degree of hearing loss is on right ordinate. ANSI = American National Standards Institute.

1. in the 26 to 45 dB HL region—have a mild hearing loss and have difficulty hearing soft or distant speech
2. in the 46 to 65 dB HL area—have difficulty hearing speech at conversational levels and are considered to have a moderate hearing loss
3. in the 66 to 85 dB HL range—have severe hearing loss, indicating difficulty hearing even loud speech
4. greater than 85 dB HL—have profound hearing loss.

The term "deaf" is reserved for people in group 4, and "hearing impaired" and "hard of hearing" are used to describe groups 1 to 3.

The threshold data shown in Figure 18–1 reveal mild hearing loss in the left ear and, in the right ear, hearing sensitivity that ranges from normal for the lower frequencies to mild hearing loss for the middle frequencies and moderate to severe hearing loss for the higher frequencies. The hearing loss in the right ear is sensorineural, as shown by the interweaving air conduction and bone conduction thresholds. In the left ear, hearing loss is conductive, as indicated by the separation of air conduction and bone conduction thresholds.

Speech Reception Thresholds

Also shown on the audiogram in Figure 18–1 are speech reception thresholds that are based on correct responses to 50% of spondaic words (spondees are words with two syllables of equal stress, such as "airplane" and "baseball") at the hearing levels indicated (40 dB HL). The values in the audiogram indicate a mild hearing loss for speech, as do pure tone thresholds in the 500-Hz to 2000-Hz frequency region.

Speech Discrimination

Ability to understand lists of monosyllables (one-syllable words, such as "ease," "hit," "pew," and "thin," presented singly) at clearly audible levels is shown by the percentage speech discrimination score. In the example shown in Figure 18–1, the lists of monosyllables were presented at a sensation level of 25 dB, that is, 25 dB above speech recognition thresholds. The score of 76% reveals less than perfect understanding of the test items through the right ear. The score for the left ear is perfect (100%), as expected for conductive hearing loss. These results demonstrate that speech loud enough to be heard easily was understood perfectly by the left ear, that is, conductive hearing loss, but not by the right ear, sensorineural hearing loss (see Fig. 18–1). Scores of 90 to 100% indicate that people should have essentially no difficulty understanding speech loud enough to be easily heard. Correct responses to 70 to 88% of the monosyllables in a given list suggest occasional difficulty, 60 to 68% definite difficulty, 40 to 58% marked difficulty, and less than 40% extreme difficulty in understanding speech.

ACOUSTIC REFLEX

Intense stimulation of either ear causes contraction of the stapedius muscle in both ears. This protective response, called the "acoustic reflex," is easily and quickly measured with a device called an "immittance unit" (Fig. 18–2). The system microphone monitors the level of a low-frequency tone (226 Hz) maintained in the space between the probe tip sealed in the ear canal and the eardrum while intense tones (500, 1000, 2000, or 4000 Hz) are presented to the opposite ear or, in some situations, to the same ear as the 226-Hz probe tone. Contraction of the stapedius muscle, which is attached to the neck of the stapes, stiffens the middle ear system, thereby altering the level of the 226-Hz tone maintained between the probe tip and the tympanic membrane. This change is the acoustic reflex response measured by the immittance unit.

Measurement of acoustic reflexes requires an intact eardrum, mobile middle ear ossicles (no conductive hearing loss) (Table 18–1), hearing adequate to allow sufficient stimulation of the ear with at least one of the above-mentioned tones, and intact CN VIII, reflex arc in the brain stem, CN VII, and stapedius muscle attachment to the stapes. Because of the complexity of this system, various response patterns emerge. In conjunction with the case history and other audiologic test results, acoustic reflex testing provides valuable diagnostic information.

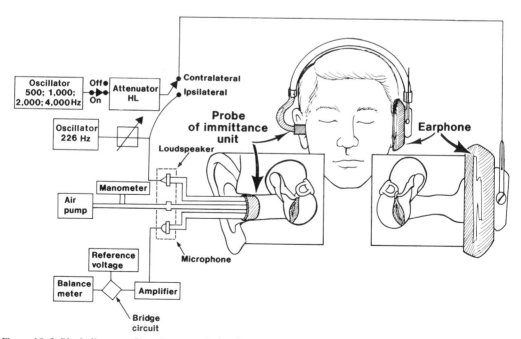

Figure 18–2. Block diagram of immittance unit showing setting for eliciting contralateral acoustic reflexes (stimulus presented through earphone).

TABLE 18–1. AUDIOGRAM AND ACOUSTIC REFLEX FINDINGS FOR VARIOUS CONDITIONS

Condition	AUDIOGRAM			ACOUSTIC REFLEXES	
	Normal Hearing	Conductive Hearing Loss	Sensorineural Hearing Loss	Normal Response	Abnormal Response
Normal	X			X	
Cerumen plug		X			X
Thickened tympanic membrane	X			X	
Perforated tympanic membrane		X			X
Otitis media fluid		X			X
Ossicular discontinuity		X			X
Otosclerosis stapes fixation		X			X
Sensorineural loss					
Cochlea			X	X	
CN VIII			X		X
Facial paralysis					
Medial to stapedial branch CN VII	X				X
Peripheral to stapedial branch CN VII	X			X	

Source. Modified from Keating and Olsen.[2]

CN VIII Versus Cochlear Findings

Absence of acoustic reflexes or response only to very intense tones in an ear with sensorineural hearing loss no worse than severe in degree makes one suspect neural (CN VIII) involvement on the side of the stimulated ear.[1] Similarly, acoustic reflex decay, that is, diminished amplitude of the acoustic reflex response to less than half within 5 seconds to a 500-Hz or 1000-Hz tone delivered at 10 dB above acoustic reflex threshold, suggests a lesion of CN VIII. Elicitation of acoustic reflexes to normal levels of stimulation and absence of reflex decay indicate that the middle and inner ears and acoustic reflex arc are normal or, in the case of a sensorineural hearing loss, indicates a sensory (cochlear) abnormality (see Table 18–1). The sensitivity and specificity levels of acoustic reflex and reflex decay tests are 85%; that is, this test combination correctly identifies lesions of CN VIII and correctly rules out such lesions 85% of the time.

CN VII

Measurement of stapedius muscle contraction on the same side as facial paralysis reveals that the CN VII lesion is distal to the branch that innervates the stapedius muscle. Absence of a reflex response on the same side as facial paralysis indicates that the involvement of CN VII is medial to the stapedius branch of the nerve (see Table 18–1).

Brain Stem

Absence of acoustic reflexes with contralateral stimulation (e.g., stimulating the right ear and measuring the acoustic reflex response in the left ear and vice versa) but their occurrence with ipsilateral stimulation (i.e., stimulating and measuring the response in the same ear) indicates a brain stem lesion that interrupts the crossing acoustic reflex tracts (Fig. 18–3).[4]

CHAPTER SUMMARY

Audiologic testing in the form of puretone air-conduction and bone-conduction audiograms provides diagnostic information about the type of hearing loss—conductive, sensorineural, or mixed—and the degree of hearing loss and attendant communication difficulties. The addition of speech tests that

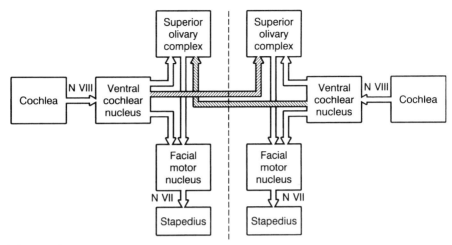

Figure 18–3. Contralateral and ipsilateral acoustic reflex arcs. Crossing tracts for contralateral reflexes are shaded. (From Wiley and Block,[4] p 388, with permission.)

use specific types of speech stimuli directly assesses the patient's ability to hear and to understand speech. Acoustic reflex and reflex decay tests are used to evaluate the integrity of a complicated neural network involving not only the auditory tracts to and through the brain stem but also crossing pathways in the brain stem and CN VII, through to its innervation of the stapedius muscle.

REFERENCES

1. Bauch, CD, Olsen, WO, and Harner, SG: Preoperative and postoperative auditory brain-stem response results for patients with eighth-nerve tumors. Arch Otolaryngol Head Neck Surg 116:1026–1029, 1990.
2. Keating, LW and Olsen, WO: Practical considerations and applications of middle-ear impedance measurements. In Rose, DE (ed): Audiological Assessment, ed 2. Prentice-Hall, Englewood Cliffs, New Jersey, 1978, pp 336–367.
3. Silman, S (ed): The Acoustic Reflex: Basic Principles and Clinical Applications. Academic Press, Orlando, Florida, 1984.
4. Wiley, TL and Block, MG: Acoustic and nonacoustic reflex patterns in audiologic diagnosis. In Silman, S (ed): The Acoustic Reflex: Basic Principles and Clinical Applications. Academic Press, Orlando, Florida, 1984, pp 387–411.

CHAPTER 19

BRAIN STEM AUDITORY EVOKED POTENTIALS IN PERIPHERAL ACOUSTIC DISORDERS

Christopher D. Bauch, Ph.D.

The otoneurologic assessment of patients who, on the basis of complaints of dizziness, hearing loss, or tinnitus, are suspected to have retrocochlear abnormality often includes brain stem auditory evoked response (BAER) testing. Because *conductive hearing losses* resulting from abnormalities of the ear canal, tympanic membrane, or middle ear can affect BAER waveform morphology and latencies, as can *sensorineural disorders* (lesions of the cochlea or CN VIII), the potential effects of hearing losses must be considered in the interpretation of BAER findings.

STIMULI

Clicks are the most effective stimuli in BAER assessment; their short duration, 50 to 100 microseconds, and abrupt onset disperse acoustic energy and provide good synchronization of neural discharges across a broad frequency range. However, the importance of high-frequency hearing sensitivity is accentuated by the spectral characteristics of the earphone and by the response characteristics of the ear canal and middle ear, resulting in greater stimulation in the 2000-Hz to 4000-Hz range. Because this region is stimulated maximally by clicks, routine pure-tone assessment is recommended before BAER evaluation. Hearing losses, particularly in the 2000-Hz to 4000-Hz range, can affect BAER results.[3] Behavioral thresholds for clicks are not an adequate screen for hearing because the click's spectral spread of energy can yield relatively good thresholds despite significant hearing loss in the 2000-Hz to 4000-Hz range.

INTERPRETATION

Three basic measurements are often made in the typical evaluation of BAER waveforms: absolute latencies, wave-V interaural latency differences, and interpeak intervals. BAER wave amplitude may be an unreliable

186

criterion for clinical testing because of marked variations among normal subjects.[3,4]

Absolute Latencies

Absolute latencies of the BAER waves may be influenced by peripheral, that is, conductive or cochlear, hearing loss. Conductive hearing loss reduces the effective stimulus intensity reaching the cochlea and causes absolute latency delays dependent on the degree of conductive impairment. On average, a 0.4-millisecond shift can be expected for each 10 dB of conductive hearing loss.

For cochlear disorders, the hearing thresholds for 2000, 3000, and 4000 Hz are important. As hearing thresholds for these three frequencies become poorer, the latencies of waves I, III, and V increase on a systematic basis. Presumably, this is because of the time delay of the traveling wave in the cochlea reaching more responsive apical (lower frequency) areas and also because of the decrease in effective intensity stimulating the defective cochlea. These factors, and the fact that the cochlea produces more synchronous responses at its high-frequency basal end, lead to latencies that depend on the integrity of high-frequency hearing. Clinical experience has shown that when the average hearing thresholds for 2000, 3000, and 4000 Hz equal or are greater than 60 dB hearing loss, at least 60% of the absolute latencies for waves I, III, and V are abnormal. Knowledge of these tendencies is important in the interpretation of absolute wave V latencies for patients suspected to have a CN VIII abnormality. Wave V latency delays for patients with cochlear hearing losses are often indistinguishable from those of patients with lesions of CN VIII.

Interaural Latency Differences

Another measure, interaural latency differences (ILD), compares wave-V latencies at the two ears of the patient. The advantage of this measure is that the patient serves as his or her own control. Normal variability for ILD is 0.2, 0.3, or 0.4 milliseconds. Larger wave-V latency differences between ears are considered indicative of CN VIII involvement. However, the degree of hearing loss in the 2000-Hz to 4000-Hz range also can influence the validity of such comparisons. When wave-V latency differences between ears exceed the predetermined criterion, the examiner must determine what influence, if any, the hearing loss has on the results. Adjustments in wave-V latency based on various levels of high-frequency hearing loss have been advocated, but the application of these corrections is often misleading and confusing.

Interpeak Intervals

Interpeak intervals reflect the time interval from one neural generator to another. The primary interpeak intervals (I to III, III to V, and I to V) separate a delayed wave-V absolute latency into its peripheral (I to III) and central (III to V) components. Normal I to III and III to V intervals are each approximately 2 milliseconds, which provide an overall I to V interval of approximately 4 milliseconds. One advantage of measuring the interpeak intervals is that they are usually unaffected by moderate-to-severe levels of cochlear or conductive hearing loss. A prolonged I to V interval (longer than 4.54 milliseconds) suggests a retrocochlear lesion, whereas conductive and cochlear losses usually have normal intervals (4.54 milliseconds or less). The main disadvantage to using the interpeak intervals is that wave I cannot always be identified when mild, moderate, or severe peripheral hearing losses exist. In these cases, the examiner must rely on the absolute latencies or ILD measures for interpretation. When all three measures—absolute latency, ILD, and interpeak intervals—are used collectively, the sensitivity of the BAER to CN VIII nerve lesions is approximately 95%, and its specificity for cochlear hearing loss is nearly 90%.

ELECTRODES

Whereas conventional mastoid or earlobe electrodes usually allow recording of waves I, III, and V from patients with normal hearing sensitivity, wave I becomes difficult or impossible to identify in patients with mild, moderate, or severe cochlear hearing loss. An ear canal electrode can enhance wave-I amplitude. The electrode is a disposable, soft, foam plug wrapped in a thin layer of conducting foil. It couples to a transducer through flexible silicon tubing. Such an electrode serves dual roles as a recording electrode and a stimulus delivery system (Fig. 19–1).

In direct comparison with mastoid electrodes, the ear canal electrode improves wave-I amplitude by nearly 100% for patients with normal hearing and from 41 to 127% for patients with mild-to-severe hearing losses. In a large sample of hearing impaired patients, wave I was easily identified 96% of the time with the ear canal electrode compared with 70% of the time with mastoid electrodes.[2]

The primary advantages of the ear canal electrode are that wave-I amplitude is improved for all degrees of hearing loss and that the foam material is compressible, fits comfortably, and prevents collapse of the ear canals. Its disadvantage is that it can be used only once.

APPLICATIONS

Absolute latencies of waves I, III, and V have been compared in patients with CN VIII tumors (tumor group) with patients with cochlear hearing loss (nontumor group) matched for audiometric pure-tone configurations. Mean wave-I absolute latencies are usually similar between the tumor and nontumor groups, but mean latencies for waves III and V are prolonged by as much as 1 millisecond for the tumor group. The range of latencies for waves I, III, and V is also considerably larger for the tumor group than for the patients in the nontumor group who have a similar degree of cochlear hearing loss.

Interaural latency differences that exceed the 0.4-millisecond criterion identify more than 90% of the patients with CN VIII tumors. If the criterion is decreased to 0.3 millisecond, the rate of tumor detection in-

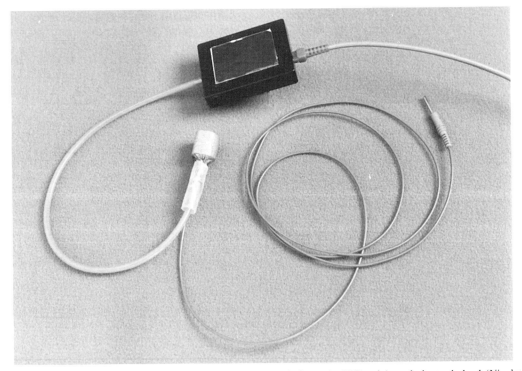

Figure 19–1. Etymotic ER-3A transducer, sound tube, ear canal electrode (TIPtrode), and electrode lead (Nicolet Biomedical Instruments, Inc., Madison, WI).

creases only slightly, and the number of cochlear hearing losses exceeding the 0.3-millisecond criterion increases substantially. The 0.4-millisecond criterion for ILD appears to be a reasonable compromise.[1]

Interpeak intervals have also been compared between the tumor and nontumor groups matched for hearing loss. The mean I to III interval for the tumor groups exceeds that of the nontumor group by approximately 0.6 milliseconds, whereas the mean III to V intervals are similar for both groups. Mean overall I to V intervals are larger by nearly a whole millisecond for the tumor group. Only rarely do I to V interpeak intervals for patients in the nontumor group exceed 4.54 milliseconds.

CHAPTER SUMMARY

In summary, the 2000-Hz to 4000-Hz range of hearing sensitivity is important in BAER assessment. Absolute latencies and interaural latency differences are often af-fected by increasing degrees of peripheral hearing loss in this range, whereas interpeak intervals are relatively stable measures, even for patients with moderate-to-severe degrees of peripheral hearing loss. However, the reduction in amplitude or the absence of wave I associated with peripheral hearing losses often makes the I to III and I to V intervals difficult or impossible to measure. In these cases, use of the TIPtrode can be extremely helpful.

REFERENCES

1. Bauch, CD and Olsen WO: Wave V interaural latency differences as a function of asymmetry in 2,000-4,000 Hz hearing sensitivity. Am J Otol 10:389–392, 1989.
2. Bauch, CD and Olsen, WO: Comparison of ABR amplitudes with TIPtrode and mastoid electrodes. Ear Hear 11:463–467, 1990.
3. Hall, JW III: Handbook of Auditory Evoked Responses. Allyn and Bacon, Boston, 1992.
4. Jiang, ZD, Zhang, L, Wu, YY, and Liu, XY: Brainstem auditory evoked responses from birth to adulthood: Development of wave amplitude. Hear Res 68:35–41, 1993.

CHAPTER 20

VISUAL EVOKED POTENTIALS

Gregory D. Cascino, M.D.

VISUAL SYSTEM ANATOMY AND
   PHYSIOLOGY
VISUAL EVOKED POTENTIALS IN
   NORMAL SUBJECTS
METHODS
INTERPRETATION
LOCALIZATION OF VISUAL CONDUCTION
   LESIONS
Prechiasmatic and Chiasmatic Lesions
Retrochiasmatic Lesions

Visual evoked potentials are highly sensitive to lesions of the optic nerve and anterior chiasm but relatively insensitive to ophthalmologic disorders. Visual evoked potentials are noninvasive studies and allow a quantitative determination of visual function. Importantly, these electrophysiologic studies are sensitive but not pathologically specific. They are usually performed by using a shift of a checkerboard pattern without changing luminance. Monocular visual stimulation is always preferred. In normal subjects, a prominent waveform with a positive polarity is recorded in the posterior head region at a mean latency of approximately 100 milliseconds. This potential, the P1 or P100 wave, is generated by striate and peristriate occipital cortex after visual stimulation. The most common transient visual evoked potential abnormality in patients with anterior visual pathway lesions is a prolonged P1-wave la-

tency. Monocular P1-wave alterations in latency are superior in diagnostic yield to the physical examination findings in patients with optic neuritis. Full-field visual stimulation usually does not reveal abnormality in patients with unilateral retrochiasmatic lesions. This chapter reviews the methodology, interpretation, and clinical applicability of transient full-field pattern-reversal visual evoked potentials in patients with neurologic disease.

VISUAL SYSTEM ANATOMY AND PHYSIOLOGY

Discussion of visual evoked potentials must begin with a review of the relevant anatomy and physiology of the visual system. The visual system functions at several levels, beginning with the retina and terminating in multiple regions of the cerebral cortex.[13,15] Each eye projects to both occipital lobes because of the decussation of the axons from the nasal half of the retina. Important structures involved in visual conduction include the macula, optic nerve, optic chiasm, lateral geniculate body in the thalamus, and thalamocortical pathways. The macula at the posterior pole of the retina is specialized for high-acuity central vision. The primary visual system projects to striate and peristriate areas of the occipital cortex (Brodmann's areas 17, 18, 19). The occipital cortex projects to the midtemporal region and the posterior parietal cortex. Cells in the visual cortex are most sensitive to movement and edges.[23,24] The retina topographically transmits visual information to the occipital cortex. The macula projects to the occipital poles, and more peripheral regions of the retina project to medial calcarine cortex. The type of visual stimulation may affect visual system activation. For example, certain neuronal groups in the retina and lateral geniculate body are primarily involved in detecting visual motion or color. Neurons in the visual cortex also appear to demonstrate these unique electrophysiologic properties.[23,24]

VISUAL EVOKED POTENTIALS IN NORMAL SUBJECTS

A normal transient visual evoked potential to a pattern-reversal checkerboard is a positive midoccipital peak that occurs at a mean latency of 100 milliseconds[13,15] (Fig. 20–1). The waveform consists of three separate phases, with an initial negative deflection (N1 or N75), a prominent positive deflection (P1 or P100), and a later negative deflection (N2 or N145).[15] The numbers used for the waveform designation refer to the approximate latency (in milliseconds) in the control population. The amplitude and latency of the N1 and N2 waveforms are too variable in normal subjects to be useful in interpreting visual evoked potentials in patients with neurologic diseases. Sedation and anesthesia abolish visual evoked potentials; therefore, these studies are not useful for intraoperative neurophysiologic monitoring.[13]

The size of the checks used in the checkerboard pattern may affect the amplitude and latency of the P1 wave.[9,13,15,29] The checks are measured by the degree of visual angle (minutes of arc [′]). Checks that are

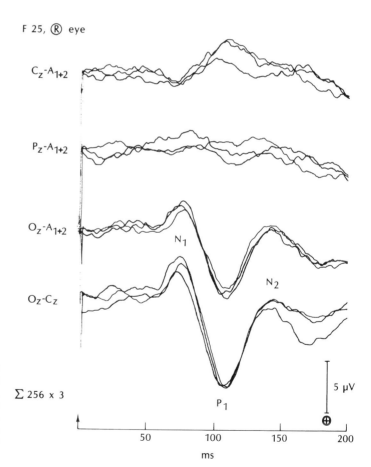

Figure 20–1. Full-field visual evoked potential after stimulation of the right eye in a 25-year-old woman. P1 waveform is maximal in the posterior head region (O_z electrode); P1 amplitude and latency are normal. (From Stockard, et al,[29] p 175, with permission.)

between 28' and 31' are associated with stimulation of the central retina and are usually satisfactory.[13,15] Larger check sizes may be necessary in patients with decreased visual acuity. A decrease in check size to 10' in people with normal acuity is associated with an increase in the amplitude and latency of the P1 wave. The fovea is most sensitive to smaller checks and makes the largest contribution to the P1 amplitude.[15] Unfortunately, smaller check sizes are inherently sensitive to ophthalmologic disorders, including poor visual acuity, and are not used routinely in performing visual evoked potentials.

Visual acuity, pupillary size, age, and gender may alter the P1 waveform in normal subjects.[1,2,9,11,13–16,27–29] In the absence of an alteration in luminance, the visual acuity must be reduced to 20/200 for the P1 latency to be abnormal[14] (Fig. 20–2). The P1

is not prolonged with a visual acuity of 20/200 if large checks, for example, greater than 35', are used.[13] Therefore, subtle changes in visual acuity, for example, 20/40, do not explain significant prolongations of the P1 latency (see Fig. 20–2). Patients who have an asymmetry in pupillary diameter may have interocular differences in the P1 latency.[22,29] Therefore, it is preferred that patients not have their pupils dilated before undergoing visual evoked potential studies.[13,22,29] A miotic pupil may reduce luminance and prolong the P1 latency and reduce the P1 amplitude.[22,29] Age is a significant variable in determining the normal P1 latency.[1,11,13,15,27–29] The P1 latency increases in older persons[1,13,29] and may be more marked in those older than 60 years. Also, women have a shorter P1 latency than men. This factor has to be considered when

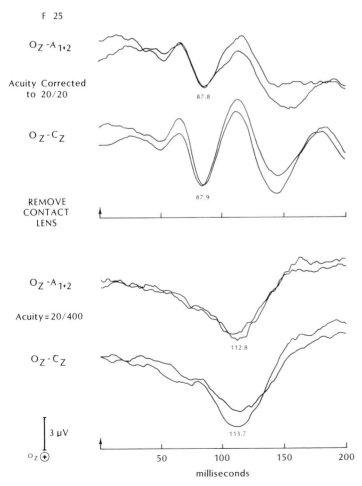

Figure 20–2. Full-field visual evoked potential obtained with and without a contact lens. The P1 amplitude and latency are normal in top tracing. With reduction of visual acuity to 20/400 (bottom tracing), P1 morphology is distorted and P1 latency is prolonged.

deciding on normative data for the P1 latency. Finally, if the normal subject chooses not to look directly at the screen used for visual stimulation, the P1 waveform may be distorted.[13,15] Patient cooperation may be extremely important in a person with psychiatric disease and in very young and old people.

METHODS

Transient visual evoked potential studies are usually performed by using a shift of a checkerboard pattern (black and white).[12,13,15,16,19–22,27,29] There is no change in luminance (total light output) with this form of stimulation.[13,15] Studies that change luminance, for example, pattern-flash or strobe light, produce more variable results in normal subjects and are not as sensitive for detecting abnormalities in visual conduction.[13,15] Flash visual evoked potentials may be appropriate for intraoperative monitoring of visual function.[8] Monocular testing is preferred because binocular stimulation may mask a unilateral visual conduction abnormality. The patient is seated a fixed distance, 70 to 100 cm, from the screen (usually a television monitor) and is asked to focus on the center of the screen. The technician may need to verify that the patient is looking at the screen in certain situations (see discussion later in this chapter). The patient should not be sedated before being examined. Electrodes are placed at C_Z, O_Z, A_1, and F_Z. Other acceptable electrode positions include midoccipital (5 cm above the inion), right and left occipital (5 cm lateral to the midoccipital electrode), and midfrontal (12 cm above the nasion).

The P1 waveform with full-field stimulation is maximal in the midoccipital region but may be well-recorded between the inion and the vertex of the head. The checkerboard pattern is reversed (black to white to black) at a rate of once or twice per second. An increase of the stimulus rate to 4 per second or greater may prolong the P1 latency.[29] Steady-state visual evoked potentials obtained with stimulus rates of 8 to 10 per second are technically more difficult to perform and are not commonly used to evaluate patients with suspected neurologic disease.[13]

The low-frequency filter may range between 0.2 Hz and 1.0 Hz; the high-frequency filter should be 200 Hz to 300 Hz. A sweep length of 200 to 250 milliseconds is used and 100 to 200 responses are averaged. At least two trials should be performed before the identification of the P1 latency. The trials should be reproducible. The American Electroencephalographic Society guidelines recommend a check size of approximately 30′.[2]

The procedure for performance of a test of visual evoked potentials should be explained to the patient before the study is initiated. It is also beneficial if the physician who requested the study indicates to the patient the rationale for the examination. Visual acuity and pupillary size should be determined in each eye of the patient before the study is performed. If appropriate, the patient should wear his or her eyeglasses or contact lenses for the study. Mydriatic drops should not be used before the procedure.

INTERPRETATION

Interpretation of visual evoked potentials in patients with suspected neurologic disease begins with identification of the amplitude and latency of the P1 waveform. Results of visual evoked potential studies in normal subjects should be available in a laboratory to determine whether an absolute P1 latency and the interocular difference in latency are abnormal (Table 20–1). Each evoked potential laboratory should preferably have its own normative data. An acceptable alternative is to use published normal values obtained at a reference laboratory. Before performing visual evoked potential studies, however, at least 20 normal subjects should be examined with methods similar to those of the reference laboratory. P1 latencies and interocular

Table 20–1. **NORMATIVE P1 LATENCY VALUES USED AT THE MAYO CLINIC**

Age	Female	Male
Less than 60 y	<115 ms	<120 ms
60 y or older	<120 ms	<125 ms

♂ Age: 42Yrs

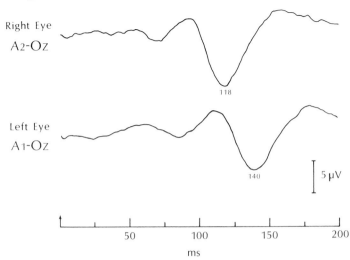

Right Eye
A₂-Oz

118

Left Eye
A₁-Oz

140

5 μV

50   100   150   200

ms

Figure 20–3. Full-field visual evoked potential obtained in a 42-year-old man with multiple sclerosis and optic neuritis in the left eye. P1 latency and amplitude are normal with stimulation of the unaffected right eye. Absolute P1 latency is prolonged and the interocular difference is abnormal with stimulation of the left eye. Note that the amplitude in the lower tracing is preserved. This study suggests an anterior conduction defect on the left. (From Daube, JR, et al: Medical Neurosciences: An Approach to Anatomy, Pathology, and Physiology by Systems and Levels, ed 2. Little, Brown & Company, Boston, 1986, p 413, with permission.)

differences in latencies greater than the mean plus three standard deviations are often used to identify abnormal studies. Absolute amplitude determinations are not particularly useful when interpreting a visual evoked potential study. Luminance and patient visual fixation may affect the amplitude.[20] An interocular difference in amplitude greater than two may be considered abnormal if the asymmetry cannot be explained by technical factors.[15] However, amplitude abnormalities usually occur with latency criteria for an abnormal visual evoked potential. Certain lesions in the visual pathway may distort amplitude more than latency (see later discussion).[18]

In reporting visual evoked potential studies, the anatomic localization of the lesion in

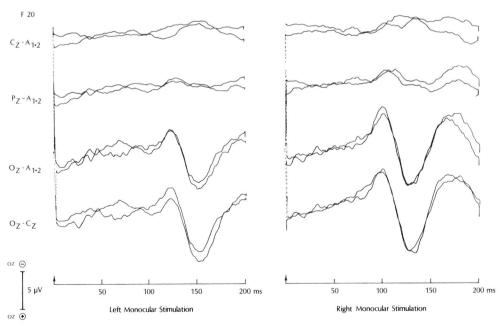

F 20

Cz-A₁₊₂

Pz-A₁₊₂

Oz-A₁₊₂

Oz-Cz

oz ⊖

5 μV

oz ⊕

50   100   150   200 ms

Left Monocular Stimulation

50   100   150   200 ms

Right Monocular Stimulation

Figure 20–4. P1 latencies are prolonged bilaterally, maximal on the left, in patient with a spinal cord lesion who subsequently was shown to have demyelinating disease. (From Stockard, et al,[29] p 193, with permission.)

the visual pathway and the lack of specificity must be emphasized. An anterior visual conduction abnormality indicated by a unilateral prolongation of the P1 latency should not be considered specific for a pathologic process. The most common neurologic disease associated with a unilateral P1 abnormality is demyelinating disease (see discussion later in this chapter); however, other lesions in the anterior visual pathway may produce an identical electrophysiologic alteration. Unilateral P1 abnormalities indicate a visual conduction defect anterior to the optic chiasm[13,15] (Fig. 20–3). Bilateral full-field P1 abnormalities can be seen in patients with bilateral anterior visual conduction lesions or retrochiasmatic disease and do not allow a determination of lesion location (Fig. 20–4).

LOCALIZATION OF VISUAL CONDUCTION LESIONS

Prechiasmatic and Chiasmatic Lesions

Transient full-field visual evoked potentials are highly sensitive to anterior visual conduction lesions. Abnormal visual evoked potentials are commonly found in patients with suspected optic neuritis, multiple sclerosis, or both.[12,17,21,25,26] The rationale for the study is to determine the presence of an electrophysiologic alteration in the anterior visual conduction pathway consistent with optic neuritis. Visual evoked potentials are highly sensitive, but not specific, in patients with optic nerve lesions associated with multiple sclerosis. Importantly, they may be abnormal in a patient without a history of optic neuritis and with normal ophthalmologic findings (see Fig. 20–4). An alteration of visual evoked potentials may be identified in patients with retrobulbar neuritis who characteristically have no abnormality on physical examination. The sensitivity of visual evoked potentials may be superior to that of magnetic resonance imaging (MRI) in patients with optic nerve lesions and demyelinating disease.[17]

Visual evoked potentials should be used to complement other neurodiagnostic studies

and should be correlated with the clinical presentation before the diagnosis of demyelinating disease is made. The results of electrophysiologic studies remain abnormal even several years after the optic neuritis has resolved. Visual evoked potentials may be useful diagnostically in demonstrating a lesion in the optic nerve in patients with suspected multiple sclerosis who have disease localized to the cerebral hemispheres or spinal cord. The incidence of abnormalities of visual evoked potentials depends on the degree of clinical confidence that a patient has multiple sclerosis.[12] Patients with clinically definite or probable multiple sclerosis are more likely to have abnormal visual evoked potentials than those with possible disease. The most common visual evoked potential in patients with optic neuritis is an ipsilateral P1-latency prolongation (Fig. 20–3 and Fig. 20–5). This may be shown by an abnormality in the absolute P1 latency or with a prolonged interocular difference. The amplitude of the P1 wave may be normal even when prolongation in latency is marked. Virtually all patients with clinically demonstrated optic neuritis have unilateral or bilateral abnormalities in visual evoked potentials. Rarely, in acute optic neuritis with a severe alteration in visual acuity, a P1 wave is not recorded.

Tumors compressing the optic nerve and optic chiasm may be associated with a unilateral P1 alteration.[13,18] The P1 latency may be prolonged; however, more commonly, the amplitude is reduced disproportionately to the latency change. The morphology of the visual evoked potential may be markedly distorted, and occasionally the P1 wave may not be recorded. The neoplasms associated with optic nerve compression include optic nerve gliomas, meningiomas, craniopharyngiomas, and pituitary tumors.[13] Giant aneurysms may produce a similar optic nerve lesion. Improvement in P1 waveforms is variable in patients examined after surgical extirpation of the tumor.

Other anterior visual pathway lesions that may be associated with an abnormality in full-field visual evoked potentials include anterior ischemic optic neuropathy, toxic (drug-induced) amblyopia, glaucoma, and Leber's optic atrophy.[6,7,13,15] Correlation of the results of electrophysiologic studies with

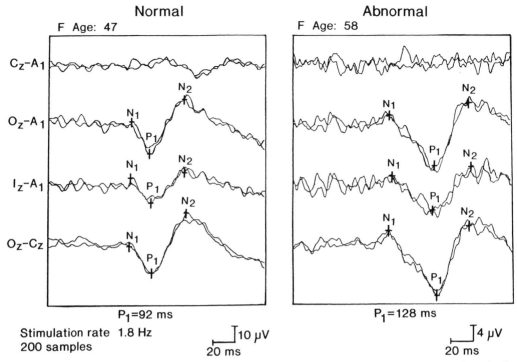

Figure 20–5. Full-field visual evoked potential in a normal subject and in a patient with an anterior visual conduction defect. Note that the P1 latency is prolonged in the patient, with a preservation of P1 amplitude.

the clinical presentation is required to confirm these diagnoses.

Retrochiasmatic Lesions

As indicated previously, recording of full-field visual evoked potentials from the mid-occipital region usually does not reveal any P1 abnormality in patients with unilateral posterior visual conduction defects.[4,10,13] MRI increasingly has been shown to be more useful than full-field transient visual evoked potentials in evaluating patients with these disorders.[25,26] Full-field visual evoked potentials may be normal in patients with abnormal neuroimaging findings retrochiasmatically or visual field defects or both. Bilateral retrochiasmatic disease commonly reveals binocular abnormalities when patients are studied with full-field transient visual evoked potentials.[3] The diagnostic yield of visual evoked potentials is increased with partial-field stimulation in patients with posterior visual conduction defects.[4] Partial-field studies are not commonly performed and require an altered methodology. Partial-field

studies require the additional placement of lateral temporal electrodes (Fig. 20–6). (See Chiappa[13] for a more complete discussion of the methodology and interpretation of partial-field studies.) The clinical applicability of partial-field visual evoked potentials is uncertain because of developments in quantitative visual perimetry and neuroimaging.

Patients with cortical blindness associated with various pathologic processes have been studied with transient visual evoked potentials.[5,10] Importantly, full-field visual evoked potentials have been reported to be normal in patients with blindness and with neuroimaging and pathologic changes confined to the visual cortex.[10] Sensitivity of visual evoked potentials in patients with cortical blindness depends on the anatomy of the cortical lesion and the methodology of the study.[5,10] Lesions involving only Brodmann's area 17 (bilaterally) may be associated with visual loss and normal visual evoked potentials.[10] The use of smaller check sizes is important to identify changes in visual evoked potentials. Patients studied with "normal size" checks, for example, 27′, may have normal visual evoked potentials,[10] but checks

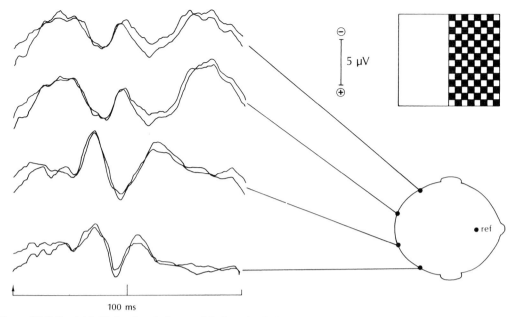

Figure 20–6. Partial-field visual evoked potential after stimulation of the right hemifield. A P1 waveform maximal on the right is present in the posterior head region. (From Stockard, et al, [29] p 200, with permission.)

less than 20′ usually reveal an alteration.[13] Normal visual evoked potentials obtained with large checks in patients with suspected cortical blindness should not be considered evidence for functional visual loss. A normal P1 latency and amplitude in a blind person is highly unusual except for patients with visual cortex disease.[15] Normal findings on a visual evoked potential study virtually exclude an optic nerve or anterior chiasm lesion as the cause of visual loss.[15] As previously noted, with small checks, a significant percentage of patients with retrochiasmatic lesions have changes in visual evoked potentials. Fortunately, most patients with cortical blindness have neuroimaging findings that indicate the anatomy and pathology of the lesion.

CHAPTER SUMMARY

An important question that remains is "What, if any, is the role of visual evoked potentials in evaluating patients with neurologic disease in an era of advanced neuroimaging techniques?" MRI is clearly superior in sensitivity and specificity to visual evoked potentials in patients with retrochiasmatic lesions. There are several important advantages of visual evoked potential studies in patients with lesions involving the optic nerve and anterior chiasm: (1) visual evoked potentials are objective and reproducible and may demonstrate a functional abnormality that is not evident on physical examination or with neuroimaging studies; (2) visual evoked potential abnormalities may persist over time even when there is clinical resolution of visual symptoms; (3) visual evoked potentials may be a more reliable indicator of disease than MRI. MRI may reveal abnormalities that do not represent a pathologic process, such as alterations of the white matter of the cerebral hemisphere; and (4) visual evoked potential studies are less expensive than MRI studies and require minimal patient cooperation. The waiting period at most institutions is also considerably shorter for performing visual evoked potentials.

REFERENCES

1. Allison, T, Hume, AL, Wood, CC, and Goff, WR: Developmental and aging changes in somatosensory, auditory and visual evoked potentials. Electroencephalogr Clin Neurophysiol 58:14–24, 1984.
2. American Electroencephalographic Society: Guide-

lines for evoked potentials. American EEG Society, Bloomfield, Connecticut, 1992.

3. Ashworth, B, Maloney, AFJ, and Townsend, HRA: Delayed visual evoked potentials with bilateral disease of the posterior visual pathway. J Neurol Neurosurg Psychiatry 41:449–451, 1978.

4. Blumhardt, LD, Barrett, G, Kriss, A, and Halliday, AM: The pattern-evoked potential in lesions of the posterior visual pathways. Ann N Y Acad Sci 388:264–289, 1982.

5. Bodis-Wollner, I, Atkin, A, Raab, E, and Wolkstein, M: Visual association cortex and vision in man: Pattern-evoked occipital potentials in a blind boy. Science 198:629–631, 1977.

6. Cappin, JM and Nissim, S: Visual evoked responses in the assessment of field defects in glaucoma. Arch Ophthalmol 93:9–18, 1975.

7. Carroll, WM and Mastaglia, FL: Leber's optic neuropathy: A clinical and visual evoked potential study of affected and asymptomatic members of a six generation family. Brain 102:559–580, 1979.

8. Cedzich, C, Schramm, J, and Fahlbusch, R: Are flash-evoked potentials useful for intraoperative monitoring of visual pathway function? Neurosurgery 21:709–715, 1987.

9. Celesia, GG: Visual evoked potentials and electroretinograms. In Niedermeyer, E and Lopes da Silva, F (eds): Electroencephalography: Basic Principles, Clinical Applications, and Related Fields, ed 3. Williams & Wilkins, Baltimore, 1993, pp 911–936.

10. Celesia, GG, Archer, CR, Kuroiwa, Y, and Goldfader, PR: Visual function of the extrageniculo-calcarine system in man: Relationship to cortical blindness. Arch Neurol 37:704–706, 1980.

11. Celesia, GG, Kaufman, D, and Cone, S: Effects of age and sex on pattern electroretinograms and visual evoked potentials. Electroencephalogr Clin Neurophysiol 68:161–171, 1987.

12. Chiappa, KH: Pattern shift visual, brainstem auditory, and short-latency somatosensory evoked potentials in multiple sclerosis. Neurology 30 (7, Part 2):110–123, 1980.

13. Chiappa, KH: Evoked Potentials in Clinical Medicine, ed 2. Raven Press, New York, 1990, pp 37–171.

14. Collins, DWK, Carroll, WM, Black, JL, and Walsh, M: Effect of refractive error on the visual evoked response. Br Med J 1:231–232, 1979.

15. Epstein, CM: Visual evoked potentials. In Daly, DD and Pedley, TA (eds): Current Practice of Clinical Electroencephalography, ed 2. Raven Press, New York, 1990, pp 593–623.

16. Erin, CW: Pattern reversal evoked potentials. Am J EEG Technol 20:161–165, 1980.

17. Farlow, MR, Markland, ON, Edwards, MK, et al: Multiple sclerosis: Magnetic resonance imaging, evoked responses, and spinal fluid electrophoresis. Neurology 36:828–831, 1986.

18. Halliday, AM, Halliday, E, Kriss, A, et al: The pattern-evoked potential in compression of the anterior visual pathways. Brain 99:357–374, 1976.

19. Halliday, AM and McDonald, WI: Visual evoked potentials. In Stålberg, E and Young, RR (eds): Clinical Neurophysiology. Butterworths, London, 1981, pp 228–258.

20. Halliday, AM, McDonald, WI, and Mushin, J: Delayed pattern-evoked responses in optic neuritis in relation to visual acuity. Trans Ophthalmol Soc UK 93:315–324, 1973.

21. Halliday, AM, McDonald, WI, and Mushin, J: Visual evoked response in diagnosis of multiple sclerosis. Br Med J 4:661–664, 1973.

22. Hawkes, CH and Stow, B: Pupil size and pattern evoked visual response. J Neurol Neurosurg Psychiatry 44:90–91, 1981.

23. Hubel, DH and Wiesel, TN: Receptive fields and functional architecture of monkey striate cortex. J Physiol (Lond) 195:215–243, 1968.

24. Hubel, DH and Wiesel, TN: Functional architecture of macaque monkey visual cortex. Proc R Soc Lond (Biol) 198:1–59, 1977.

25. Paty, DW, Isaac, CD, Grochowski, E, et al: Magnetic resonance imaging (MRI) in multiple sclerosis (MS): A serial study in relapsing and remitting patients with quantitative measurements of lesion size (abstr). Neurology 36 Suppl 1:177, 1986.

26. Paty, DW, Oger, JJF, Kastrukoff, LF, et al: MRI in the diagnosis of MS: A prospective study with comparison of clinical evaluation, evoked potentials, oligoclonal banding, and CT. Neurology 38:180–185, 1988.

27. Sokol, S and Jones, K: Implicit time of pattern evoked potentials in infants: An index of maturation of spatial vision. Vision Res 19:747–755, 1979.

28. Sokol, S, Moskowitz, A, and Towle, VL: Age-related changes in the latency of the visual evoked potential: Influence of check size. Electroencephalogr Clin Neurophysiol 51:559–562, 1981.

29. Stockard, JJ, Hughes, JF, and Sharbrough, FW: Visual evoked potentials to electronic pattern reversal: Latency variations with gender, age, and technical factors. American Journal Electroencephalography Technology 19:171–204, 1979.

CHAPTER 21

COMPOUND MUSCLE ACTION POTENTIALS

Jasper R. Daube, M.D.

The chapters from Chapter 15 to this point have described the assessment of sensory pathways. This chapter begins the discussion of the assessment of motor pathways. For an introduction to this topic, see page 235.

A *compound muscle action potential (CMAP)* is the action potential recorded from muscle when stimulation anywhere along the motor pathway is sufficient to activate some or all the muscle fibers in that muscle. The CMAP is the summated activity of the synchro-nously activated muscle fibers in the muscle innervated by the axons and motor units represented in that muscle. Therefore, a CMAP provides a physiologic assessment of (1) the descending motor axons in the pathway below the level of stimulation; (2) the neuromuscular junction between the axons and the muscle; and (3) the muscle fibers activated by the stimulus. Because disease of the axons, neuromuscular junction, or muscle fiber can alter the CMAP, CMAP recording can be used to assess disease at each of these locations. CMAP recordings are least useful for assessing muscle disease because the potentials are not altered until the disease is either severe or late in its course, when marked atrophy and loss of muscle tissue occur. CMAP assessment for disease of the neuromuscular junction is discussed in Chapter 22 (Repetitive Stimulation Studies). CMAPs are also recorded with motor evoked potentials to assess central motor pathways (Chapter 23). The major application of CMAP recording is in motor nerve conduction studies.

Motor nerve conduction studies and CMAP recordings are equivalent. The rest of this chapter focuses on several aspects of CMAP recording as part of motor nerve conduction studies and their application. The chapter begins with a review of the techniques of stimulation and recording, including technical problems. The next section discusses modifications of the techniques of

199

stimulation and recording to obtain F-wave latencies and is followed by a general discussion of the approach to selecting motor nerve conduction studies and CMAP recording in different clinical entities. The final section summarizes the technical aspects of the recording of individual motor nerves.

CLINICAL APPLICATION

Recording of CMAPs in motor nerve conduction studies is used for several purposes to assist clinicians in assessing neuromuscular disease. CMAPs are particularly useful in providing objective measurements of the extent and type of weakness, particularly when the apparent weakness may be due to hysteria, malingering, or upper motor neuron disease. In these situations, the CMAP is normal. If the weakness is due to a peripheral neuromuscular disease, motor nerve conduction studies can identify and localize the sites of damage, whether from compression, ischemia, or other focal lesion. These studies can also characterize the type of abnormality as a conduction block with neurapraxia or as slowing of conduction at a localized area. They can identify the changes associated with wallerian degeneration and regeneration in the motor nerve. Measurement of CMAP can assist in distinguishing peripheral nerve disease from lower motor neuron disease, neuromuscular junction disease, and myopathies.

In some situations, CMAP recordings can go beyond confirming the presence of disease and the definition of severity by identifying disease that may not be clinically apparent. For example, in patients with clinical evidence of a mononeuropathy, CMAP recording may show signs of multiple mononeuropathies or widespread peripheral nerve damage that may not be apparent clinically. In patients with inherited neuropathies, motor conduction studies can identify the process early in the disease or when there is mild involvement and no clinical evidence of the neuropathy. Motor conduction studies can also identify disease early in its evolution, for example, diabetes mellitus, when a mild peripheral neuropathy may not yet be clinically apparent.[5] In patients with an atypical distribution of deficits, the presence of anomalous innervation can be traced. This is particularly useful for Martin-Gruber anastomosis in the forearm, Riech-Cannieux anastomosis in the hand, the deep accessory branch of the superficial peroneal nerve in the leg, and crossed innervation after reinnervation.

A less common application of CMAP recording is to identify and to measure transient loss of function in primary muscle disease such as periodic paralysis. The recordings can also be used to study abnormal reflex responses in upper motor neuron lesions. In selected patients who have primary muscle disease, a study of the mechanical twitch and its relationship to electrical events may be useful as part of a CMAP recording.

RECORDING CMAPs

Type of Recording Electrode

CMAPs may be recorded with different electrodes that define the size, shape, and, to a lesser extent, latency of the response. Large surface electrodes, small surface electrodes, subcutaneous electrodes, and intramuscular electrodes each have advantages and disadvantages. A 5-mm to 10-mm surface electrode held in place with tape or electrode jelly is most commonly used because of the ease of application, the reproducibility of responses, and the availability of well-defined normal values. Normal values for CMAP recording depend heavily on the type of electrode used for recording; therefore, these values can be used reliably only when the electrodes are identical.

Larger surface electrodes provide a better sample of large muscles or multiple muscles with a common innervation underlying the electrode. However, they also record less well from small muscles and increase the likelihood of recording from more distant muscles. Their use is generally limited to laboratories in which normal values have been developed specifically for them. These electrodes are also used in some laboratories in measuring the number of motor units in a muscle (Chapter 26).

Subcutaneous needle electrodes have the

advantage of being placed closer to the muscle tissue; therefore, they sometimes record a higher amplitude CMAP. For some muscles, these electrodes are easier to apply. The disadvantages are those of any invasive technique, including greater discomfort for the patient and the extremely low risk of infection. The subcutaneous needle recording electrode occasionally has higher impedance, resulting in greater noise, shock artifact, or both.

Intramuscular needle or wire electrodes have the advantage of recording from well-defined small areas of muscle, thereby better isolating the CMAP of individual muscles. Intramuscular recordings are able to record small potentials, particularly of deep muscles that may not be recordable on the skin surface. However, the configuration of the potential varies markedly with the precise location of the recording electrode, which may shift during movement produced by the stimulation. Thus, amplitude and area measurements are not sufficiently reliable to be useful clinically. Latencies may be more difficult to measure because of irregular initiation of the CMAP.

Location of Recording Electrode

CMAP recording is made with an active electrode and a reference electrode whose locations are critical for the size, shape, and, to a lesser extent, latency of the CMAP. Normal values must be determined for specific recording electrode locations. The CMAP is maximal directly over the muscle generating it. The amplitude and area of the action potential decrease with the distance from the muscle. Nonetheless, a CMAP from any limb muscle can be recorded with electrodes far from the muscle but with much lower amplitude.

When recording a CMAP with the active electrode directly over the muscle, the potential is a well-defined, large negative waveform. When recording the CMAP with electrodes that are either off the muscle or at some distance from the muscle that generates the CMAP, the potential is predominantly or initially positive in polarity and much smaller, with a significantly slower rise time to the negative peak. The presence of an initial positivity on a CMAP is evidence that the active electrode is not over the end-plate region of the muscle generating the CMAP and may be entirely off the muscle. The slope or rate of rise of the positive-to-negative peak of the CMAP is a rough gauge of the distance between the recording electrode and the muscle generating the CMAP.

The active electrode ideally is placed over the end-plate region of the muscle or group of muscles being recorded. In some large muscles, the end-plate region is not well defined and such electrode placement is more difficult to obtain. Electrode placement over the end-plate region where the CMAP is initiated results in an initial negativity with a sharp inflection and maximum amplitude that facilitates recording and measuring the CMAP.

Location of the reference electrode, sometimes referred to as the *inactive* or *G2* or *terminal-two electrode*, also has an effect on the amplitude and configuration of the CMAP and, thus, must be the same as the location used to obtain the normal values. Maximum amplitude is generally obtained with the reference electrode over the tendon of the muscle being recorded, especially at the junction of the tendon with the muscle. Placement of the reference recording electrode differs in certain laboratories in one of two ways.

The first variation uses a fixed distance between the active and reference electrodes by using a bar electrode. When the active electrode is over the end-plate region, the reference electrode is over the muscle generating the CMAP. CMAPs recorded in this fashion often have a shorter duration and may have a lower amplitude and smaller area; occasionally, they have a different configuration.

The second variation places the reference electrode at some distance from the active muscle, such as on the tip of a digit. These recordings also significantly alter the amplitude and configuration of the CMAP. Because of the effect of volume change, recordings on the ends of digits are comparable to placement of the recording electrode at the base of the same digit. These recordings include potentials from surrounding muscles, further distorting the CMAP.

STIMULATION

CMAP recording requires stimulating a nerve somewhere along its length. Stimuli can be applied in several ways. The stimulation technique used to activate a nerve affects the normal values. This must be kept in mind when selecting a stimulation technique. The two major types of stimulation are magnetic and electrical. Magnetic stimulation can activate some but not all peripheral nerves (their activation may not be supramaximal). The location of the site of stimulation cannot be defined as precisely with magnetic stimulation as with electrical stimulation. Thus, the measurement of distances and conduction velocity is significantly less reliable with magnetic than with electrical stimulation. The advantage of magnetic stimulation is that it is sometimes easier to obtain and record potentials from deep nerves with less discomfort to the patient. With magnetic stimulation, CMAPs can be evoked by stimulating through clothing and at some distance from the nerve. However, it is not supramaximal for some nerves, and it eliminates the clinical examination that is often helpful.

Type of Stimulating Electrode

Stimulation is applied through a cathode (+) and an anode (−) of various sizes and shapes. Surface electrodes that are placed over the nerve and in parallel with it evoke the most reproducible responses with lowest stimulus intensity. Other types of stimulating electrodes have different advantages and disadvantages, which must be understood to select the optimal electrode for each motor nerve tested. A commonly used electrode is a surface electrode in a handheld stimulator, which allows the electrode to be moved easily in search of the nerve. A smooth, rounded, 5-cm to 10-cm electrode mounted on the end of a curved, removable stimulating pole permits rapid change of the anode and cathode position. This stimulator is most convenient for stimulating nerves that may require pressure on the overlying skin to place the electrode closer to the nerve and for rotating the position of the anode to reduce shock artifact. This type of electrode is optimal for standard motor nerve conduction studies. When stimuli have to be applied for longer periods of time, such as in the operating room for intraoperative monitoring and for measuring motor unit estimates, flat disk electrodes taped on the skin over the nerve or a pair of electrodes mounted in a bar with the electrode protruding from the bar allow more stable positioning of the electrode.

Needle electrodes, either a 1-cm or 2-cm needle that is entirely bare or a longer needle that is bare for 1 to 2 mm at its tip, can also be used to stimulate motor fibers. These electrodes are particularly useful for stimulating deep nerves such as the median and ulnar nerves in the forearm and the tibial and sciatic nerves in the leg. The disadvantages of needle electrodes are that their placement is invasive, and that they are more difficult to move medially and laterally when attempting to find the optimal location for nerve stimulation.

Position of Stimulating Electrode

Depolarization of motor axons occurs at the cathode. The anode hyperpolarizes the nerve and may block conduction of an action potential through the area of hyperpolarization. Activation of a motor axon requires areas of both depolarization and hyperpolarization along the length of the axon, with current flow through the axon between the two locations. Therefore, the optimal position of stimulating electrodes is for the cathode to be as close as possible to the nerve between the anode and the recording site so that the activated action potential does not traverse the area of hyperpolarization at the anode. The optimal location of the anode is longitudinally along the course of the axon away from the recording electrode. Ideally, the anode and cathode are adjacent to the nerve and only a few millimeters apart so that all current flow is directed through the nerve being tested and not into surrounding muscle, another nerve, or other tissue.

Needle electrodes can be placed immediately adjacent to the nerve, but this may require considerable probing in the tissue.

The optimal location of a needle electrode can be obtained with sequential tests of threshold by repeated stimulation. A fixed stimulus is applied twice at two locations. If amplitude increases with the second position, the threshold is reduced, indicating that the electrode is closer to the nerve. The electrode is then moved further in that direction and retested. If amplitude decreases, the threshold has increased because the electrode has been moved away from the nerve. Thus, the electrode has to be moved in a different direction to bring it closer to the nerve. When the anode and cathode are both immediately adjacent to the nerve, stimuli of less than 2 mA are adequate for activating all the motor axons. An anode at some distance from the nerve, either on the surface or elsewhere in the tissue, may be used with the needle cathode near the nerve. This arrangement results in a somewhat higher threshold for activation, a greater risk of current spreading to the surrounding nerves, and a less accurate site of stimulation. These disadvantages are generally outweighed by the advantage of not having to probe the tissue with the anode to find the optimal location near the nerve. The invasive nature of needle stimulation and the time it takes to achieve optimal location of the stimulating electrode have made it less generally accepted than surface stimulation.

The ideal position of surface stimulating electrodes is along the length of the nerve, with the cathode closest to the recording electrode. The anode and cathode must be farther apart than for needle electrode stimulation. If the anode and cathode are too close, current flow passes directly between them without entering the tissue to the depth of the nerve. Thus, activation of all motor axons may not occur despite use of high voltage and the passage of a large current. For most motor nerves, a distance of 3 to 5 cm between the anode and cathode is sufficient for adequate current to penetrate the tissues to the depth of the motor axons. This increases the likelihood of stimulating surrounding nerves, which must be kept in mind as a potential technical problem. For nerves that are very deep in the tissue, a greater distance between the anode and cathode may be necessary, thus further increasing the risk of inadvertent stimulation

of other nerves and muscles. The anode may also be placed perpendicular to the course of the nerve and laterally from it. This position requires a higher current intensity to obtain depolarization, increasing the possibility that adjacent nerves will be stimulated. A lateral position is used for the anode only when this position is necessary for other purposes. The most common need for the lateral position is when the stimulating and recording electrodes are placed so close that there is a prominent shock artifact in the recording. The shock artifact occurs because the current flow from the stimulating electrode spreads through the tissue directly to the recording electrode and charges the capacitance of the intervening tissue, which then discharges over 2 to 20 milliseconds with a waveform superimposed on the CMAP. This occurs especially with stimulation of the tibial nerve at the ankle, the sural nerve at the ankle, and the facial nerve at the angle of the mandible. In these situations, it may become necessary to locate the anode perpendicular to the nerve as the anode is rotated to find a position of minimal shock artifact.

The location of most nerves can be identified reasonably well from anatomic landmarks for each nerve. However, it must always be remembered that the exact location of a nerve can vary significantly among normal subjects. The most striking example is the peroneal nerve at the ankle; its position can vary from 0.5 cm to 4 cm lateral to the tibia. Therefore when attempting to stimulate a motor nerve, the nerve must be localized to minimize stimulus intensity for lessened patient discomfort and to reduce the likelihood of current spread to other nerves. This is best accomplished by placing the stimulating electrodes at the location judged to be the nerve and then obtaining an initial low-amplitude CMAP. The stimulating electrode is then moved medially or laterally perpendicular to the nerve without changing the stimulus intensity. If the subsequent CMAPs have increasing amplitude, the electrode is being moved closer to the nerve. However, if the amplitude decreases, the electrode is being moved away from the nerve. The electrode continues to be moved until the maximal amplitude is obtained with the original stimulus intensity. The voltage is then increased until the CMAP does not increase further with a 25% to 30% increase in applied voltage or current.

In normal subjects, these techniques allow supramaximal or full amplitude CMAPs to be obtained with stimulus intensities less than 20 mA (100 V) in the arm and less than 40 mA (200 V) in the leg. In obese subjects or with particularly deep nerves and in patients with peripheral nerve disease, a greater intensity of current may be needed to activate motor nerves.

The intensity of a stimulus applied to a motor nerve is defined by *total current flow*, which is a function of the intensity of the applied voltage, the resistance to current flow, and the duration of the stimulus. Pulses of 0.1 to 0.2 millisecond are usually adequate for motor nerve stimulation, but longer durations of up to 1 millisecond may be necessary for deep or diseased nerves.

CMAP MEASUREMENTS

The CMAP recorded over any muscle in response to the stimulation of its peripheral nerve is sometimes called the *M wave* (M for motor). It is characterized by several specific measurements. The most valuable measurement is the size of the CMAP, measured either as the amplitude or area of the CMAP. Both these variables reflect the total number of muscle fibers that contribute to the potential. In most laboratories, amplitude is measured from the baseline to the peak, although the measurement can be made from the positive peak to the negative peak. The duration of the response of the CMAP is a function of the synchrony of firing of the components contributing to the potential. A loss of synchrony results in longer duration and lower amplitude. Thus, the area of the CMAP is related most directly to the number of muscle fibers or motor units that contribute to the CMAP.

The latency of the CMAP from the time of stimulation is best measured to the onset of the initial negativity. The latency defines the time it takes the action potential to travel from the stimulation site to the recording site and depends mainly on the conduction time in the peripheral axons. A small amount of time is needed to traverse the neuromuscular junction. If the electrodes are not over the end-plates, latency also includes the time for conduction along the muscle fiber to the recording electrode. In this case, the CMAP initially is positive rather than negative, with the elapsed time to reach the end-plate being the latency of the initial positive deflection. Initial positive deflections may also be due to the recording of a CMAP of a distant muscle, for example, a contribution from the anterior compartment muscles with stimulation of the peroneal nerve at the knee when recording from the extensor digitorum brevis. This initial positivity should not be measured.

Distal latency is the onset of CMAP at the most distal site of stimulation and is best measured as an absolute value. Attempts have been made to correct for slowing in the nerve terminal and at the neuromuscular junction, a measurement called *residual latency*. This method has been reported to be of value in diagnosing early carpal tunnel syndrome. Residual latency is calculated with the formula:

$$RL = DML - \left(\frac{\text{distal distance}}{\text{conduction velocity}} \right)$$

where *RL* is residual latency and *DML* is distal motor latency.

In one study,[11] a normal range for distal motor latency was determined in 100 asymptomatic subjects and applied to patients with "clinical and electrodiagnostic evidence of carpal tunnel syndrome." Residual latency was prolonged in more patients than other measures. However, residual latency was not compared with the more sensitive palmar sensory conduction studies.

Latency measurements should be made from the point at which the negative action potential is initiated, as defined by the nerve action potentials seen with stimulation of the distal site. Latency measurement reproducibility can be enhanced by automated measurement at a fixed voltage above baseline.

The difference in CMAP latency with stimulation at two points along a nerve is a function of the distance between the two points and the rate of conduction of the action potentials in that nerve between the two points.

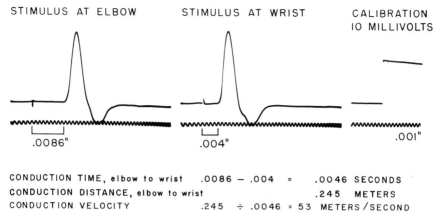

STIMULUS AT ELBOW   STIMULUS AT WRIST   CALIBRATION
10 MILLIVOLTS

.0086"   .004"   .001"

CONDUCTION TIME, elbow to wrist   .0086 − .004 =   .0046 SECONDS

CONDUCTION DISTANCE, elbow to wrist   .245   METERS

CONDUCTION VELOCITY   .245 ÷ .0046 = 53 METERS/SECOND

Figure 21–1. Calculation of conduction velocity from latency and distance measurements on standard nerve conduction studies. (From Daube, JR: Nerve conduction studies. In Aminoff, MJ [ed]: Electrodiagnosis in Clinical Neurology, ed 3. Churchill Livingstone, New York, 1992, p 289, with permission.)

Dividing the difference in CMAP latencies by the distance between the two points measures the conduction velocity of the nerve fibers (Fig. 21–1). Because the latency measurements are made to the initial negativity, the conduction velocity measurement is that of the fastest conducting fibers. Paired stimulation techniques, in which the action potentials in the fast conducting fibers are obliterated by collision, have been used to measure conduction velocity in slower conducting axons. However, the additional clinical data provided by paired stimulation is not sufficiently useful clinically to make it a standard procedure.

CMAP size is also measured as part of repetitive stimulation in response to exercise or drugs used in disorders such as periodic

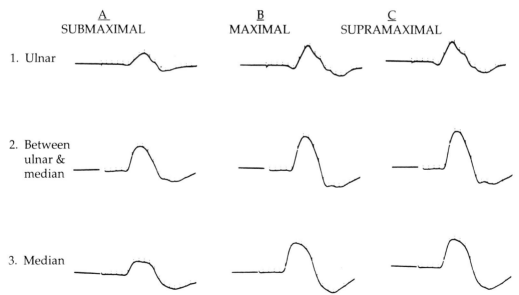

<u>A</u>
SUBMAXIMAL

<u>B</u>
MAXIMAL

<u>C</u>
SUPRAMAXIMAL

1. Ulnar

2. Between
ulnar &
median

3. Median

Figure 21–2. Summation of CMAPs recorded from thenar muscles with stimulation of median and ulnar nerves. Rows 1 and 3 show the CMAP obtained with isolated stimulation of each of the two nerves. Note the initial positivity with ulnar stimulation that results from recording CMAP from ulnar-innervated muscles at a distance from the thenar recording electrode. Row 2 shows the effect of simultaneous stimulation of both median and ulnar nerves, with summation of the potentials recorded in rows 1 and 3.

paralysis and myasthenia gravis (see Chapter 22).

Several potential sources of error must be kept in mind in measuring CMAPs during nerve conduction studies. The commonest one is incorrect measurement of the distance between the two points of stimulation, which may be due to (1) distortion of skin when applying the stimulating electrodes or when making the measurement; (2) non-standard position of the body during the measurement, such as having the elbow extended rather than flexed during ulnar nerve conduction studies; (3) erroneous polarity of the stimulating electrode; and (4) simultaneous stimulation of adjacent nerves (Fig. 21–2). Sources of error in latency measurements include (1) failure to note the sweep speed correctly; (2) a poorly defined shock artifact that interferes with the take-

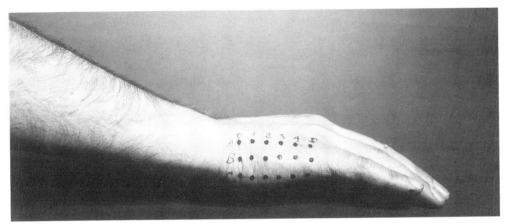

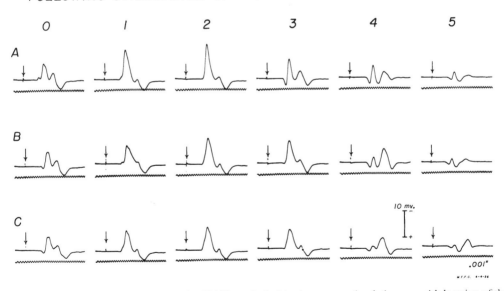

Figure 21–3. The size and configuration of the CMAPs evoked with ulnar nerve stimulation vary with location of the hypothenar recording electrodes. (*Top*) Location of the active recording electrode with the reference electrode on the fifth digit. (*Bottom*) The corresponding CMAPs. (From Carpendale, MTF: Conduction Time in the Terminal Portion of the Motor Fibers of the Ulnar, Median and Peroneal Nerves in Healthy Subjects and in Patients With Neuropathy. Thesis, Mayo Graduate School of Medicine [University of Minnesota], Rochester, 1956, with permission.)

off of the CMAP; (3) incorrect electrode location, resulting in an initial positivity or a poorly defined onset of the negative CMAP (Fig. 21–3); and (4) failure to select the same point on the inflection of the CMAP for the measurement at two points of stimulation. When a CMAP is recorded at two sites of stimulation, it should be very similar at both sites unless there is disease present or anomalous innervation. If the two responses are not similar, technical or physiologic errors must be excluded before the difference is attributed to localized disease. Technical errors can be due to submaximal stimulation at one location or excessive stimulation with activation of an adjacent nerve in another location (see Fig. 21–2). Excessive stimulation at one site may also shorten the latency because of current spread along the nerve.

Values in CMAP Recordings

For clinical reports, values in CMAP recordings should include the actual value measured and the appropriate associated normal values. The normal value can be presented as the range of normal for that age group or three standard deviations from the mean for that age. Because of the skewed distribution of the normal values, the latter method is the less satisfactory description of normal.[7]

F WAVES

F waves are small CMAPs recorded from the muscle fibers of a single motor unit or a small number of motor units activated by antidromic action potentials that travel centrally along motor axons to anterior horn cells.[6] Consequently, the latency of an F wave includes the time required for the action potential to travel antidromically from the site of stimulation to the spinal cord and the time to travel orthodromically from the spinal cord to the muscle. In most muscles, only a small proportion of the motor units is activated antidromically by any one supramaximal stimulus, and the motor units that are activated vary from stimulus to stimulus. Therefore, F waves typically are much lower in amplitude than the directly evoked CMAP. F-wave latencies vary with each stimulus, because axons with different conduction velocities are activated from stimulus to stimulus (Fig. 21–4). As the site of stimulation is moved proximally on a limb, F-wave latency decreases (because the distance the action potential travels decreases) and the M-wave latency increases until the two potentials merge, usually with stimulation at the elbow or just proximal to it (Fig. 21–5). The F-wave latency varies with the distance from the spinal cord to the site of stimulation and to the muscle and with the conduction velocity of the motor fibers (Fig. 21–6).[9]

AMPLIFICATION OF F WAVES
(Tibial nerve/abductor hallucis muscle)

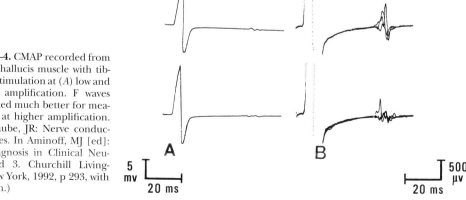

Figure 21–4. CMAP recorded from abductor hallucis muscle with tibial nerve stimulation at (*A*) low and (*B*) high amplification. F waves are depicted much better for measurement at higher amplification. (From Daube, JR: Nerve conduction studies. In Aminoff, MJ [ed]: Electrodiagnosis in Clinical Neurology, ed 3. Churchill Livingstone, New York, 1992, p 293, with permission.)

5 mv

20 ms

A

B

500 µv

20 ms

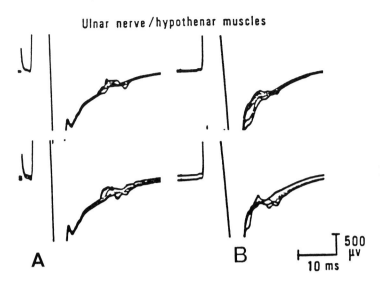

Ulnar nerve/hypothenar muscles

A   **B**

⊤ **500**
| **µV**
├─────┤
10 ms

Figure 21–5. CMAP recorded from hypothenar muscles with ulnar nerve stimulation at (*A*) wrist and (*B*) elbow. The CMAP latency increases whereas the F-wave latency decreases with more proximal stimulation. F waves summate with the late component of the CMAP with proximal stimulation.

Methods

Stimulation applied to the median, ulnar, tibial, or peroneal nerve at the wrist or ankle evokes an F wave that is clearly separated from the M wave. The cathode should be proximal to the anode, and the stimulus should be supramaximal to ensure antidromic activation of all the axons. A series of stimuli is applied until a minimum of eight F waves have been obtained. In some nerves, particularly the peroneal, F waves may be too infrequent for an adequate number to be obtained to make reliable measurements.

Recording electrodes for F waves are placed over the muscle in the standard locations used for motor nerve conduction studies so that F-wave recordings can be made immediately after standard nerve conduction studies are completed. Higher amplification is needed than for standard nerve conduction studies; gains of 200 or 500 mV/cm are usually adequate. The longer latencies of F waves require slower sweep speeds than needed for standard nerve conduction studies. The latency is measured to the earliest reproducible potential in the series recorded. The latency of each of the F waves can be measured and the values plotted as a histogram, but this is time-consuming and does not have any additional value clinically (Fig. 21–7). Different laboratories use different distance measurements. Normal values must be recorded using the same techniques. For this textbook, arm measurements are made from the site of stimulation at the wrist (cathode) to the sternoclavicular joint and leg measurements from the cathode to the xiphoid process.

In measuring F-wave latencies, it is particularly important to pay attention to potential errors. A poorly relaxed muscle may produce deflections throughout the sweep, making it difficult to identify F waves (Fig. 21–8*B*). Late components or satellite potentials on a dispersed compound action potential may incorrectly be identified as F waves. Satellite potentials can be recognized by their constant location and configuration, in contrast to the variable F waves. Also, the latency of satellite potentials increases with more proximal stimulation, whereas F-wave latencies decrease. Axon reflexes (A waves) may resemble F waves in that they decrease in latency with more proximal stimulation (Fig. 21–9, *small arrow*). However, axon reflexes have a more constant configuration and latency than F waves. The criterion for identifying F waves is a response that is variable in latency, amplitude, and configuration but that occurs grouped with a consistent range of latencies. The latency increases with more distal sites of stimulation and has a well-defined onset. F-wave latency measurements are made to the reproducible responses of shortest latency. F-wave latencies

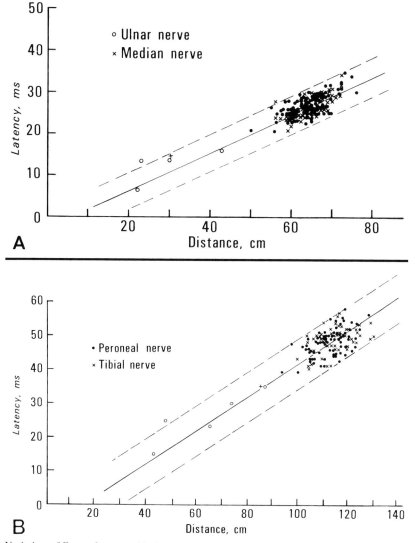

Figure 21–6. Variation of F-wave latency with distance in normal subjects for (A) arm conduction studies and (B) leg conduction studies.

obtained in normal subjects 18 to 88 years old are shown in Table 21–1. The F-wave latency should not be more than 4 milliseconds longer than the estimated F-wave latency (see later discussion).

Several methods have been suggested for assessing F waves, including comparing the latency with normal values corrected for age and distance, calculating the conduction velocity in the central segment, and calculating a central latency and comparing it with an estimated latency based on known conduc-

tion velocity (see Fig. 21–7). The most convenient and readily applied method is to compare F-wave latency with normal values corrected for distance. F waves can be elicited by stimulating any nerve, but they are more prominent in some nerves, for example, in the tibial nerve when recording from foot muscles (hence, "F" for foot).

The rate of stimulation does not affect the F waves, but minimal muscle contraction may enhance them. However, such contraction can make it more difficult to recognize

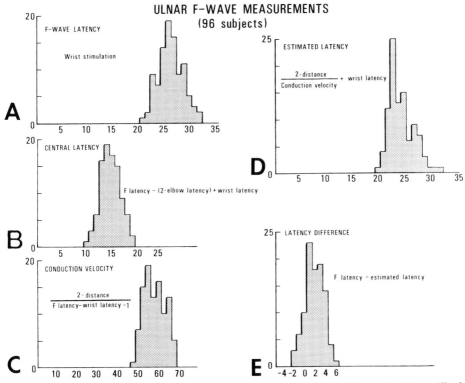

ULNAR F-WAVE MEASUREMENTS
(96 subjects)

A F-WAVE LATENCY — Wrist stimulation

B CENTRAL LATENCY — F latency – (2·elbow latency) + wrist latency

C CONDUCTION VELOCITY — $\dfrac{2 \cdot \text{distance}}{\text{F latency} - \text{wrist latency} - 1}$

D ESTIMATED LATENCY — $\dfrac{2 \cdot \text{distance}}{\text{Conduction velocity}} + \text{wrist latency}$

E LATENCY DIFFERENCE — F latency – estimated latency

Figure 21–7. Calculated values for ulnar F-wave latency based on recordings from 96 normal subjects. All values are derived from the F-wave latencies shown in (*A*) by the calculations shown with each histogram. (*B*) *Central latency* estimates the time from elbow stimulation to return of the F-wave response to the elbow location. (*C*) *Conduction velocity* is the velocity of the F waves over the length of the nerve from wrist to spinal cord. (*D*) *Estimated latency* for the F wave is based on peripheral conduction and distance from the wrist to the sternal notch. (*E*) *Latency difference* compares estimated latency with measured F-wave latency. Proximal slowing alone results in positive differences; distal slowing alone results in negative differences.

Ulnar nerve/hypothenar muscles

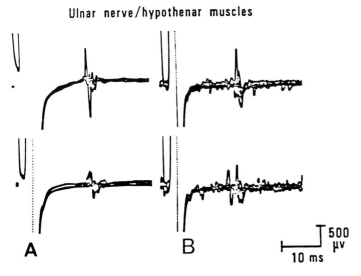

A **B** 500 µV 10 ms

Figure 21–8. F-wave recordings made (*A*) with the muscle at rest and (*B*) with muscle contraction. Reliable measurements are not possible with poor relaxation.

STIMULATE

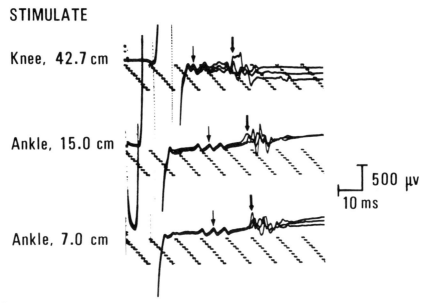

Figure 21–9. Superimposed abductor hallucis CMAP recorded in response to tibial nerve stimulation. Late responses are variable latency and amplitude F waves (*heavy arrow*) and stable axon reflexes (*light arrow*). Both have shorter latency with more proximal stimulation. (From Daube, JR: Nerve conduction studies. In Aminoff, MJ [ed]: Electrodiagnosis in Clinical Neurology, ed 3. Churchill Livingstone, New York, 1992, p 294, with permission.)

F waves. Contraction and relaxation of muscles in another limb or in the jaw may enhance F waves without obscuring them. F waves should be recorded with only supramaximal stimulation; otherwise, they may be confused with H reflexes (Fig. 21–10). F waves are seen best with stimulation at distal sites. The decrease in latency with more proximal stimulation is an important test to ensure that the responses are late responses.

Because F-wave latency varies with distance, it depends on limb length. Measurements of limb length should be made as described for each nerve whenever F waves are recorded. An estimated F-wave latency (F_{EST}) can be calculated on the basis of the distance and conduction velocity in the distal segment with the formula (see Fig. 21–7D):

$$F_{EST} = \frac{2 \times distance}{conduction\ velocity} + distal\ latency$$

F-wave latencies should be within the normal range of F-wave estimates. If they are shorter, proximal conduction is *faster* than distal conduction. If they are longer, proximal conduction is *slower* than distal conduction.

Table 21–1. **F-WAVE LATENCY IN NORMAL SUBJECTS 18 TO 88 YEARS OLD**

	Mean, ms	Range, ms	Distance, cm	Contralateral Difference
Ulnar/ADM	26.6	21–32	50–76	0–3
Median/APB	26.4	22–31	57–73	0–3
Tibial/AHB	48.6	41–57	106–125	0–4
Peroneal/EDB	47.4	38–57	102–128	0–4

ADM = abductor digiti minimi, AHB = abductor hallucis brevis, APB = abductor pollicis brevis, EDB = extensor digitorum brevis.

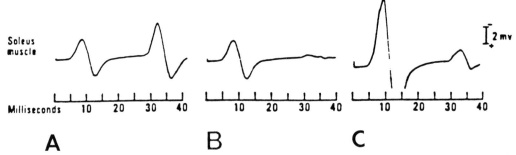

Figure 21–10. Soleus CMAP with tibial nerve stimulation at the knee. (*A*) Low stimulus intensity evokes an H reflex (second wave of higher amplitude than the M wave). (*B*) Stimulation more frequent than 1/s results in reduction of the H reflex. (*C*) Higher stimulus intensities result in lower amplitude H reflexes.

PHYSIOLOGIC VARIABLES

In normal subjects, CMAP measurements made during motor nerve conduction studies vary with several factors. The temperature of the limb is the most significant factor and produces a 2.0 m/s slowing per degree centigrade. Calculations can be made to correct for a cool limb, but it is more reliable and effective to warm a cool limb, which can be done by immersing the limb in a water bath at 40°C for 5 minutes. Arm temperature measured on the surface over the hand should be at least 31°C, and leg temperature measured on the surface over the lateral malleolus should be at least 29°C. More distal sites have lower temperatures. CMAP measurements in patients should always be performed in the same temperature range in which the normal values were determined. Temperature also changes CMAP amplitude and area; both increase with lower temperature.

A second factor is the age of the patient. Conduction velocities are slower in children younger than 3 years than in adults. Also, elderly people, particularly over age 65, on the average show slowing of conduction. This varies with patient population and, therefore, should be compared with the normal population for each patient. CMAP recordings show no significant differences between men and women. The range of normal values is wide, making the measurement of a single value less reliable in identifying mild disease. For example, the range of normal peroneal/extensor digitorum brevis CMAP amplitudes from 2 to 12 mV means that a patient who had a normal amplitude of 10 mV may lose 80% of the response before the value is in the abnormal range. The range of normal values for conduction velocity, latency, and F-wave latency is narrower and, thus, better for identifying mild changes in disease.

There are also significant differences in the normal values for amplitudes and rates of conduction between different nerves, particularly between the upper and lower extremities. Therefore, the significance of any value is evaluated by comparing it with values obtained in the same nerve in a limb at the same temperature of subjects of the same age who participated in the study in which the normal values were determined using the same methods.

The reproducibility of CMAP recordings must also be considered both in identifying abnormality and in comparing values over time when a patient is being monitored. These range from 5% for F-wave latencies in the arm to 15% for CMAP amplitudes in the foot.

CMAP CHANGES IN DISEASE

Pathophysiology

The techniques routinely used to study nerve conduction test large diameter afferent fibers and alpha motor fibers. The compound action potential recorded from peripheral nerves is the action potential from activation of large myelinated fibers. Even the nerve action potential from a mixed nerve is predominantly from large afferent fibers. Components resulting from activation of small myelinated (delta) fibers and C fibers cannot be identified.

MECHANISMS OF CONDUCTION IN MYELINATED FIBERS

Conduction in myelinated fibers is *saltatory*. The action potential jumps from one node of Ranvier to the next. The action potential of one node provides the current that excites the subsequent nodes. Conduction velocity is determined by the time required for one node to excite the next. Thus, if the distance between two nodes (internodal distance) is 1 mm and the nodal conduction time is 20 microseconds, the conduction velocity is 50 m/s. The time required for one node to excite the next node is determined by several factors:

1. The faster the rate of rise of the action potential at node 1, the more rapidly node 2 will be activated.
2. The smaller the amount of current required to neutralize the charge held by the membrane capacitance of node 2 and to depolarize the nodal membrane to threshold, the more rapidly an action potential will appear at node 2.
3. The more current that is lost in neutralizing the charge across the axonal membrane in the internode and by leakage through the myelin, the longer it will take to activate the next node.
4. The higher the resistance to current flow in the axoplasm from node 1 to node 2, the longer it will take to activate node 2.

MECHANISMS OF SLOWING CONDUCTION

Paranodal demyelination increases the capacitance of the internodal membrane. More current is needed to neutralize the charge across the internodal membrane and less is available to discharge node 2. Thus, it takes longer to initiate an action potential at node 2.

Segmental demyelination results in a more profound increase in capacitance and decrease of resistance across the internodal membrane. In large diameter fibers, conduction may be blocked. In smaller diameter fibers, conduction may become continuous, as in unmyelinated fibers, instead of saltatory.[1]

Decreased diameter of fibers occurs with axonal atrophy or compression. It has been observed in a compressed zone and 1 to 2 cm proximal to it. Decreased diameter increases the resistance to flow of current from node 1 to node 2. Reduction in current flow increases the time required to excite node 2. Simultaneous reduction in internodal membrane capacitance because of reduced membrane area does not compensate for the higher resistance.

Loss of large diameter fibers by conduction block or degeneration results in measurement of conduction from smaller diameter, more slowly conducting fibers rather than slowing of conduction in individual fibers. This may account for the slow conduction observed in segments proximal to a focal lesion rather than the abnormality extending proximal to the compression site.

With *decreased myelin thickness*, particularly during remyelination, the number of myelin lamellae is small in proportion to fiber diameter. The capacitance and conductance of the internodal membrane are high, the loss of current through the internode is more than normal, and a longer time is required to excite the next node. Other possible factors in the slowing of conduction are altered characteristics of the nodal membrane which affect the generation of the action potential. No such factor has been identified in focal lesions.

EFFECTS OF NERVE LENGTH, TEMPERATURE, AND AGE

Conduction velocity slows with axonal length. The effect is particularly noteworthy in persons more than 6 feet tall, in whom normal values are significantly slower than in shorter subjects.

Temperature is a greater cause of variation in measurements of conduction velocity than errors in measurement of latency or distance. Between 22°C and 38°C, conduction velocity is related approximately linearly to temperature, increasing about 50% when the temperature is increased 10°C (Q_{10} = 1.5). Thus, a nerve with a conduction velocity of 60 m/s at 36°C conducts at 40 m/s at 26°C (i.e., a decrease of 2 m/s per degree centigrade). The change per degree centigrade is proportionally less for nerves that have a lower conduction velocity.

Age must also be considered in determining the significance of prolonged latencies, slow conduction velocities, and low amplitudes of compound action potentials. Conduction velocity slows progressively between 20 to 30 years of age and by age 80 is about 10 m/s slower.

OTHER EFFECTS OF DEMYELINATION

Demyelination increases the refractory period, decreases the ability of the fiber to conduct impulses at high frequency, and increases the susceptibility to block of conduction with increasing temperature.

FINDINGS IN PERIPHERAL NERVE DISORDERS

The only electrophysiologic findings in peripheral nerve disorders are *conduction slowing*, *conduction block*, and *reduced* or *absent CMAPs*. Each may have a focal or a diffuse distribution. Conduction block can result from a metabolic alteration in the axonal membrane, such as local anesthetic block, or the myelin, such as structural changes of telescoping or segmental demyelination. Segmental demyelination and the narrowing of axons both slow conduction. Reduced or absent responses are the result of wallerian degeneration (after axonal disruption) or axonal degeneration, as in "dying-back" neuropathies.

Large diameter myelinated fibers are the nerve fibers most sensitive to damage by localized pressure. The largest ones are the afferent fibers that mediate touch-pressure, vibration, and proprioception. In a mixed nerve, these fibers generally have larger diameters than alpha motor fibers, as evidenced by their 10% to 15% faster conduction velocity. In a chronic compression lesion, measurement of conduction velocity in the sensory fibers often demonstrates an abnormality before it is evident in the motor fibers.

In a conduction block, the CMAP area obtained with stimulating just proximal to the site of the block is reduced compared with that just distal to the block. Conduction slowing, in contrast, is seen as a prolonged latency. Although conduction block and slowing may be seen together, they often occur independently. Conduction block is more common in rapidly developing disorders, and conduction slowing is more common in chronic disorders. Loss of strength is most accurately quantitated by reduction in CMAP to stimulation at a proximal site. CMAP amplitude and area changes can help in categorizing nerve damage into broad groups. For instance, in traumatic injuries of a nerve, there is usually conduction block (with an area or amplitude change) or axonal disruption (with low amplitude at all stimulation points) or some combination of the two. The clinical deficit caused by either conduction block or axonal disruption may have a variable duration (Table 21–2). The results of nerve conduction studies are a function of the underlying pathologic change, not of the duration of the disorder.

No single change in nerve conduction studies is typical of the clinical phenomenon of *neurapraxia*, in which there is a transient weakness without atrophy. If neurapraxia is due to a metabolic alteration, it lasts only a few minutes; however, if it is caused by axonal distortion with telescoping of internodes, it may persist for weeks or months.[17]

Table 21–2. DURATION OF DEFICIT AFTER PERIPHERAL NERVE INJURY

Injury	Duration of Deficit
Conduction block (amplitude change)	
Metabolic	Seconds to minutes
Myelin loss	Days to weeks
Axonal distortion	Weeks to months
Axonal disruption (fibrillation potentials)	
Few axons	No deficit
Many axons	Weeks to months
All axons	Months to years

Source. From Daube, JR: Nerve conduction studies. In Aminoff, MJ [ed]: Electrodiagnosis in Clinical Neurology, ed 3. Churchill Livingstone, New York, 1992, p 308, with permission.

Table 21–3. **PATTERNS OF ABNORMALITY IN NERVE CONDUCTION STUDIES OF PERIPHERAL NEUROMUSCULAR DISORDERS**

Disorder	MOTOR NERVE STUDIES				SENSORY NERVE STUDIES		
	ACTION POTENTIAL		Conduction Velocity	F-Wave Latency	ACTION POTENTIAL		Conduction Velocity
	Amplitude	Duration			Amplitude	Duration	
Axonal neuropathy	↓	Normal	>70%	Mild ↑	↓ ↓	Normal	>70%
Demyelinating neuropathy	↓ Proximal	↑ Proximal	<50%	↑	↓	↑ Proximal	<50%
Mononeuropathy	↓	↑	↓	↑	↓ ↓	↑	↓
Regenerated nerve	↓	↑	↓	↑	↓	↓	↓
Motor neuron disease	↓ ↓	Normal	>70%	Mild ↑	Normal	Normal	Normal
Neuromuscular transmission defect	(↓)	Normal	Normal	Normal	Normal	Normal	Normal
Myopathy	(↓)	Normal	Normal	Normal	Normal	Normal	Normal

Symbols: ↑ = increase, ↓ = decrease, ↓ ↓ = greater decrease, (↓) = occasional decrease.

Source. From Daube, JR: Nerve conduction studies. In Aminoff, MJ [ed]: Electrodiagnosis in Clinical Neurology, ed 3. Churchill Livingstone, New York, 1992, p 309, with permission.

Some patterns of abnormality are summarized in Table 21–3. CMAP measurements cannot predict how long these abnormalities will last.

Findings in Focal Lesions

Localized peripheral nerve damage is characterized by a low-amplitude CMAP, slow conduction, or a conduction block. The amplitude of the CMAP may be low at all sites of stimulation if wallerian degeneration has occurred. In conduction block, some axons are unable to transmit action potentials through the damaged segment but are functioning distal to it.[3] If all axons are blocked, no response is obtained proximal to the site of the lesion. A block in conduction must be distinguished from slowing in conduction, which may resemble it. In conduction block, there is an abrupt decrease in CMAP amplitude and area over a short segment (Fig. 21–11). Slowing of nerve conduction in some axons in the nerve is associated with a gradual decrease in amplitude as stimulation is moved proximally, because of dispersion of the CMAP. The area of the evoked response remains constant with conduction slowing unless there is also phase cancellation.

Liability to Pressure Palsies

In patients with hereditary neuropathy with liability to pressure palsies, even slight traction or compression of a nerve may cause motor and sensory disturbances. Nerves of the patients' relatives who are unaffected clinically may, however, also have EMG and histologic abnormalities. An increased incidence of pressure palsies has been observed in patients with diabetes mellitus.[16]

Slowing of conduction in Guillain-Barré syndrome (inflammatory polyradiculoneuropathy) is often most marked at sites commonly affected by pressure lesions (e.g., the median nerve at the wrist, the ulnar nerve at the elbow, and the peroneal nerve at the knee). A similar phenomenon has been observed in diphtheritic neuropathy in guinea pigs (plantar nerves).[12] Other conditions, including renal failure, alcoholism, and mal-

PRESSURE PALSY: Amplitude of Muscle Action Potential Evoked by Maximal Stimulation of Nerves and Conduction Velocity of Nerves During Recovery

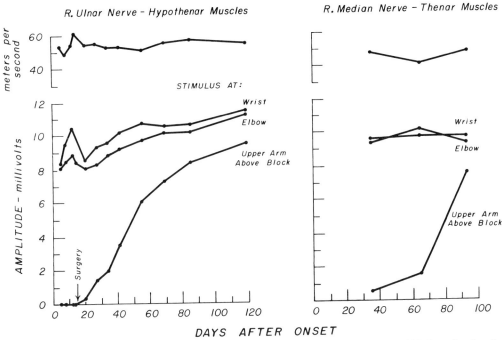

Figure 21–11. Median and ulnar CMAP amplitude and conduction velocity changes over 100 days after localized compression neuropathy. The upper arm conduction block is seen as the marked difference in response amplitude with elbow and upper arm stimulation. Note the normal conduction velocities. (From Daube, JR: Nerve conduction studies. In Aminoff, MJ [ed]: Electrodiagnosis in Clinical Neurology, ed 3. Churchill Livingstone, New York, 1992, p 308, with permission.)

nutrition, have been reported to increase susceptibility to focal compression lesions.

Evaluation of Focal Neuropathies

If a nerve is conducting slowly, it is important to identify whether the abnormality is localized or diffuse. Lesions are best localized by measuring latencies and amplitudes obtained with stimulation (or recording) *over short distances.*

If there is slowing of conduction velocity over any length of nerve (e.g., the median and ulnar nerves in the forearm or the tibial and peroneal nerves in the leg), the severity of slowing must be compared with that of other nerves in the patient. If the slowing is out of proportion to the slowing elsewhere or if there is more than the normal ampli-

tude reduction for that nerve, a localized abnormality must be sought. Localized lesions can be identified by stimulating proximally and distally to the suspected area of localized abnormality (e.g., knee or elbow). If a conduction block or slowing is found between two points of stimulation, the method of *inching* should also be used. Inching begins with supramaximal stimulation in the normal segment just distal to the area of abnormality. The response is saved and stored on the oscilloscope. The point of stimulation is noted, and the same voltage and duration are then applied 2 cm proximal along the nerve. The response is superimposed and compared with the more distal response. Similar stimulations are applied at 2-cm intervals along the nerve to above the area of abnormality. Stimulation with near-nerve needle electrodes can be used for inching in nerves deep in the tissue, for example, the median nerve in the forearm.

A localized area of abnormality is indicated by a greater reduction in amplitude or a greater increase in latency between two adjacent points of stimulation than between other sites. The anatomic location of this point is measured from a fixed landmark (e.g., the medial epicondyle).

DIFFUSE PERIPHERAL NERVE DAMAGE

Nerve conduction studies can distinguish primarily axonal loss from primarily demyelinating nerve disorders.[15] A disorder associated primarily with axonal destruction, as in axonal dystrophies and dying-back neuropathies, is associated predominantly with low-amplitude CMAP at all sites of stimulation, with no more than about 30% slowing in conduction velocity. Segmental demyelination is associated with pronounced slowing of conduction, usually to less than 50% of normal, and with a progressive decrease in the CMAP amplitude at proximal stimulation sites (Fig. 21–12). Slowing of conduction also occurs in severe, chronic axonal disorders with axonal narrowing.

Peripheral Neuropathies

The severity of peripheral nerve damage can be well defined by nerve conduction studies, but the pathologic alteration cannot be predicted for mixed patterns or mild changes. In diabetes mellitus, there is commonly a mild generalized distal neuropathy due to multiple small additive lesions along the nerves. There may also be mononeuropathies of the median nerve at the wrist, the ulnar nerve at the elbow, or the peroneal nerve at the knee, with localized slowing or conduction block superimposed on a generalized reduction in the CMAP amplitude and mild generalized slowing of conduction. Needle examination usually shows only mild changes distally. However, some patients with diabetes mellitus have a lumbosacral polyradiculopathy with diffuse fibrillation potentials in lumbar paraspinal muscles. This pattern may be associated with prolongation of F-wave latencies due to proximal slowing of conduction.

Guillain-Barré syndrome has a spectrum of electrophysiologic changes.[14] There may be no abnormalities on nerve conduction studies or the abnormalities may be limited to proximal slowing with prolongation of the

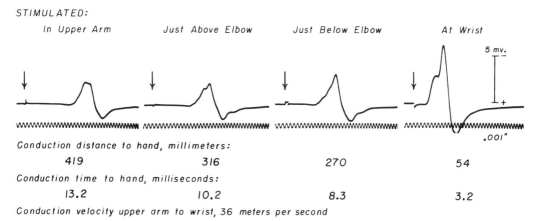

RESPONSE OF HYPOTHENAR MUSCLES TO STIMULATION OF THE ULNAR NERVE

Polyneuropathy

STIMULATED:

| In Upper Arm | Just Above Elbow | Just Below Elbow | At Wrist |

Conduction distance to hand, millimeters:

| 419 | 316 | 270 | 54 |

Conduction time to hand, milliseconds:

| 13.2 | 10.2 | 8.3 | 3.2 |

Conduction velocity upper arm to wrist, 36 meters per second

Figure 21–12. Hypothenar CMAP recorded with ulnar nerve stimulation in a patient with a generalized peripheral neuropathy. Velocity is slow throughout, with gradual dispersion of the CMAP to produce lower amplitudes with proximal stimulation. ↓ = stimulation. (From Daube, JR: Nerve conduction studies. In Aminoff, MJ [ed]: Electrodiagnosis in Clinical Neurology, ed 3. Churchill Livingstone, New York, 1992, p 311, with permission.)

F-wave latency or H-reflex latency. Distal recording with stimulation at proximal sites, such as a spinal nerve or brachial plexus, also may be abnormal. More commonly, however, Guillain-Barré syndrome is associated with prolonged distal latencies of a mild-to-moderate degree, dispersion of CMAP with proximal stimulation, and symmetrical or asymmetrical slowing of conduction velocities. Facial nerves or other cranial nerves may be involved, with abnormalities on blink reflex testing or on facial nerve stimulation. Patients with mild nerve conduction abnormalities and only mild changes on needle examination have a good prognosis; others with low-amplitude CMAPs and prominent fibrillation potentials have severe axonal destruction and a poor prognosis for rapid recovery.

Table 21–4. **PATTERNS OF ELECTROMYOGRAPHIC ABNORMALITY IN CONDITIONS ASSOCIATED WITH PERIPHERAL NEUROPATHY**

Predominant EMG/NCS changes of "axonal degeneration"
    Diabetes mellitus (some patients)
    Guillain-Barré syndrome (some patients)
    Toxic—vincristine, acrylamide, others
    Alcohol
    Uremia
    Acute intermittent porphyria
    Collagen-vascular diseases
    Carcinoma
    Amyloid
Predominant EMG/NCS changes of "segmental demyelination"
    Diabetes mellitus (some patients)
    Guillain-Barré syndrome (some patients)
    Dejerine-Sottas disease
    Diphtheritic neuropathy
    Chronic inflammatory neuropathy
    Refsum disease
    Leukodystrophies
    Neuropathy with monoclonal protein

EMG = electromyographic, NCS = nerve conduction studies.

Source. From Daube, JR: Nerve conduction studies. In Aminoff, MJ [ed]: Electrodiagnosis in Clinical Neurology, ed 3. Churchill Livingstone, New York, 1992, p 312, with permission.

Although many patients with neuropathy show mixed findings on nerve conduction studies, some patients may have a predominantly axonal or segmental demyelinating neuropathy (Table 21–4).

AXONAL NEUROPATHIES

Axonal neuropathies primarily affect the axon and produce either diffuse degeneration or dying-back of the distal portion of the axon. Axonal damage is particularly common with toxic and metabolic disorders. The major change on nerve conduction studies is a decrease in the amplitude of the CMAP or the compound nerve action potential (or both). This reduction is proportional to the severity of the disease. Some axonal neuropathies, such as those associated with vitamin B_{12} deficiency, carcinoma, and Friedreich's ataxia, chiefly affect sensory fibers; others, such as lead neuropathies, have a greater effect on motor fibers. Sensory axons are commonly involved earlier and more severely than motor axons. Occasionally, sensory potentials can be low amplitude and associated with only mild sensory symptoms. In contrast to the change in amplitude, there is usually little change in latency or conduction velocity in axonal neuropathies; conduction in individual axons generally is normal until the axon has degenerated. Therefore, normal conduction velocities should not be considered to be evidence against the presence of a neuropathy. Often, the only finding in a case of axonal neuropathy will be fibrillation potentials on needle examination of distal muscles, especially intrinsic foot muscles.

If many large axons are lost because of axonal neuropathy, conduction velocity may be decreased but not to less than 70% of normal. Axonal neuropathies typically affect the longer axons earlier and are first seen in the lower extremities. Nerves that are more susceptible to local trauma because of their superficial location are also more sensitive to axonal damage; typically, axonal neuropathies are first manifested as peroneal neuropathies with low-amplitude or absent responses, whereas other motor nerves remain intact. Axonal neuropathies may be associated with a change in the refractory period of the nerve and with a relative resistance to ischemia.

SEGMENTAL DEMYELINATING NEUROPATHIES

Segmental demyelinating neuropathies are usually subacute inflammatory disorders, such as Guillain-Barré syndrome, chronic inflammatory demyelinating polyradiculoneuropathy, and diphtheritic neuropathies. However, similar patterns may be seen in chronic hypertrophic neuropathies, for example, Dejerine-Sottas disease and hereditary motor sensory neuropathy. Demyelinating neuropathies typically are associated with prolonged latencies and a pronounced slowing of conduction, often in the range of 10 to 20 m/s.[10] Commonly, the amplitude is relatively preserved on distal stimulation but is decreased proximally because of dispersion of the CMAP on proximal stimulation. In some hereditary disorders, such as Dejerine-Sottas disease, the velocity may be only a few meters per second. Acquired demyelinating neuropathies commonly affect sites of nerve compression early and produce asymmetrical neuropathies of the peroneal, ulnar, or median nerves at the knee, elbow, or wrist, respectively. One form of demyelinating neuropathy with multifocal motor conduction block may superficially resemble amyotrophic lateral sclerosis. The refractory period in demyelinating neuropathies is decreased, often to the extent that repetitive stimulation at rates as low as 5 Hz causes a decrement. The decrement usually does not appear until rates of 10 or 20 Hz are used.

At times, the pattern of abnormality in demyelinating neuropathies helps differentiate an acquired from a hereditary process.[13] An acquired demyelinating neuropathy has scattered areas of slowing, with some areas being much more abnormal than others; hereditary disorders generally have a symmetrical pattern. Acquired demyelinating disorders often show more dispersion with proximal stimulation than hereditary disorders do. This distinction is not always reliable because some patients who have a hereditary demyelinating neuropathy with a low-amplitude CMAP may have pronounced dispersion at proximal sites of stimulation.

Focal Neuropathies

Changes on nerve conduction studies in mononeuropathies vary with the rapidity of development, the duration of damage, and the severity of damage as well as with the underlying disorder.[8] With a chronic compressive lesion, localized narrowing or paranodal or internodal demyelination produces localized slowing of conduction. Narrowing of axons distal to a chronic compression slows conduction along the entire length of the nerve. Telescoping of axons with intussusception of one internode into another distorts and obliterates the nodes of Ranvier and thus blocks conduction. Moderate segmental demyelination and local metabolic alterations are often associated with conduction block. The segment of nerve with dis-

Table 21–5. COMPOUND ACTION POTENTIAL AMPLITUDE AFTER PERIPHERAL NERVE INJURY*

	AMPLITUDE		
Injury	0–5 Days	After 5 Days	Recovery
Conduction block			
Proximal stimulation	Low	Low	Increases
Distal stimulation	Normal	Normal	Normal
Axonal disruption			
Proximal stimulation	Low	Low	Increases
Distal stimulation	Normal	Low	Increases

* = Supramaximal stimulation.

Source. From Daube, JR: Nerve conduction studies. In Aminoff, MJ [ed]: Electrodiagnosis in Clinical Neurology, ed 3. Churchill Livingstone, New York, 1992, p 314, with permission.

Table 21–6. **FINDINGS ON NEEDLE EXAMINATION AFTER PERIPHERAL NERVE INJURY**

	0–15 Days	After 15 Days	Recovery
Conduction block			
Fibrillation potentials	None	None	None
Motor unit potentials	↓ Recruitment	↓ Recruitment	↓ Recruitment
Axonal disruption			
Fibrillation potentials	None	Present	Reduced
Motor unit potentials	↓ Recruitment	↓ Recruitment	Nascent

↓ = decrease.

Source. From Daube, JR: Nerve conduction studies. In Aminoff, MJ [ed]: Electrodiagnosis in Clinical Neurology, ed 3. Churchill Livingstone, New York, 1992, p 314, with permission.

ruption of the axons distal to an acute lesion may continue to function normally for as long as 5 days. Then, as the axons undergo wallerian degeneration, their conduction ceases, and the amplitude of the evoked response diminishes and finally disappears. One week after an acute injury, the amplitude of the evoked response can be used as an approximation of the number of intact, viable axons (Table 21–5).

The evolution of electrophysiologic changes after peripheral nerve injury is also seen on needle examination and is an aid in characterizing mononeuropathies (Table 21–6). Therefore, adequate assessment of a peripheral nerve injury should include both needle examination and nerve conduction studies. The significance of changes with time after injury is outlined in Table 21–7. The sequence of changes shows that nerve

Table 21–7. **ELECTROMYOGRAPHIC INTERPRETATIONS AFTER PERIPHERAL NERVE INJURY**

Finding	Interpretation
0–5 days	
Motor unit potentials present	Nerve intact, functioning axons
Fibrillations present	Old lesion
Low compound action potential	Old lesion
5–15 days	
Compound action potential distal only	Conduction block
Low compound action potential	Amount of axonal disruption
Motor unit potentials present	Nerve intact
After 15 days	
Compound action potential distal only	Conduction block
Motor unit potentials present	Nerve intact
Fibrillation potentials	Amount of axonal disruption
	Distribution of damage
Recovery	
Increasing compound action potential	Block clearing
Increasing number of motor unit potentials	Block clearing
Decreasing number of fibrillation potentials	Reinnervation
"Nascent" motor unit potentials	Reinnervation

Source. From Daube, JR: Nerve conduction studies. In Aminoff, MJ [ed]: Electrodiagnosis in Clinical Neurology, ed 3. Churchill Livingstone, New York, 1992, p 315, with permission.

conduction studies can be important in assessing localized nerve injury within the first few days after injury.

MEDIAN NEUROPATHIES

The most common focal mononeuropathy is *carpal tunnel syndrome*, in which the median nerve is compressed in the space formed by the wrist bones and the carpal ligament. Early or mild compression of the median nerve in the carpal tunnel may not show any electrophysiologic abnormalities, especially of CMAP. However, more than 90% of symptomatic patients have localized slowing of conduction in sensory fibers. The sensory latency through the carpal tunnel is the most sensitive measurement for identifying the earliest abnormality. This so-called *palmar latency* may be compared with normal values but is more reliable when compared with the latency in ulnar sensory fibers over the same distance. More severe nerve compression decreases the amplitude of the sensory nerve action potential and prolongs the latency to a greater extent and over a longer distance. Severe median neuropathy at the wrist also increases the distal motor latency to the thenar muscles and decreases the thenar CMAP (Fig. 21–13). Reduction of the CMAP is often associated with mild slowing

of motor conduction velocity in the forearm and with fibrillation potentials in the thenar muscles. Carpal tunnel syndrome of moderate severity is often associated with anomalous innervation of the thenar muscles, with the amplitude of the response being higher on elbow stimulation than on wrist stimulation.

Many patients with carpal tunnel syndrome have bilateral abnormalities on nerve conduction studies, even though the symptoms may be unilateral. Therefore, the conduction in the opposite extremity should be measured if a median neuropathy at the wrist is identified. A few patients have a normal sensory response and a prolonged distal motor latency. Chronic neurogenic atrophy from a proximal lesion, such as damage to a spinal nerve or anterior horn cells, can result in distal motor slowing and a normal sensory response. A radial sensory response may also be evoked inadvertently by high-voltage stimulation of the median nerve and recorded as an apparent median sensory potential. Occasionally, patients have sensory branches that innervate one or more fingers which are anatomically separated from the motor fibers and relatively spared. The severity of compression may also not be the same for all the fascicles of the median nerve, which would result in greater slowing

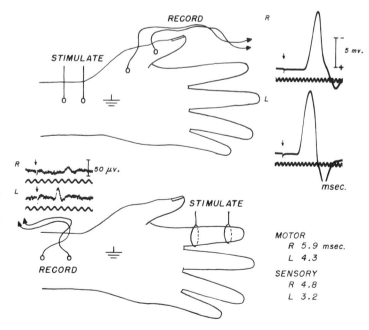

Figure 21–13. Right and left median nerve conduction studies with carpal tunnel syndrome. Distal latency is prolonged on the right. (From Department of Neurology, Mayo Clinic and Mayo Foundation for Medical Education and Research: Clinical Examinations in Neurology, ed 6, Mosby-Year Book, St. Louis, 1991, p 434, with permission.)

MOTOR
  R 5.9 msec.
  L 4.3

SENSORY
  R 4.8
  L 3.2

in the axons to some digital nerves than to others. A median neuropathy may be an early finding in patients with more diffuse neuropathies. To exclude this possibility, it is necessary to assess other nerves.

Median neuropathies in the forearm are much less common and only rarely show abnormality on nerve conduction studies, other than slightly low-amplitude sensory or motor responses (or both).[4] Anterior interosseous neuropathy and pronator syndrome are usually manifested electrophysiologically by fibrillation potentials in the appropriate muscles. Infrequently, patients have localized slowing of conduction in the damaged segment of nerve.

ULNAR NEUROPATHIES

Findings in ulnar neuropathies vary with the severity and location of the lesion.[19] In most patients with an ulnar neuropathy, the abnormality is at the elbow (Fig. 21–14). As in carpal tunnel syndrome, sensory fibers are more likely to be damaged than motor fibers, so that the sensory nerve action potential is commonly lost early. In some patients, focal slowing in ulnar sensory fibers across the elbow can be demonstrated. The most common localizing finding in ulnar neuropathy of recent onset is conduction block at the elbow that can be localized precisely by using the inching procedure (see Fig. 21–14). This conduction block may be associated with local slowing. Chronic ulnar neuropathy usually results in slowing of conduction, which may extend distal to the elbow. Rarely, when the ulnar nerve is compressed in the hand, there is a prolonged latency only in the branch to the first dorsal interosseous muscle. In such a patient, the hypothenar muscles should not be the only recording sites for ulnar nerve conduction studies. Although there may be slowing of conduction to the flexor carpi ulnaris, this muscle usually shows little or no change on nerve conduction studies and needle examination. In ulnar and median neuropathies, F-wave latency is prolonged proportional to the slowing in the peripheral segments. Ulnar neuropathies are commonly bilateral; if an ulnar neuropathy is evident on one side, the opposite extremity should also be tested.

PERONEAL NEUROPATHIES

Neuropathy of the peroneal nerve at the head of the fibula is another common focal lesion. Peroneal neuropathy of recent onset due to compression typically is associated with a conduction block that can be localized precisely using the inching procedure to identify the area where the evoked response decreases (Fig. 21–15). Conduction across this segment is usually not slowed, although in lesions of longer standing the slowing becomes prominent.[20]

Nerve conduction studies of the superfi-

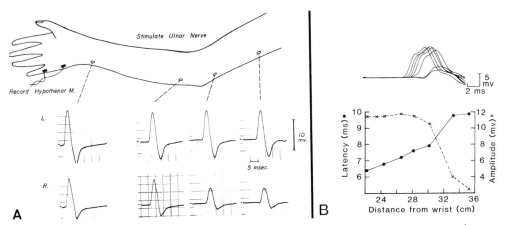

Figure 21–14. (*A*) CMAP recordings with the right ulnar neuropathy at the elbow. (*B*) The inching procedure across the elbow demonstrates localized slowing and conduction block at the elbow. (From Daube, JR: Nerve conduction studies. In Aminoff, MJ [ed]: Electrodiagnosis in Clinical Neurology, ed 3. Churchill Livingstone, New York, 1992, p 310, with permission.)

RIGHT PERONEAL (CROSS-LEG) PALSY:

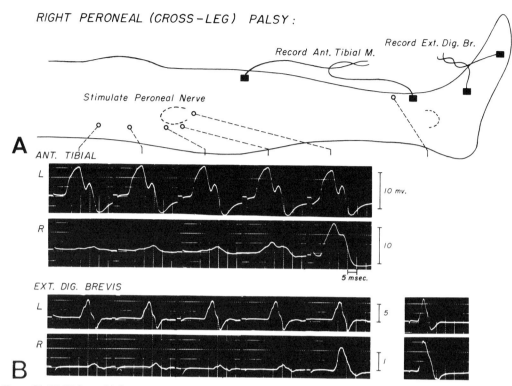

Figure 21–15. Right and left peroneal nerve conduction studies with compression peroneal neuropathy at the fibula. (*A*) Sites of stimulation and recording. (*B*) CMAPs. Note the decrease in amplitude (conduction block) with stimulation at the fibula on the right. (From Daube, JR: Nerve conduction studies. In Aminoff, MJ [ed]: Electrodiagnosis in Clinical Neurology, ed 3. Churchill Livingstone, New York, 1992, p 318, with permission.)

cial peroneal nerve may be of value in differentiating a peroneal neuropathy from an L-5 radiculopathy with some motor slowing. In the latter condition, the superficial peroneal sensory nerve is normal. Patients with a moderately severe peroneal neuropathy often do not have an evoked response from the extensor digitorum brevis. Recordings from the anterior tibial muscle and other anterior compartment muscles on stimulation at the head of the fibula and the knee may still demonstrate a block or slowing of conduction in the nerve. Anomalous innervation of the extensor digitorum brevis muscle by a deep accessory branch of the superficial peroneal nerve may make it more difficult to recognize a peroneal neuropathy. In apparent peroneal neuropathies without localized slowing of conduction, the short head of the biceps femoris muscle should be tested for fibrillation potentials to exclude a sciatic nerve lesion. Sciatic nerve lesions may present with only peroneal deficit, and sciatic

nerve conduction studies are technically difficult to perform.

OTHER NEUROPATHIES

Neuropathies of the *radial and tibial nerves* may similarly be localized by nerve conduction studies.[21] Evaluation of most other nerves is not aided by nerve conduction studies because these neuropathies do not show localized slowing. In *facial neuropathies* like Bell's palsy, stimulation cannot be applied proximal to the site of the lesion. The usual findings in Bell's palsy with a neurapraxia are normal amplitudes and latencies; in axonal degeneration, the amplitude of the evoked response is decreased in proportion to the axonal destruction.[18] Blink reflexes can be used to measure conduction across the involved segment, but they are commonly absent in Bell's palsy. Conduction studies can help differentiate *hemifacial spasm* from other facial movements by demonstrating ephaptic activation of lower facial muscles during periods of spasm, called the *lateral spread response.*

Most *brachial plexus lesions* are traumatic, and nerve conduction studies are of limited value. Generally, the amplitude of the CMAP is reduced and sensory responses are absent in the distribution of the damaged fibers. In patients with lower trunk lesions, the ulnar sensory response is absent, and in those with upper trunk lesions, the median sensory response of the index finger is reduced or absent. In patients with slowly evolving or compressive lesions of the plexus, such as tumors, a localized slowing of conduction of motor fibers and occasionally conduction block may be identified on stimulation at the supraclavicular or nerve root level. Thoracic outlet syndrome, which has been reported to show abnormalities on nerve conduction studies, is usually a vascular syndrome with a change, if any, only in sensory potential amplitudes and little or no slowing of nerve conduction.

RADICULOPATHIES

Cervical and lumbosacral *radiculopathies* are not usually associated with changes in nerve conduction studies; however, if there is sufficient destruction of axons and wallerian degeneration in the distribution of the nerve being tested, the amplitude of the CMAP may be decreased. For example, in an L-5 radiculopathy with weakness, the response of the extensor digitorum brevis muscle on peroneal nerve stimulation is often of low amplitude or absent. In the presence of atrophy and a low CMAP, there may be mild slowing of conduction in the motor axons innervating the atrophic muscle. In a few patients with lumbosacral radiculopathy, measurements of F-wave or H-reflex latencies in mild lesions have been valuable in identifying proximal slowing of conduction.[2] Because most lesions of the spinal nerve and nerve root are proximal to the dorsal root ganglion, the sensory potentials usually are normal, even in the distribution of a sensory deficit. This phenomenon is valuable in identifying avulsion of a nerve root in which there is total anesthesia and loss of motor function with normal sensory potentials.

SPECIFIC NERVE CONDUCTION STUDIES

Cranial Nerve Conduction Techniques

FACIAL NERVE MOTOR FIBER CONDUCTION STUDIES

Electrode Placement

1. Recording electrodes:
   A. Active electrode (G1): over the nasalis muscle just lateral to and 1 cm above the external nares, directly beneath the pupil
   B. Reference electrode (G2): in the same position as G1, but on the opposite side of the face
   C. Recording can also be made from the frontalis, orbicularis oris, orbicularis oculi, or mentalis muscles when indicated
2. Stimulating electrodes:
   A. The cathode is placed just below and anterior to the lower tip of the mastoid, beneath the earlobe.
   B. The anode is inferior to the cathode.
   C. In some patients, the nerve is deep and difficult to stimulate at this location. The cathode can then be placed just anterior to the tragus of the earlobe. Often, the anode must be rotated to reduce artifact or to eliminate masseter contraction.
3. Ground: electrode placed on the chin

Measurements

1. Distance from the cathode to the active recording electrode—*this must be identical on both sides.*
2. Latency from the shock artifact to the initial negative deflection of the CMAP

Table 21–8. **NORMAL VALUES FOR FACIAL NERVE CONDUCTION**

	Amplitude, mV	Latency, ms	Distance, cm
Range	1.8–4.0	1.5–4.0	8–14
Mean	2.2	2.7	—

3. Amplitude of the CMAP from baseline to negative peak
4. Normal values for facial nerve conduction are given in Table 21–8. The values must always be compared with those from the opposite facial nerve (≤ 0.6 millisecond).

ACCESSORY NERVE MOTOR FIBER CONDUCTION STUDIES

Electrode Placement

1. Recording electrodes:
   A. Active electrode (G1): on the trapezius muscle one half the distance along a straight line from the C-7 spinous process to the prominence of the acromioclavicular joint
   B. Reference electrode (G2): over the acromion
2. Stimulating electrodes:
   A. The cathode is placed immediately behind the posterior border of the sternocleidomastoid muscle, 10 cm from the active (G1) electrode.
   B. The anode is placed proximal and parallel to the course of the nerve.
3. Ground: electrode is placed between the active recording electrode and the cathode.
4. Subject position: the head of the bed is elevated 45 degrees from horizontal, with the patient's head turned 45 degrees to the opposite side.
5. Repetitive stimulation: the arm can be immobilized by having the patient grasp a strap held by the patient's foot.

Measurements

1. Distance from the cathode to the active recording electrode, following the contour of the neck
2. Latency from the shock artifact to the initial negative deflection of the CMAP
3. Amplitude of the CMAP from baseline to negative peak

JAW JERK

Electrode Placement

1. Recording electrodes:
   A. Active electrode (G1): over the bony mandible inferior to the anterior border of the masseter muscle
   B. Reference electrode (G2): over the masseter muscle just below the zygoma
   C. Place electrodes symmetrically on the two sides.
2. Stimulating electrodes:
   A. Stimulation with a triggering hammer plugged into the Trigger In input. No electrical stimuli are given.
   B. The examiner's finger is tapped gently with the hammer as the finger rests on the patient's chin.
   C. Both sides are activated simultaneously with a single tap.
3. Ground: electrode on the neck

Measurements

1. No distance measurements
2. Latency to initial reproducible deflection from the baseline with four repeated taps
3. Difference in latency on the two sides when recorded simultaneously
4. Amplitude is variable and need not be measured
   Normal Values. Normal mean is 8.4 milliseconds (range, 6 to 10.5 milliseconds). The difference in latency on the two sides should be no more than 1 millisecond.

Upper Limb Nerve Conduction Techniques

AXILLARY NERVE MOTOR FIBER CONDUCTION STUDIES

Electrode Placement

1. Recording electrodes:
   A. Active electrode (G1): one half the distance along a line that bisects the deltoid muscle on a line extending from the tip of the acromion to the insertion of the deltoid muscle
   B. Reference electrode (G2): just distal to the insertion of the deltoid muscle over the humerus laterally
2. Stimulating electrodes:
   A. The cathode is placed behind and deep to the clavicle, one half the distance between the sternum and the acromion.
   B. The anode is placed posterior.
   C. If the stimulus does not produce a supramaximal response or there is ex-

cessive artifact, the electrodes should be moved around the area. If an adequate response cannot be produced, then a monopolar surface electrode should be tried as the cathode in the same location, with a surface electrode placed posteriorly over the trapezius muscle as the anode.
3. Ground: large electrode over the acromion
4. Repetitive stimulation: the arm can be immobilized with a sheet around it tucked under the torso or with a self-adhering shoulder stabilizer strap.

Measurements

1. Distance from the cathode to the active recording electrode is measured *with calipers.*
2. Latency from the shock artifact to the initial deflection, positive or negative, of the action potential. If the initial deflection is positive, move the active electrode (G1) to try to eliminate the positive deflection.
3. Amplitude of the CMAP from baseline to negative peak

MEDIAN NERVE MOTOR FIBER CONDUCTION STUDIES

Electrode Placement

1. Recording electrodes:
    A. Active electrode (G1): over the center of the abductor pollicis brevis muscle on the thenar eminence, one third the distance between the major creases at the metacarpal-carpal and metacarpal-phalangeal joints of the thumb
    B. Reference electrode (G2): just distal to the metacarpal-phalangeal joint on the lateral side of the thumb
2. Stimulating electrodes:
    A. Distal: the cathode is 2 cm proximal to the crease on the volar surface of the wrist, between the flexor carpi radialis and palmaris longus tendons, 7.0 cm from G1. The anode is 2 cm proximal to the cathode.
    B. Proximal: the cathode is on the anterior surface of the upper arm between the biceps tendon and the medial epicondyle, directly over the brachial artery. The anode is 2 cm proximal.
    C. Other: the nerve may also be stimulated along its course on the medial surface of the upper arm, beneath the biceps muscle, along the brachial artery as high as the axilla. When stimulating here, it is easy to activate the ulnar nerve. ***This must be avoided.***
    D. To record F waves, recording electrode locations are the same. The cathode is in the same location used for distal stimulation. The anode of the stimulating electrode at the wrist is rotated off the nerve away from the ulnar nerve.
3. Ground: electrode placed on the dorsum of the wrist

Table 21–9. NORMAL VALUES FOR MEDIAN NERVE MOTOR FIBER CONDUCTION IN ADULTS 16 TO 65 YEARS OLD

	Amplitude, mV, Stimulus at Elbow	Amplitude Difference, mV,* Wrist to Elbow	Conduction Velocity, m/s,[†] Elbow to Wrist	Distal Latency, ms,[‡] 6–8 cm
Range	3.8–21.4	0–40	49–74	2.3–4.6
2 SD	4–18	0–20	49–70	2.5–4.4
Mean	10.7	10	59.1	3.4

*Amplitude difference between arms should be < 10 mV.
[†]Conduction velocity difference between arms should be < 5 m/s.
[‡]Distal motor latency for the median nerve should not be more than 1.8 milliseconds greater than the distal motor latency of the ulnar nerve.

Table 21–10. MEDIAN NERVE MOTOR FIBER CONDUCTION IN CHILDREN 0 TO 15 YEARS OLD

	0–6 mo	6–12 mo	1–2 y	2–4 y	5–15 y
Conduction velocity, m/s	30–44	38–45	39–48	43–57	48–70
Amplitude, mV	—	5.3–8.6	3.7–7.1	4.1–12.4	4.3–16.6
Distal latency, ms	—	1.9–2.8	2.3–3.0	2.0–3.0	2.0–4.2

Measurements

1. Distance from distal to proximal cathode points of stimulation at the elbow and wrist
2. Distance from the cathode at the wrist to the active recording electrode
3. Distance from any other stimulation points (cathode) to the wrist
4. F-wave distance measured from the cathode at the wrist to the sternoclavicular joint; it is measured as a straight line with the arm at a right angle to the body.
5. Latency from the shock artifact to the initial negative deflection of the CMAP for each stimulation site
6. Latency to earliest reproducible F wave from a minimum of eight responses
7. Amplitude of the CMAP from baseline to negative peak at all stimulation sites
8. The normal values for median nerve motor fiber conduction in adults 18 to 65 years old are given in Table 21–9 and in children 0 to 15 years old, in Table 21–10. The normal values for F waves in median nerve motor fiber conduction in subjects 12 to 73 years old are given in Table 21–11.

MUSCULOCUTANEOUS NERVE MOTOR FIBER CONDUCTION STUDIES

Electrode Placement

1. Recording electrodes:
   A. Active electrode (G1): halfway between the tendons of origin and insertion of the biceps muscle over the center of the muscle belly
   B. Reference electrode (G2): over the biceps tendon distally, 3 cm proximal to the antecubital fossa

2. Stimulating electrodes:
   A. Upper arm: cathode high in the axilla immediately beneath the tendon of the short head of the biceps muscle near its insertion, anode proximal. If unsuccessful, two alternative types of stimulation may be tried:
      (1) "Monopolar" cathode prong and large surface electrode anode on the posterior deltoid muscle
      (2) "Monopolar" needle stimulating electrode inserted near the nerve and a large electrode anode on the posterior deltoid muscle
   B. Supraclavicular: cathode behind and deep to the clavicle, one half the distance from the sternum to the acromion. Anode is placed posterior. If a response is not obtained, the electrodes should be moved or a "monopolar" cathode prong should be tried in the same location with a large surface electrode anode posteriorly over the trapezius muscle.

3. Ground electrode: large surface electrode between the active recording elec-

Table 21–11. NORMAL VALUES FOR F WAVES IN MEDIAN NERVE MOTOR FIBER CONDUCTION IN SUBJECTS 12 TO 73 YEARS OLD

F Wave (12–73 y)	Latency, ms*	Distance, cm	Differences From Estimated F Wave, ms
Range	22–31	57–73	−3 to +4
2 SD	22–31	57–71	—
Mean	26.4	64.2	0.8

*Ulnar/median nerve F-wave latency difference: <3 milliseconds.

trode and the cathode on the lateral side of the arm

Measurements

1. Distance from the cathode to the active recording electrode
2. Distance between proximal and distal sites is measured *with calipers.*
3. Latency from the shock artifact to the initial negative deflection of the CMAP at all stimulation sites
4. Amplitude of the CMAP from baseline to negative peak at all stimulation sites
5. Normal values for musculocutaneous nerve motor fiber conduction in subjects 18 to 65 years old are given in Table 21–12.

PHRENIC NERVE MOTOR FIBER CONDUCTION STUDIES

Electrode Placement

1. Recording electrodes:
   A. Active electrode (G1): in the sixth intercostal space at the anterior axillary line
   B. Reference electrode (G2): at the lower border of the rib cage on the anterior axillary line
2. Stimulating electrodes:
   A. The cathode is placed posterior to the sternocleidomastoid muscle and on the scalene muscles at the level of the thyroid cartilage. The anode is placed cephalad to the cathode. The stimulator can be moved to ensure selective stimulation of the phrenic nerve.
3. Ground: electrode in place on the upper third of the sternum between the active recording electrode and the cathode

Table 21–12. NORMAL VALUES FOR MUSCULOCUTANEOUS NERVE MOTOR FIBER CONDUCTION IN SUBJECTS 18 TO 65 YEARS OLD

	Amplitude, mV	Conduction Velocity, m/s	Distal Latency, ms, 5.0–10.8 cm
Range	4–16	53–85	1.5–3.3
Mean	6.0	—	2.3

4. Patient position:
   A. The patient should be supine if possible.
   B. If the patient is unable to lie flat, a 45 degree position should be used.

Measurements

1. Distance from the cathode to the active recording electrode on each side is measured *with calipers.*
2. Latency from the shock artifact to the initial negative deflection of the CMAP. The potential may be positive initially, although the initial deflection is usually negative.
3. Amplitude of the CMAP from baseline to negative peak
4. Stimulation should be performed on each side in the same fashion and then the two sides should be compared.
5. Procedure:
   A. Stimulate the phrenic nerve until a maximal response is obtained.
   B. With a maximal response obtained, ask the patient to take a deep breath and then exhale. At maximal exhalation, obtain the response. If the patient is unable to do this, then obtain a response at maximal inhalation.
6. The normal values for phrenic nerve motor fiber conduction in adults 20 to 80 years old are given in Table 21–13.

ULNAR NERVE MOTOR FIBER CONDUCTION STUDIES

Electrode Placement

1. Arm position:
   A. Standard (bent arm position with hand under head)
      (1) Arm (shoulder) abducted 90 degrees
      (2) Forearm supinated (palm up)
      (3) Elbow flexed 45 degrees
   B. Repetitive stimulation: arm at side with hand pronated
   C. Proximal (supraclavicular) stimulation: arm at side with hand supinated
   D. Spinal nerve ("root") stimulation: arm at side with patient in lateral decubitus position
2. Recording electrodes:
   A. Active electrode (G1): over the hypothenar muscles at the dorsal border

Table 21–13. **NORMAL VALUES FOR PHRENIC NERVE MOTOR CONDUCTION IN ADULTS 20 TO 80 YEARS OLD**

96 Nerves	Amplitude,* mV	Latency,* ms
Range	0.3–1.3	5.0–10.6
2 SD	0.3–1.1	5.3–9.7
Mean	0.7	7.5

*With subject supine and with maximal expiration.

(just below the fifth metacarpal bone) halfway between the wrist crease and the crease at the fifth metacarpal-pha-langeal joint

B. Reference electrode (G2): just distal to the fifth metacarpal-phalangeal joint on the lateral surface of the fifth digit

3. Stimulating electrodes:
   A. Distal: the cathode is 2 cm proximal to the crease on the volar surface of the wrist, just medial to the flexor carpi ulnaris tendon, approximately 6.5 cm from G1. The anode is 2 cm proximal to the cathode.
   B. Proximal: the cathode is 5 cm proximal to the medial epicondyle over the palpable nerve trunk; the anode is 2 cm proximal.
   C. When standard studies suggest localized damage at the elbow, stimulation should be performed at three proximal sites:
      (1) with the cathode > 5 cm distal to the medial epicondyle

(2) >5 cm proximal to the medial epicondyle
(3) in the axilla, 10 to 15 cm proximal to the second site. If these are abnormal, proceed to segmental stimulation ("inching").

D. For suspected brachial plexus lesions, stimulation is also performed in the supraclavicular fossa, just lateral to the scalenus anticus muscle beneath the clavicle. The cathode may have to be moved in this region to obtain maximal stimulation of the fibers in the ulnar nerve; a monopolar surface prong with a large surface electrode as the anode may be required.

E. To record F waves, the recording electrode locations are the same. The cathode is on the nerve in the same location used for distal stimulation. The anode is rotated off the nerve toward the ulnar side.

4. Ground: electrode on the dorsum of the wrist.

Measurements

1. Distance from the distal to each proximal point of stimulation (cathode) measuring *around the elbow* along the ulnar groove. The tape should be anchored in the groove by the examiner's finger. Conduction velocity is measured to the wrist from each stimulation site, *not segmentally.*
2. Distance from the cathode at the wrist to the active recording electrode
3. F-wave distance from the cathode at the wrist to the sternoclavicular joint, measured as a straight line with the arm at a right angle from the body

Table 21–14. **NORMAL VALUES FOR ULNAR NERVE MOTOR CONDUCTION IN ADULTS 22 TO 69 YEARS OLD**

	Amplitude, mV, Stimulus at Elbow	Amplitude Difference,* %, Below to Above Elbow	Conduction Velocity, m/s, Elbow to Wrist	Distal Latency, ms, 5.5–7.5 cm
Range	6.3–16.0	90–100	51.5–75.9	2.0–3.5
2 SD	6.6–15.0	90–100	53.6–72.4	2.1–3.3
Mean	10.8	96	63.0	2.7

*Amplitude difference wrist to elbow, <20%.

Table 21–15. **NORMAL VALUES FOR MOTOR FIBER CONDUCTION VELOCITY ALONG SEGMENTS OF THE ULNAR NERVE ABOVE AND BELOW THE ELBOW**

	Conduction Velocity, m/s*			
	Below Elbow to Wrist	**Upper Arm to Wrist**	**Differences Below to Above**	**Differences Above to Axilla**
Range	49.5–76.7	51.8–73.6	−9.9 to +8.4[†]	−4.3 to +4.8[†]
2 SD	52.8–73.6	54.4–70.8	−5.5 to +6.6[†]	−3.9 to +3.3[†]
Mean	63.2	62.6	0.3	−0.3

*Normal differences between arms: conduction velocity, <7 m/s; amplitude, <5 mV.
[†]This value defines the upper limit of normal.

4. Latency from the shock artifact to the initial negative deflection of the CMAP at all stimulation sites
5. Latency to the earliest reproducible F wave from a minimum of eight responses
6. Amplitude of the CMAP from baseline to negative peak at all stimulation sites
7. Normal values for ulnar nerve motor fiber conduction studies are given in Tables 21–14, 21–15, and 21–16.

Inching. When standard and 4-point stimulation indicate the presence of an ulnar nerve lesion at or near the elbow, inching should be used to define the precise location of the damage.
1. Recording: as above, with arm up in standard position
2. Stimulation: cathode at most distal elbow stimulation site and at 2-cm intervals along the course of the nerve to the most proximal site
   A. Obtain supramaximal stimulation at each site, and store the response.

Table 21–16. **NORMAL VALUES FOR F WAVE IN ULNAR NERVE MOTOR FIBER CONDUCTION IN SUBJECTS 12 TO 73 YEARS OLD**

	Latency, ms	Distance, cm	Difference From Estimated F Wave
Range	22–31	55–73	−2 to +5
2 SD	21–32	50–76	−2 to +3
Mean	24.6	64.3	2.0

B. Graph the latency difference between each successive stimulation site against the distance from the wrist.
C. Graph the amplitude against distance.

Lower Limb Nerve Conduction Techniques

Peroneal Nerve Motor Fiber Conduction Studies

Electrode Placement

1. Recording electrodes:
   A. Active electrode (G1): over the center of the palpable portion of the extensor digitorum brevis muscle on the lateral aspect of the dorsum of the foot, 1 cm distal to the bony prominence from which the muscle takes origin.
   B. Reference electrode (G2): over the fifth metatarsal-phalangeal joint on the dorsum of the foot, lateral to the long extensor tendons of the fifth toe
   C. Anterior tibial recording can be used when the extensor digitorum brevis muscle provides no evoked response.
      (1) Active electrode (G1): over the anterior tibial muscle one third of the distance from the tibial tubercle to the lateral malleolus (approximately halfway down the muscle), immediately lateral to the tibia
      (2) Reference electrode (G2): on the anterior surface of the ankle at the level of the medial malleolus

2. Stimulating electrodes:
   A. Distal: cathode is on the anterior aspect of the ankle 2 to 5 cm lateral to the tendon of the anterior tibial muscle, 5 cm proximal to the lateral malleolus, *8.5 cm from G1* recording electrode. The anode is proximal to the cathode.
   B. Proximal: cathode over the palpable portion of the peroneal nerve in the lateral part of the popliteal fossa approximately 10 cm proximal to the head of the fibula
   C. If a localized conduction abnormality at the knee is suspected, additional sites should be used:
      (1) Immediately below the head of the fibula where the nerve circles the bone as it enters the anterior compartment
      (2) 10 cm proximal to the head of the fibula in the popliteal fossa
      (3) If there is a difference in CMAP size between stimulation sites, "inching" (segmental stimulation) should be performed along the peroneal nerve as described for the ulnar nerve.
      (4) The same sites of stimulation are used for recordings from either

the anterior tibial or extensor digitorum brevis muscles.
   D. If there is anomalous innervation of the extensor digitorum brevis muscle, the amplitude of the response with stimulation at the knee will be higher than at the ankle. The stimulus should then be applied immediately behind the lateral malleolus to identify such an anomaly.
   E. Late responses occur infrequently but can be obtained with peroneal nerve stimulation. The cathode should be proximal. Other electrode locations are the same.
3. Ground electrodes:
   A. Extensor digitorum brevis recording: electrode on the dorsum of the foot at the insertion of the anterior tibial tendon
   B. Anterior tibial recording: over the anterior tibial muscle just distal to the tibial tubercle

Measurements

1. Distance from the distal point of stimulation to each proximal point of stimulation (cathode)
2. Distance from the cathode at the ankle to the active recording electrode

Table 21–17. **NORMAL VALUES FOR PERONEAL NERVE MOTOR CONDUCTION IN SUBJECTS FROM NEWBORN TO 65 YEARS OLD**

	Amplitude,* mV, Stimulus at Knee	Amplitude Difference, %, Ankle/Knee	Conduction Velocity,[†] m/s	Distal Latency, ms	F-Wave Latency, ms,[‡] 86–125 cm
Newborn					
Range	2.2–4.0	—	22–35	1.8–3.0 (2.0–3.0 cm)	—
Mean	2.9	—	29	2.6	—
6–12 mo	2.6–5.2	—	40–53	1.8–2.5	—
1–3 y	2.3–5.8	—	44–58	2.3–4.9	—
3–15 y	1.6–8.2	—	48–64	2.4–6.2 (6–11 cm)	—
16–65 y					
Range	2.2–14.6	0–22	41–59	2.4–7.0	38–57
2 SD	2–12	0–21	44–57	3.3–6.5	38–57
Mean	7.3	10	50.7	4.9	47.4

*Amplitude difference between legs should be < 5 mV.
[†]Conduction velocity difference between legs should be < 5 m/s.
[‡]Tibial/peroneal F-wave latency difference should be < 7 milliseconds.

Table 21–18. **SEGMENTAL CONDUCTION VELOCITY ACROSS FIBULA TO EXTENSOR DIGITORUM BREVIS IN ADULTS**

	Below Fibula, m/s	Above Fibula, m/s	Difference, m/s
Range	41–56	42–56	−7 to +13*
Mean	50	53	—

*This value defines the upper limit of normal.

Table 21–19. **NORMAL VALUES FOR F WAVE IN PERONEAL NERVE MOTOR FIBER CONDUCTION IN SUBJECTS 10 TO 79 YEARS OLD**

	Latency, ms	Distance, cm	Difference From Estimated F Wave*
Range	40–59	93–133	−10 to +3†
2 SD	40–56	96–126	−9 to +3†
Mean	47.8	111	−3.1

*Measured F-wave latency should not be more than 3 milliseconds longer than estimated.
†This value defines the limits of normal.

3. F-wave distance from the cathode at the ankle to the xiphoid process
4. Latency from the shock artifact to the initial negative deflection of the CMAP at all stimulation sites
5. Latency to the earliest reproducible F wave from a minimum of eight responses
6. Amplitude of the CMAP from baseline to negative peak at all stimulation sites
7. The normal values for peroneal nerve motor fiber conduction in subjects from newborn to 65 years old are given in Table 21–17; the values for stimulation across the fibula to the extensor digitorum brevis, in Table 21–18; and F-wave values, in Table 21–19.

POSTERIOR TIBIAL NERVE MOTOR FIBER CONDUCTION STUDIES

Electrode Placement

1. Recording electrodes:
   A. Active electrode (G1): over the abductor hallucis muscle 1 cm below and 1 cm behind the navicular tubercle, approximately 8 cm from the distal stimulating cathode. End-plate location varies and G1 may need to be moved to obtain clear onset.
   B. Reference electrode (G2): over the medial aspect of the first metatarsal-phalangeal joint
   C. If a tarsal tunnel syndrome is suspected, recording should also be made from the abductor digiti minimi pedis muscle.
      (1) Active electrode (G1): 3 cm proximal to the head of the fifth metatarsal on the lateral aspect of the foot

      (2) Reference electrode (G2): over the lateral aspect of the fifth metatarsal-phalangeal joint
2. Stimulating electrodes:
   A. Distal: cathode is behind and 1 to 2 cm above the medial malleolus. Anode is proximal to cathode
   B. Proximal: cathode is at the lower border of the popliteal space near the popliteal artery (in the center behind the knee). It may be easier to stimulate with the knee partially flexed.
   C. May also stimulate with the cathode behind and 1 to 2 cm below the medial malleolus if a block in this area is suspected.
   D. Stimulation for F waves is at the ankle with the cathode proximal.
3. Ground: electrode on the dorsum of the foot

Measurements

1. Distance from the cathode at the ankle to each of the active recording electrodes
2. Distance from the distal to the proximal point of stimulation (cathode)
3. F-wave distance from cathode at the ankle to the xiphoid process
4. Latency from the shock artifact to the initial negative deflection of the CMAP at all stimulation sites
5. Latency to the earliest reproducible F wave from a minimum of eight responses
6. Amplitude of the CMAP from baseline to negative peak at all stimulation sites
7. The normal values for posterior tibial nerve conduction in adults 16 to 65 years

Table 21–20. **NORMAL VALUES FOR POSTERIOR TIBIAL NERVE CONDUCTION IN ADULTS 16 TO 65 YEARS OLD**

	Amplitude,* mV, Knee Stimulus	Conduction Velocity, m/s	Distal Latency, ms, 6–8 cm	F-Wave Latency,[†] ms, 90–128 cm
Recording AH				
Range	4–25	56–58	3.0–6.0	41–57
2 SD	4–19	41–58	3.0–5.8	41–56
Mean	10.1	49.0	4.2	48.6
Recording ADMP				
Range	1.4–15.6	41–53	4.6–7.3	—

*Amplitude difference, ankle to knee, should be < 50%.
[†]Differences between tibial and peroneal F-wave latency, <7 ms.
ADMP = abductor digiti minimi pedis muscle, AH = abductor hallucis muscle.

old are given in Table 21–20 and for F wave in subjects 12 to 79 years old, in Table 21–21.

CHAPTER SUMMARY

Compound muscle action potentials (CMAPs) are among the most helpful recordings in the electrophysiologic assessment of peripheral neuromuscular disease. CMAPs are the recordings made for all motor conduction studies, both of the directly recorded M-wave used for peripheral conduction and the F-wave late response used for testing proximal conduction. Reliable CMAP recordings require the use of standard stimulating and recording electrode types and locations and standard measurement criteria. The sensitivity and specificity of motor conduction studies depend on comparing the results obtained in a patient with the normal values obtained by using exactly the same methods. The methods and normal values developed and used in the Mayo Clinic EMG Laboratory are presented in this chapter. Motor conduction study normal values vary with physiologic factors such as age and temperature, which must be controlled and adjusted.

Motor nerve conduction studies with CMAP localize focal lesions in a nerve by identifying either localized conduction block or localized slowing of conduction. Conduction block is a change in size of CMAP when stimulating at two points near each other along the nerve. Both conduction block and slowing of conduction represent pathophysiologic changes in the nerve which can sometimes be predicted by the changes on nerve conduction studies. These changes can be helpful in defining prognosis for improvement after nerve damage.

Table 21–21. **NORMAL VALUES FOR F WAVE IN POSTERIOR TIBIAL NERVE MOTOR FIBER CONDUCTION IN SUBJECTS 12 TO 79 YEARS OLD**

	Latency, ms	Distance, cm	Difference From Estimated F Wave,* ms
Range	41–57	74–99	−14 to +5[†]
2 SD	41–56	76–99	−13 to +5[†]
Mean	48.6	86.9	−2.4

*Measured F-wave latency should not be more than 5 milliseconds longer than estimate.
[†]This value defines the limits of normal.

REFERENCES

1. Bostock, H and Sears, TA: Continuous conduction in demyelinated mammalian nerve fibres. Nature 263:786–787, 1976.
2. Braddom, RL and Johnson, EW: H reflex: Review and classification with suggested clinical uses. Arch Phys Med Rehabil 55:412–417, 1974.
3. Brown, WF and Yates, SK: Percutaneous localization of conduction abnormalities in human entrapment neuropathies. Can J Neurol Sci 9:391–400, 1982.
4. Buchthal, F, Rosenfalck, A, and Trojaborg, W: Electrophysiological findings in entrapment of the median nerve at wrist and elbow. J Neurol Neurosurg Psychiatry 37:340–360, 1974.
5. Dyck PJ, Karnes JL, O'Brien PC, et al: The Rochester Diabetic Neuropathy Study: Reassessment of tests and criteria for diagnosis and staged severity. Neurology 42:1164–1170, 1992.
6. Eisen, A, Schomer, D, and Melmed, C: The application of F-wave measurements in the differentiation of proximal and distal upper limb entrapments. Neurology 27:662–668, 1977.
7. Falco, FJE, Hennessey, WJ, Goldberg, G, and Braddom, RL: Standardized nerve conduction studies in the lower limb of the healthy elderly. Am J Phys Med Rehabil 73:168–174, 1994.
8. Fowler, TJ, Danta, G, and Gilliatt, RW: Recovery of nerve conduction after a pneumatic tourniquet: Observations on the hind-limb of the baboon. J Neurol Neurosurg Psychiatry 35:638–647, 1972.
9. Guiloff, RJ, and Modarres-Sadeghi, H: Preferential generation of recurrent responses by groups of motor neurons in man: Conventional and single unit F wave studies. Brain 114:1771–1801, 1991.
10. Gutmann, L, Fakadej, A, and Riggs, JE: Evolution of nerve conduction abnormalities in children with dominant hypertrophic neuropathy of the Charcot-Marie-Tooth type. Muscle Nerve 6:515–519, 1983.
11. Halvorson, GA and Kraft, GH: Median nerve residual latency in carpal tunnel syndrome (abstr). Muscle Nerve 6:535–536, 1983.
12. Hopkins, AP and Morgan-Hughes, JA: The effect of local pressure in diphtheritic neuropathy. J Physiol (Lond) 189:81P–82P, 1967.
13. Lewis, RA and Sumner, AJ: The electrodiagnostic distinctions between chronic familial and acquired demyelinative neuropathies. Neurology 32:592–596, 1982.
14. McLeod, JG: Electrophysiological studies in the Guillain-Barré syndrome. Ann Neurol 9 Suppl:20–27, 1981.
15. McLeod, JG, Prineas, JW, and Walsh, JC: The relationship of conduction velocity to pathology in peripheral nerves: A study of the sural nerve in 90 patients. In Desmedt, JE (ed): New Developments in Electromyography and Clinical Neurophysiology, vol 2, S Karger, Basel, 1973, pp 248–258.
16. Mulder, DW, Lambert, EH, Bastron, JA, and Sprague, RG: The neuropathies associated with diabetes mellitus: A clinical and electromyographic study of 103 unselected diabetic patients. Neurology 11:275–284, 1961.
17. Ochoa, J, Fowler, TJ, and Gilliatt, RW: Anatomical changes in peripheral nerves compressed by a pneumatic tourniquet. J Anat 113:433–455, 1972.
18. Olsen, PZ: Prediction of recovery in Bell's palsy. Acta Neurol Scand Suppl 61:1–121, 1975.
19. Pickett, JB and Coleman, LL: Localizing ulnar nerve lesions to the elbow by motor conduction studies. Electromyogr Clin Neurophysiol 24:343–360, 1984.
20. Singh, N, Behse, F, and Buchthal, F: Electrophysiological study of peroneal palsy. J Neurol Neurosurg Psychiatry 37:1202–1213, 1974.
21. Trojaborg, W: Rate of recovery in motor and sensory fibres of the radial nerve: Clinical and electrophysiological aspects. J Neurol Neurosurg Psychiatry 33:625–638, 1970.

SECTION 2
Electro-physiologic Assessment of Neural Function
PART C
Motor Pathways

Weakness, fatigue, loss of strength, and loss of power are among the major symptoms of neurologic disease that can be assessed with neurophysiologic testing. Strength and movement are under the control of the motor system, which includes the central mechanisms for integrating motor activity and the output pathways. Reflexes and other central motor control systems are discussed in Part E of this section. The electrophysiologic assessment of peripheral motor pathways is reviewed in this part and Part D. As with the sensory pathways, the most direct assessment of the motor pathways can be obtained with stimulation along the motor pathway and measurement of the response evoked by the stimulation. These measurements can include the threshold for activation, the conduction time or velocity (or both) between the points of stimulation and recording, and the size and shape of the evoked response.

Compound muscle action potentials recorded directly from a muscle are the response measured for each of the assessments of motor pathways, whether activated centrally or peripherally. Method of application, strength, and type of stimulus vary with the site along the motor pathway that is being stimulated. Stimulation at the cortical level requires a much higher intensity electrical stimulus or magnetic stimulation to produce adequate responses than stimulation at the spinal level does. Deep-lying motor nerves, such as the spinal nerves, may require needle electrodes for stimulation. Surface electrical stimulation is adequate for stimulation of most peripheral motor nerves.

Motor nerve function in peripheral neuromuscular disorders is assessed by the recording of compound muscle action potentials described in Chapter 21. Repetitive activation of compound muscle action potentials, described in Chapter 22, assesses the function of the neuromuscular junction. Central stimulation of motor pathways at the spinal cord or cortical level evokes compound muscle action potentials that are called "motor evoked potentials," which are described in Chapter 23. The distinction between the terms "compound muscle action potential" and "motor evoked potential" is made based on the site of stimulation. Stimulation of motor nerve fibers anywhere along their course after they leave the spinal cord produces "compound muscle action potentials." Stimulation along the motor pathways in the spinal cord or at the cortical level produces "motor evoked potentials."

CHAPTER 22

REPETITIVE STIMULATION STUDIES

Robert C. Hermann, Jr., M.D.

ANATOMY AND PHYSIOLOGY OF THE
  NEUROMUSCULAR JUNCTION
TECHNIQUE
CRITERIA OF ABNORMALITY
RAPID RATES OF STIMULATION
SELECTION OF NERVE-MUSCLE
  COMBINATIONS
CLINICAL CORRELATIONS

Repetitive stimulation is a clinical neurophysiologic technique designed to evaluate the function of the neuromuscular junction, the point at which a motor nerve axon connects functionally with a voluntary (striated) muscle fiber. The function of the neuromuscular junction is disturbed and the test results are abnormal in a group of rare diseases, including myasthenia gravis, myasthenic syndrome of Lambert and Eaton, botulism, and congenital myasthenia. Repetitive stimulation is often important in the detection, clarification, and follow-up evaluation of these unusual diseases. More commonly, this test is useful in excluding these rare disorders in patients with the common symptoms of fatigue, vague weakness, diplopia, ptosis, and malaise, or with the finding of weakness of uncertain origin.

On the other hand, repetitive stimulation demands an unusual degree of attention to detail and technical expertise to avoid misleading false-positive and false-negative results.

This chapter includes a brief review of the anatomy and physiology of the neuromuscular junction as it applies to repetitive stimulation, a detailed discussion of the technique involved, criteria used to classify the results as normal or abnormal, and the patterns of abnormality seen in various diseases.

ANATOMY AND PHYSIOLOGY OF THE NEUROMUSCULAR JUNCTION

Knowledge of the anatomy and function of the neuromuscular junction is important in understanding the indications for, techniques of, and results of repetitive stimulation.[5]

Each muscle fiber is innervated by a motor neuron whose axon loses its myelin sheath near the muscle and divides into numerous branches, each of which proceeds to join a muscle fiber midway along its length. As the axonal branch nears the muscle fiber, it expands into a presynaptic terminal bouton that lies in a depression in the muscle cell membrane. The muscle cell membrane beneath the nerve terminal undergoes specialized changes, with the development of a highly folded postsynaptic membrane. These presynaptic neural and postsynaptic

237

muscle membrane specializations constitute the *neuromuscular junction*, which is the synapse between nerve and muscle (Fig. 22–1).

The presynaptic nerve terminal has specialized anatomic and metabolic features for the formation, storage, release, and reuptake of acetylcholine, which is required for chemical synaptic transmission (see Fig. 22–1). Acetylcholine is stored in synaptic vesicles that release their contents into the synaptic cleft under appropriate conditions. The contents of a vesicle are all released together as a quantum.

The postsynaptic membrane contains acetylcholine receptor protein concentrated on the crest of the junctional folds (see Fig. 22–1). When acetylcholine binds to the postsynaptic membrane receptors, it causes a membrane channel to open, which in turn causes depolarization of the muscle cell membrane.

Randomly, presynaptic vesicles containing acetylcholine join the presynaptic membrane and release their contents into the neuromuscular junction. The acetylcholine joins with the acetylcholine receptor and produces a small depolarization of the muscle membrane in the area around the neuromuscular junction. This local change is called a *miniature end-plate potential* (MEPP).

When an action potential travels along a motor axon and reaches the nerve terminal, it causes the release of a large number of vesicles (quanta) of acetylcholine in a short time. This acetylcholine binds with the postsynaptic receptors to produce a depolarization that is much larger than a miniature end-plate potential. This larger depolarization is termed an *end-plate potential* (EPP). An end-plate potential is large enough to depolarize the muscle membrane around the end plate sufficiently to generate an action potential, which then spreads in all-or-none fashion over the membrane of the entire muscle fiber. Through excitation-contraction coupling, the action potential causes contraction of the muscle fibers. In normal subjects, the end-plate potential is much greater in amplitude than necessary for the muscle cell membrane to reach threshold, and each generates a muscle contraction. The size of the end-plate potential is determined by the amount of acetylcholine in each vesicle, the number of vesicles released, and the number of acetylcholine receptors stimulated.

The amount of acetylcholine released at the neuromuscular junction varies under different conditions. The mechanisms involved in the release of acetylcholine by an action potential are such that if another action potential occurs within 200 milliseconds of the first one, the amount of acetylcholine released is greater with the second action potential (Fig. 22–2). If a second action potential arrives more than 200 milliseconds after the first one, less acetylcholine is released with the second action potential.

Thus, if a nerve is stimulated repetitively, the amount of acetylcholine released varies

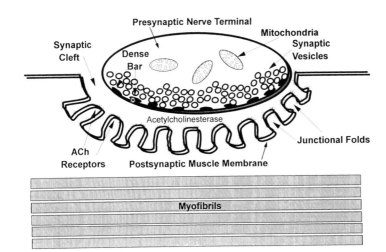

Figure 22–1. Functional anatomy of the neuromuscular junction. ACh = acetylcholine.

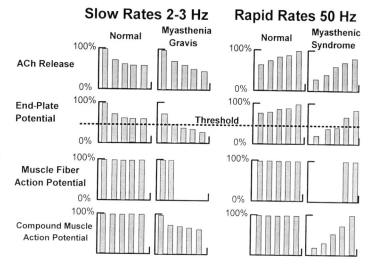

Figure 22–2. The effects of repetitive stimulation at slow and fast rates on the release of acetylcholine (ACh), end-plate potential, individual muscle fiber action potentials, and the compound muscle action potential. This is a comparison of normal subjects with patients with myasthenia gravis or myasthenic syndrome.

depending on the rate of stimulation. At fast rates of repetition, that is, more than 10 per second (short interval between successive stimuli), the amount of acetylcholine released either increases or is potentiated. After a series of stimuli (tetanic), potentiation of acetylcholine release may persist for 30 to 60 seconds. With slow rates of repetitive stimulation, that is, fewer than 5 per second (long intervals between stimuli), the amount of acetylcholine released is less with each of the first four stimuli. This decrease is much greater for 2 to 5 minutes after a period of exercise or after repetitive (tetanic) stimulation.

In normal subjects the amplitude of the end-plate potential is so much greater than required to reach threshold ("safety factor") that small decreases in amplitude of the end-plate potential with repetitive firing have no effect. Each nerve action potential produces a muscle contraction.

However, in disorders in which the size of the end-plate potential is decreased to the point when the amplitude falls just above or just below threshold, minor physiologic fluctuations in the amplitude may assume major significance. If the end-plate potential is marginally above threshold, repetitive stimulation at slow rates results in a lower amplitude that may not reach threshold and neuromuscular transmission may fail, decreasing the number of muscle fibers that contract (decrement). If the end-plate po-

tential is just below threshold for generating an action potential, repetitive stimulation at rapid rates may result in increased release of acetylcholine, thus causing an increased amplitude that may exceed threshold and result in an increase (increment) or facilitation of neuromuscular transmission, with an increment in the number of muscle fibers responding.

TECHNIQUE

Technique is important in repetitive stimulation, because, on the one hand, poor technique can result in "abnormal" findings in patients with normal neuromuscular transmission, leading to erroneous diagnosis of a disorder of neuromuscular transmission; on the other hand, poor technique may result in "normal" findings in patients with abnormal neuromuscular transmission. This gives a false-negative result and a missed diagnosis.

The basic techniques required are those used for routine motor nerve conduction studies. These basic techniques must be mastered before repetitive stimulation is attempted. Attention to these basic technical details results in reliable and rapid testing. Ignoring these technical details can produce unreliable results, requiring repetition of the procedure and causing both great expenditure of time and unnecessary patient discomfort.

Recording electrode placement is the same as for routine motor nerve conduction studies, with the active, or G1, electrode placed over the end plate or motor point, and the G2 electrode placed over the tendon. Repetitive stimulation requires that the G1 electrode be positioned so there is no initial downward deflection (i.e., positivity at G1) and that the initial negative deflection is very sharp. The recording electrodes and wires must be attached securely so that no movement or loosening occurs during the study. The ground electrode should be positioned to minimize stimulus artifact.

Immobilization is important in repetitive stimulation studies. Movement results in broken connectors and wires, loose electrodes, alterations in the relationship between the recording electrode and the muscles, introduction of unwanted activity from neighboring muscles, and shifts in the location of the stimulator, among many other effects. These effects cause problems of varying severity. Movement by the patient should be avoided. The physician must: (1) instruct patients about the importance of remaining as relaxed as possible; and (2) explain that their cooperation will result in a shorter, less painful study and will produce more reliable results. The patient should let the movement produced by the stimulation occur but should avoid the contraction of other muscles, especially between stimuli.

In addition, some form of physical restraint should be used when possible to minimize movement. These immobilization devices include clamps, boards, straps, sheets, and towels. Basically, any device that limits shifts between the electrodes and the recording or stimulating sites without harming the patient or recording setup could be useful.

Stimulation technique is based on the usual techniques of supramaximal stimulation with the smallest stimulus possible. Therefore, the stimulator must be as close as possible to the nerve. The longer the duration and the greater the intensity of the stimulus, the more uncomfortable the patient will be, the more movement by the patient that will occur, the more stimulus artifact that will be created, and thus the greater the chance of stimulating unwanted surrounding nerves. For these reasons, the nerve must be carefully localized. In repetitive stimulation, the stimulus site must be stable. This is a challenge because movements produced by the stimulation and the reaction of patients tend to cause movement of the stimulating electrode in relation to the nerve. Stimulation of the nerve distally limits contraction of unwanted muscles. A near-nerve needle electrode may result in shorter duration, lower intensity, and more localized stimulation. This may produce less stimulus artifact, less patient discomfort, and more reliable results.

The stimulus must be strong enough to excite all motor axons, and the intensity must be increased 10% to 25% above the level that produces a maximal response. This supramaximal stimulus cancels the effect of small degrees of movement of the stimulating electrode away from the nerve. Watch the response carefully for any change in amplitude or configuration. If any such change is noted, the adequacy of the stimulus intensity should be checked immediately.

The stimulus rate and the number of stimuli vary depending on the clinical problem. In most situations, slow rates of stimulation of 2 per second with an interstimulus interval of 500 milliseconds maximize any potential decrement. The greatest decrease in acetylcholine release at slow rates of stimulation occurs during the first four stimuli. For these reasons, four or five stimuli at 2 Hz are usually satisfactory. With a slower rate and fewer stimuli, the patient will better tolerate the procedure.

The train of four stimuli should be repeated, with at least 15 to 30 seconds of rest between trains. Trains are repeated to check for reproducible amplitudes, areas, and configurations as well as for the stability of the baseline, presence of stimulus artifact, patient relaxation or movement, and stability of recording and stimulating electrodes. If any changes are found, it is wise to consider the problem to be technical and to proceed with a systematic checklist to eliminate artifact. When all are checked, three sets of four stimuli at 2 Hz with 15 to 30 seconds between sets should be obtained as baseline.

Depending on the clinical problem and the results of the baseline 2-Hz repetitive stimulation, a decision must be made about the usefulness of further testing of neuromuscular transmission with repetitive stimulation after exercise or tetanic stimulation. Generally, the patient performs an exercise for a brief (i.e., 10 seconds) or an intermediate (i.e., 1 minute) period.

Brief (10-second) periods of exercise have almost the same effect as rapid stimulation at 20 to 50 Hz for 10 seconds but are not nearly as uncomfortable for the patient. After 10 seconds of exercise, the release of acetylcholine with each action potential is potentiated for 30 to 60 seconds. During this period of postactivation (i.e., post-tetanic) potentiation, amplitude of the end-plate potential is increased, and amplitude of the evoked potential may be markedly increased in myasthenic syndrome or botulism. In myasthenia gravis, the decrement at slow rates of stimulation may be decreased or absent during this period.

With no decrement or only a very questionable decrement on baseline testing, and if every attempt has been made to unmask a borderline or mild defect of neuromuscular transmission, the patient should be exercised for 1 minute (exercise for 20 seconds, rest 5 seconds, exercise 20 seconds, rest 5 seconds, and exercise 20 seconds) to simulate prolonged stimulation. The amount of acetylcholine released with each stimulus should be minimal 2 to 5 minutes after exercise, thus providing the greatest chance

for detecting any defect of neuromuscular transmission. Usually, after 1 minute of exercise, 4 stimuli are given at 2 Hz immediately after exercise and at 30, 60, 120, 180, and 240 seconds after exercise. As previously emphasized any change in amplitude, configuration, and other factors should be considered to represent a technical problem, and technical factors, including strength of stimulation, should be checked.

Display of the results varies with the machine, hardware, software, and display devices. In general, sensitivity should be adjusted to display the potentials as large as possible without causing overflow or blocking. The sweep speed should be fast enough to spread the potential out so that it can be analyzed visually, but slow enough so that the entire potential is displayed, including any late components. The repetitive potentials should be displayed in an x-shifted fashion (Fig. 22–3). This means that the onset of the sweep for each successive stimulus is shifted to the right on the horizontal, or x, axis or delayed so that every potential can be analyzed individually. Thus, if there are changes, it is possible to determine which potential in the sequence of four changed and what the order of change was. Superimposition of successive stimuli allows closer inspection and detection of changes in amplitude, area, or configuration, if needed.

It is most important that the amplitude,

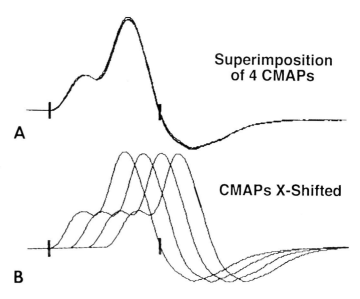

Figure 22–3. (*A*) Superimposition of compound muscle action potentials (CMAPs) evoked by repetitive stimulation allows easy visual identification of small decrements. (*B*) Also, CMAPs can be displayed in a staggered, or x-shifted, fashion to allow inspection of each potential and determination of the sequence of any changes.

duration, and area of the negative waveform of the compound muscle action potential be measured. This assumes that the other standard nerve conduction studies are complete. Measurements of amplitude can be made either by hand or with manual or automated markers. Measurements of area are made quickly and reliably by computer-driven digital machines, provided that the markers are accurately placed. Such computer-based digital machines can quickly measure changes in amplitude and area between the first and subsequent responses.

Decreases in amplitude or area are expressed as a percentage decrement, as derived from the following formula:

$$\frac{\text{Percentage}}{\text{decrement}} = \frac{\left(\begin{array}{c}\text{amplitude of}\\\text{first response}\end{array}\right) - \left(\begin{array}{c}\text{amplitude of}\\\text{subsequent response}\end{array}\right)}{\text{amplitude of first response}} \times 100$$

or

$$\frac{\text{Percentage}}{\text{decrement}} = \left(1 - \frac{\text{amplitude of last response}}{\text{amplitude of first response}}\right) \times 100$$

Increases in amplitude or area are expressed as percentage increments or facilitation, as in the following:

$$\frac{\text{Percentage}}{\text{increment}} = \frac{\left(\begin{array}{c}\text{amplitude of}\\\text{last response}\end{array}\right) - \left(\begin{array}{c}\text{amplitude of}\\\text{first response}\end{array}\right)}{\text{amplitude of first response}} \times 100$$

$$\text{Facilitation} = \frac{\text{facilitated amplitude}}{\text{baseline amplitude}} \times 100$$

Note that no change from baseline is a facilitation of 100% and a 100% increment is a facilitation of 200%.

A small increment in amplitude (less than 40%) may be seen in normal subjects. The increment is usually accompanied by a decrease in duration and by little or no change in area. This is best explained by an increased synchronization of firing of the motor units rather than an increase in the number of units or an increase in the amplitude of the response of individual fibers.

A change in CMAP that is not due to disease may be obtained in several situations:

1. There may be movement of the stimulating electrode in relation to the nerve, producing a random variation in amplitude or, less frequently, a sequential decrement. This is more common just after exercise and is more likely to occur when the stimulus is not supramaximal. Submaximal stimulation is suggested by a loss of amplitude of the initial response in the train from the baseline responses.
2. Movement of the recording electrode relative to the underlying muscle produces a change in configuration and amplitude that is usually random but occasionally may be in a decremental or incremental pattern (Fig. 22–4). Movement may produce a shift in the baseline or visible muscle activity between the stimuli.
3. Technical factors should be suspected if (a) the results are not reproducible; (b) the pattern of decrement, increment, postexercise potentiation, or exhaustion is unusual; (c) there are baseline shifts or changes in configuration; or (d) there is evidence of muscle activity or movement between stimuli.

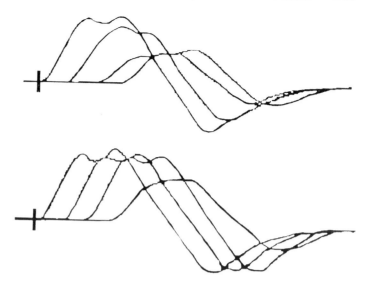

Figure 22–4. Technical problems such as poor relaxation or movement during repetitive stimulation can produce apparent decrements (pseudodecrements) in normal subjects.

If technical factors cannot be excluded and are suspected to be the cause of abnormality, the study should be considered technically inadequate and thus not diagnostic. This is preferable to making a serious diagnosis on the basis of questionable data. Repeating the study on another nerve, at a later date, or under other circumstances may be helpful.

False-negative results or normal results in a patient with a disorder of neuromuscular transmission can be caused by low temperatures, which mask mild defects of neuromuscular transmission. The patient must be warm before and during the study. This requires that temperatures are monitored and that hand skin temperature is above 32°C and foot skin temperature 30°C. If temperatures are cooler, the patient should be warmed before further studies are undertaken.

False-negative results may occur if ongoing treatment with anticholinesterase agents, immunosuppressant drugs, or plasma exchange is successful. The test should be conducted when the patient is most symptomatic, usually late in the day when fatigued. When possible, treatment with acetylcholinesterase inhibitors such as pyridostigmine (Mestinon) or neostigmine should be discontinued for at least 4 to 6 hours before the test. The risk to the patient from discontinuing treatment must be weighed against the importance of the test.

Patients for whom the diagnosis is in question probably should be tested when the effects of treatments such as plasma exchange, intravenous gamma globulin, and corticosteroids are minimal.

CRITERIA OF ABNORMALITY

In normal subjects, there should be no decrement with 2-Hz stimulation; however, technical problems often result in small decrements. A conservative criterion of abnormality is to require a decrement of at least 10% in two different muscle/nerve preparations that meet the following criteria:

1. Reproducible results on repeated testing.
2. The pattern of decrement or the shape of the envelope of the responses should be similar to that seen in neuromuscular disorders. In myasthenia gravis, the decrement is usually greatest between the first and second responses in the train of four responses.
3. The changes induced by exercise with potentiation and exhaustion should be compatible with what is seen in the disease under question.
4. Edrophonium (Tensilon) or neostigmine (Prostigmin) should decrease or correct the abnormalities seen in myasthenia gravis.

RAPID RATES OF STIMULATION

Rapid rates of stimulation of 10 to 50 Hz are helpful in some disorders such as Lambert-Eaton myasthenic syndrome and botulism in which there may be a marked increment with rapid rates of stimulation. Unfortunately, rapid repetitive stimulation is painful. However, brief exercise with voluntary, strong contraction produces the same effect without pain, although some patients are unable to produce a strong contraction because of extreme weakness, lack of understanding, or inability to cooperate. If a disorder such as the myasthenic syndrome is suspected, rapid rates of stimulation are necessary. Explain the details of the test to the patient before proceeding. Stress the importance of remaining as still and as relaxed as possible so that the results of the test will be reliable. Before beginning, check the setup carefully to exclude any technical errors that might make the results uninterpretable. If a large facilitation is expected, the gain or sensitivity should be set so a much larger response can be recorded without blocking. Stimulate at 20 to 50 Hz for 2 to 10 seconds, depending on the situation. After 3 seconds of stimulation, an increment greater than 40% in adults or 20% in infants is considered abnormal.

SELECTION OF NERVE-MUSCLE COMBINATIONS

Most diseases studied affect certain muscles more than others. Proximal muscles are usually more involved than distal muscles. A basic set of rules can be developed for selecting the nerve-muscle preparations (Table 22–1) to be studied in searching for defects of neuromuscular transmission:

1. The clinically weakest muscles are most likely to show a defect on repetitive stimulation.
2. In most cases, proximal muscles are more likely to show abnormalities.
3. A clinically uninvolved muscle is unlikely to reveal abnormal results.
4. Testing distal muscles is technically easier and causes less discomfort, and the results are more reliable.
5. Generally, it is probably wise to begin by testing a distal muscle in the most affected extremity and then moving more proximally if indicated.

CLINICAL CORRELATIONS

Myasthenia gravis[3] is the classic disease of the neuromuscular junction. It is usually the result of an autoimmune-mediated attack on acetylcholine receptors on the muscle (postsynaptic) cell membrane. This results in fewer functional receptors and fluctuating, fatigable weakness involving proximal muscles more than distal muscles, particularly affecting bulbar muscles and often the extraocular muscles. Experimentally, the amplitudes of miniature end-plate potentials and end-plate potentials are low (Fig. 22–5). The resting compound muscle action potential is in the normal or low-normal range except in the most severe cases. With repetitive stimulation at 2 Hz, there is a decrement, with the greatest decrease between the first and second response and smaller decreases after that. By the fifth response, the decrement levels off. After 10 seconds of exercise, the decrement is partially or completely repaired and the increment in the amplitude of the compound muscle action potential can be as much as 94% (i.e., facilitation of 194%). At 2 to 4 minutes after 1 minute of exercise, the decrement is larger than when measured at rest. Similar findings are seen in cases of myasthenia gravis associated with the use of the drug D-penicillamine.

Lambert-Eaton myasthenic syndrome[3,4] is a rare entity, often found in association with systemic malignancies such as small cell carcinoma of the lung. It seems to be an autoimmune disorder produced by an antibody that alters the function of the calcium channels in the axon terminal of the neuromuscular junction. The result is decreased release of acetylcholine with each action potential. The amount of acetylcholine released increases rapidly with rapid rates of activation or stimulation. The weakness is more generalized than in myasthenia gravis but involves proximal muscles more than distal muscles. The bulbar muscles are not involved as prominently as they are in myasthenia gravis. Baseline studies usually re-

Table 22–1. **SPECIFIC NERVE-MUSCLE COMBINATIONS**

Nerve	Muscle	Stimulation Site	Advantages	Disadvantages	Immobilization
Ulnar	Abductor digiti quinti manus	Wrist	Reliably immobilized, well tolerated	Distal muscle may be spared	Velcro strap, clamp
Median	Abductor pollicis brevis	Wrist	Well tolerated	Distal muscle may be spared, difficult to immobilize	Thumb restrained with towel
Musculocutaneous	Biceps	Lower edge axilla	Proximal muscle	Unstable stimulus, difficult to immobilize, painful	Board
Axillary	Deltoid	Supraclavicular	Proximal muscle	Unstable stimulus, difficult to immobilize, painful	Large Velcro strap or sheet
Spinal accessory	Trapezius	Posterior border upper sternocleidomastoid	Proximal muscle, well tolerated	Difficult to immobilize	Strap
Facial	Nasalis	Between mastoid and tragus	Proximal muscle	Painful, unstable stimulation, shock artifact, cannot immobilize. Masseter contraction	None
Peroneal	Anterior tibial	Knee	Leg muscle	Distal muscle may be spared	Board
Femoral	Rectus femoris	Femoral triangle	Proximal leg muscle	Painful, difficult to immobilize	Restrained manually

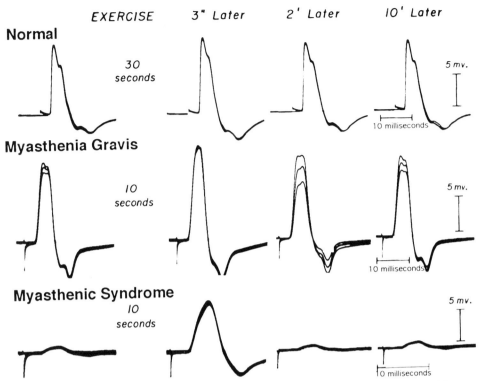

Figure 22–5. Examples of supramaximal repetitive stimulation at 3 Hz of the ulnar nerve at the wrist while recording over the hypothenar muscle at rest and at 3 seconds, 2 minutes, and 10 minutes after exercise. Each waveform consists of three compound muscle action potentials superimposed in a normal subject, in a patient with myasthenia gravis, and in another patient with myasthenic syndrome. In the patient who has myasthenia gravis, the decrease in amplitude from the first to third response recovers after 10 seconds of exercise and is accentuated 2 minutes after exercise. In a patient with the Lambert-Eaton myasthenic syndrome, the CMAP at rest is very low in amplitude; there is a decrement that is not appreciable at this sensitivity. After brief exercise, there is a transient facilitation in amplitude of the CMAP. (From Lambert, EH, et al: Myasthenic syndrome occasionally associated with bronchial neoplasm: Neurophysiologic studies. In Viets, HR [ed]: Myasthenia Gravis [The Second International Symposium Proceedings.] Charles C Thomas, Springfield, IL, 1961, pp 362–410. By permission of the publisher.)

veal low-amplitude compound muscle action potentials at rest. Repetitive stimulation at 2 Hz produces decrements similar to, but often more prominent than, those found in myasthenia gravis. Brief (10-second) exercise or rapid stimulation at 50 Hz produces a marked increment or facilitation of the amplitude of the compound muscle action potential to 2 to 20 times the baseline amplitude. This effect is transient and must be looked for in a well-rested, warm muscle immediately after brief exercise. After 60 to 120 seconds, the amplitude returns to baseline and the decrement returns and may be more prominent 3 to 4 minutes after exercise.

Botulism[1,2,6] is a rare disorder caused by

exposure to one of the types of botulinum toxin. In adults, the toxin is ingested with inadequately preserved or prepared food. In infants, botulism is due to ingestion of spores that germinate into bacteria that produce botulinum toxin in the gut. Between 12 and 36 hours after ingestion of the toxin, blurred vision, dysarthria, dyspnea, and generalized weakness develop. The toxin markedly decreases the quanta of acetylcholine released by an action potential, thus also reducing the amplitude of the end-plate potential. Clinically, results of nerve conduction studies are normal except for the presence of low-amplitude compound muscle action potentials. Rapid repetitive stimulation at 50 Hz or brief exercise produces an increment in most cases (in 62% of adults and 92% of infants). This may not be found in very severe cases; although when found, the decrement ranges from 30% to 200%. There is only a small decrement at slow rates of repetitive stimulation in 56% of the cases in infants and in 8% of the cases in adults. Postexercise exhaustion is not seen.

Congenital myasthenia is the general term applied to a group of rare inherited disorders of neuromuscular transmission. Results of repetitive stimulation are abnormal, but the pattern of abnormality varies among the different disorders. One should look for repetitive firing of the CMAP that occurs in some congenital myasthenic disorders.

In myotonic disorders, repetitive stimulation may produce a small decrement at rest that is more prominent as the rate of stimulation is increased. The decrement occurs at lower frequencies of stimulation in myotonia congenita than in myotonic dystrophy. In paramyotonia congenita, the decrement may progress to electrical silence with cooling of the muscle. In the myotonic disorders, the amplitude of compound muscle action potentials is reduced for several minutes after exercise.

Periodic paralysis may reveal a gradual decrease in the amplitude of the compound muscle action potential 20 to 30 minutes after exercising the muscle for 3 to 5 minutes. Usually, there is no decrement at slow rates of repetitive stimulation.

In motor neuron disorders such as amyotrophic lateral sclerosis, slow rates of stimulation may produce a decrement that is less prominent after brief exercise and more prominent after prolonged exercise. Such abnormalities are seen more frequently when the disease is rapidly progressive.

Disordered neuromuscular transmission is rarely found in cases of peripheral neuropathy or inflammatory myopathy.

CHAPTER SUMMARY

The electrophysiologic technique of repetitive stimulation is an important component of the evaluation of patients with fatigue, weakness, ptosis, diplopia, dysphagia, and dysarthria. For proper application and interpretation, the technique requires a knowledge of the physiology and pathophysiology of neuromuscular transmission and the basics of nerve conduction studies. Errors in technique can produce false-positive or false-negative results. The test must be individualized for each patient, with the proper selection of nerve-muscle combinations, rates of stimulation, and types of exercise. The examiner must be aware of the various abnormalities that occur with different disease entities.

REFERENCES

1. Cherington, M: Electrophysiologic methods as an aid in the diagnosis of botulism: A review. Muscle Nerve 5:S28–S29, 1982.
2. Cornblath, DR, Sladky, JT, and Sumner, AJ: Clinical electrophysiology of infantile botulism. Muscle Nerve 6:448–452, 1983.
3. Howard, JF Jr, Sanders, DB, and Massey, JM: The electrodiagnosis of myasthenia gravis and the Lambert-Eaton myasthenic syndrome. Neurol Clin 12:305–330, 1994.
4. Jablecki, CK: Electrodiagnostic evaluation of patients with myasthenia gravis and related disorders. Neurol Clin 3:557–572, 1985.
5. Lambert, EH, Okihiro, M, and Rooke, ED: Clinical physiology of the neuromuscular junction. In Paul, WM, Daniel, EE, Kay, CM, and Monckton, G (eds): Muscle. Pergamon Press, Oxford, 1965, pp 487–497.
6. Pickett, JB III: AAEE case report 16: Botulism. Muscle Nerve 11:1201–1205, 1988.

CHAPTER 23

MOTOR EVOKED POTENTIALS

Jasper R. Daube, M.D.

The term *motor evoked potentials* (MEPs) has been used to describe potentials from muscle or nerve elicited by several modes of stimulation. Stimulating a peripheral nerve and recording a compound muscle action potential (see Chapter 21) are, strictly speaking, a form of MEP but are not referred to as such. MEP usually describes potentials recorded from muscles after stimulation of motor structures in the central nervous system and is used in this sense in this book. MEPs may be elicited by stimulating either the spinal cord or the cerebral hemispheres with electrical or magnetic stimuli. Magnetic stimulation of the brain is not approved for general use in the United States, where it is considered experimental. However, studies of the efficacy and safety of MEP have been performed in other countries. MEP testing may include various locations for both stimulation and recording (described later in this chapter). The clinical application of MEPs parallels the application of somatosensory evoked potentials. The major value of both methods has been in two areas: (1) identifying slowing of impulse conduction in demyelinating diseases, particularly multiple sclerosis; and (2) monitoring motor pathways intraoperatively. The former is the commonest clinical diagnostic application. Intraoperative monitoring is increasingly used as the reliability, validity, and range of application of this method become better known.[1]

MEP STIMULATION

Theoretically, MEPs can be obtained by stimulating anywhere along motor pathways. For clinical purposes, these potentials are elicited with stimulation of either the cerebral cortex or the cervical spinal cord.[9] At each location, motor pathways can be activated by electrical or magnetic stimulation, each of which is described below.

Electrical Stimulation

Direct electrical activation of the motor pathways at the level of the cerebral cortex has been used in experimental animals for many years to study the motor systems and their pathways. The first extensive study of stimulation of the motor cortex in humans was conducted 50 years ago by Wilder Penfield in Montreal during surgical procedures for epilepsy.[7] He noted that the responses were significantly attenuated by anesthesia and so conducted many of his operations on patients under local or light anesthesia. He also recognized that the cerebral cortex was more readily activated with rapid repetitive stimulation, ranging from 20 to 50 Hz; and that optimal activation occurred with the anode, rather than the cathode, placed over the motor area. Thus, stimulation of the motor cortex differs from that of peripheral nerves or motor fiber tracts in the spinal cord in both the polarity of stimulation and the rate of repetitive stimulation.

MEP augmentation with repetitive stimulation occurs at the cortical level, but it is also a function of the physiology of activation of the pool of motor neurons in the anterior horn. Direct stimulation of the cerebral cortex may activate the dendrites and cell bodies or the axon hillocks of the motor neurons in the precentral gyrus.

Cathodal activation at the axon hillock is obtained with the anode over the region of the neuron and the cathode at a distal site. This orientation enhances current entry and depolarization in the deep layers of the cortical gray matter and hyperpolarization on the cortical surface. Repetitive activation of cortical motor neurons drives their membrane threshold closer to threshold for firing, thereby increasing the likelihood of activation. However, as found in experimental testing of the motor pathways, activation of an anterior horn cell is more likely to occur when multiple depolarizations arrive at the neuron within a few milliseconds of one another. Normal activation during motor behavior is due to a series of descending impulses in the motor pathways occurring at intervals of 3 to 5 milliseconds. Microelectrode recordings of the depolarizations in anterior horn cells in response to cortical stimulation show an initial large *direct*, or *D*, *wave*, followed by a series of three to eight smaller depolarizations occurring at intervals of 3 to 5 milliseconds and referred to as *indirect*, or *I*, *waves*. The occurrence and number of I waves increase with repetitive stimulation.

Consequently, optimal activation of MEPs for diagnostic purposes would be expected to require a series of 3 to 10 stimuli to the brain. At some medical institutions, such repetitive stimulation is used to enhance MEPs. Comparable enhancement of these potentials can be obtained by low-level voluntary activation, which partially depolarizes the cell body and axon hillock.

In some monitoring of MEPs intraoperatively, electrodes are placed directly on the dura mater through a burr hole. However, diagnostic MEPs and most monitoring MEPs use transcranial stimulation of the motor cortex through the intact bony skull. The high resistance of bone to electric current requires much higher applied voltages to drive the few milliamps of current to the cortex needed for cortical stimulation. Transcranial stimulation in normal adults often requires stimuli of 1000 to 1500 V. They are applied in short pulses, with durations of less than half a millisecond. The intensity of the stimulus produces considerable contraction of the cranial musculature. Some patients report it quite uncomfortable, but others tolerate it without difficulty. Pulse duration and configuration appear to make less difference for activation at the cortical level than they do with peripheral stimulation. Because of the high voltages required, MEPs are typically performed with specially designed equipment to generate high voltages without risk to the patient. Electrical activation of the motor pathways in the spinal cord occurs by depolarization at the cathode, with hyperpolarization at the anode. Therefore, activation of a spinal cord pathway is similar to that of a peripheral nerve except that responses are enhanced in most patients when there is a low level of voluntary contraction. Such background contraction partially depolarizes the anterior horn cells before the stimulus is applied.

Electrical stimulation at the spinal cord level relies on the depolarization of the motor axons in the corticospinal tract. This can be obtained percutaneously with stimuli greater than 1000 V, similar to those for transcranial stimulation. However, in the operating room, activation of the axons in the cord can be obtained with lower intensity stimuli because the cathode can be in the form of a flat electrode placed directly on the dura mater in the surgical field or a needle electrode placed either in the interspinous ligament or over the vertebral lamina. Anode placement for activation of the spinal cord may be epidural, laminar, esophageal, subcutaneous, or on the surface of the skin at some distance. Stimulation of the cervical spinal cord is enhanced by repetitive stimulation at intervals of 3 to 5 milliseconds. Repetitive stimulation enhances activation of the anterior horn cells, as with cortical stimulation. Stimulation of motor pathways in the cervical spinal cord is easier than cortical stimulation in surgical settings because anesthetics have less effect on the spinal cord than on the cerebral cortex. However, volatile anesthetics suppress transsynaptic activity in the anterior horn cells, thereby partially reducing MEPs with spinal cord stimulation but to a lesser extent than with cortical stimulation.

Stimulation of the motor pathways at either the cortical or spinal cord level activates multiple descending pathways and, therefore, is associated with widespread muscle contraction. Recording electrodes can be placed over either multiple muscles or muscles of particular clinical concern. Selection of the recording electrode defines which components of the motor pathways are tested. Typically, the ankle extensor and flexor muscles and quadriceps muscles are tested in the lower extremity, and intrinsic hand muscles, forearm extensors, and arm flexors are tested in the upper extremities.

Although some reports have cited normative data, most have not. Consequently, generally accepted values for MEP amplitude and latencies do not exist. For clinical studies, amplitude measures usually are not used because of the marked differences among normal subjects. Latency measures can show marked slowing or dispersion of the response. Such a finding is strong evidence for a demyelinating process, as in multiple sclerosis.

Magnetic Stimulation

Activation of the motor pathways with magnetic stimulation is generally more difficult and less successful than with electrical stimulation. Magnetic stimulation is better tolerated than electrical stimulation by patients who are awake. Although the strong electric shock is not felt by the patient being magnetically stimulated, the discomfort caused by the activation of the axons in a nerve remains.

The activation of either the axons in the corticospinal tract at the level of the cervical cord or the neurons in the motor cortex depends on the same mechanism. In both cases, electrical current flowing through the neuron or axon causes depolarization, and this depolarization initiates descending action potentials in the corticospinal pathway. Magnetic stimulation differs from electrical stimulation in that no current flow occurs in the superficial levels of the skin, muscle, or bone. All current flow is induced intracranially by the rapid change in current in the stimulating coil. An increase or decrease of current or a change in direction of current always establishes a magnetic field around the wire through which current is flowing. In transcranial magnetic stimulation, the electrically induced pulse of the magnetic field induces current flow in the underlying tissue.[3] When this current flow reaches threshold, it activates either a nearby neuron or a motor axon. These in turn generate action potentials that travel to the anterior horn cells and elicit compound muscle action potentials.

With optimal recording and no interference, supramaximal or nearly supramaximal compound muscle action potentials can be recorded with electrical and magnetic stimulation of the brain or spinal cord. At any given site, current flow is greater and activation is more readily obtained and more effective with electrical than with magnetic stimulation.

Recording

MEPs elicited by stimulation of the cerebral cortex or the cervical spinal cord may be recorded from any of the limb muscles. Attempts at recording MEPs from muscles innervated by cranial nerves have been limited to recordings from facial muscles with intracranial stimulation of the facial nerve.

The selection of muscles for recording is determined by the clinical problem. Specific recordings for muscles innervated by individual roots, such as C-6, C-7, and C-8 in patients undergoing cervical spine operations, are not often needed. Commonly, surface electrodes placed over the major muscle groups, including the extensors and flexors of the ankle, knee, wrist, and elbow, provide satisfactory measures of central motor conduction deficits. The thenar muscle in the hand and the anterior tibial muscle in the leg are the sites most frequently used and for which the most definitive normal data are available. The potentials may be recorded with subcutaneous electrodes, intramuscular wires, or other electrodes, which for most clinical needs add little to surface electrodes. The standard gains, filter settings, and sweep speeds that are used for peripheral motor conduction studies are usually satisfactory, although with some MEPs recorded intraoperatively the amplitudes are small enough to require averaging.

APPLICATIONS

MEPs are not widely used clinically because of several restricting factors. The equipment needed for either magnetic or transcranial electrical stimulation has not been widely available and is not approved for clinical use in the United States, mainly because the safety of the techniques has not been sufficiently defined to allow their adoption. Also, normal data have been obtained in only a few laboratories (and for specific stimulation and recording methods that have not been widely adopted).[4] Furthermore, variability of the response—with and without voluntary muscle contraction—has raised concern about the reliability of the responses for clinical interpretation.[11] Finally, MEPs have been shown to have clinical value primarily in patients with multiple sclerosis.

MEPs have been tested and reported on in various neurologic disorders, including cerebral infarcts, Parkinson's disease and other degenerative movement disorders, motor neuron disease, bony spondylosis of the cervical and lumbar spine, and other less common disorders.[6,8]

Demyelinating Disease

In clinical diagnostic neurology, the commonest use of MEPs is the search for evidence of demyelinating disease.[5] MEPs show marked slowing and dispersion in the presence of demyelinating disease in the spinal cord or brain stem. Such changes may be present even in the absence of any major or clear-cut neurologic deficit. Therefore, MEPs can be used to identify subclinical lesions in patients with multiple sclerosis. Many reports have shown the sensitivity of MEPs in recognizing and characterizing slowing of conduction in the motor pathways which may not produce a measurable motor deficit. Virtually all patients with multiple sclerosis who have weakness in a limb also have abnormal MEPs in that limb. Patients with multiple sclerosis who do not have a neurologic deficit in a limb nonetheless have up to 50% abnormal findings on MEP testing conducted in that limb.

Bony Spine Disease

MEPs are frequently abnormal in patients with cervical or lumbar spondylosis.[10] The changes may be a prolongation of latency in that segment of the recording or a change in configuration between sites of recording. In cervical spondylosis, abnormality on MEP testing may be caused by bony compression of the spinal cord or of the spinal nerves in the intervertebral foramina. The abnormalities are prolongation of latency and reduction in amplitude. Generally a change in latency is more prominent than a reduction in amplitude. Amplitude reduction is particularly difficult to assess in MEPs because it is difficult to obtain a supramaximal response in many normal subjects. Assessment of amplitude depends on the ability to evoke a supramaximal response. This difficulty is particularly true for the lower extremities, and it may also be the case in testing the upper extremities of elderly subjects.

Other Disorders

Case reports or reports of small numbers of patients with other disorders who underwent MEP testing are available, but their clinical significance is difficult to determine. Amyotrophic lateral sclerosis is reported to show abnormalities of latency, but these are difficult to distinguish from the changes that would be expected solely from loss of amplitude and loss of facilitation. None of these reports provide clear evidence of actual slowing of conduction in the motor pathways. MEP testing has also been extensively applied in analyzing the underlying pathophysiology of such disorders as Parkinson's disease, chorea, and myoclonus. MEPs have helped define the localization of motor cortex in humans. Attempts to apply measurements of threshold to MEPs have not been successful in distinguishing specific clinical disorders.

Complications

Reports of complications of MEP testing are few. Most studies that have specifically assessed safety have not found side effects. There is one report of damage to the hearing mechanism caused by a magnetic pulse applied near the ear. Another report described the initiation of epileptic discharge after repeated MEP testing in an epileptic patient. No evidence has been found that MEP testing causes increased frequency of seizures in patients with epilepsy.

Measurement Variables

MEPs are measured by the same variables used for peripherally evoked compound muscle action potentials, that is, latency and amplitude. Because MEPs traverse the peripheral segment of the motor pathway before being recorded, several methods have been developed to distinguish conduction over the central portion of the motor pathways from that over the peripheral portion. The most direct and reliable method of measuring MEP latency is by direct measurements that are compared with those of normal subjects matched for age and gender; however, according to some reports, this has not been sufficiently sensitive for identifying mild involvement by disease.

Central conduction time is a measure designed to better identify central disorders.[2]

It is calculated by subtracting the time needed for the signal to travel over the peripheral segment from the spinal cord to the muscle from the total latency of the MEP from the site of stimulation to the muscle. The latency of the peripheral segment may be obtained by evoking an MEP at the level of the spine and measuring latency at that point. Peripheral latency can also be obtained indirectly from F-wave measurements made during standard nerve conduction studies. Because the F wave traverses the peripheral motor pathway from the anterior horn cell to the muscle, F-wave latency minus the distal latency at the site of stimulation divided by two gives the time from the spinal cord to the distal site of stimulation. The distal latency is added to this to obtain the spinal latency for the MEP. The spinal latency subtracted from the latency obtained with cortical stimulation gives the central conduction time.

Peripheral Nerve Disease

Peripheral segments of the motor pathways are included in standard MEP latency and amplitude measurements; therefore, MEPs are often abnormal in cases of peripheral neuropathy. MEP technical problems and lack of reliability make this method less satisfactory for assessing peripheral neuropathies than standard nerve conduction studies. However, in some patients with primarily proximal involvement, as in acute or chronic demyelinating polyradiculopathy, MEP testing may demonstrate the slowing at the level of the spinal nerve and nerve root when it is not readily apparent on standard nerve conduction studies.

CHAPTER SUMMARY

MEPs are difficult to perform reliably, but they offer a direct method of assessing descending motor pathways. Both electrical and magnetic stimulation can elicit MEPs, but they have just been approved for clinical use in the United States. In Europe, MEP testing has proved most useful in identifying subclinical demyelination in multiple sclerosis and in monitoring motor pathways intraoperatively.

REFERENCES

1. Adams, DC, Emerson, RG, Heyer, EJ, et al: Monitoring of intraoperative motor-evoked potentials under conditions of controlled neuromuscular blockade. Anesth Analg 77:913–918, 1993.
2. Claus, D: Central motor conduction: Method and normal results. Muscle Nerve 13:1125–1132, 1990.
3. Counter, SA and Borg, E: Analysis of the coil generated impulse noise in extracranial magnetic stimulation. Electroencephalogr Clin Neurophysiol 85:280–288, 1992.
4. Eisen, A, Siejka, S, Schulzer, M, and Calne, D: Age-dependent decline in motor evoked potential (MEP) amplitude: With a comment on changes in Parkinson's disease. Electroencephalogr Clin Neurophysiol 81:209–215, 1991.
5. Jones, SM, Streletz, LJ, Raab, VE, et al: Lower extremity motor evoked potentials in multiple sclerosis. Arch Neurol 48:944–948, 1991.
6. Murray, NMF: Magnetic stimulation of cortex: Clinical applications. J Clin Neurophysiol 8:66–76, 1991.
7. Penfield, WP, and Jasper, HR: Epilepsy and the Functional Anatomy of the Human Brain. Little, Brown, Boston, 1954.
8. Pillai, JJ, Markind, S, Streletz, LJ, et al: Motor evoked potentials in psychogenic paralysis. Neurology 42:935–936, 1992.
9. Rossini, PM: Methodological and physiological aspects of motor evoked potentials. Electroencephalogr Clin Neurophysiol Suppl 41:124–133, 1990.
10. Travlos, A, Pant, B, and Eisen, A: Transcranial magnetic stimulation for detection of preclinical cervical spondylotic myelopathy. Arch Phys Med Rehabil 73:442–446, 1992.
11. Triggs, WJ, Kiers, L, Cros, D, et al: Facilitation of magnetic motor evoked potentials during the cortical stimulation silent period. Neurology 43:2615–2620, 1993.

SECTION 2
Electro-physiologic Assessment of Neural Function

PART D
Assessing the Motor Unit

Contraction of somatic muscles underlies all human movement. Therefore, motor pathways that control movement must act directly or indirectly on muscle. Motor units form the connection between motor pathways in the central nervous system and the muscles. For this reason, the motor neuron has been called the *final common pathway*. Lower motor neurons are the connection between the central motor pathways and the muscles.

Each motor unit consists of a cell body in the central nervous system, an axon in the peripheral nervous system, and all the muscle fibers innervated by that axon. The cell body is that of a motor neuron located in one of the cranial nerve motor nuclei in the brain stem or in the anterior horn of the spinal cord. The peripheral axon is a myelinated fiber that courses in one of the cranial nerves or peripheral nerves to the muscle, where the axon branches into multiple nerve terminals. The number of these terminal branches determines the innervation ratio: the number of muscle fibers in a motor unit. Innervation ratios are as small as 50 (in extraocular muscles and other small muscles requiring fine control) and as large as 2000 (in large, powerful muscles such as the gastrocnemius).

Most peripheral neuromuscular diseases involve one or more components of a motor unit. Primary myopathies and such disorders as myasthenia gravis affect the muscles and neuromuscular junctions. Mononeuropathies, peripheral neuropathies, and radiculopathies involve the peripheral axons,

255

whereas motor neuron diseases, such as amyotrophic lateral sclerosis, involve the anterior horn cells.

Nerve conduction studies measure alterations in function in the peripheral motor axons but only indirectly measure the extent of axonal destruction or anterior horn cell loss. The electrophysiologic assessments described in the three chapters in this section complement nerve conduction studies in defining the character, severity, and distribution of neuromuscular diseases. To distinguish among primary muscle disease, disorders of the neuromuscular junction, and neurogenic disorders often depends on needle electromyography. In addition, needle electromyography helps characterize the defects of neuromuscular transmission and the number of functioning axons or anterior horn cells. These two aspects of neuromuscular disease can be quantitated more precisely with the special techniques of single fiber electromyography and motor unit estimates that are reviewed in this section. By quantitating the number of motor units in a muscle or in a group of muscles, motor unit estimates can provide the most critical information about the severity and progression of a neurogenic process.

CHAPTER 24

ASSESSING THE MOTOR UNIT WITH NEEDLE ELECTROMYOGRAPHY*

Jasper R. Daube, M.D.

*Reprinted from Daube JR: AAEM minimonograph 11: Needle examination in clinical electromyography. Muscle Nerve 14:685–700, 1991. By permission of Mayo Foundation.

The electrical activity of motor units recorded with a needle electrode in a muscle is derived from the action potentials of the muscle fibers that are firing singly or in groups near the electrode. The muscle fiber and motor unit action potentials have a triphasic configuration. If the motor unit potential is recorded from a region of a muscle fiber that is unable to generate a negative potential (e.g., if the membrane of the muscle fiber is damaged), the potential will be recorded as a large positivity followed by a long low negativity.[23] For example, potentials recorded from damaged areas of fibrillating muscle fibers are recorded as positive waves (Fig. 24–1). The amplitude of the externally recorded action potential and the rate of rise of the positive-negative inflection (*rise time*) fall off exponentially in proportion to the distance of the recording electrode from the muscle fibers. Normally, the size and shape of the potential are constant each time it occurs.

257

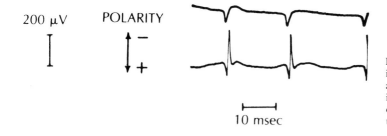

200 μV   POLARITY

10 msec

Figure 24–1. Simultaneous recording from one electrode that initiates the potentials when inserted into the muscle and from another electrode, a few millimeters distant, on the surface of the muscle.

Action potentials in individual muscle fibers may occur spontaneously or they may be initiated by external excitation. Normally, muscle fibers are under neural control and fire only in response to an end-plate potential that reaches threshold. Therefore, the rate and pattern of firing of all muscle fibers in the motor unit are under neural control. These are seen as *motor unit potentials*. Muscle fibers that are not innervated by an axon have an unstable muscle fiber membrane potential and fire individually without external stimulation, usually with a regular rhythm. These are *fibrillation potentials*.

NORMAL ELECTROMYOGRAPHIC ACTIVITY

Spontaneous Activity

Normal muscle fibers show no spontaneous electrical activity outside the end-plate region. In the end-plate area, spontaneous

miniature end-plate potentials occur randomly. These may be recorded with needle electrodes in the end-plate region as many monophasic negative waves that have amplitudes less than 10 μV and durations of 1 to 3 milliseconds or less. Individual potentials occur irregularly but usually cannot be distinguished. This activity is usually seen as an irregular baseline called *end-plate noise*; it has a typical "seashell sound" (Fig. 24–2).

The action potentials of some muscle fibers may be recorded in the end-plate region as brief spike discharges called *end-plate spikes*. They have a rapid irregular pattern, with an initial negative deflection. End-plate spikes sound like "sputtering fat in a frying pan." They are due to mechanical activation of a nerve terminal, with a secondary discharge of a muscle fiber. End-plate spikes often have interspike intervals of less than 50 milliseconds. If the muscle fibers have been damaged by the needle electrode, end-plate spikes will be recorded as rapid, irregularly firing, positive waves. Normal muscle fibers that are not innervated, for example, muscle

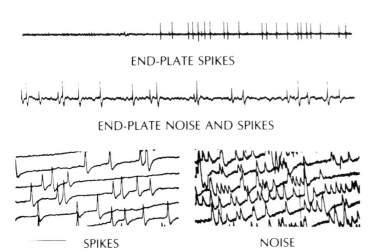

END-PLATE SPIKES

END-PLATE NOISE AND SPIKES

SPIKES   NOISE

Figure 24–2. Normal spontaneous activity in the end-plate region.

fibers in tissue culture, generate rhythmic spontaneous action potentials. These potentials are associated with regular muscle fiber twitches called *fibrillations.*

Voluntary Activity

All voluntary muscle activity is mediated by lower motor neurons and the muscle fibers they innervate (motor units) and is recorded electrically as motor unit potentials. The *motor unit potential* is the sum of the potentials of muscle fibers innervated by a single anterior horn cell. The muscle fibers in the region of the needle electrode discharge in near synchrony; the motor unit potential has a more complex configuration of higher amplitude and longer duration than a single fiber action potential. Motor unit potentials may be characterized by their *firing pattern* and by their *appearance.* Only a small proportion of the fibers in a motor unit are near the electrode; those at a distance contribute little to the motor unit potential. Thus, the appearance of a motor unit potential from one motor unit varies with electrode position. No single motor unit potential characterizes a motor unit but rather a multiplicity of them recorded from different sites (Fig. 24–3).

Motor unit potentials are under voluntary control and have a characteristic semirhythmic firing pattern in which the rate is continuously changing by small amounts. The firing pattern of the motor unit potential is assessed in terms of rate and recruitment.

Recruitment is the initiation of the firing of additional motor units as the rate of discharge of the active motor unit potential increases. Normally, recruitment of additional motor unit potentials occurs at low levels of effort and at slow rates of firing (Fig. 24–4). Recruitment can be characterized by the *recruitment frequency*, the frequency of firing of a unit when the next unit is recruited (i.e., begins to discharge). This is a function of the number of units capable of firing and is 7 to 16 Hz for motor units in a normal muscle during mild contraction.[17] Recruitment frequencies vary in different muscles and for different types of motor units. Recruitment may also be characterized by the ratio of the rate of firing of the individual motor units to the number that are active. Normally, this ratio averages less than 5 (e.g., 3 units firing at about 15 Hz each). If the ratio is greater than 5 (e.g., 2 units firing at 16 Hz), a loss of motor units is indicated. Motor unit firing can also be characterized in terms of the number of motor unit potentials firing in relation to the force being exerted.

A motor unit potential is also characterized by its appearance, including its duration (onset from the baseline to the final return to baseline), number of phases (baseline crossings plus one), amplitude (peak to peak), turns (potential reversals), area (under the curve), and rate of rise of the fast component, that is, rise time (Fig. 24–5).

Each of these characteristics has multiple

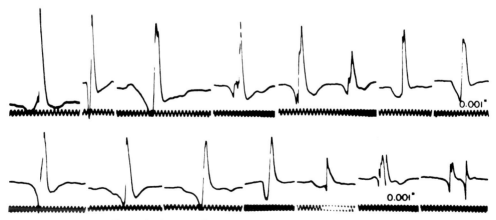

Figure 24–3. Normal motor unit potentials in biceps muscle.

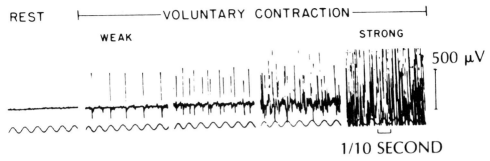

Figure 24–4. Recruitment of motor unit potentials.

determinants. Technical factors have a major influence on the appearance of motor unit potentials.[16] These factors include the type of needle electrode, the area of exposed surface of the active leads, the characteristics of the metal recording surfaces, and the electrical characteristics of the cables, preamplifier, and amplifier.

Appearance of motor unit potentials also changes with a number of normal physiologic variables, including the subject's age, the muscle being studied, the location of the needle in the muscle, the manner of activation of the potentials (minimal voluntary contraction, maximal voluntary contraction, reflex activation, or electric stimulation), and the temperature of the muscle.[2,4,8,9]

If these technical and physiologic factors

are controlled, the characteristics of the motor unit potentials will be determined by the normal anatomic and histologic features of the motor unit and any pathologic changes that may affect these features. These anatomic and histologic features include the innervation ratio (number of muscle fibers in the motor unit), the fiber density (number of muscle fibers per given cross-sectional area), the distance of the needle tip from the muscle fibers and from the end-plate region, and the direction of the muscle fiber axis. The capacitance and resistance of the tissue between the electrode and the discharging muscle fibers depend on the amount of connective tissue, blood vessels, and fat. The characteristics of the action potentials generated by individual muscle fibers depend

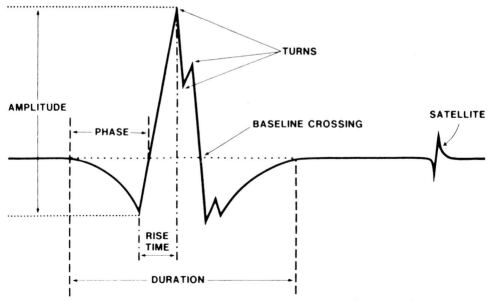

Figure 24–5. Motor unit potential with characteristics that can be measured.

on muscle fiber membrane resistance and capacitance, intracellular and extracellular ionic concentrations, muscle fiber diameter, and conduction velocity. The synchrony of firing of the muscle fibers in a motor unit depends on the length, diameter, and conduction velocity of the nerve terminals; the diameter of the muscle fibers; and the relative location of the end-plates on the muscle fibers. The firing characteristics of the motor unit depend on the amount of overlap with other motor units, the number of motor units in the muscle (or per given area), the differential response to sources of activation (monosynaptic, local spinal cord, higher centers), and the rates and patterns of discharge of the anterior horn cell.

The configuration of a motor unit potential may be monophasic, biphasic, triphasic, or have multiple phases. The configuration depends on the synchrony of firing of the muscle fibers in the region of the electrode. Usually, only a small proportion of motor unit potentials have more than four phases; those that do are called *polyphasic potentials.* The percentage varies with the muscle being tested and the age of the patient. A late spike, distinct from the main potential, that is time locked to the main potential may occur; this is called a *satellite potential* (see Fig. 24–5). The satellite potential is generated by a muscle fiber in a motor unit that has a long nerve terminal, narrow diameter, or distant end-plate region. If a motor unit potential is recorded from damaged muscle fibers or from the end of the muscle fibers, it may have the configuration of a positive wave with low, long, late negativity.

The *rise time* is the duration of the rapid positive-negative inflection and is a function of the distance of the muscle fibers from the electrode. It is less than 500 microseconds if the electrode is near muscle fibers in the active motor unit. The *duration* of the motor unit potential is the time from the initial deflection away from baseline to the final return to baseline. It varies with the muscle, muscle temperature, and the patient's age. The *amplitude* of the motor unit potential is the maximum peak-to-peak amplitude of the potential and varies with the size and density of the muscle fibers in the region of the recording electrode and with their synchrony of firing. It also differs with the muscle, muscle temperature, and the patient's age (Fig. 24–6). Decreased muscle temperature produces higher amplitude, longer duration motor unit potentials (see Fig. 24–6). *Variability* of a motor unit potential is any change in its configuration, or amplitude, or both in the absence of movement of the recording electrode as the motor unit fires repetitively. Normally, there is no such variation.

The polarity of all potentials recorded on needle electromyography (EMG) depends on recording the potential with the active (G1 amplifier input) electrode. If a motor unit potential is recorded with the shaft of a standard concentric electrode or with the reference of a "monopolar" electrode, it will be displayed as an inverted triphasic potential (apparently negative-positive-negative).

ELECTRODE TYPES

Various types of electrodes have been used to record the electrical activity of normal and diseased muscles. Surface electrodes

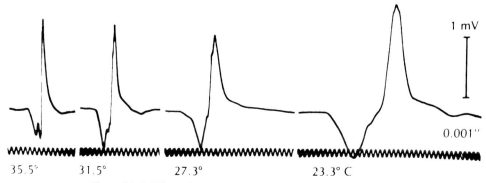

Figure 24–6. Effect of muscle temperature on motor unit potential.

35.5°   31.5°   27.3°   23.3° C

1 mV

0.001″

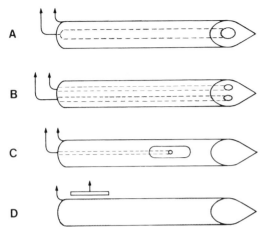

Figure 24–7. Electrode types used in recording electromyographic signals. (*A*) The standard concentric and (*D*) monopolar electrodes are the ones commonly used for clinical diagnostic recordings. (*B*) Bipolar concentric and (*C*) single-fiber electromyographic needle electrodes are used to isolate potentials.

can depict the extent of EMG activity and measure the conduction velocity of muscle fibers, but they are not helpful in distinguishing neuromuscular diseases, because of distortion of the electrical signals by intervening skin, subcutaneous tissue, and other muscles. Needle electrodes inserted into the muscle can accurately depict the electrical signals, but depending on needle type, needle electrodes record from different numbers of muscle fibers and from muscle fibers in different locations (Fig. 24–7).

Single-Fiber Electrode

Recordings made with small recording surfaces (25 μm) referenced to the shaft of the needle, with filtering of the low-frequency components, focus on a small number of muscle fibers in the immediate vicinity of the electrode.[29] *Single-fiber EMG needles* record from small areas of muscle and cannot be used to characterize motor unit potential size. This method has been used

primarily in disorders of neuromuscular transmission, because it can detect variation in motor units (*jitter* between single fiber potentials) not seen with other needle electrodes. Single-fiber EMG can also be used to quantitate the density of muscle fibers in a motor unit (*fiber density*), a measurement closely related to the percentage of motor unit potentials that are polyphasic and to the number of turns on the motor unit potential.

A larger area of muscle is sampled with a *bipolar concentric electrode*, in which two recording surfaces (80 × 320 μm) are side-by-side and insulated from one another in the beveled open end of the needle (Fig. 24–8). This electrode has not been used for standard clinical diagnostic studies; thus, standard values for the size and shape of motor unit potentials obtained with this electrode have not been defined. This type of electrode has been used to record the patterns of firing of single motor units at high levels of force because it is able to isolate individual motor unit potentials better than other electrodes.

Standard Concentric Electrodes

Standard concentric electrodes are one of the two types of electrodes commonly used for diagnostic clinical EMG. A bare, 24-gauge to 26-gauge hollow needle with a fine wire down the center is bevelled at its tip to expose an active, oval recording surface of 125 × 580 μm. The electrode is referenced to the shaft of the needle, thereby cancelling unwanted activity from surrounding muscle. Although these electrodes previously were expensive, inexpensive disposable models are now available. This type of electrode has several advantages: (1) its ability to record EMG activity with a minimum of interference from surrounding muscles; (2) its fixed-size recording surface; (3) the absence of a separate reference electrode; and (4) the extensively defined quantitation of nor-

Figure 24–8. Normal insertional activity.

mal motor unit potential sizes for various ages and muscles.

Monopolar Electrode

The monopolar electrode is a solid 22-gauge to 30-gauge Teflon-coated needle with a bare tip of approximately 500 μm in diameter. Monopolar electrodes record essentially the same activity as recorded with a standard concentric electrode, but motor unit potentials are slightly longer in duration and have a higher amplitude.[10] Monopolar electrodes are preferred by electromyographers because they are less expensive and, at times, less uncomfortable for patients.

Macroelectrode

A larger needle electrode is the *"macro" needle*, or macroelectrode.[28,30] A macroelectrode recording is made from 15 mm of the shaft of a needle electrode referenced to a surface electrode. The macroelectrode records from a large number of muscle fibers of multiple motor units in a cylinder along the shaft of the needle. This recording summates the activity of many motor unit potentials, which of course cannot be differentiated from one another. The potential from a single motor unit potential is isolated with the help of simultaneous recording of potentials from single muscle fibers with a 25-μm-diameter electrode halfway along the shaft of the macroelectrode on a second channel. The second channel is used to identify the firing pattern of a single motor unit. The electrical activity recorded from the macroelectrode at the time of the firing of a single fiber potential on the small electrode is averaged over multiple discharges. This results in an averaged potential from all muscle fibers along the macroelectrode which are innervated by the same motor unit as the single muscle fiber. Thus, the averaged potential gives an estimate of the activity in a larger portion of the muscle fibers of the motor unit. Occasionally, macrorecordings are able to identify changes in the whole motor unit which are not apparent with smaller electrodes.[15]

SIGNAL ANALYSIS

The spontaneous activity and motor unit potentials recorded from muscle can be displayed and analyzed in several different ways.[12] The usual method in clinical studies is to display and measure isolated potentials (described later in this chapter); however, other approaches that analyze the entire sequence of waveforms in an interference pattern when multiple motor units are firing have also been used. Such analyses are applied almost exclusively to motor unit activity, but not to spontaneous activity. These methods are not widely used clinically.

Frequency analysis, most commonly performed by a fast Fourier transformation, determines the frequency components in a limited segment of a recording, usually a few seconds.[1] These frequencies range from very low (less than 1 Hz) to more than 10,000 Hz. Small shifts in the frequency components occur in some diseases.[14] A more common approach to interference pattern analysis is to measure the number of reversals of potential (turns) per unit time (or the number of baseline crossings) and the amplitude of the overall signal.[13,20] The ratios of these to each other can be defined for normal muscles, and the ratios change with disease.

Measurement of Motor Unit Potentials

Assessment of the variables of a motor unit potential may be performed subjectively, but this requires experience and knowledge of the normal variations from muscle to muscle and with age. It is more reliable to make a quantitative measurement of the characteristics of motor unit potentials; this is needed in questionable cases to increase the certainty of a diagnosis.[25] In mild myopathies, such objective measurement may be a necessity.

Measurements may be made in two ways: by isolation and measurement of a single motor unit potential and by interference pattern analysis. Because of the number and variety of normal motor unit potentials, both of these methods require multiple measurements and a statistical description of the results obtained from different areas of a muscle.

The classic method of measuring a motor unit potential is to isolate and record at least 20 single potentials and then manually measure the duration, number of phases, and amplitude. These measurements must be compared with the values recorded from the same muscle in normal subjects of the same age. This method provides no quantitative assessment of recruitment and makes the measurements only at minimal-to-moderate levels of contraction. Recently developed digital EMG machines have automated the measurements.

Interference pattern analysis summates the effect of recruitment with the duration and amplitude of the potentials and records the number of turns and total amplitude of the electrical activity during a fixed period of time with an automatic counting device.[24] This method varies with patient effort, which must be accounted for in measurements. Recording EMGs by isolation of single motor unit potentials and by interference pattern provide reliable estimates of the electrical activity in a muscle. The results of these two methods correlate well with each other and with muscle histology. Neither method has been shown to be superior; they may complement one another.

Recording Technique

Needle electrode recording must sample three types of activity: *insertional activity, spontaneous activity*, and *voluntary activity*. Because needle electrodes primarily record activity from a small area in a muscle, the electrode must be moved in the muscle to record each type of activity in multiple areas. Adequate control during needle manipulation can only be obtained manually, with small advances of the needle. With the examiner's hand resting on the patient, the needle should be held firmly and steadily in the hand without release. Small movements are less painful than large movements with insertional bursts longer than 500 milliseconds.

Each electromyographer develops a preference for how to display the electrical activity; however, certain variables should be familiar to all examiners because of their common use and advantages in certain situations. Oscilloscope sweep speeds of 5 to 10 milliseconds per centimeter are best for characterizing the appearance of motor units, but slower speeds of 50 or 100 milliseconds per centimeter are needed to characterize firing patterns. Amplification settings of 50 μV/cm and 200 μV/cm are most useful for examining spontaneous and voluntary activity, respectively. Filter settings of approximately 30 Hz and 10,000 Hz or more should be used for routine studies. Measurements of motor unit potential duration should be made with a gain of 100 μV/cm at a sweep speed of 5 milliseconds per centimeter (10 milliseconds per centimeter if long duration) and with a low filter at 2 to 3 Hz. The convention of displaying negative potentials at the active electrode as upward deflections is now generally accepted in clinical EMG.

Insertional activity is the electrical response of the muscle to the mechanical damage by a small movement of the needle (see Fig. 24–8). Evaluation of insertional activity requires a pause of 0.5 to 1 second or more to see any repetitive potentials that may be activated. Insertional activity may be increased, decreased, or show specific wave forms, such as myotonic discharges.

Spontaneous activity occurs with the needle and muscle both at rest but does not include motor unit potentials occurring because of upper motor neuron activity, as in poor relaxation, tremor, or spasticity (these can be distinguished by their firing patterns and are discussed later in this chapter). Spontaneous activity is assessed during the pauses between needle movements. It decreases when muscle temperature is low. A search for spontaneous activity requires adequate relaxation of the muscle, which may be difficult to obtain in some patients. Relaxation is achieved readily by gentle manipulation or passive positioning of the extremity. Occasionally, it can be obtained by activation of an antagonist, for example, by neck flexion in testing cervical paraspinal muscles.

The recruitment and appearance of motor unit potentials are examined during *voluntary activity*. Several different motor unit potentials (a minimum of 20) in different areas of the muscle must be assessed. They are studied most efficiently by having the patient maintain a *minimal voluntary contraction* while the needle is advanced through different areas of the muscle. It is difficult to identify individual motor unit potentials when there is tremor or strong contractions, and attempts should be made to minimize these conditions. Motor unit potentials cannot be measured reliably during a strong voluntary contraction, which normally produces a dense pattern of multiple superimposed potentials called an *interference pattern*. Less dense patterns may occur with a loss of motor units, poor effort, an upper motor lesion, or in a powerful muscle. The latter three conditions can be distinguished from a loss of motor units only by estimates of firing rates. The combination of rapid firing rates of individual potentials and few motor units occurs only with a loss of motor units.

Care must be used to prevent muscle damage, bleeding, and needle damage during strong contractions. During contraction, it is best to start with the needle tip in a subcutaneous location and to insert it into the muscle after the muscle is contracted. Otherwise, muscle damage and pain are likely, the needle electrode may be bent, and the electrode may be dislodged from the muscle into surrounding tissue. The presence of polyphasic potentials may also mask a loss of motor unit potentials in an interference pattern. The most reliable method of judging a loss of motor unit potentials is by assessing recruitment from the ratio of rate to number of motor unit potentials during mild-to-moderate contraction rather than during strong contraction.

ABNORMAL ELECTRICAL ACTIVITY

Neuromuscular diseases are best described by a combination of clinical findings, histologic changes, and the pattern of abnormal findings on needle EMG.[5] Needle EMG findings are combinations of different specific types of abnormal electrical wave forms described later in this chapter. The clinical electromyographer must recognize specific discharges and know what diseases are associated with them. In most cases, a specific discharge may be associated with several different diseases. In the following discussion, the types of abnormal electrical activity recorded with a needle electrode and the diseases associated with them are described.

Neuromuscular diseases may show abnormal spontaneous discharges or abnormal voluntary motor unit potentials or both. Abnormal spontaneous activity includes fibrillation potentials, fasciculation potentials, myotonic discharges, complex repetitive discharges, myokymic discharges, cramps, and neuromyotonic discharges. Motor unit potentials may have an abnormal duration, be polyphasic, or vary in size. The recruitment pattern of motor unit potentials may be altered or there may be abnormal patterns of activation, as in tremor and synkinesis.

Recognition of all normal and abnormal EMG discharges is made most accurately and reliably by auditory pattern recognition. The trained ear of an electromyographer can define the discharge frequency, rise time, duration, and number of turns/phases of EMG potentials. This skill is called *semiquantitation*.

Insertional Activity

Insertional activity may be prolonged (Fig. 24–9) or reduced from the brief burst seen in normal subjects. Prolonged insertional activity occurs in two types of normal variants and in denervated muscle and myotonic discharges.

Normal variants are recognized by their widespread distribution. One type is composed of short trains of regularly firing positive waves. It is sometimes familial and may be a subclinical myotonia. The second type has short recurrent bursts of irregularly firing potentials. This type is seen most often in muscular individuals, especially in their calf muscles. Reduced insertional activity occurs in periodic paralysis (during paralysis) and with replacement of muscle by connective tissue or fat in myopathies and neurogenic disorders.

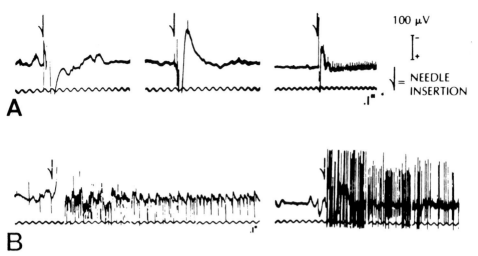

Figure 24–9. (*A*) Normal insertional activity. (*B*) Increased insertional activity.

Fibrillation Potentials

Fibrillation potentials are the action potentials of single muscle fibers that are twitching spontaneously in the absence of innervation. These potentials typically fire in a *regular* pattern at rates of 0.5 to 15 per second (Fig. 24–10). Infrequently, they may be intermittent or irregular, but if so, the interspike interval is longer than 70 milliseconds. Fibrillation potentials have one of two forms: either a brief spike or a positive wave. When seen as brief spikes, fibrillation potentials are triphasic or biphasic, 1 to 5 milliseconds in duration, and 20 to 200 μV in amplitude, with an initial positivity (unless recorded at their site of origin). When seen as positive waves, fibrillation potentials are of long du-

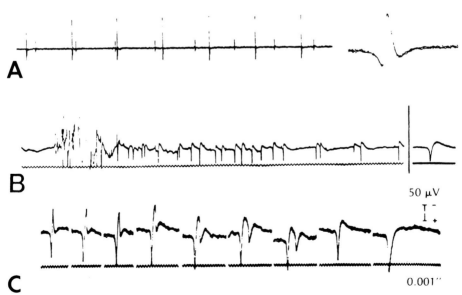

Figure 24–10. Fibrillation potentials. (*A*) Spike form. (*B*) Positive waveform. (*C*) Development of a positive waveform from a spike form (serial photographs taken after insertion of needle electrode).

ration and biphasic, with an initial sharp positivity followed by a long-duration negative phase. Amplitudes are from 20 to 200 μV, and durations are from 10 to 30 milliseconds. Amplitude is proportional to muscle fiber diameter and decreases with muscle atrophy.[21] The positive waveforms are muscle fiber action potentials recorded from an injured portion of the muscle fiber. The spike form and the positive waveform are both recognized as fibrillation potentials by their slow, regular firing pattern, which is like the "ticking of a clock."

Any muscle fiber that is not innervated can be expected to fibrillate; thus, many neurogenic and myopathic disorders show fibrillation potentials (Table 24–1). These potentials may occur in muscle fibers that (1) have lost their innervation, (2) have been sectioned transversely or divided longitudinally, (3) are regenerating, and (4) have never been innervated. The density of fibrillation potentials is a rough estimate of the number of denervated muscle fibers (Fig. 24–11).

Other forms of electrical activity may be mistaken for fibrillation potentials. These include the spontaneous activity in the region

Table 24–1. **DISEASES WITH FIBRILLATION POTENTIALS**

Lower motor neuron diseases
   Anterior horn cell diseases
   Polyradiculopathies
   Radiculopathies
   Plexopathies
   Peripheral neuropathies, especially axonal
   Mononeuropathies
Neuromuscular junction diseases
   Myasthenia gravis
   Botulinum intoxication
Muscle diseases
   Myositis
   Duchenne dystrophy
   Myotonic dystrophy
   Myotubular myopathy
   Late-onset rod myopathy
   Toxic myopathy
   Hyperkalemic periodic paralysis
   Acid maltase deficiency
   Rhabdomyolysis
   Trichinosis
   Muscle trauma

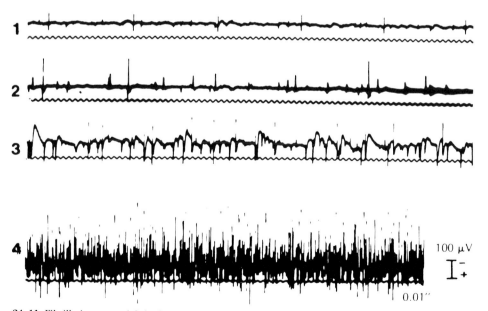

Figure 24–11. Fibrillation potentials in denervated muscle. Grades of activity: 1+, fibrillation potentials persistent in at least two areas; 2+, moderate number of persistent fibrillation potentials in three or more areas; 3+, large number of persistent discharges in all areas; 4+, profuse, widespread, persistent discharges that fill the baseline.

Figure 24–12. Myotonic discharge.

of the end-plate (end-plate noise and end-plate spikes), short-duration motor unit potentials, and motor unit potentials with a positive configuration. All of them are best distinguished from fibrillation potentials by their firing patterns.

Myotonic Discharges

Myotonic discharges are the action potentials of muscle fibers that are firing spontaneously in a prolonged fashion after external excitation. They are less readily elicited in a muscle that has just been active. The potentials wax and wane in amplitude and frequency because of an abnormality in the membrane of the muscle fiber. Myotonic discharges are regular in rhythm, but vary in frequency between 40 and 100 per second, which makes them sound like a "dive-bomber." Myotonic discharges occur as brief spikes or positive waveforms, depending on the relationship of the recording electrode to the muscle fiber. When initiated by insertion of the needle, myotonic potentials have the configuration of a positive wave, with an initial sharp positivity followed by a long-duration negative component. These are action potentials recorded from an injured area of the fiber. Both amplitude and frequency may increase or decrease as the discharge continues (Fig. 24–12). Myotonic discharges that occur after a voluntary contraction are brief, biphasic or triphasic,

Table 24–2. MUSCLE DISEASES WITH MYOTONIC DISCHARGES

Myotonic dystrophy
Myotonia congenita
Paramyotonia
Hyperkalemic periodic paralysis
Polymyositis
Acid maltase deficiency

initially positive spikes of 20 to 300 μV that resemble the spikes of fibrillation potentials. They wax and wane, similar to the mechanically induced myotonic discharges. This afterdischarge corresponds to the clinically evident poor relaxation.

Myotonic discharges may occur with or without clinical myotonia in several disorders (Table 24–2). Rarely, similar discharges may be seen with fibrillation potentials in chronic denervating disorders and with some drugs (e.g., 20,25-diazocholesterol, triparanol, 2,4-dichlorophenoxyacetic acid, clofibrate, propranolol).

Complex Repetitive Discharges

Complex repetitive discharges, referred to previously as bizarre repetitive (or high-frequency) potentials or as pseudomyotonic discharges, are the action potentials of

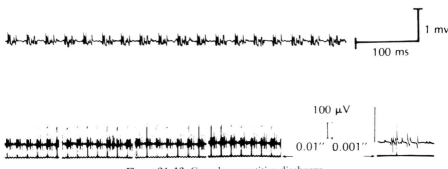

Figure 24–13. Complex repetitive discharge.

Table 24–3. **DISORDERS ASSOCIATED WITH COMPLEX REPETITIVE DISCHARGES**

Neuropathies
    Polymyositis
    Amyotrophic lateral sclerosis
    Spinal muscular atrophy
    Chronic radiculopathies
    Charcot-Marie-Tooth disease
    Chronic neuropathies
Myopathies
    Poliomyelitis
    Duchenne dystrophy
    Limb-girdle dystrophy
    Myxedema
    Schwartz-Jampel syndrome

groups of muscle fibers discharging spontaneously in near synchrony. Standard and single-fiber EMG recordings suggest that they are the result of ephaptic activation of groups of adjacent muscle fibers.[26] Complex repetitive discharges are characterized by abrupt onset and cessation. During the discharge, they may have abrupt changes in their configuration. They have a uniform frequency that ranges from 3 to 40 per second (Fig. 24–13). Although their form is variable, it typically is polyphasic, with 3 to 10 spike components with amplitudes from 50 to 500 μV and durations of up to 50 milliseconds. Complex repetitive discharges sound like "a motor boat that misfires." They occur in several chronic disorders, both myopathic and neurogenic (Table 24–3).

Complex repetitive discharges may be confused with other repetitive discharges, such as myokymic discharges, cramps, neuromyotonia, tremor, and synkinesis. However, each of these has a characteristic pattern of firing best recognized by its sound and distinct from that of complex repetitive discharges.

Fasciculation Potentials

Fasciculation potentials are the action potentials of a group of muscle fibers innervated by an anterior horn cell that discharges in a random fashion (Fig. 24–14). The rates of discharge of an individual potential may vary from a few per second to less than 1 per minute. The sum of all fasciculations in a muscle may reach 500 per minute. Fasciculation potentials may be of any size and shape, depending on the character of the motor unit from which they arise and their relationship with the recording electrode, and may have the appearance of normal or abnormal motor unit potentials. They can be identified only by their firing pattern. The discharges may arise from any portion of the lower motor neuron but usually from spontaneous firing of the nerve terminal. The random occurrence sounds like "large raindrops on a roof."

Fasciculation potentials may occur in normal persons and in many diseases. They are especially common in chronic neurogenic disorders but have been seen in all neuromuscular disorders (Table 24–4). It has not been clearly shown that they occur more of-

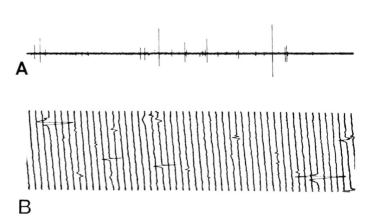

Figure 24–14. Fasciculation potentials. (*A*) Slow sweep speed, continuous. (*B*) Fast sweep speed, raster.

Table 24–4. COMMON SOURCES OF FASCICULATION POTENTIALS

Normal
   Benign (fatigue)
   Benign with cramps
Metabolic disorders
   Tetany
   Thyrotoxicosis
   Anticholinesterase medication
Lower motor neuron diseases
   Amyotrophic lateral sclerosis
   Root compression
   Peripheral neuropathy
   Jakob-Creutzfeldt disease

ten in patients with myopathy than in normal persons. Fasciculations usually occur in an overworked muscle, especially if there is underlying neurogenic disease.

No reliable method exists for distinguishing benign fasciculations from those associated with specific diseases. However, in normal persons, fasciculations occur more rapidly, on the average. Patients who have large motor units caused by chronic neurogenic diseases may have visible twitching during voluntary contractions. Such *contraction fasciculations* must be differentiated from true fasciculations by the pattern of firing.

Myokymic Discharges

Spontaneous muscle potentials associated with the fine, worm-like quivering of facial myokymia are called *myokymic discharges.* These potentials have the appearance of normal motor unit potentials that fire with a fixed pattern and rhythm. They occur in bursts of 2 to 10 potentials that fire at 40 to 60 Hz. The bursts recur at regular intervals of 0.1 to 10 seconds (Fig. 24–15). The firing pattern of any one potential is unrelated to the myokymic discharge pattern of other potentials and is unaffected by voluntary activity. Myokymic discharges sound like "marching soldiers."

Some forms of clinical myokymia—especially the form in the syndrome of continuous muscle fiber activity (Isaac's syndrome)—are associated with neurotonic discharges and not with myokymic discharges. Although discharges that have regular patterns of recurrence but fire at different rates or with a regularly changing rate of discharge may have similar mechanisms, they are better classified with the broad group of iterative discharges. Some investigators consider iterative discharges and myokymic discharges to be forms of fasciculation because they arise in the lower motor neuron or axon. It is best to separate these discharges from fasciculation potentials because of their distinct patterns and different clinical significance. Diseases associated with myokymic discharges are listed in Table 24–5.

Table 24–5. DISORDERS ASSOCIATED WITH MYOKYMIC DISCHARGES

Facial muscles
   Multiple sclerosis
   Brain stem neoplasm
   Polyradiculopathy
   Facial palsy
Extremity muscles
   Radiation plexopathy
   Chronic nerve compression (e.g., carpal
      tunnel syndrome)

100 μv
200 ms

Figure 24–15. Facial myokymia in orbicularis oris muscle.

Neurotonic Discharges (Neuromyotonia)

Motor unit potentials that are associated with some forms of continuous muscle fiber activity (Isaac's syndrome) and that fire at frequencies of 100 to 300 Hz are called *neurotonic discharges*, or *neuromyotonia* (Figs. 24–16, 24–17, 24–18). These potentials may decrease in amplitude because of the inability of muscle fibers to maintain discharges at rates greater than 100 Hz. The discharges may be continuous for long intervals or recur in bursts. They are unaffected by voluntary activity and are commonly seen in neurogenic disorders (Table 24–6).[18]

Neuromyotonia occurring with tetany may be distinguished by its precipitation or augmentation with ischemia. Neurotonic dis-

Table 24–6. **DISORDERS ASSOCIATED WITH NEUROTONIC DISCHARGES**

Syndrome of continuous muscle fiber activity
    Spontaneous (see Fig. 24–16)
    Hereditary (see Fig. 24–17)
Anticholinesterase poisoning
Tetany
Chronic spinal muscular atrophy (see Fig. 24–18)

charges also occur intraoperatively with the mechanical irritation of cranial or peripheral nerves and, thus, are valuable in alerting surgeons to possible nerve damage (Fig. 24–19).

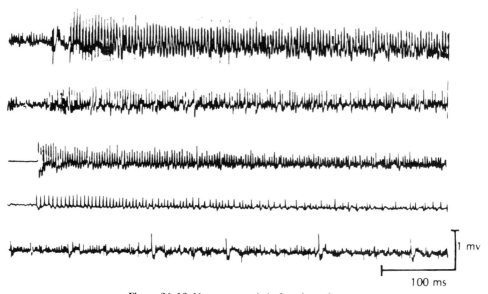

Figure 24–16. Neuromyotonia in Isaac's syndrome.

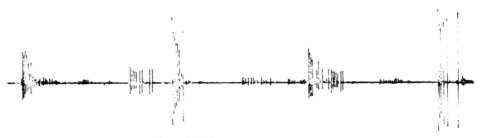

Figure 24–17. Familial neuromyotonia.

First dorsal interosseous

1 mv

500 ms

Abductor digiti quinti

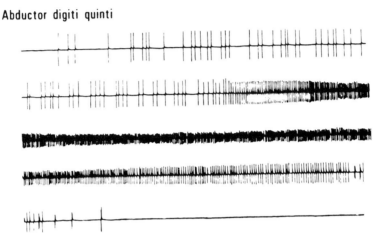

Figure 24–18. Neuromyotonia in spinal muscular atrophy (two forms).

EMG

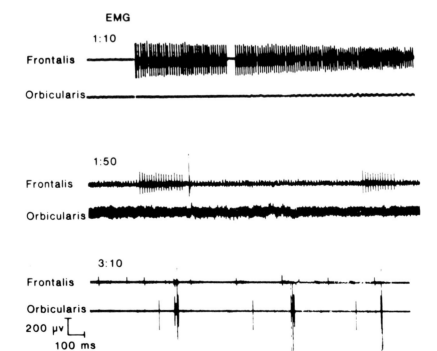

Figure 24–19. Neurotonic discharges in facial muscles during acoustic neuroma surgery. The times of recordings were at 1:10 PM, 1:50 PM, and 3:10 PM.

Cramp Potentials

The potentials associated with a muscle cramp do not have a specific name but can be distinguished from other spontaneous activity by their firing pattern. Individual potentials are not distinctive and resemble motor unit potentials. They fire rapidly at 40 to 60 Hz, usually with abrupt onset and cessation; however, during their discharge, they may fire irregularly in a sputtering fashion, especially just before termination (Fig. 24–20). Typically, increasing numbers of potentials that fire at similar rates are recruited as the cramp develops and then drop out as the cramp subsides. Cramps are a common phenomenon in normal persons and usually occur when a muscle is activated strongly in a shortened position (Table 24–7).

Short-Duration Motor Unit Potentials

Single potentials that are outside the normal range or groups of motor unit potentials that have a mean duration less than the normal range for the same muscle in a patient of the same age are called *short-duration motor unit potentials.* Commonly, these potentials are also low in amplitude and show rapid re-

Table 24–7. **DISORDERS ASSOCIATED WITH CRAMP DISCHARGES**

Salt depletion
Chronic neurogenic atrophy
Benign nocturnal cramps
Myxedema
Pregnancy
Uremia (dialysis)

cruitment with minimal effort, but they may have reduced recruitment or normal amplitude (or both). They may be as short in duration as a fibrillation potential if only a single muscle fiber is in the recording area of the needle electrode. Short-duration motor unit potentials occur in diseases in which there is (1) physiologic or anatomical loss of muscle fibers from the motor unit, or (2) atrophy of component muscle fibers (Table 24–8).

Some myopathies, such as metabolic and endocrine disorders, show few short-duration motor unit potentials or none. Technical errors (e.g., incorrect filter settings, electrical short in the recording electrode or connecting cables, and reduced recording surface area) that may produce similar find-

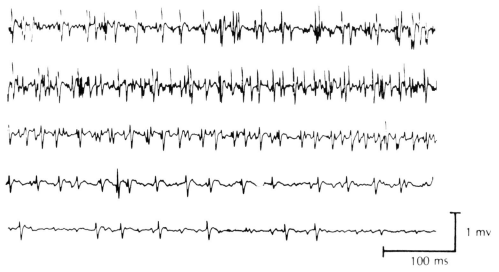

1 mv

100 ms

Figure 24–20. Muscle cramp (30 to 50 Hz).

Table 24–8. DISORDERS ASSOCIATED WITH SHORT-DURATION MOTOR UNIT POTENTIALS

Myasthenia gravis
Myasthenic syndrome
Botulinum intoxication
Early reinnervation after nerve damage
Late-stage neurogenic atrophy
Muscular dystrophies (all forms)
Periodic paralysis
Polymyositis
Toxic myopathies
Congenital myopathies (not all forms)

ings must be excluded to identify true short-duration motor unit potentials.

Short-duration motor unit potentials may be intermingled with long-duration potentials in rapidly progressing neurogenic atrophies such as amyotrophic lateral sclerosis and Werdnig-Hoffman disease.

Long-Duration Motor Unit Potentials

Individual motor unit potentials that are outside the normal range or groups of motor unit potentials that have a mean duration greater than the normal range for the same muscle in a patient of the same age are

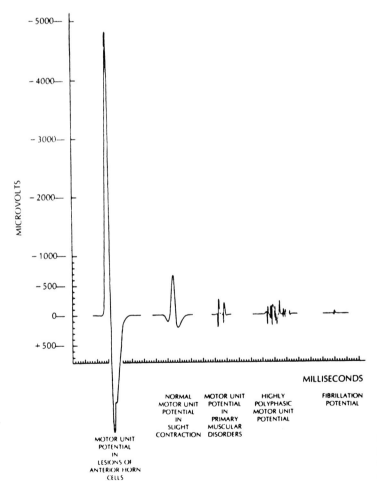

Figure 24–21. Relative average durations and amplitudes of some electric potentials observed in electromyography of human muscle.

<table>
Table 24–9. **DISORDERS ASSOCIATED WITH LONG-DURATION MOTOR UNIT POTENTIALS**

Motor neuron diseases (all types)
Axonal neuropathies with collateral sprouting
Chronic radiculopathies
Chronic mononeuropathies
Residual of a neuropathy (polyneuropathy or mononeuropathy)
Chronic myositis
</table>

called *long-duration motor unit potentials* (Fig. 24–21). They generally have high amplitude and show poor recruitment, but they may have normal or low amplitude. Motor unit potentials recorded from damaged muscle fibers are preponderantly positive and have a long late negativity, which is a recording artifact that should not be measured. Long-duration motor unit potentials occur in diseases in which there is increased fiber density in a motor unit, an increased number of fibers in a motor unit, or loss of synchronous firing of fibers in a motor unit (Table 24–9). Long-duration motor unit potentials are a frequent finding in inclusion body myositis and in some cases of chronic polymyositis.

Polyphasic Motor Unit Potentials

A polyphasic motor unit potential is one that has five or more phases. (Phase is defined as the area of a potential on either side of the baseline; it is equal to the number of baseline crossings plus one.) Polyphasic potentials may be of any duration. Some may have late, satellite components, sometimes called *linked potentials* or *parasites*, that give the total unit a long duration (Fig. 24–21 and 24–22).[22] Such long-duration motor unit potentials with satellites may occur in myopathies or neurogenic diseases. The individual components of a polyphasic potential are action potentials recorded from single muscle fibers. Thus, they are commonly seen in myopathies in which there is fiber regeneration and increased fiber density. Regeneration of axons in neurogenic diseases can also produce low-amplitude polyphasic motor unit potentials that vary in size and shape (*nascent motor unit potential*). If a motor unit potential is large, an increase in fiber density produces more turns without extra phases, a *complex motor unit potential.* Polyphasic motor unit potentials may occur in any of the myopathies or the neurogenic atrophies listed above and are identified as abnormal by an increase in the proportion of polyphasic potentials over that which is normal for that muscle in a patient of the same age. These potentials must be differ-

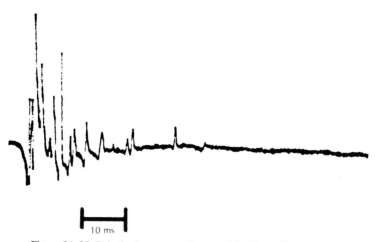

Figure 24–22. Polyphasic motor unit potential with satellite potential.

entiated from doublets, multiplets, tremor, and motor unit potentials that vary in configuration.

Mixed Patterns: Long-Duration and Short-Duration Motor Unit Potentials

Occasionally, patients have a combination of the abnormalities described for short, long, and polyphasic motor unit potentials, but instead of having the usual pattern of an excess of either long-duration or short-duration potentials, both types occur. The quantitative distribution becomes broad rather than shifting to long or short. Rarely, the distribution of durations may be bimodal (Fig. 24–23). These combinations commonly occur in chronic myositis or rapidly progressing motor neuron disease.

Motor Unit Potential Variation

As they fire repetitively under voluntary control, motor unit potentials normally have the same amplitude, duration, and configuration. Fluctuation in time of any of these variables during repeated discharge of a motor unit potential is abnormal and is usually caused by blocking the discharge of action potentials of the individual muscle fibers in the motor unit.

The disorders in which motor unit potentials fluctuate from moment to moment (Fig. 24–24) are listed in Table 24–10. In disorders of muscle membrane, such as myotonia, there may be a slower progressive decrease or increase in a motor unit potential (Fig. 24–25). In myasthenia gravis or in cases of active reinnervation, amplitude may initially decline, but in the myasthenic syndrome, it may increase (Fig. 24–26).

Doublets (Multiplets)

Motor units under voluntary control normally discharge as single potentials in a semirhythmic fashion. In some disorders, they fire two or more times at short intervals of 10 to 30 milliseconds (Fig. 24–27) (Table 24–11). These are called *doublets, triplets,* or *multiplets*. The bursts of two or more recur in a semirhythmic pattern under voluntary control. They are often increased by ischemia.

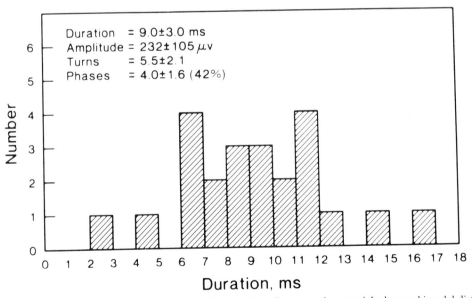

Figure 24–23. Patient with inclusion body myositis. Quantitation of motor unit potentials shows a bimodal distribution.

NORMAL

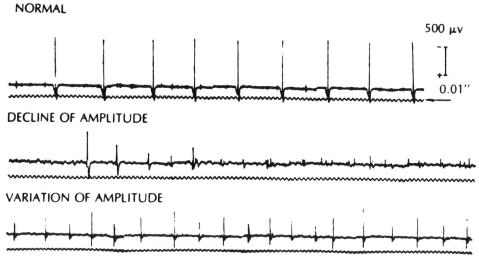

DECLINE OF AMPLITUDE

VARIATION OF AMPLITUDE

Figure 24–24. Voluntary motor unit potentials.

Table 24–10. DISORDERS ASSOCIATED WITH RAPID MOTOR UNIT POTENTIAL VARIATION (UNSTABLE MOTOR UNIT POTENTIAL)

Myasthenia gravis

Myasthenic syndrome

Botulinum intoxication

Myositis

Muscle trauma

Reinnervation after nerve injury

Rapidly progressing neurogenic atrophy (e.g., amyotrophic lateral sclerosis)

Abnormal Recruitment

In normal muscle, increasing voluntary effort causes an increase in the rate of firing of individual motor unit potentials and initiation of the discharge of additional motor unit potentials. The relationship of the rate of firing of individual potentials to the number of potentials firing is constant for a particular muscle and is called the *recruitment pattern*. If there is a loss of motor unit potentials in any disease process, the rate of firing of individual potentials will be disproportionate to the number firing; this is referred to as *reduced recruitment* (Fig. 24–28). It may

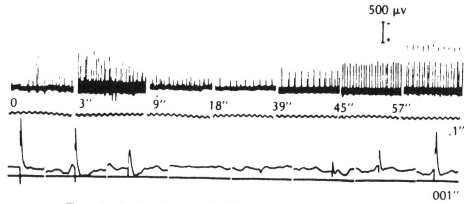

Figure 24–25. Myotonia congenita with slow motor unit potential variation.

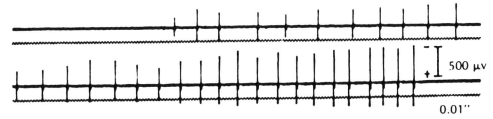

Figure 24-26. Increase in amplitude and variation of a motor unit potential.

be seen in any disease process that destroys or blocks conduction in the axons innervating the muscle or that destroys a sufficient proportion of the muscle so that whole motor units are lost. This pattern is seen in association with all neurogenic disorders and may be the only finding in a neurapraxia in which the sole abnormality is a localized axonal conduction block or in cases of acute axonal loss in which fibrillation has not yet developed. In myopathies, however, more motor units are activated than would be expected for the force exerted in disorders in which the force that a single motor unit can generate is decreased. This is called *rapid recruitment*. The recruitment frequency and rate of firing in relation to number are normal with rapid recruitment.

Disorders of Central Control

One pattern of motor unit firing that is caused by disorders of the central nervous system must be recognized because it may resemble the changes seen with lower motor neuron disease.[11] In muscle *tremor*, which

Table 24-11. **DISORDERS ASSOCIATED WITH DOUBLETS**

Hyperventilation
Tetany
Motor neuron disease (infrequent)
Other metabolic diseases

may not be apparent clinically, motor unit potentials fire in groups but not in a fixed relationship. The potentials of these motor units are superimposed and may resemble polyphasic, complex, or long-duration motor unit potentials (Fig. 24-29). They are recognized by their rhythmic pattern and their changing appearance. Minimal activation, with slightly increasing and decreasing effort, often allows single motor unit potentials to be resolved and characterized. Motor unit potential firing patterns in stiff-man syndrome, rigidity, and spasticity resemble normal patterns, but with loss of voluntary control. In upper motor neuron weakness, motor unit firing cannot be maintained by patients.

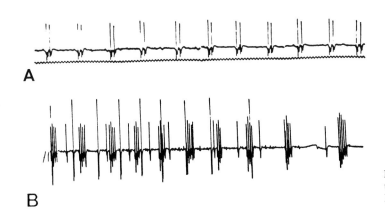

A

B

Figure 24-27. Voluntary motor unit potentials. (*A*) Doublets. (*B*) Multiplets.

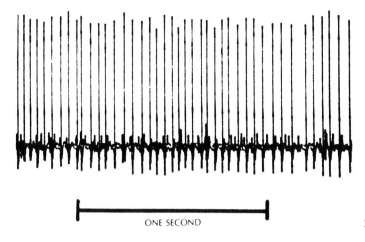

ONE SECOND

Figure 24–28. Poor recruitment.

Synkinesis

The aberrant regeneration of axons after nerve injury may result in two different muscles being innervated by the same axon. In such cases, voluntary potentials may be mistaken for spontaneous activity. Examples include potentials in facial muscles in association with blinking and potentials in shoulder girdle muscles in association with respiration.

Patterns of Abnormality

The types of needle EMG abnormalities described previously may occur in different combinations.[3] It is only through knowledge of these combinations that reliable interpretations can be made. No single finding allows the identification of a specific disease.[19] The combinations of particular forms of spontaneous activity and changes in motor unit potentials in neuromuscular diseases are too varied to be included in this review, but some general comments about patterns of abnormality of motor unit potentials can be made. Motor unit potential changes have been broadly divided into *neuropathic* and *myopathic*.[6,7] (*SSAP*—**s**mall, **s**hort, **a**bundant **p**otentials—is an alternative term for myo-

pathic.) The concept that motor unit potential changes must be either one or the other of these two types is incorrect and can lead to misinterpretations.

Each of the variables—recruitment, duration, amplitude, configuration, and stability—of motor unit potentials may be altered separately or in combination with one or more of the others in different disorders.[27] Each must be judged individually, quantitated if necessary, and compared with normal values. The result should then be interpreted on the basis of known pathophysiologic mechanisms or by common association with known disorders. Recruitment, duration, and stability are the important features of motor unit potentials in determining the underlying pathologic factors. With these three criteria, it is possible to distinguish most patterns of motor unit potential abnormality.

Each pattern of abnormality changes with the severity and duration of the disease. Careful attention to the independent changes of the variables of motor unit potentials can allow an electromyographer to comment on the severity, duration, and prognosis of a disease. Because of the various patterns that may be seen, description of the abnormalities should always include comments about each of the variables. The find-

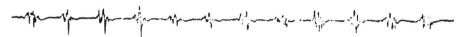

Figure 24–29. Tremor with superimposed motor unit potentials.

Table 24–12. **PATTERNS OF ABNORMALITY SEEN WITH NEEDLE ELECTROMYOGRAPHY**

Recruitment	Appearance	Variation	Disorders
Normal	Normal	No	Normal
			Some endocrine and metabolic myopathies
		Yes	Myasthenia gravis
			Myasthenic syndrome
	Short-duration, polyphasic	No	Primary myopathies
		Yes	Severe myasthenia
			Botulinum intoxication
			Reinnervation (neurogenic or myositis)
	Mixed short-duration and long-duration	Yes or no	Chronic myositis and inclusion body myositis
			Rapidly progressing neurogenic disorder, such as amyotrophic lateral sclerosis and Werdnig-Hoffman disease
Poor	Normal	No	Acute neurogenic lesion
	Long-duration, polyphasic	No	Chronic neurogenic atrophy
		Yes	Progressing neurogenic atrophy
	Short-duration, polyphasic	No	Severe myopathy (end-stage, neurogenic atrophy)
		Yes	Early reinnervation after severe nerve damage

ings can then be interpreted most reliably by listing the disorders that may be seen with the pattern of abnormality found, as in Table 24–12.

CHAPTER SUMMARY

Virtually all primary neuromuscular diseases result in changes in the electrical activity recorded from muscle fibers. These changes can best be depicted using fine needle electrodes inserted into the muscle to record spontaneous and voluntary EMG. Thus, EMG can be used to distinguish lower motor neuron, peripheral nerve, neuromuscular junction, and muscle disease with great sensitivity and some specificity. The sensitivity is usually greater than clinical measures; specificity in identifying the cause of the disease often requires muscle biopsy or other clinical measures. Although EMG is somewhat uncomfortable because of the need to insert needles into the muscles, it is generally well tolerated by patients and provides a rapid, efficient means of testing the motor unit.

The application of techniques of clinical neurophysiology in the evaluation of peripheral neuromuscular disorders relies heavily on needle EMG. It was the first of the electrophysiologic techniques to be applied in this way and has remained a mainstay of electrodiagnosis. The collection of data in needle EMG and the interpretation for clinical purposes require a firm understanding of the physiology of muscle fibers and motor units. It is heavily dependent on controlling various technical factors, mastering the skills of the collection, and understanding the changes that occur with the many disorders that may affect peripheral nerves, neuromuscular junctions, and muscle.

The essence of quality needle EMG rests with the ability to isolate, recognize, and interpret the wide range of specific waveforms and their variation that occur in normal and diseased muscle. The nature and meaning of fibrillation potentials and of the alteration in appearance and firing pattern of motor unit potentials in each muscle tested are the data on which clinical interpretation is based. The extent and distribution of these abnormalities allow conclusions to be drawn about the type and severity of disease, its age or stage of evolution, and the likely anatomic location of the process.

REFERENCES

1. Bergmans, J: Clinical applications: Applications to EMG. In Rémond A (ed): EEG Informatics: A Didactic Review of Methods and Applications of EEG Data Processing. Elsevier-North Holland Biomedical Press, Amsterdam, 1977, pp 263–280.
2. Bischoff, C, Machetanz, J, and Conrad, B: Is there an age-dependent continuous increase in the duration of the motor unit action potential? Electroencephalogr Clin Neurophysiol 81:304–311, 1991.
3. Black JT, Bhatt GP, Dejesus PV, et al: Diagnostic accuracy of clinical data, quantitative electromyography and histochemistry in neuromuscular disease: A study of 105 cases. J Neurol Sci 21:59–70, 1974.
4. Brandstater, ME and Lambert, EH: Motor unit anatomy: Type and spatial arrangement of muscle fibers. In Desmedt, JE (ed): New Developments in EMG and Clinical Neurophysiology. Vol 1. New Concepts of the Motor Unit, Neuromuscular Disorders, Electromyographic Kinesiology. S. Karger, Basel, Switzerland, 1973, pp 14–22.
5. Buchthal, F: Electrophysiologic abnormalities in metabolic myopathies and neuropathies. Acta Neurol Scand 46 Suppl 43:129–176, 1970.
6. Buchthal, F: Diagnostic significance of the myopathic EMG. Excerpta Medica International Congress Series No. 404, 1976, pp 205–218.
7. Buchthal, F: Electrophysiological signs of myopathy as related with muscle biopsy (1). Acta Neurol (Napoli) 32:1–29, 1977.
8. Buchthal, F, Guld, C, and Rosenfalck, P: Action potential parameters in normal human muscle and their dependence on physical variables. Acta Physiol Scand 32:200–218, 1954.
9. Buchthal, F, Pinelli, P, and Rosenfalck, P: Action potential parameters in normal human muscle and their physiological determinants. Acta Physiol Scand 32:219–229, 1954.
10. Chan, RC, and Hsu, TC: Quantitative comparison of motor unit potential parameters between monopolar and concentric needles. Muscle Nerve 14:1028–1032, 1991.
11. Freund, H-J: Motor unit and muscle activity in voluntary motor control. Physiol Rev 63:387–436, 1983.
12. Fuglsang-Frederiksen, A: Quantitative electromyography. I. Comparison of different methods. Electromyogr Clin Neurophysiol 27:327–333, 1987.
13. Fuglsang-Frederiksen, A: Quantitative electromyography. II. Modifications of the turns analysis. Electromyogr Clin Neurophysiol 27:335–338, 1987.
14. Fuglsang-Frederiksen, A and Ronager, J: EMG power spectrum, turns-amplitude analysis and motor unit potential duration in neuromuscular disorders. J Neurol Sci 97:81–91, 1990.
15. Gan, R and Jabre, JF: The spectrum of concentric macro EMG correlations. Part II. Patients with diseases of muscle and nerve. Muscle Nerve 15:1085–1088, 1992.
16. Guld, C, Rosenfalck, A, and Willison, RG: Report of the Committee on EMG Instrumentation: Technical factors in recording electrical activity of muscle and nerve in man. Electroencephalogr Clin Neurophysiol 28:399–413, 1970.
17. Gunreben, G and Schulte-Mattler, W: Evaluation of motor unit firing rates by standard concentric needle electromyography. Electromyogr Clin Neurophysiol 32:103–111, 1992.
18. Hahn, AF, Parkes, AW, Bolton, CF, and Stewart, SA: Neuromyotonia in hereditary motor neuropathy. J Neurol Neurosurg Psychiatry 54:230–235, 1991.
19. Hausmanowa-Petrusewicz, I, and Jedrzejowska, H: Correlation between electromyographic findings and muscle biopsy in cases of neuromuscular disease. J Neurol Sci 13:85–106, 1971.
20. Hayward, M and Willison, RG: Automatic analysis of the electromyogram in patients with chronic partial denervation. J Neurol Sci 33:415–423, 1977.
21. Kraft GH: Fibrillation potential amplitude and muscle atrophy following peripheral nerve injury. Muscle Nerve 13:814–821, 1990.
22. Lang, A and Partanen, VSJ: "Satellite potentials" and the duration of motor unit potentials in normal, neuropathic and myopathic muscles. J Neurol Sci 27:513–524, 1976.
23. Lorente De Nó, R: A Study of Nerve Physiology. Studies from The Rockefeller Institute for Medical Research. Vol 132. The Rockefeller Institute for Medical Research, New York, 1947, pp 466–470.
24. Pfeiffer, G and Kunze, K: Turn and phase counts of individual motor unit potentials: Correlation and reliability. Electroencephalogr Clin Neurophysiol 85:161–165, 1992.
25. Pinelli, P: The value of motor unit parameter measurement. In Van Duijl, H, Donker, DNJ, and Van Huffelen, A (eds): Current Concepts in Clinical Neurophysiology. Didactic Lectures of the 9th International Congress of EEG and Clinical Neurophysiology. NV Drukkerij Trio, The Hague, 1977, pp 111–122.
26. Roth G: Repetitive discharge due to self-ephaptic excitation of a motor unit. Electroencephalogr Clin Neurophysiol 93:1–6, 1994.
27. Sonoo, M and Stålberg, E: The ability of MUP parameters to discriminate between normal and neurogenic MUPs in concentric EMG: Analysis of the MUP "thickness" and the proposal of "size index." Electroencephalogr Clin Neurophysiol 89:291–303, 1993.
28. Stålberg, E: Macro EMG, a new recording technique. J Neurol Neurosurg Psychiatry 43:475–482, 1980.
29. Stålberg, E and Trontelj, JV: Single Fibre Electromyography. Mirvalle Press, Old Woking, Surrey, England, 1979, pp 1–244.
30. Stålberg, EV: Macro EMG. Minimonograph 20. American Association of Electromyography and Electrodiagnosis, Rochester, Minnesota, 1983.

CHAPTER 25

QUANTITATIVE ELECTROMYOGRAPHY AND SINGLE-FIBER ELECTROMYOGRAPHY

W. J. Litchy, M.D.

QUANTITATIVE METHODS IN NEUROMUSCULAR DISORDERS

Much of the early application of clinical electromyography (EMG) was descriptive and subjective. However, during the past 30 years, there has been a steady evolution toward more quantitation, especially in nerve conduction studies but also for needle EMG.

Quantitation has several advantages. It allows precise statements about the severity of a disease process and allows comparison of EMG findings in a patient over time as a disease evolves. Quantitation also allows results obtained by one physician to be compared with those obtained by another. Most important, quantitation can demonstrate changes that are not recognized using subjective analysis. The major disadvantage of quantitation is that it requires extra time. Occasionally, it may also require special equipment, but additional training for performing it correctly is rarely needed.

Each of the electrical signals recorded in nerve conduction studies and needle EMG can be precisely quantitated, although when the signal is complex, it becomes more difficult to do so.[3,7,8] Compound muscle action potentials in motor nerve conduction studies are the simplest and easiest to quantitate. Sensory nerve action potentials are more variable and, thus, more difficult to quantitate. The spontaneous and voluntary activity recorded with needle EMG is the most difficult to quantitate because of the complexity and rapid recurrence of the waveforms. Each of these types of waveforms has similar basic characteristics. In this chapter, these characteristics are considered together before features unique to motor unit potentials are discussed.

ATTRIBUTES OF ELECTROMYOGRAPHIC WAVEFORMS

Each of the waveforms measured in clinical EMG is a voltage fluctuation over time for which the extent, direction, and duration of the voltage change are described. By convention, an upward deflection represents a relative negativity at the active electrode and a downward deflection represents a relative positivity. The positions of the reference electrode and active electrode must also be carefully considered, because some electrical activity virtually always occurs at both electrodes with voltage changes of opposite direction that can cancel or distort the signal. The voltage changes can be described by the attributes discussed later in this chapter. Each of these attributes can be measured manually from film or automatically by a computer. Quantitation by computer requires more precise definitions of the measurements.

Negative Components

Negative components are upward deflections from the baseline when the active electrode is relatively negative to the reference elec-

trode.[11] The presence of a negativity indicates that the area of depolarization is immediately adjacent to the active electrode. Negative components can be reliably recorded only if the active electrode is near the generator. Distances greater than 1 to 2 mm, or intervening tissue such as bone, distorts the potential into a complex waveform with latencies that are different from the actual latencies measured near the generator. Such recordings are called *far-field recordings* and are not reliable for standard studies; if they are used, they must be interpreted with caution. Electrodes that record far-field potentials record the negativity earlier than electrodes near the generator because of the field distribution of the voltages. Thus, the time of occurrence of a waveform is measured less reliably at a distance from a generator.

Positive Components

Positive components are downward deflections from the baseline when the active electrode is positive compared with the reference electrode. The presence of a positivity indicates that the major area of depolarization is at some distance from the active electrode.[11] Positivity can distort negative waves in two ways. First, if two generators near each other are not in synchrony, the positive component of one may occur before the negative component of the other and either distort the latter or end it sooner. A common example is the positivity from anterior compartment muscles occurring before the negativity from the extensor digitorum brevis muscle during peroneal nerve conduction studies. Second, placement of the reference electrode too close to the active electrode can superimpose the negativity at the reference electrode on the negativity at the active electrode to produce distortion, latency changes, or early termination of the negative component. Such distortions are common in both motor and sensory nerve conduction studies. Reference electrodes are best placed distant from the active electrode so that the negative component ends at the active electrode before it begins at the reference electrode. Activity at the reference electrode is also less likely to distort the potential if the electrode is placed longitudinally further along a nerve or muscle, because the potential with a laterally placed reference electrode occurs during or even before that at the active electrode.

Onset

Onset is the first deflection from the baseline in either direction (Fig. 25–1). This length of time may be difficult to measure if it has a gradual onset. Adequate amplification is needed to measure it. Time of onset becomes measurably earlier with higher amplification of the signal, because earlier components that are below the level of resolution at lower amplification can be recorded. When automatically measuring the onset of the potential with a computer, the criteria must be defined in terms of the length of time that the potentials must remain outside a defined voltage or the slope of the waveform as it leaves the baseline or both. Time from a stimulus to the onset is the latency of a waveform.

Termination

Termination is the final return to the baseline and is more difficult to measure than

onset because it often ends gradually. Measurement is also complicated by components of the potential that occur after the main component has returned to the baseline. These can occur with compound muscle action potentials, sensory nerve action potentials, and motor unit potentials (e.g., late components, satellite potentials, linked potentials, axon reflexes, and repetitive discharges). If these other components are present, the duration of the waveform should be measured with and without them.

Baseline Crossing

Baseline crossing is when the voltage changes from positive to negative or from negative to positive (see Fig. 25–1). A baseline crossing designates the end of a phase. Measurement should include the time of the baseline crossing and the number of baseline crossings in the waveform. The number of total phases is one more than the number of baseline crossings. Individual muscle fiber potentials and normal compound muscle action potentials and sensory nerve action potentials are simple biphasic and triphasic waveforms. The potential is biphasic if it is initiated in the region of the active electrode and is triphasic if it is initiated elsewhere and

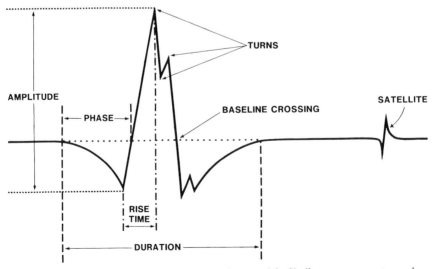

Figure 25–1. Standard measurements made from motor unit potentials. Similar measurements can be made from compound muscle or sensory nerve action potentials. (From Daube, JR: AAEM minimonograph #11: Needle examination in clinical electromyography. Muscle Nerve 14:685–700, 1991. By permission of Mayo Foundation.)

moves toward the active electrode. The presence of more than two baseline crossings is evidence of a loss of synchronization of the discharges of individual components in the potential. The time of occurrence of baseline crossings is measured to determine durations (see later discussion).

Turns

Turns are changes in direction of a potential. Measuring turns requires defining the amount of voltage change necessary to call a direction reversal a turn. This amount must be significantly more than the noise level in the signal. A turn represents the discharge of a distinct component in the potential, such as an individual muscle fiber, an individual motor unit, groups of nerve fibers, or a separate muscle. The number of turns measures the extent of dispersion or the number of components contributing to a waveform. The time of occurrence of turns does not provide useful diagnostic information and so is not measured. The time after a stimulus of the first major positive or negative turn can be used as a measure of latency.

Duration of Waveform

Duration of a waveform is the time from onset to termination. *Total duration* measures the dispersion of all components. Analysis of the shape and duration of a waveform can provide an estimate of the *distribution of conduction velocities* of the fibers contributing to the potential. Estimates of the number of fast and slow conducting axons can also be made from the change in size with paired stimuli at decreasing intervals between the stimuli. In specific waveforms, such as motor unit potentials, it is also related to other factors such as the location of end-plates and disease (Fig. 25–2). In waveforms with late components, the duration is best measured to the end of the main segment and to the end of the late component.

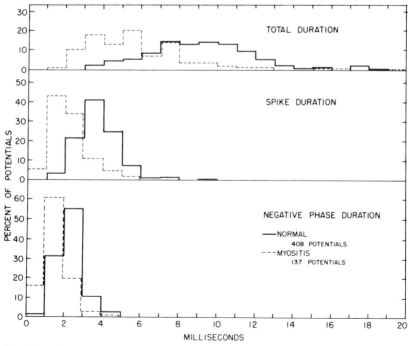

Figure 25–2. Durations of motor unit potentials in normal and myositic biceps brachii muscle. Durations of different components of motor unit potentials are altered to different extents by disease.

Duration of Individual Phases

This is measured from the onset of a phase to its termination (baseline crossings). It is commonly used to measure the duration of the major negative component (see Fig. 25–1). Phase duration reflects the dispersion of conduction of components that have a similar range of conduction velocities.

Rise Time

Rise time is the duration of the rising phase (positive peak to negative peak). The moving region of depolarization in nerve or muscle fibers has reached the active electrode when this reversal of potential from positive to negative occurs. The rate of rise (slope) is related directly to the distance of the generator from the electrode and, to a lesser extent, to fiber conduction velocity. Intervening tissue acts as a combined resistance and capacitance to slow the rate of rise of the potential. Rise time has no direct diagnostic significance, but it provides a useful criterion for the proximity of the generator in measuring the amplitude or area (see later in this chapter).

Amplitude

Amplitude is the voltage from the baseline to the maximum negative peak or from the maximum positive peak to the maximum negative peak (or both). The amplitude of the negative component is proportional to the number and size of fibers that are depolarized; therefore, it provides an estimate of the amount of functioning tissue. The amplitude also depends on the distance of the recording electrode from the generator. This distance can be judged from the rise time. When a waveform has multiple components or when it is abnormally prolonged, the area provides a more accurate estimate of the amount of functioning tissue than the amplitude does. The positive components of late occurring potentials can cancel the negative components of early potentials. Thus, amplitude measurements are the most reliable estimates of the number of active fibers when amplitude is recorded close to the site of stimulation, where dispersion is less.

Area

Area is the space under the curve of the waveform (usually measuring only the negative components). Although area cannot be measured reliably by hand or with standard equipment, it can be done easily with digital computers. Area provides the most direct estimate of the functioning tissue that generates the waveform. Positive components are not useful in this measurement because they are depolarizations that occur at a distance. Area measurements, similar to amplitude measurements, provide the most information about the number of fibers close to the site of activation.

Stability/Variation

The waveforms recorded in clinical EMG usually repeat spontaneously, under voluntary control, or with stimulation. Normal waveforms are identical when measured in the usual fashion, because each fiber is activated with each stimulus and summates with other fibers in the same manner. Small (< 50 microseconds) variations in latency occur because of changes in end-plate potential amplitude, but these are recorded only with single-fiber electromyography (SFEMG). Increases in interpotential interval variation or loss of some fibers on consecutive discharges occur with block of conduction in axons, neuromuscular junctions, or muscle fibers or with changes in conduction velocity in individual fibers. Stability can be measured as change in amplitude or area (*decrement*) or as variation in latency of individual components (*jitter*). Stability largely depends on the rate of discharge of the waveform. Stability is greatest at slow rates of discharge in some disorders and at high rates in others.

MOTOR UNIT POTENTIALS

Each of the waveform attributes already listed can be reliably and usefully measured on motor unit potentials but requires consideration of one factor not present with either the compound muscle action potential or sensory nerve action potential. The latter two potentials occur under the control of the stimulation by the electromyographer and, therefore, occur singly and without other interfering potentials. However, a motor unit potential is under the control of the patient and is usually intermingled with other motor unit potentials. When single motor unit potentials can be isolated either with low rates of activation or by using a selective delay line, the attributes can be reliably measured. However, if motor unit potentials overlap, they summate and cannot be differentiated. In this case, meaurements cannot be reliably performed on a single potential, and other methods need to be used. Even when single motor unit potentials are isolated, their amplitude is often low enough that the noise in the system may distort individual potentials, in which case averaging is necessary, just as for sensory nerve action potentials.

Motor unit potentials are generated by muscle fibers in one motor unit, with the major contributions coming from fibers within a 200-μm radius of the electrode. Therefore, a single motor unit has many motor unit potentials recorded in different areas of the motor unit territory. At any one location, there may be 10 to 20 nearby units, but motor unit potentials can be measured reliably from only two to four of them before they summate excessively with one another. Reliable estimates of the characteristics of motor unit potentials require measurement of at least 20 potentials, with calculation of the means and standard deviations of each attribute.

Measurement of amplitudes, durations, turns, and phases of a motor unit potential are made best with a delay line to isolate 20 or more single potentials. However, reliable estimates can be made visually with repeated storage of many waveforms on the oscilloscope or film. Computer-controlled automatic measurement of motor unit potentials is being perfected for routine clinical use.

All the attributes of a motor unit potential, whether obtained singly or by averaging with a delay line or storage methods, provide valuable information about the muscle and neuromuscular junction.[41] The amplitude and area reflect the number and size of muscle fibers contributing to the potential and their density in the area of the needle electrode.[4] The duration is also related to these variables, but it is affected more directly by the location and dispersion of the end-plates and the conduction in muscle fibers. The number of turns and, less directly, the number of phases reflect the number of muscle fibers in the area of the electrode which contribute to the potential. This is referred to as the *fiber density*. The time interval between these components is related primarily to the rate of conduction of action potentials in muscle fibers and, thus, to their size and, to a lesser extent, the size, length, and conduction velocity in nerve terminals. Each of these measurements must be made on 20 or more motor unit potentials to provide a reliable estimate of the population of motor unit potentials. Quantitation of these potentials can identify otherwise unrecognized disease and mixed disorders.[16–18]

A larger proportion of the motor unit potential can be quantitated by macrorecordings that sample from a cylindrical core of muscle rather than at just the tip of the electrode. This is referred to as *macro EMG*. Macrorecordings average all electrical activity recorded from 15 mm of the shaft of a needle electrode by using a single fiber potential to trigger the average. Amplitudes in macrorecordings of the motor unit potential can distinguish different groups of neurogenic and myopathic processes.

The stability of the potential depends primarily on the neuromuscular junction, although some instability can result from nerve or muscle disease. Instability can be measured as percentage decrement, as with the compound muscle action potential, or as variation of the motor unit potential. It is most reliably measured on SFEMG as *jitter*, the variation in interval between individual spike components of the motor unit potential, or *blocking*, the frequency with which individual components are not present. Quantitation of the stability is of importance in identifying mild myasthenia gravis and in observing patients who have myasthenia gravis.

Because motor unit potentials are under voluntary control, they have an additional attribute, which is not present with the compound muscle action potential or sensory nerve action potential that can be measured: recruitment. As a motor unit potential is activated under voluntary control, it fires at increasingly rapid rates. In normal muscle, there is a simultaneous activation of other motor unit potentials that is proportional to the rate of firing of the individual potentials. This relationship is called *recruitment*, which can be quantitated as *onset frequency* or *recruitment frequency*. The former is the rate of firing when a motor unit potential begins to fire, and the latter is the frequency of firing of the first unit when the next unit begins to fire. Although recruitment is more difficult to measure than other attributes, it can be measured.

When multiple motor unit potentials overlap, an interference pattern results and the attributes of the individual potentials cannot be measured. Interference pattern measurements can be made using the summated signal. The simplest of these measures are turns and amplitude. Turns, as defined above, are measured over a prescribed length of time. Amplitude of the interference pattern can be measured as the maximum amplitude or, more commonly, as the total amplitude of all deflections over a unit of time. Similar information can be obtained by measuring baseline crossings or the frequency content of the signal. These measurements require specialized equipment and the development of normal values for different ages and dif-

Motor Unit Potentials in
Inclusion Body Myositis
Automated Quantitation– Biceps Muscle

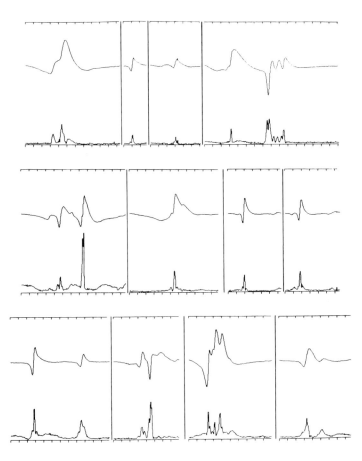

Figure 25–3. Automated computer recordings of averaged motor unit potentials in the biceps muscle of a patient with myositis. Waveforms are reproduced above, and the variance of each point averaged is shown below.

ferent muscles. They also may vary with the level of effort and require some control of patient voluntary activity. Measurements of the interference pattern can distinguish neurogenic and myopathic processes, complementing single motor unit potential measurements, but they cannot provide the detailed information about individual motor unit potentials, such as variation and dual populations, that single motor unit potential measurements can (Fig. 25–3).

QUANTITATION OF NEEDLE ELECTROMYOGRAPHY

Electrical activity recorded on needle EMG during voluntary activity is the sum of multiple motor units under voluntary control. Motor unit potential generated by each motor unit is the spatially summated activity of all muscle fibers in the motor unit firing together in response to activation by the lower motor neuron. Characteristics of each motor unit potential include duration, amplitude, number of phases, interval between the initial positive and initial negative peaks, rate of firing, pattern of firing, recruitment frequency, area of distribution, and stability of size and configuration.[27] In addition, the pattern of firing of different motor unit potentials in relation to one another, the extent of voluntary control, the number of motor units that can be activated, and the character of the units active at different force levels all combine to produce potentials that make a recording distinctive. Subjective assessment of these attributes by an experienced electromyographer is satisfactory for identifying the patterns of abnormalities seen in most patients during routine clinical EMG. However, in patients with mild changes, in those with unusual neuromuscular diseases, and in those with a mixed pattern of abnormality, subjective assessment may be inadequate for providing a definite description. In these situations, quantitative measurements of the electrical activity are needed. In addition, reliable comparison of recordings made at different times or made by different electromyographers requires some method of quantitation.

Three general methods of quantitation of

needle EMG that can be used for routine clinical applications have been described. One of these, SFEMG, uses a special recording electrode and recording techniques to be described later. The other methods differ only in the level of voluntary activation made by the patient during the recording. In the first of these, attributes of individual motor unit potentials are measured at low levels of voluntary activation and are referred to as *single motor unit potential quantitation.* More recently, computer-assisted approaches have been used to try to quantitate motor unit potentials at stronger levels of contraction.[12] In the other method, the *interference pattern* generated during moderate or strong contraction is quantitated.

Technical Factors

Quantitation of either single motor unit potentials or the interference pattern requires attention to several technical factors that can alter the measurements. The type and size of the recording electrode have a major effect on the measurements. The standard concentric needle electrode with an exposed surface of 0.07 mm^2 is the one usually used to make these measurements. The use of other electrodes requires different sets of normal values.[6] The characteristics of the recording amplifiers and their input cable may also alter the characteristics of the motor unit potential. In particular, the frequency response of the equipment should extend from 1 Hz to at least 10 kHz. Motor unit potentials also vary in amplitude and duration with the temperature of the muscle. A standard temperature is needed for accurate quantitation.

To compare quantitative measurements, normal values must first be determined. The techniques of electrical recording and measurements of potentials must be the same for both the normal population and the patients. Because there are significant differences in the characteristics of the motor unit potential activity among muscles and in patients with different ages, normal values must be available for different age groups and for each muscle that may be tested.

The motor unit potentials in any single muscle differ from area to area in the muscle as the recording electrode selectively records from different groups of muscle fibers. Therefore, an adequate description of the electrical characteristics of any one muscle requires the recording of multiple motor unit potentials in multiple areas in each muscle and a statistical description of the values, including the mean, standard deviation, and range of values for each of the attributes for each muscle.[14]

Single Motor Unit Potential Measurement

The method of measurement of single motor unit potentials was first described by Buchthal, Guld, and Rosenfalck[5] who subsequently published normal values for a wide range of muscles in subjects of different ages. The original technique used photographs of a large series of sequential sweeps from which individual recurring motor unit potentials were selected for measurement. More recently, computer software applications have made their identification and measurement more efficient.[14,28]

Motor unit potentials are measured by moving the needle electrode during minimal to moderate contraction of the muscle to a site at which one to three motor unit potentials are firing repetitively under voluntary control. The measurements must be made at standard gain settings and filter settings, as noted above. The motor unit potentials must have a rise time of less than 500 microseconds to be certain that the recording electrode is in the vicinity of the muscle fibers generating a potential. Also, the motor unit potentials must have a negative component, because recording from damaged areas of the fiber is seen as a large positive wave with a long, slow return to baseline that cannot be measured.

Duration is measured from the initial deflection of the potential from the baseline to its final return to the baseline. These points may be ill defined in some potentials and, therefore, must be selected consistently both in normal subjects and in patients. Amplitude is measured from maximal positive to maximal negative peak. After the return to baseline, some motor unit potentials have a time-locked component that is seen as a separate potential.[21] Measurements may be made either including or excluding such a late or satellite potential as long as normal values are measured in the same fashion. The work of Buchthal, Guld, and Rosenfalck[5] excluded these late components from the duration measurement. Configuration may be measured in one of two ways. The number of baseline crossings of the potential plus one can be used to identify the number of phases, or the number of reversals of the direction of the potential of over 50 μV may be used to determine the number of turns.

Interference Pattern Quantitation

Although several authors have described methods for assessing frequency content of the electrical activity in needle EMG, the method reported by Willison[39] and other authors[23] is the only available clinical method for quantitating the interference pattern. This method measures the attributes of turns and amplitude during a strong voluntary contraction. The method depends on a fixed level of activation for the muscle being tested. The measurements of interference pattern turns and amplitude are made at 10 to 15 different sites in the muscle to define mean values of turns and amplitude. These values then can be compared with measurements in the same muscles in subjects of the same age who make the same effort.

Interpretation

Both techniques have advantages and disadvantages. The measurement of single motor unit potentials gives a precise assessment of the characteristics of the electrical activity produced by individual motor unit potentials and, thus, can distinguish alterations that are limited to a small group of motor units or one type of motor unit. It permits assessment of highly complex potentials and their differentiation from many short-duration potentials. It also permits assessment of variability in the configuration of the motor unit potential from moment to moment, but does not assess the pattern of recruitment of motor unit potentials with increasing effort. The measurement of interference pattern includes information about the pattern of recruitment and the number of motor unit potentials recruited. However, it is unable to identify changes of long duration or distinguish highly complex motor unit potentials from many short-duration motor unit potentials. It also cannot identify variation in the configuration of a motor unit potential. Because recruitment depends on voluntary effort, it requires strict control of the extent of voluntary activation by measurement of force exerted. Force measurements are cumbersome in the clinical setting, limiting the application of recruitment analysis.

SINGLE-FIBER ELECTROMYOGRAPHY: PHYSIOLOGY AND METHODS

Contrast Between Concentric Needle Electromyography and Single-Fiber Electromyography

Standard concentric needle examination of voluntary muscle contraction displays single or multiple motor unit potentials, depending on the strength of contraction. Average motor units contain 50 to 400 muscle fibers (range, 9 to 1900 fibers). The recording surface of standard concentric needles is oblong, 150×580 μm. During a minimal voluntary contraction, a motor unit potential is observed which reflects the activity of 3 to 20 muscle fibers of a single motor unit lying within a few millimeters of the electrode. Tissue acts as a low-pass filter, damping the high-frequency components of potentials generated by more distant muscle fibers. The major negative components of motor unit potentials recorded with standard concentric needle electrodes are produced by three to eight muscle fibers within 500 μm of the electrode; there is only a small contribution to the early and late positive phases from the more distant fibers.

SFEMG needle examination is designed to record from a single muscle fiber or a small number of muscle fibers, enhancing the recording from the fibers near the electrode while suppressing the activity of more distant fibers. The two essential steps in this enhancement are equally important. The first is the suppression of activity from more distant fibers by filtering low-frequency components of the electrical signal. A 500-Hz filter effectively eliminates the contribution from fibers more than 500 μm from the electrode. The second step is a small recording surface (25 μm), which reduces the number of muscle fibers in direct contact with the electrode to one or two and decreases the effective distance over which the recordings from individual muscle fibers can be made to approximately 200 μm. The combination of these methods provides precise recordings from individual muscle fibers which can be accurately measured and quantitated. Four types of measurement can be made during SFEMG: *fiber density, jitter, blocking,* and *duration*. The techniques for recording are similar for each of these measurements.

Equipment Requirements

NEEDLE

For standard SFEMG recordings, the needle consists of a stainless steel shaft with a single platinum wire down the center opening onto a sideport opposite the beveled edge 3 mm proximal to the electrode tip. The active recording surface is 25 μm in diameter. The shaft of the needle serves as reference electrode.

AMPLIFIER

Standard clinical EMG preamplifiers and amplifiers are sufficient for SFEMG recording as long as the low-frequency filter can be set to approximately 500 Hz.

TRIGGER AND DELAY LINE

For quantitating the SFEMG potential, the equipment must be able to isolate an individual potential so that it can be recorded and superimposed prior to quantitative measurements.

SFEMG potentials are studied during slight activation of a muscle by either minimal voluntary contraction or electrical or re-flex activation.

Common Single-Fiber Electromyographic Measurements

FIBER DENSITY

In most areas of normal muscle, an SFEMG electrode records an action potential from only one muscle fiber in a motor unit, because only that is found within 200 μm of the electrode in most areas of a muscle. Approximately 60% of recording sites in normal muscle have such single potentials, but the percentage varies with age and from muscle to muscle. At some locations, two single-fiber potentials are recorded together, time-locked to one another. Pairs of potentials are found in approximately 35% of sites, whereas three or more time-locked potentials are found in the remaining small percentage of recording sites.

Fiber density is measured by recording the number of single-fiber potentials generated by a single motor unit at 30 different recording sites and then taking the average. The needle is positioned to maximize the amplitude of the triggering potential, and any time-locked potentials larger than 200 μV with peak-to-peak rise times less than 500 microseconds are counted. Fiber density ranges from 1.3 to 1.8 in normal subjects less than 70 years old. Fiber density reflects the density of muscle fibers in one motor unit within the recording area and corresponds most directly to the number of turns seen on standard EMG. This feature of the motor unit potential is sometimes called *complexity*, because each of these turns usually represents a separate fiber that contributes to the motor unit potential. Fiber density has a less direct relationship to the percentage of polyphasic motor unit potentials, because in a polyphasic potential a phase may include more than one turn. Satellite potentials seen in standard recordings also are recorded as separate single fiber potentials in SFEMG.

Fiber density is related directly to the number of muscle fibers innervated by a single motor unit in an area; consequently, it increases with any disorder that results in re-innervation. Thus, increased fiber density is seen in many disorders of peripheral nerve or anterior horn cells. It is particularly striking early in the course of reinnervation when the differences in conduction along regenerating nerve terminals cause marked asynchrony of firing of newly reinnervated muscle fibers and dispersion of the single-fiber potentials. However, reinnervation of muscle fibers also occurs in myopathies as a result of fiber splitting or fiber regeneration. Therefore, increases in fiber density are also seen in a number of myopathies. Although fiber density can be used to quantitate the severity of some diseases and their evolution, it cannot be used to differentiate the two broad categories of neurogenic and myopathic diseases.

JITTER

As already noted, *jitter* is defined as the variation in the interpotential interval between two single fiber potentials from a single motor unit. This results from end-plate potential variation at the neuromuscular junction. The amount of neurotransmitter (i.e., acetylcholine at the neuromuscular junction) released varies from moment to moment and produces end-plate potentials of varying size. A lower amplitude end-plate potential has a slower rise time to peak and reaches threshold later than a higher amplitude end-plate potential so that the interval of time from the action potential in the nerve terminal to the action potential in the muscle fiber varies by as much as 50 microseconds. The presence of this variation is evidence of a synapse between the sites of activation and recording. Two single fiber potentials with no jitter are time-locked by ephaptic (electrical) activation of each other or are recorded from a single fiber that has been split or otherwise distorted.

A measurement of jitter is the measurement of the variability of the interval between two single-fiber potentials. Jitter can be measured as the standard deviation of the interpotential interval, but because of the occasional occurrence of a gradual change in the mean interpotential interval over time, it is more reliably measured as the mean consecutive difference:

$$MCD = \frac{(IPI_1 - IPI_2) + (IPI_2 - IPI_3) + \cdots + (IPI_N - 1 - IPI_N)}{N - 1}$$

wherein MCD is mean consecutive difference, IPI is interpotential interval, and N is the number of intervals measured.

For maximum reliability, the mean consecutive difference should be calculated from 50 or more consecutive interpotential intervals.

Normal jitter of less than 70 microseconds varies with age and the muscle from which the recording is being made. In the extensor digitorum communis, the most commonly recorded muscle, normal mean jitter for 20 fiber pairs is between 18 and 34 microseconds (upper limit of normal for single pair is 55 microseconds) for people younger than 60. Similar normal values have been defined for several other muscles.

Technical problems with jitter relate primarily to an unstable triggering point. This can be caused by an unstable position of the needle electrode, excessive background activity leading to spurious triggers, or to amplitudes that are too similar between the two potentials, leading to alternation in the triggering point and poor superimposition.

In normal muscle, jitter is not identifiable on standard needle EMG. However, if a motor unit potential is recorded with a standard concentric needle electrode with a low-frequency cutoff of 500 Hz at a sweep speed of 100 or 200 μs/cm, the jitter can be identified. Quantitative measurements of jitter with a standard concentric needle electrode are somewhat larger than those recorded with a single fiber electrode.

Because jitter results from fluctuations in end-plate potential amplitudes, any disorder that produces a reduction in the end-plate potential also increases jitter. This most often is seen in disorders of neuromuscular transmission, such as myasthenia gravis, but it may also be seen in disorders with ongoing reinnervation or regeneration, such as amyotrophic lateral sclerosis and polymyositis. Thus, abnormalities of jitter are not diagnostic of specific disease of the neuromuscular junction but must be considered in relation to findings seen with standard electrophysiologic recordings. Abnormalities of jitter can occur without clinical weakness in the muscle.

Jitter is a function of the variation in synaptic potential size; therefore, it is present in recordings that include other synapses. F waves are due to antidromic activation of the anterior horn without central synapse, so that F-wave jitter is approximately the same magnitude as with voluntary motor unit potentials. In contrast, H reflexes, which traverse a synapse in the spinal cord in addition to the neuromuscular junction, have normal jitter of two to three times that of voluntary motor unit potentials. Other more complex reflex phenomena, such as the blink and flexion reflexes, have correspondingly larger amounts of jitter.

BLOCKING

In a normal muscle, the end-plate potential always reaches threshold and initiates a single-fiber action potential; therefore, when multiple single-fiber potentials are found, they occur with each discharge of the motor unit potential. However, if an end-plate potential does not reach threshold or if conduction fails in a nerve terminal, one or more single-fiber potentials would be missing with some discharges of the motor unit. Such potentials are said to be blocked. *Blocking* is measured as the percentage of discharges of a motor unit in which a single-fiber potential is missing. A motor unit in which a single-fiber potential does not fire half the time has 50% blocking. Quantitation of blocking for a muscle can be either as the percentage of 20 fiber pairs that show any blocking or as the percentage of blocking in all discharges of all motor unit potentials. For example, if 20 potential pairs each discharge 50 times and blocking occurs a total of 20 times in two of the pairs, 10% of the pairs show blocking and 2% of all discharges have blocking.

Normal elderly persons may have occasional blocking in some muscles. In fact, in a study of 20 pairs of single fiber potentials, if a single pair exceeds the limit of normal jitter or displays blocking (or both), many electromyographers do not use those changes alone to call the study abnormal. Blocking begins to occur when the jitter in a pair has increased between 80 and 100 microseconds. Amplitude decrement of a compound muscle action potential to 2-Hz stimulation is the nerve conduction study equivalent, and moment-to-moment variation is the concentric needle EMG equivalent to blocking.

Blocking is seen most commonly in disorders of neuromuscular transmission such as myasthenia gravis and is evidence of a defect severe enough to be associated with weakness. As with jitter, however, blocking can occur in other disorders in which neuromuscular transmission may be impaired, such as amyotrophic lateral sclerosis, polymyositis, and ongoing reinnervation. For blocking to be considered diagnostic evidence of a disorder such as myasthenia gravis, it should be found in the absence of other electrophysiologic signs of neurogenic or myopathic disease.

DURATION

The time interval between the first and last potentials of multiple single-fiber potentials recorded from a motor unit has been measured as duration. *Duration* is the total elapsed time from the first to last potential averaged for all multiple potentials recorded. An alternative measure is *mean interspike interval*, which divides the duration by the number of single-fiber potentials in each discharge. Mean interspike interval and duration might be expected to increase with any phenomenon that increases dispersion in the time of activation of individual single-fiber potentials and reduces their synchrony of firing, such as differences in conduction in nerve terminals, spread of the sites of endplates along muscle fibers, and differences in conduction velocity among muscle fibers because of differences in diameter. Therefore, duration is also nonspecific but can provide useful evidence for the presence of peripheral nerve or muscle abnormality. Duration measurements are used less frequently than are the other attributes mentioned above.

Artifacts

Technical artifacts can suggest falsely low jitter, falsely high jitter, or blocking. These phenomena need to be recognized so they are not reported as abnormal.

Excessively low jitter between potentials (less than 8 microseconds) indicates the absence of a neuromuscular junction or a technical problem. Blocking the amplifier from too high a gain results in an extra phase, which may be mistaken for a second fiber with low jitter. An artifact from the delay line may have the same result, but it can be recognized by its fixed location after each potential.

An entirely positive potential with a slow rise time, long duration, and high jitter is due to damage to a muscle fiber and should not be measured. In some instances, a bimodal distribution of the interpotential interval is seen. When this occurs, the pair should be excluded from measurement.

Apparent blocking can be observed in normal muscle. If two motor unit potentials are firing, with similar amplitude single-fiber potentials being triggered, and one single-fiber potential is a singlet while the other is one of a pair, alternation of triggering between the two potentials gives the appearance of blocking of the second potential. Firing patterns allow this to be recognized.

Stimulated Single-Fiber Electromyography

SFEMG studies are usually performed on muscle fibers that are voluntarily activated by the patient.[35] Sometimes, however, the patient is unable to cooperate adequately and reliable measurements cannot be obtained. Patients with tremors or cognitive impairment and young children may not be able to activate muscles adequately for a good evaluation.

Stimulated SFEMG is a technique that does not require effort by the patient. Intramuscular nerves are electrically stimulated with a needle placed close to the nerve. Recordings from single muscle fibers are made and the variation in the time between the stimulus and the muscle fiber action potential is measured as the jitter.[36] Generally, the jitter using stimulated SFEMG is about 50% of that recorded with the voluntary technique.

The stimulated single-fiber EMG methods should be used cautiously. Although activation of the nerve and recording of the muscle fiber potentials may appear straightforward, there is potential for misinterpretation of the data.[35]

In summary, SFEMG is a sensitive method for identifying and quantitating various disorders of peripheral nerve or muscle. The changes seen with SFEMG are nonspecific and must always be assessed in conjunction with standard electrophysiologic tests. The precise quantitative measurement of small numbers of muscle fibers in a motor unit is particularly useful for identifying mild or early disease, especially in disorders of neuromuscular transmission, before they are manifested on standard electrophysiologic testing. In most of the disorders in which there are electrophysiologic abnormalities on standard testing, SFEMG will show changes that are not specific. SFEMG can also provide quantitative measures that can be followed serially during the course of a disease.

ABNORMALITIES IN DISEASE STATES

Diseases of the Neuromuscular Junction

The sensitivity of SFEMG allows an abnormality of the neuromuscular junction to be recognized in the absence of clinical weakness or abnormality revealed by other physiologic tests.[2] The clinical usefulness of SFEMG in identifying and quantitating defects of neuromuscular transmission that Stålberg and colleagues[26,29,33] showed has been confirmed by several authors. Although the method is more complicated and time-consuming than repetitive stimulation studies of decrement, it can be readily learned and applied with a minimum of specialized equipment. However, in uncooperative or tremulous patients, reliable SFEMG can be time-consuming, so the selection of

patients and muscles for SFEMG requires consideration.

The abnormalities found on SFEMG in patients with myasthenia gravis were clearly demonstrated in the early studies of Ekstedt and Stålberg,[13] as well as in the textbook by Stålberg and Trontelj,[32] *Single Fibre Electromyography.* Both jitter and blocking are increased in proportion to the severity of clinical involvement, with greater abnormality in weaker muscles. However, in a muscle, and even among the end-plates of a single-motor unit, there is marked variation in the amount of jitter in different fiber pairs. In a single muscle, there may be some entirely normal fiber pairs and others that are grossly abnormal with frequent blocking.

Most important among the features that Ekstedt, Stålberg, and Broman[13,29] noted was the presence of abnormal jitter even in clinically normal muscles. In their early experience with 70 patients with myasthenia gravis, jitter was always abnormal. They concluded that if jitter was normal in the presence of weakness, the weakness was not due to myasthenia gravis. They also noted that jitter was increased with motor unit potential firing rate and muscle activity and decreased with rest or edrophonium. Frequent blocking in patients with severe myasthenia gravis made single-fiber EMG recordings difficult to perform.

Stålberg's work was soon confirmed by Blom and Ringqvist,[1] who examined small numbers of fiber pairs in 12 patients with myasthenia gravis and found them abnormal, with blocking more prominent with greater weakness. Since then, studies by Sanders, Howard, and Johns,[24] Konishi and associates,[20] Martinez and colleagues,[22] Kelly and co-workers,[19] and others have further confirmed and amplified the initial reports. Even though the patient populations studied by these authors differed, all of them found that 77% to 95% of patients with myasthenia gravis have abnormal SFEMG, with increased jitter or blocking (or both). Patients with generalized myasthenia gravis have a higher proportion (86% to 98%) of abnormal jitter, and if there is weakness due to myasthenia gravis in the muscle being tested, all authors agreed that the jitter in that muscle is abnormal.

The frequency of SFEMG abnormalities in

patients with ocular myasthenia gravis has been less consistent among the authors mentioned above, ranging from 20% to 70% abnormal in the extensor digitorum communis muscle and from 60% to 100% in proximal and facial muscles. Slight increases in fiber density have been reported for up to 25% of patients with myasthenia gravis; however, this usually is not found, and if it is present, it is minimal and of little clinical significance.

Remarkable amounts of SFEMG abnormality have also been reported in studies of patients in clinical remission (66% to 83%) and in relatives of patients with myasthenia gravis (33%). These abnormalities are usually mild increases in jitter. Stålberg[26] also reported SFEMG abnormalities in patients with thymoma and no clinical weakness. Stålberg and Hansson[30] demonstrated the clinical application of SFEMG in children for identifying the presence of disease and for following its course.

SFEMG findings have been compared with those of other diagnostic studies. All of the comparisons have shown a much greater frequency of abnormality on SFEMG than on repetitive stimulation. These studies found a slightly higher percentage of abnormality for SFEMG than for acetylcholine receptor antibody levels. However, the studies of Stålberg, Trontelj, and Schwartz[33] and Kelly and associates[19] demonstrated that SFEMG and acetylcholine receptor antibodies are complementary tests: one test identifies abnormality in some patients in whom the results of the other test are normal.

SFEMG has been used in the Mayo Clinic EMG Laboratory since 1976 and has been a useful adjunct to standard EMG and nerve conduction studies in patients thought to have a defect of neuromuscular transmission. Patients with other disorders have been assessed with SFEMG only as part of special studies and not routinely. SFEMG is applied regularly only to patients in whom defects of neuromuscular transmission are suspected. Furthermore, it is limited to patients who have no evidence of disease on other electrophysiologic testing.

Each patient in whom myasthenia gravis or a similar disorder is suspected first undergoes repetitive stimulation of distal and proximal muscles before and after exercise. Repetitive stimulation is a reliable, readily performed method of obtaining a quantitative measure of a defect of neuromuscular transmission. The presence of a reproducible decrement with slow rates of stimulation that partially repairs after exercise and is enhanced late after exercise provides enough evidence of abnormality that SFEMG is not required unless further quantitation of the defect is needed. Moreover, most patients who have a decrement with repetitive stimulation have marked SFEMG abnormalities that make measurement of the jitter and blocking more difficult and less reliable.

In addition to repetitive stimulation, each patient undergoes standard needle EMG of the proximal and distal muscles, including clinically weak muscles, paraspinal muscles, and bulbar muscles. Needle examination can provide evidence of other diseases that may be associated with abnormal findings on SFEMG, such as myopathies and anterior horn cell disease. These must be identified before SFEMG is performed, not only to save the time and effort of performing SFEMG, but also to ensure that they are not mistaken for myasthenia gravis. Standard needle EMG can also rapidly sample a number of muscles on which it may be difficult to perform SFEMG. In these muscles, single motor unit potentials are studied for the variation in size or shape that can provide clear evidence of a defect of neuromuscular transmission. The presence of abnormal motor unit potentials or motor unit potential variation on standard needle EMG makes SFEMG unnecessary unless further quantitation is needed.

Therefore, in our practice, SFEMG is limited to patients whose standard studies are normal. SFEMG studies begin on the extensor digitorum communis muscle because it is easy to examine and well defined and because age-controlled normal values are available. Twenty or more fiber pairs are measured for mean consecutive difference, blocking, and mean duration of the total interpotential interval. This muscle is tested even in patients in whom there are no symptoms or signs of weakness in the extremities, because a high proportion of them show abnormalities. If the extensor digitorum communis muscle is normal, SFEMG is performed on other muscles selected on the basis of clinical weakness. The frontalis muscle is commonly used for patients with ocular or bulbar symptoms.[37]

In more than 2 years, 115 patients (mean age, 44 years; range, 11 to 74 years) in whom myasthenia gravis was suspected were studied with an automated SFEMG system. All had normal findings on nerve conduction studies, repetitive stimulation, and standard concentric needle examination. Of these patients, 108 had the extensor digitorum communis muscle tested alone or with other muscles, and seven had only the frontalis muscle tested.

The 81 patients (age range, 11 to 59 years) had mean consecutive differences in the extensor digitorum communis muscle ranging from 20.5 to 177.8 microseconds, with an abnormal mean greater than 34 microseconds in 32 patients. In the 27 patients 60 or more years old, the mean consecutive differences ranged from 28.3 to 139.7 microseconds, with 14 abnormal. Eight patients from the entire group of 115 had an abnormal mean for the mean consecutive differences but normal numbers of fiber pairs with a mean consecutive difference greater than 60. None had more than two abnormal fiber pairs with a normal mean consecutive difference. Among the seven patients in whom the frontalis muscle was tested, the muscle was abnormal in four whose extensor digitorum communis muscles were normal on SFEMG. One patient with ocular myasthenia gravis had normal facial jitter with abnormal limb jitter.

Use of single-fiber techniques to study patients with congenital forms of myasthenia gravis is limited by the difficulty in obtaining cooperation from young patients. Presynaptic disorders of neuromuscular transmission have also been studied with single-fiber techniques. Abnormalities (both with standard EMG and SFEMG) are readily seen in disorders such as the myasthenic syndrome, because they are most marked with the patient at rest. Increased jitter and blocking that decrease at higher innervation rates are characteristic of myasthenic syndrome.[32]

Primary Neuropathic Diseases With Denervation and Reinnervation

Because reinnervating nerve terminals have immature neuromuscular junctions, SFEMG demonstrates abnormalities. The specific type and degree of abnormality depend on the magnitude and rate of progression of the neurogenic process. Abnormalities on standard nerve conduction studies and EMG differentiate these from disorders of the neuromuscular junction. In studies of severed nerves, increased fiber density is the first sign of reinnervation. Increases in fiber density are seen as early as 3 to 4 weeks after nerve injury (usually before changes can be detected on muscle biopsy). Fiber density increases rapidly for the first 3 months after injury and more slowly thereafter. Increased jitter and blocking are seen for 3 to 6 months after the lesion but rarely longer than that. Most clinical neuropathic disease presents with more complex findings because the process is progressive rather than a single insult and disease may affect the ability to reinnervate.

Stålberg, Schwartz, and Trontelj[31] reported the SFEMG findings in 21 patients with anterior horn cell disease and in 3 with syringomyelia. All patients demonstrated increased fiber density. The increase was greatest in anterior horn cell disorders that were slowly progressive (fiber density, 5.4). The increase (fiber density, 3.3) was less in rapidly progressive amyotrophic lateral sclerosis. Increased jitter and blocking were observed in all these conditions: the largest increase was in amyotrophic lateral sclerosis, and the increase in the spinal muscular atrophies and syringomyelia was less. In chronic conditions, the complexes (particularly the initial part) were more stable. Duration of single-fiber potentials varied considerably. However, the longest durations were seen in the more chronic conditions. The authors concluded that the dual findings of moderately increased fiber density and unstable complexes of varying configuration represent a rapidly progressive process with active reinnervation, such as amyotrophic lateral sclerosis. Markedly increased fiber density and relatively stable complexes (particularly of the initial part) indicate a slowly progressive disease or burned-out process with long-standing reinnervation. The combination of markedly increased fiber density and unstable potentials was believed to reflect reactivation of a long-standing process.

Schwartz, Stålberg, and Swash[25] reported similar conclusions in 10 patients with long-established syringomyelia. SFEMG abnormalities (and clinical changes) were maximum in muscles innervated by spinal segments C-8 and T-1. In the first dorsal interosseous muscle, mean fiber density was 4.1, with 21% of potential pairs demonstrating increased jitter and 7% demonstrating blocking. The distribution of abnormalities, rather than the type, differentiated these patients from those with anterior horn cell disease. Patients with chronic nonprogressive clinical conditions demonstrated complex, but stable, motor unit action potentials and increased fiber density. Patients with recent clinical progression demonstrated more blocking.

Daube and Mulder[9] reported mildly increased fiber density and more markedly increased jitter and blocking in 31 unselected patients with amyotrophic lateral sclerosis. Patient age, clinical severity of disease, compound muscle action potential amplitude, and presence of a decrement to slow repetitive stimulation were valuable predictors of longevity. However, SFEMG findings did not add to the prognostic accuracy. Single-fiber study of spontaneously recorded fasciculations in patients with amyotrophic lateral sclerosis has documented increased jitter and blocking in those discharges.[38]

Thiele and Stålberg[34] reported SFEMG findings in 54 patients with a polyneuropathy associated with uremia, diabetes mellitus, or alcohol abuse. Findings of increased fiber density, jitter, and blocking were seen in alcoholic patients who had only mildly slowed nerve conduction velocities but evidence of denervation on concentric needle examination. Patients with diabetic and uremic polyneuropathies had slower nerve conduction velocities, relatively normal concentric needle examinations, and mild abnormalities on SFEMG (mildly increased jitter without blocking and normal fiber density and duration). SFEMG corroborated standard EMG and pathologic data about the neuropathies of these patients.

In the presence of neuropathic disease, the abnormalities on SFEMG complement those seen during conventional needle EMG. The single-fiber profile can delineate the rate of progression and longevity of the neuropathic disease. Whether SFEMG will become clinically useful in establishing prognosis in certain disease states or in se-

rially following the disease state in treatment trials or for evaluating reinnervation potential is not known.

Primary Disorders of Muscle

Most muscle disorders that cause abnormalities on SFEMG do so by causing muscle necrosis with secondary regeneration and reinnervation. However, muscle fiber membrane abnormalities that cause electrophysiologic failure can also lead to changes. Decreased fiber density and blocking without much increase in jitter were observed during a paralytic attack in a patient with hypokalemic periodic paralysis.[10]

Muscular dystrophies and inflammatory myopathies are the muscle diseases that show the most marked abnormalities on SFEMG. Foote, O'Fallon, and Daube[15] found significantly increased jitter, blocking, duration of the single-fiber complex, and fiber density in 18 patients with inflammatory myopathy. A significant negative correlation existed between motor unit potential duration (measured with concentric needle electrode) and the degree of jitter and blocking. This was believed to provide evidence that the myopathic process was involved in the genesis of both electrophysiologic abnormalities. Duchenne muscular dystrophy causes increased fiber density and a markedly increased duration as well as moderately increased jitter and blocking. Limb girdle muscular dystrophy demonstrates quantitatively less marked changes.

SFEMG is not a good method for primary diagnosis of myopathies. However, it adds complementary information to the concentric needle examination about the degree and timing of associated reinnervation and may detect early mild abnormalities not recognized on standard EMG. The usefulness of SFEMG for serial study of myopathies during treatment trials and for investigation of membrane abnormalities has not been determined.

CHAPTER SUMMARY

Quantitation of nerve conduction studies is widely used in clinical neurophysiology laboratories. Quantitation of needle EMG has been used less widely because the techniques have required special equipment, special expertise, and extra effort.

In this chapter, the attributes of motor unit potentials are presented and techniques for quantitating these attributes are discussed. Methods for measuring single-motor unit potentials, interference pattern of motor unit potentials, and SFEMG are reviewed.

REFERENCES

1. Blom, S and Ringqvist, I: Neurophysiological findings in myasthenia gravis: Single muscle fibre activity in relation to muscular fatiguability and response to anticholinesterase. Electroencephalogr Clin Neurophysiol 30:477–487, 1971.
2. Bromberg, MB and Scott, DM: Single fiber EMG reference values: Reformatted in tabular form. Muscle Nerve 17:820–821, 1994.
3. Brown, WF: The Physiological and Technical Basis of Electromyography. Butterworth, Boston, 1984, pp 95–168; 287–316.
4. Buchthal, F: Electrophysiological signs of myopathy as related with muscle biopsy. Acta Neurol (Napoli) 32:1–29, 1977.
5. Buchthal, F, Guld, C, and Rosenfalck, P: Action potential parameters in normal human muscle and their dependence on physical variables. Acta Physiol Scand 32:200–229, 1954.
6. Chan, RC and Hsu, TC: Quantitative comparison of motor unit potential parameters between monopolar and concentric needles. Muscle Nerve 14:1028–1032, 1991.
7. Daube, JR: Nerve conduction studies. In Aminoff, MJ (ed): Electrodiagnosis in Clinical Neurology. Churchill Livingstone, New York, 1980, pp 229–264.
8. Daube, JR: Quantitative EMG in nerve-muscle disorders. In Stålberg, E and Young, RR (eds): Clinical Neurophysiology. Butterworth, London, 1981, pp 33–65.
9. Daube, JR and Mulder, DW: Prognostic factors in amyotrophic lateral sclerosis (abstr). Muscle Nerve 5 Suppl:S107, 1982.
10. DeGrandis, D, Fiaschi, A, Tomelleri, G, and Orrico, D: Hypokalemic periodic paralysis: A single fiber electromyographic study. J Neurol Sci 37:107–112, 1978.
11. deWeerd, JPC: Volume conduction and electromyography. In Notermans, SLH (ed): Current Practice of Clinical Electromyography. Elsevier Science Publishers, New York, 1984, pp 9–28.
12. Dorfman, LJ and McGill, KC: AAEE minimonograph #29: Automatic quantitative electromyography. Muscle Nerve 11:804–818, 1988.
13. Ekstedt, J and Stålberg, E: Single muscle fibre electromyography in myasthenia gravis. In Kunze, K and Desmedt, JE (eds): Studies on Neuromuscular Diseases. Karger, Basel, 1975, pp 157–161.
14. Engstrom, JW and Olney, RK: Quantitative motor unit analysis: The effect of sample size. Muscle Nerve 15:277–281, 1992.
15. Foote, RA, O'Fallon, WM, and Daube, JR: A comparison of single fiber and routine EMG in normal subjects and patients with inflammatory myopathy. Bull Los Angeles Neurol Soc 43:95–103, 1978.
16. Fuglsang-Frederiksen, A, Scheel, U, and Buchthal, F: Diagnostic yield of analysis of the pattern of electrical activity and of individual motor unit potentials in myopathy. J Neurol Neurosurg Psychiatry 39:742–750, 1976.
17. Fuglsang-Frederiksen, A, Scheel, U, and Buchthal, F: Diagnostic yield in the analysis of the pattern of electrical activity of muscle and of individual motor unit potentials in neurogenic involvement. J Neurol Neurosurg Psychiatry 40:544–554, 1977.
18. Hausmanowa-Petrusewicz, I and Jedrzejowska, H: Correlation between electromyographic findings and muscle biopsy in cases of neuromuscular disease. J Neurol Sci 13:85–106, 1971.
19. Kelly, JJ Jr, Daube, JR, Lennon, VA, et al: The laboratory diagnosis of mild myasthenia gravis. Ann Neurol 12:238–242, 1982.
20. Konishi, T, Nishitani, H, Matsubara, F, and Ohta, M: Myasthenia gravis: Relation between jitter in single-fiber EMG and antibody to acetylcholine receptor. Neurology 31:386–392, 1981.
21. Lang, AH and Partanen, VSJ: "Satellite potentials" and the duration of motor unit potentials in normal, neuropathic and myopathic muscles. J Neurol Sci 27:513–524, 1976.
22. Martinez, AC, Ferrer, MT, Perez Conde, MC, et al: Diagnostic yield of single fiber electromyography and other electrophysiological techniques in myasthenia gravis. II. Jitter and motor unit fiber density studies. Clinical remission and thymectomy evaluation. Electromyogr Clin Neurophysiol 22:395–417, 1982.
23. Nandedkar, SD, Sanders, DB, and Stålberg, EV: On the shape of the normal turns-amplitude cloud. Muscle Nerve 14:8–13, 1991.
24. Sanders, DB, Howard, JF Jr, and Johns, TR: Single-fiber electromyography in myasthenia gravis. Neurology 29:68–76, 1979.
25. Schwartz, MS, Stålberg, E, and Swash, M: Pattern of segmental motor involvement in syringomyelia: A single fibre EMG study. J Neurol Neurosurg Psychiatry 43:150–155, 1980.
26. Stålberg, E: Clinical electrophysiology in myasthenia gravis. J Neurol Neurosurg Psychiatry 43:622–633, 1980.
27. Stålberg, E, Andreassen, S, Falck, B, et al: Quantitative analysis of individual motor unit potentials: A proposition for standardized terminology and criteria for measurement. J Clin Neurophysiol 3:313–348, 1986.
28. Stålberg, E, Bischoff, C, and Falck, B: Outliers, a way to detect abnormality in quantitative EMG. Muscle Nerve 17:392–399, 1994.
29. Stålberg, E, Ekstedt, J, and Broman, A: The electromyographic jitter in normal human muscles. Electroencephalogr Clin Neurophysiol 31:429–438, 1971.
30. Stålberg, E and Hansson, O: Single fibre EMG in

juvenile myasthenia gravis. Neuropaediatrie 4:20–29, 1973.

31. Stålberg, E, Schwartz, MS, and Trontelj, JV: Single fibre electromyography in various processes affecting the anterior horn cell. J Neurol Sci 24:403–415, 1975.

32. Stålberg, E and Trontelj, JV: Single Fibre Electromyography. Mirvalle Press, Old Woking, Surrey, England, 1979.

33. Stålberg, E, Trontelj, JV, and Schwartz, MS: Single-muscle-fiber recording of the jitter phenomenon in patients with myasthenia gravis and in members of their families. Ann N Y Acad Sci 274:189–202, 1976.

34. Thiele, B and Stålberg, E: Single fibre EMG findings in polyneuropathies of different aetiology. J Neurol Neurosurg Psychiatry 38:881–887, 1975.

35. Trontelj, JV and Stålberg, E: Jitter measurement by axonal micro-stimulation. Guidelines and technical notes. Electroencephalogr Clin Neurophysiol 85:30–37, 1992.

36. Trontelj, JV, Stålberg, E, Mihelin, M, and Khuraibet, A: Jitter of the stimulated motor axon. Muscle Nerve 15:449–454, 1992.

37. Ukachoke, C, Ashby, P, Basinski, A, and Sharpe, JA: Usefulness of single fiber EMG for distinguishing neuromuscular from other causes of ocular muscle weakness. Can J Neurol Sci 21:125–128, 1994.

38. Wiechers, D and Johnson, EW: Characteristics of malignant fasciculations (abstr). Electroencephalogr Clin Neurophysiol 49 Suppl:17P, 1980.

39. Willison, RG: A method of measuring motor unit activity in human muscle. J Physiol (Lond) 168:35P–36P, 1963.

CHAPTER 26

ESTIMATING THE NUMBER OF MOTOR UNITS IN A MUSCLE

Jasper R. Daube, M.D.

Peripheral neuromuscular diseases are broadly divided into neurogenic and myopathic disorders, with diseases of the neuromuscular junction often being grouped with myopathic diseases. Neurogenic and myopathic diseases both result in the clinical problem of muscle weakness. The task of clinicians is first to distinguish between these two broad groups of diseases.

The critical difference between neurogenic and myopathic diseases is that motor neurons are involved in neurogenic diseases.

Damage to either the anterior horn cell or its peripheral axon is the essence of a neurogenic process, even though the associated change in muscle may be more apparent clinically. Diseases of motor neurons may produce histologic changes or physiologic changes without histopathologic correlates. An example of a physiologic abnormality is the irritability with excessive activity in the motor nerve seen as fasciculation. Physiologic disorders may cause a loss of function (as in conduction block) or no change in function (as in slowing of conduction). Degenerative or destructive processes result in the loss of an entire motor neuron or peripheral axon. Loss of either motor axons or neurons and conduction block are the bases of the weakness found in most patients with neurogenic diseases. The severity of the clinical deficit is related directly to the number of motor neurons or axons (or both) that are lost or blocked. Therefore, an important part of the assessment of neuromuscular disease is determining the number of functioning motor units. Although it would be ideal to have an actual measure of the number of motor units, current methods—physiologic, histologic, clinical, histopathologic—are not able to provide it. Only electrophysiologic

301

methods can be used to estimate the number of motor units in a muscle. This chapter describes the methods that have been developed for estimating the number of motor units and for determining the number that have been lost through neurogenic disease.

Some of the terms that are used in this chapter are defined as follows: *motor unit* refers to a single anterior horn cell or brain stem motor neuron, its peripheral axon (which travels in a cranial or peripheral nerve), and each of the muscle fibers innervated by that axon. *Number of motor units* refers to the number of functioning motor neurons or functioning motor axons innervating a muscle or group of muscles. A physiologic determination of the number of motor units is called *motor unit number estimate* (MUNE).

Physiologic estimates of the number of motor units have been hampered by the lack of a standard determination of the number of motor units in a muscle. At best, histopathologic and anatomic determinations are only estimates. Two such studies have attempted to measure the number of motor units in human fetal and newborn tissue.[10] In both studies, individual muscles and the motor nerves innervating them were dissected and counted. The number of large myelinated axons in the motor nerve was counted and divided by two to estimate what proportion of these fibers were motor rather than sensory. This proportion was based on animal studies in which degeneration of the peripheral sensory axons after section of the dorsal root indicated that approximately half of the axons in a motor nerve are sensory. This proportion of large myelinated axons then served as the estimate for the number of motor units innervating the muscle. These studies also counted the number of muscle fibers in a muscle to determine the innervation ratio of muscle fibers to axons. Although the results of the two studies were similar, the values were sufficiently different to preclude the designation of a true standard measurement of the number of motor units innervating individual human muscles. However, these studies serve as a baseline comparison for the physiologic methods that have been developed for MUNE. The absence of a standard makes the direct comparison of the values obtained by different methods an equally important part of the as-

sessment of the validity of individual methods of MUNE.[2]

Two distinctly different methods of MUNE have been developed: one uses needle electromyography (EMG), the other motor nerve stimulation and surface recordings. They are described in subsequent sections.

MOTOR UNIT NUMBER ESTIMATE BY STANDARD ELECTROMYOGRAPHY

Standard diagnostic clinical EMG has always included a subjective estimate of the number of motor units in a muscle.[5] Electromyographers have used *recruitment analysis* or *interference pattern analysis* (or both) with voluntarily activated EMG recordings to judge the number of motor units in a muscle. The EMG recorded with either surface or intramuscular electrodes during strong voluntary contractions summates the activity of the muscle fibers in the activated motor units to produce an interference pattern. The greater the number of motor units that are activated, the greater the density of the EMG pattern. The increased density with increased effort is the result of an increase in the number of motor units activated and an increase in the firing rate of individual motor units. The addition of firing rate to number significantly reduces the reliability of density measures to determine the number of motor units. Nonetheless, various measures of density, whether done by subjective or automated methods, have been used to make judgments about the loss of motor units. When the loss is moderate to severe, these methods can clearly identify a loss of motor units, for example, as in a *reduced interference pattern* or a *single motor unit* firing pattern.[15] EMG density measures of the number of motor units are further beset by the problem of relying on patient effort for obtaining activation. Uncooperative patient effort clearly alters the number of motor units activated.

Interference Pattern

The methods that have been described for analysis of interference pattern can be broadly classified into two approaches. The more common one measures the number of turns (i.e., reversals of potential per unit time) and the amplitude of the spikes in the EMG pattern. The other method analyzes the frequency components of the EMG pattern. Both provide some measure of density and, indirectly, of the number of motor units. Neither is reliable enough to be used clinically. Alterations in motor unit potentials in disease states reduce the reliability of both methods. For example, in some disease processes, the potential generated by individual motor units becomes more complex with multiple phases. These additional phases contribute to both the high-frequency components and the number of turns but do not reflect any change in the number of motor units. Because of these features, interference pattern analysis is unsatisfactory for MUNE.

Recruitment Analysis

The second method used by clinical electromyographers to judge the number of motor units in a muscle is recruitment analysis. *Recruitment* refers to the initiation of firing of additional motor units as the effort of voluntary contraction increases. Activation of motor neurons in the brain stem and anterior horn of the spinal cord depends on input to the motor neurons from local reflex pathways and from the descending direct and indirect motor pathways. The relationship of firing of different motor units is determined by the connections of these motor pathways with motor neurons and their individual thresholds for activation. These intrinsic anatomic and physiologic properties of motor neurons result in a fixed pattern of activation in response to voluntary effort. During voluntary activation, low-threshold motor neurons begin firing at rates of 5 to 7 Hz and increase their firing frequency with increasing effort. As effort increases, additional motor units are activated, and they, in turn, increase their frequency of firing with further increases in effort. The recruitment of motor units during activation is a fixed relationship between the number of activated motor units and their firing rates. In recruitment analysis, the number of motor unit potentials activated at any given level of

effort is compared with the rate of firing of individual motor units. This ratio provides an indirect measure of the number of motor units in the muscle.

Clinical EMG judgments about the number of motor units compare the rate of firing of single units with the total number of motor units. The determination of the rate of firing is one of the more difficult steps to make in standard EMG because of difficulty in obtaining sufficient patient control of motor units to isolate one or two units. When it is possible, the rate of firing of the motor unit initially activated is measured at the time the second unit begins to fire. In most muscles, this occurs at 6 to 8 Hz. If the number of motor units in a muscle is decreased, the second motor unit does not begin to fire until the first has reached a higher rate of firing. Less commonly, the first unit begins firing at a higher rate than normal (more than 10 Hz). For most normal muscles, there will be two motor unit potentials firing if one of them is firing at 10 Hz, three at 15 Hz, and four at 20 Hz. The ratio of the number of units firing to the rate of firing can provide a rough gauge of the number of motor units. If the ratio is greater than 5, there is virtually always some decrease in the number of motor units. This semiquantitative method of determining reduced recruitment provides a more accurate and reproducible estimate of the number of motor units than interference pattern analysis, because it does not depend on patient effort.

Although recruitment analysis is reasonably reproducible and clinically reliable, it is usually a subjective judgment made by electromyographers on the basis of experience. It requires taking into account differences in recruitment in different areas of individual muscles and the even greater differences among different muscles. Automated methods for formally quantitating the recruitment pattern have been developed. In these studies, individual motor unit potentials were isolated in human muscles under voluntary control in an experimental setting. The interpotential interval (inverse of frequency of firing) was determined for a population of normal subjects and for patients with amyotrophic lateral sclerosis. The normal onset frequency in the biceps muscle ranged from 6 to 8 Hz, with the recruitment frequency of the second motor unit at 7 to

12 Hz. In patients with amyotrophic lateral sclerosis, onset frequency was from 8 to 20 Hz, with recruitment frequencies of 12 to 50 Hz. These studies provided quantitative measures of MUNE. Further attempts at this form of quantitation with decomposition of the EMG using the techniques developed by Dorfman, Howard, and McGill[9] and by Guiheneuc[12] are providing additional automation and quantitation of these measures. These formal quantitative measures can provide evidence of the reliability of the clinical methods; however, they are so time-consuming and complex that they have not been applied clinically. Further studies and technical developments may eventually allow recruitment analysis to provide more accurate estimates of the number of motor units in a muscle.

Both these methods of MUNE by needle EMG are commonly used in standard diagnostic EMG. Further refinement of needle EMG in combination with surface recording techniques has provided more accurate MUNE. These techniques are described subsequently.

STANDARD NERVE CONDUCTION STUDIES

Motor nerve conduction studies are an important part of the electrophysiologic analysis of peripheral neuromuscular disease. The amplitudes of *compound muscle action potentials* are directly related to the number and size of muscle fibers in a muscle group and indirectly to the number of motor units in the muscle group. The amplitude of a compound muscle action potential is a rough estimate of the number of motor units if a disease is known to be neurogenic and acute. Its value in judging the number of motor units is limited by two factors. First, the amplitude is decreased in myopathies with loss of muscle fiber tissue. Second, the loss of muscle fiber activation that occurs if there is destruction of axons can be partially or fully compensated for by reinnervation from collateral sprouting of intact axons. Estimates of the number of motor units made on the basis of the amplitude of the compound muscle action potential are further

hampered by the wide range of normal amplitudes. For example, even in the case of acute traumatic section of a peripheral nerve that disrupts one half of the motor axons, the amplitude of the ulnar/hypothenar compound muscle action potential may decrease from 12 to 6 mV and still remain within normal limits. In the clinical setting in which the baseline amplitude is not known, the compound muscle action potential is only a rough guide to the number of motor units. Therefore, the amplitude of the compound muscle action potential cannot be used to obtain a reliable MUNE.

BASIC ASSUMPTIONS OF QUANTITATIVE MUNE

Quantitative MUNE can be obtained with needle EMG and motor nerve stimulation methods. Both needle EMG and motor nerve stimulation approaches to MUNE make basic assumptions about the electrical characteristics of motor unit potentials that are described in this section. Four methods of making quantitative MUNE have been developed. Three use nerve stimulation and recording of compound muscle action potentials, and the fourth relies on needle EMG. Variations of each of these four basic techniques continue to evolve with attempts at improving the accuracy and reliability of MUNE. Quantitative methods of MUNE were developed to improve on the accuracy and reliability of the subjective approaches described above. The newly developed techniques are based on either intramuscular needle EMG recordings or surface recordings of compound muscle action potentials of motor axon activation. Both of these techniques rely on several basic assumptions that must be understood to fully appreciate the advantages and drawbacks of quantitative MUNE.

Each method measures both the average size of the potentials generated by single motor units—*single motor unit potentials* (SMUPs)—and the size of the compound muscle action potential obtained with supramaximal stimulation of a motor nerve. The MUNE is then obtained by dividing the supramaximal compound muscle action potential by the size of the SMUP. The techniques differ in how the average size of the single motor unit potentials is obtained. These have been reviewed by Stein and Yang.[18] The underlying assumptions about the measurement of the supramaximal compound muscle action potential and the measurement of the average SMUP need to be understood to apply MUNE clinically.

Supramaximal stimulation of any peripheral motor nerve activates all the muscles innervated by that nerve distal to the point of stimulation. Therefore, measurements of the compound muscle action potential are the summation of activity from multiple muscles. For example, the median/thenar compound muscle action potential is the summation of the activity of the opponens pollicis, abductor pollicis brevis, flexor pollicis brevis, and, to a lesser extent, the lumbrical muscles. The ulnar/hypothenar compound muscle action potential is the summation of all the other intrinsic muscles of the hand. Thus, MUNE is more accurately an estimate of the number of motor units in groups of muscles rather than in a single muscle. Also, although the assumption that all motor axons are activated by supramaximal stimulation is generally true, it may not be the case in the presence of disease in which there are high-threshold axons (as in severe demyelination and regenerated axons). In these situations, the supramaximal compound muscle action potential may be difficult to obtain.

Whereas the methods of measuring the average SMUP differ, they have common assumptions that need review. The most critical issue in determining the size of individual motor unit potentials is the adequacy of sampling of the entire population of SMUPs in the muscle. In patients with severe neurogenic disease, it is possible to identify each motor unit and to obtain a reliable, reproducible, direct count of the number of motor units up to 10. This is the practical maximum of such direct counts. With more than 10 motor units, no method allows reliable measures of each motor unit, so it is uncertain whether a true count has been made. In these cases, it is necessary to select a subset of the total population of motor units and then measure their size and calculate an average size for that subgroup. If all the motor

units are nearly identical in size, then sampling a subset of motor units to obtain an average size of the SMUPs is a valid approach. With greater variation in the size of SMUPs, particularly if the range of sizes is not a normal gaussian distribution, estimates of the true size become less reliable. Each of the methods of measuring the size of SMUPs must address the bias in the selection of the SMUPs that are measured and the adequacy of the sampling to determine whether a reasonable representation of the total population of motor unit sizes has been obtained. Decreased accuracy because of variation of the size of SMUPs occurs particularly in severe chronic neurogenic processes. However, as the severity of damage of a neurogenic process increases and the variation in the size of SMUPs increases, the ease of measuring these potentials increases as well, thereby reducing the potential error from selection bias.

Spike-Triggered Motor Unit Potential Averaging

Spike-triggered averaging relies on the ability to isolate SMUPs by voluntary activation on needle EMG on a two-channel EMG machine.[4] In this method, intramuscular motor unit potentials are measured with any one of several electrodes, including single-fiber EMG, bipolar concentric, standard concentric, or fine-wire electrodes. In each of these recording methods, individual motor unit potentials are isolated on the first channel, usually by an amplitude trigger window that selects potentials on the basis of peak amplitudes. Other criteria can also be used to select motor unit potentials. The accuracy of the method depends on the ability of both patient and electromyographer to activate, identify, and trigger individual motor unit potentials for a period long enough to measure the size of SMUPs.

In spike-triggered averaging methods, the size of SMUPs is measured on a second channel of the EMG machine that is triggered by the needle-recorded motor unit potential on the first channel (Fig. 26–1). The activity averaged on the second channel from the sur-

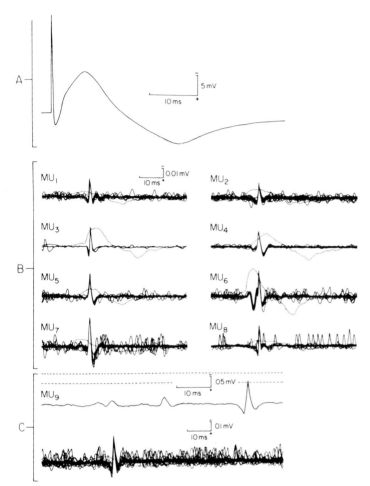

Figure 26–1. Spike-triggered averaging for motor unit number estimate with compound muscle action potential on the left. Each of the other tracks shows the triggered motor unit (*MU*) potential recorded with a needle and the averaged surface compound mucle action potential (*dotted line*). (From Brown, et al,[4] p 427, with permission of John Wiley & Sons.)

face electrode gives the size of the SMUP. The technique requires the isolation of at least 10 and preferably 20 SMUPs whose spike-triggered average can be recorded on the surface. The amplitude or area of the surface-recorded potentials is then used to calculate the average size of the SMUPs in the muscle.

The same surface recording electrodes are used to record the supramaximal compound muscle action potential evoked with stimulation of the motor nerve to the muscle. MUNE is then determined by dividing the size (area or amplitude) of the supramaximal compound muscle action potential by the average size of the SMUPs. Several assumptions made in using this technique are possible sources of error. One, the method assumes that all motor unit potentials can be recorded at the surface. Studies by Brown, Strong, and Snow[4] suggest that this is true in superficial muscles. Two, it assumes that voluntary activation recruits the full range of sizes of motor units. It is likely that this is not the case and that larger motor units are not activated with standard voluntary contraction. Despite these issues, the values obtained with the method are comparable to those expected on the basis of animal studies and those obtained with other methods of recording. De Koning and coworkers[6] have modified the technique to provide a better representation by using macro-EMG needles to record the compound muscle action potential. It is assumed that a macroneedle provides a better representation of the full range of motor units, particularly those deeper in the muscle. Milner-Brown and

Brown[14] used the technique of microstimulation of nerve terminals in the end-plate region to activate motor units recorded with a needle electrode. This reduces the bias in the selection of motor unit sizes that occurs with voluntary activation. Each of these methods gives comparable MUNE.

The spike-triggered averaging methods are generally more time-consuming and more complex to perform because of the need for two channels of recording: a motor unit potential triggering channel and a SMUP averaging channel.

All-or-None Increments in the Compound Muscle Action Potential

All-or-none increment measurement, introduced by McComas,[13] was the first method used for quantitative MUNE. The method is deceptively simple and provides the easiest, most direct, and reliable method of obtaining MUNE. It is based on the well-known all-or-none characteristic of the activation of peripheral motor axons with electrical stimulation. In the incremental method, the stimulus current is finely controlled in small steps designed to allow isolated stimulation of individual motor units in a progressive fashion (Fig. 26–2). For example, if a muscle contains only two motor units, the compound muscle action potential consists of the SMUPs of only these two potentials. Incremental testing with slow, gradually increasing current will show no responses with stimuli below the threshold of the axons of both motor units. When the threshold of one of the axons is reached, the axon is fully activated and the compound muscle action potential suddenly changes from no response to the response of the SMUP. This SMUP is in the range of 50% of the total area. When the threshold of the second axon is reached, this axon also fires and the supramaximal compound muscle action potential is obtained. Moving the stimulus current up and down across these thresholds produces stepwise activation of two steps, that is, the first single motor unit potential and then the full compound muscle action potential. If there are three motor units in the muscle, three steps would be recorded,

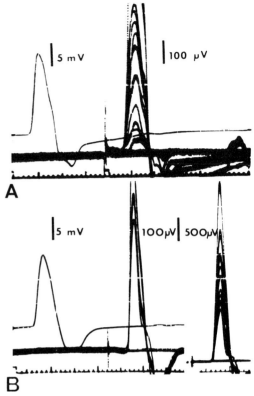

Figure 26–2. Incremental method of motor unit number estimate (MUNE). (*A*) Ten response increments in 500 μV gives a MUNE of 100. (*B*) Ten response increments in 2000 μV results in a MUNE of 25. (From Brown, WF and Feasby, TE: Estimates of functional motor axon loss in diabetics. J Neurol Sci 23:275–293, 1974, with permission.)

and similarly for larger numbers of motor units. In this technique, the size of the SMUP is estimated from the incremental change in the compound muscle action potential, with control of the stimulus current and progressively increasing numbers of motor units. The more of these distinct steps of the total compound muscle action potential that can be measured, the more reliable the MUNE becomes with incremental measurement.

If the stimulation selectively activates larger or smaller motor units, there could be a selection bias in the SMUPs used to determine the average size. Also, the occurrence of alternation may result in another error in the measurement. Alternation is best illustrated by an example of a muscle containing three motor unit potentials of slightly differing size but of nearly the same threshold. In this situation, there could be as many as seven different size steps. Because of small variations in threshold of individual axons, an axon that is activated first in one trial may be activated second or third in subsequent trials. Thus, the sizes of potentials that could be obtained when there are three motor units of different sizes, A, B, and C, are those generated by A alone, B alone, C alone, by A and C, B and C, A and B, and by A, B, and C together. Thus, three single motor unit potentials might be recorded as three to seven steps (Fig. 26–3).

Several modifications have been developed to minimize these errors: (1) use of automated computer measurement of the templates of different single motor unit potentials to reduce the likelihood of measuring alternation,[1] (2) stimulation at different points along the nerve (multipoint stimulation) to isolate only SMUPs,[7] (3) use of recording electrodes of different sizes and shapes, (4) use of an automated method of incrementing the stimulus size,[8,11] and (5) microstimulation of single nerve terminals at the end-plate region.

The McComas[13] incremental compound muscle action potential technique uses whatever average values are obtained for the size (amplitude or area) of the single motor unit potential and compares them with the supramaximal size of the compound muscle action potential. Normal value determinations by several authors[1,3,5,13] have shown that the mean normal MUNE for median/ thenar muscles is about 300, with a lower limit of normal around 100. The other well-studied muscle, the peroneal/extensor digitorum brevis, has been estimated to have about 200 motor units, with a normal lower limit of 50.

Incremental technique becomes unreliable when the motor unit potentials are smaller, as in severe myopathies or with nascent motor unit potentials. The inability to record the steps results in underestimation of MUNE.

The method of incremental compound muscle action potential is direct enough that it should be learned by every electromyographer. It can be applied to patients in whom loss of motor units allows the remaining axons to be stimulated selectively and recorded as incremental steps by stimulus control.

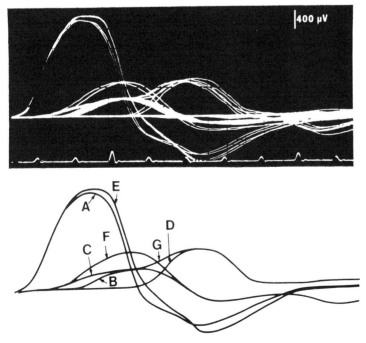

Figure 26–3. Alternation of motor unit firing during F-wave recording. (From Feasby, TE and Brown, WF: Variation of motor unit size in human extensor digitorum brevis and thenar muscles. J Neurol Neurosurg Psychiatry 37:916–926, 1974, with permission.)

F-Wave Measurements

F waves have been suggested as a method for determining the size of the single motor unit potential. Supramaximal activation of all the motor axons in a peripheral nerve is associated with antidromic activation of some of the anterior horn cells. The small proportion of anterior horn cells activated antidromically produces small late potentials, *F waves*. Repeated supramaximal stimuli activate different anterior horn cells and produce different F waves. Recording a range of sizes of F waves can be used to estimate the average size of the SMUP. This average size can then be divided into the supramaximal compound muscle action potential to obtain MUNE. Brown and associates[17] have shown that the drawbacks of this method are the alternation (see Fig. 26–3) described above for the increment method and the common activation of multiple rather than single motor units. Both of these drawbacks result in overestimating the average size of the SMUP and in underestimating the MUNE.[17] Figure 26–3 from the study of Brown[14] shows how three motor unit potentials of different sizes can summate by alternation into seven different potentials. Automated correction of these drawbacks by submaximal stimulation and template matching may make the method clinically useful.[16]

Statistical Measurements

The fourth method of estimating the size of the SMUP uses direct stimulation of the motor nerve, and although similar to the all-or-none incremental method, is conceptually different. With the statistical method, no attempt is made to identify the potentials associated with individual motor units. The method relies on the known relationship between the variance of multiple measures of step functions and the size of the individual steps when the steps have a Poisson distribution. Poisson statistics are used to calculate the number of quanta released from a nerve terminal at the neuromuscular junction when individual quanta are too small to be distinguished, as in myasthenia gravis. Because the statistical method looks only at variance of the compound muscle action potential and does not require identification of individual components, it can be used when the sizes of single motor unit potentials are too small to be isolated, which is often the case in normal muscles and myopathies. It can also be used with high-amplitude compound muscle action potentials that require gains at which the SMUP cannot be isolated. In pure Poisson statistics, the sizes of a series of measurements are multiples of the size of a single component. As shown in Figure 26–4, where steps can be seen, a Poisson distribution has discrete values at which responses are found. A pure Poisson distribution has decreasing numbers at higher values. In this distribution, the variance of that series of measurements is equal to the size of the individual components making up each measurement and can provide an estimate of the average size of the SMUP.

In common with other methods of estimating the size of the SMUP, the statistical method assumes that each motor unit has a similar size and that it is the same size each time it is activated. Defects of neuromuscular transmission that result in varying sizes of the SMUP cause inaccuracies in this measurement. With a larger number of components making up the summated potential, the distributions typically shift from a Poisson to a normal distribution. This produces an error of up to 10% in the MUNE.

In the statistical method, recording electrodes are applied as in standard nerve conduction studies, with the stimulating electrode taped firmly in place over the appropriate nerve. A sequence of 30 or more submaximal stimuli is given. The inherent variability of the threshold of individual axons causes their intermittent firing and all-or-none variations in the size of the compound muscle action potential. The occurrence of alternation with changing units that are activated does not lessen the accuracy of the statistical method. Because the method is a statistical measurement, a somewhat different result is obtained each time. Therefore, multiple trials are needed to obtain the most accurate measurement. Experimental testing with trials of more than 300 stimuli has shown that repeated measurement of

STATISTICAL ESTIMATES OF MOTOR UNIT NUMBERS
Peroneal/EDB, neuropathy, 60%

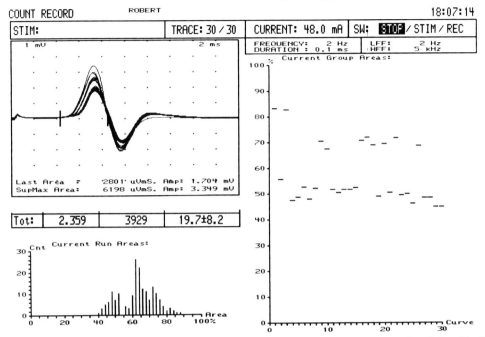

Figure 26–4. Statistical method of motor unit number estimate in a normal subject. Steps can be distinguished and an approximate Poisson distribution is obtained. *EDB* = extensor digitorum brevis.

groups of 30 until the standard deviation of the repeated trials is less than 10% provides a close estimate of the number obtained with many more stimuli (see Fig. 26–4).

The statistical method includes a scan of the compound muscle action potential in which stimulation is increased in equal increments from threshold to maximal. The scan on the left in Figure 26–5 is normal. The scan on the right, from a patient with amyotrophic lateral sclerosis, shows several large motor units and one that is very large.

CHAPTER SUMMARY

The all-or-none incremental, the spike-triggered averaging, and the statistical methods of obtaining MUNE give similar values in different muscles in normal subjects. This suggests that each of these methods can be used whenever and wherever it is most feasible and that different methods may be appropriate in different settings. For example,

when each of the motor units can be identified with small increments of stimuli in a severe neurogenic process, MUNE is most rapidly and accurately defined by actually counting the total number of increments. When the number of motor units is too large to do this or their size is too small to identify them accurately in the compound muscle action potential, the statistical method is appropriate. In those muscles in which compound muscle action potentials cannot be reliably obtained, as in proximal muscles that are difficult to immobilize during stimulation of the motor nerve, the spike-triggered averaging method is most appropriate. Applications of the methods of MUNE to diseases have demonstrated the expected decrease in the number of motor units in localized mononeuropathies (carpal tunnel syndrome, lumbar radiculopathies, and peripheral neuropathies) and in motor neuron disease. They are particularly helpful in disorders in which collateral sprouting results in a compound muscle action potential of

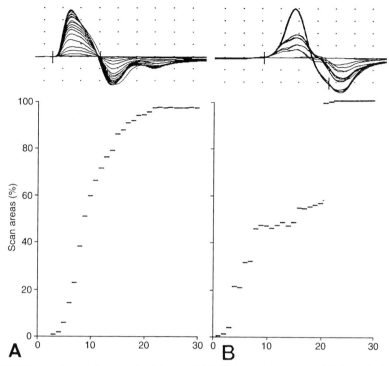

Figure 26–5. MUNE scan from (*A*) normal subject and (*B*) patient with amyotrophic lateral sclerosis. Between threshold and maximal compound muscle action potentials, 30 equal increments in stimulus intensity were applied to the nerve. The elicited compound muscle action potentials are superimposed above the histograms. The histograms depict the area of each of the 30 responses. In (*A*) note the smooth curve with small increments. In (*B*) the increments are larger, with a particularly large increment just before the maximal compound muscle action potential. The latter is due to activation of a single large motor unit.

normal amplitude despite a loss of motor units. Serial measurements of MUNE are helpful in following the course of a disease, especially in treatment trials or studies of the evolution of the disease.

REFERENCES

1. Ballantyne, JP and Hansen, S: A new method for the estimation of the numbers of motor units in muscles. 1. Control subjects and patients with myasthenia gravis. J Neurol Neurosurg Psychiatry 37:907–915, 1974.
2. Bromberg, MB, Forshew, DA, Nau, KL, et al: Motor unit number estimation, isometric strength, and electromyographic measures in amyotrophic lateral sclerosis. Muscle Nerve 16:1213–1219, 1993.
3. Brown, WF: F-wave and multipoint methods for estimating numbers of motor units. Presented at the 46th Annual Meeting of the American Academy of Neurology, Workshop 266, Washington, DC, May 1 to 7, 1994.
4. Brown, WF, Strong, MJ, and Snow, R: Methods for estimating numbers of motor units in biceps-brachialis muscles and losses of motor units with aging. Muscle Nerve 11:423–432, 1988.
5. Daube, JR: AAEM Minimonograph #11: Needle examination in clinical electromyography. Muscle Nerve 14:685–700, 1991.
6. de Koning, P, Wieneke, GH, van der Most van Spijk D, et al: Estimation of the number of motor units based on macro-EMG. J Neurol Neurosurg Psychiatry 51:403–411, 1988.
7. Doherty, TJ and Brown, WF: The estimated numbers and relative sizes of thenar motor units as selected by multiple point stimulation in young and older adults. Muscle Nerve 16:355–366, 1993.
8. Doherty, TJ, Stashuk, DW, and Brown, WF: Determinants of mean motor unit size: Impact on estimates of motor unit number. Muscle Nerve 16:1326–1331, 1993.
9. Dorfman, LJ, Howard, JE, and McGill, KC: Motor unit firing rates and firing rate variability in the detection of neuromuscular disorders. Electroencephalogr Clin Neurophysiol 73:215–224, 1989.
10. Feinstein, B, Lindegård, B, Nyman, E, and Wohlfart, G: Morphologic studies of motor units in normal human muscles. Acta Anat Scand 23:127–142, 1955.
11. Galea, V, de Bruin, H, Cavasin, R, and McComas, AJ: The numbers and relative sizes of motor units

estimated by computer. Muscle Nerve 14:1123–1130, 1991.

12. Guiheneuc, P: Automatic detection and pattern recognition of motor unit potentials. Electroencephalogr Clin Neurophysiol Suppl 39:21–26, 1987.

13. McComas, AJ: Motor unit estimation: Methods, results, and present status. Muscle Nerve 14:585–597, 1991.

14. Milner-Brown, HS and Brown, WF: New methods of estimating the number of motor units in a muscle. J Neurol Neurosurg Psychiatry 39:258–265, 1976.

15. Nomenclature Committee, American Association of Electromyography and Electrodiagnosis: AAEE glossary of terms in clinical electromyography. Muscle Nerve 10 Suppl:G1–G60, 1987.

16. Stashuk, DW, Doherty, TJ, Kassam, A, and Brown, WF: Motor unit number estimates based on the automated analysis of F-responses. Muscle Nerve 17:881–890, 1994.

17. Stashuk, DW, Komori, T, Doherty, TJ, and Brown, WF: Automatic analysis of F-responses: New methods for deriving motor unit estimates and analyzing relative latencies and conduction velocities in single motor fibers (abstr). Muscle Nerve 15:1204, 1992.

18. Stein, RB and Yang, JF: Methods for estimating the number of motor units in human muscles. Ann Neurol 28:487–495, 1990.

Electro-physiologic Assessment of Neural Function

PART E

Reflexes and Central Motor Control

Motor function is controlled by a complex combination of central nervous system circuits that include all levels of the neuraxis. Local reflexes at the levels of the spinal cord and brain stem mediate and integrate local sensory input and input from the descending motor pathways to the motor unit. Descending motor pathways from the cerebral hemisphere to the spinal cord include the rapidly conducting direct corticospinal pathways and several indirect pathways arising in the spinal cord and brain stem. Descending motor activity in these pathways is directed and controlled by motor areas in the cerebral cortex, basal ganglia, and cerebellum. The cerebellum and basal ganglia form feedback loops that extend through the thalamus to the cerebral cortex and control motor activities. Many of these pathways and functions can be monitored electrically, as described in the chapters of this section. H reflexes and cranial nerve reflexes are localized responses of the motor neurons in the brain stem and spinal cord to localized sensory input. Both groups of reflexes can be used to assess peripheral sensory and motor function as well as their central connections in the brain stem and spinal cord. In contrast, long-loop reflexes and the silent period depend more on the descending motor pathways from the brain to the spinal cord. Therefore, these reflexes are primarily useful in elucidating central disorders of motor function or neuronal excitability.

Multichannel surface electromyographic recordings from agonist and antagonist muscles in the limbs and trunk can be used to characterize several motor disorders on the basis of the patterns of activation and the timing of activity in different muscles, either in one limb or longitudinally in the body.

Surface electromyographic recordings in posturography and electronystagmography are also used in measuring the motor control of posture and vestibular function. These measurements, described in Chapters 30 and 31, assess the long pathways controlling motor function and their integration in the neuronal pools. Posturography and electronystagmography are useful in evaluating many disorders of both the vestibular pathways and motor control pathways.

CHAPTER 27

H REFLEXES

Kathryn A. Stolp-Smith, M.D.

PHYSIOLOGIC BASIS
TECHNIQUE
CLINICAL APPLICATION
Proximal Conduction
Central Nervous System Excitability

In 1918, Hoffmann[13] recorded from the soleus muscle a late response that occurred with submaximal electrical stimulation of the tibial nerve, similar to the response recorded as a result of a tendon tap. Magladery and McDougal[18] later termed this response the "Hoffmann reflex," or "H reflex." This reflex has been widely studied and has been found useful in assessing experimental and clinical aspects of disorders of the central and peripheral nervous systems. The H reflex may be obtained from other muscles, including the gastrocnemius and flexor carpi radialis. Technically, the H reflex is a study that is relatively simple to perform in a basic electrophysiology laboratory. A thorough understanding of the physiologic basis, sources of error, and clinical applications and limitations enhances the usefulness of the H reflex.

PHYSIOLOGIC BASIS

Stimulation of the tibial nerve in the popliteal fossa at intensities less than needed to generate an early M response, or compound muscle action potential, from the soleus muscle elicits the H reflex. Magladery and McDougal[18] demonstrated that the H reflex is a monosynaptic reflex response produced by activation of a small proportion of soleus motor neurons (Fig. 27–1). Although similar to a tendon reflex, the H reflex bypasses the muscle spindle component. Muscle spindle afferents are preferentially more sensitive to low-intensity stimulation than large soleus motor axons are.[12] The amplitude of the H reflex increases with gradually increasing stimulus intensity as more spindle afferents are activated. However, as stimulus intensity increases further, amplitude begins to decrease as more and more of the orthodromic reflex volley is blocked by antidromically conducted impulses. A stimulus of long duration allows for more selective activation of afferent axons, whereas a stimulus of short duration increases the likelihood of motor axon activation.[14]

Latency of the H reflex depends on several factors. These include the time to activate the primary spindle afferents, conduction velocity of the primary spindle afferents, central conduction delay, conduction velocity of motor axons and terminal conduction delay, neuromuscular delay, distance from the site of stimulation to the spinal cord, and the time to detect a compound muscle action potential by the recording electrode.[4a,10]

The amplitude of the H reflex is extremely variable. Factors that affect the H reflex and that should be considered in performing this study are listed in Table 27–1.

315

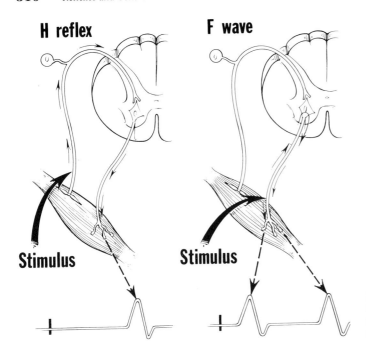

H reflex

F wave

Stimulus

Stimulus

Figure 27–1. Physiology of the H reflex: selective activation of muscle spindle afferents and monosynaptic reflex response of soleus motor axons.

TECHNIQUE

The H reflex can be recorded from many muscles in normal neonates. In normal adults, it is obtained reliably from the soleus and flexor carpi radialis muscles and has been reported in the palmaris longus, flexor carpi ulnaris, anterior tibial, vastus medialis, masseter, extensor digitorum communis, and ulnar-innervated hand intrinsic muscles.[8,12,15] The H reflex is elicited more easily in the soleus muscle after stimulation of the tibial nerve. Electrical stimulation is most commonly used; however, an Achilles ten-

Table 27–1. **FACTORS THAT AFFECT THE PRESENCE OR AMPLITUDE OF THE H REFLEX**[4b,5,11,13,19]

Suppression	Facilitation
Contraction of antagonist muscles	Mild contraction of agonists
Strong contraction of agonists	Passive stretch
Passive shortening	Labyrinthine vestibular stimulation
Strong electrical stimulus	Bite, grasp
Sleep	
Vibration	
Drugs (nicotine, pentobarbital, diazepam)	
Stimulation rate > 1/s	
Tendon tap	
Strong flexion/extension of neck muscles	
Proximal ischemia	
Spinal anesthesia	

don tap or quick Achilles tendon stretch may also be used to elicit the H reflex.[16]

The patient is placed in either a prone or a supine position to avoid excessive lengthening or shortening of the soleus muscle. The tibial nerve immediately lateral to the popliteal artery at the level of the popliteal crease is stimulated. To avoid erroneous stimulation of the peroneal nerve, the resultant twitch should be observed for gastrocnemius-soleus contraction without associated contraction of the anterior tibial or peroneal muscle. The cathode is placed proximally to the anode. Square-wave pulses of 0.5 to 1.0 millisecond in duration with an intensity of 25 to 30 V are delivered at a rate of 0.5 Hz. Intensity should be varied, and from an initial low level, it should be increased by small increments.

The active recording electrode may be placed medially over the soleus muscle approximately midway between the site of stimulation and the medial malleolus if the patient is supine and posteriorly, if prone, over the medial gastrocnemius. Recording over the soleus results in a waveform with an initial negative deflection, because the soleus is the source of the reflex compound muscle action potential.

With gradually increasing stimulus intensity, the H reflex usually appears before the M response or shortly thereafter. As stimulus voltage is increased, amplitude of the H reflex reaches a maximum before a maximum M response is obtained. With further increase in voltage, amplitude declines and eventually disappears as the M response reaches maximal amplitude at supramaximal stimulus intensity (Fig. 27–2). As recorded from the soleus muscle, the H reflex appears as a biphasic wave with a large negative deflection, which is normally 50% to 100% of the maximum M amplitude. If recorded over the gastrocnemius muscle, an initial positivity may be present.[14] Given this variability in amplitude, the minimum latency of the H reflex provides a more reliable measure as applied to clinical situations. Stimulus intensity should be increased and decreased as needed until a reproducible maximum H reflex is obtained.[6,16] This technique is important in differentiating the H reflex from the F wave (Table 27–2).

Latency is measured to the initial deflec-

H REFLEX

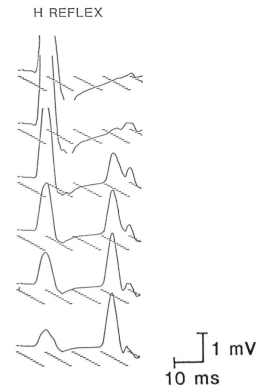

1 mV

10 ms

Figure 27–2. H reflex recorded from soleus muscle with decreasing stimulus intensity (*from top to bottom*). Note maximal amplitude of the H reflex with submaximal stimulation.

tion from baseline. The H reflex usually occurs at 28 to 35 milliseconds in adults and varies with leg length and age. Braddom and Johnson[3] developed a nomogram and formula that allow for age and leg-length differences. However, the side-to-side H-reflex latency comparison is probably more widely used than the absolute value and should vary by no more than 3 milliseconds (Table 27–3).[3,6,9,16,20]

Amplitude of the H reflex is rarely measured. Kimura[16] measured the amplitude from the baseline to the negative peak and reported a normal value of 2.4 ± 1.4 mV. It is important to distinguish the H reflex from the F wave, which has a similar latency. The F wave is not a reflex but represents recurrent discharge of multiple motor neurons, with alpha motor neurons and axons serving as both the afferent and efferent pathways. This produces a smaller amplitude wave of varying latency and morphology obtained with supramaximal stimulation. The F wave

Table 27–2. **COMPARISON OF THE H REFLEX AND F WAVE IN NORMAL SUBJECTS**[12,22]

H Reflex	F Wave
Suppressed by supramaximal stimulation	Maximal with supramaximal stimulation
Optimal with submaximal stimulation	Infrequent with submaximal stimulation
Amplitude 50% to 100% of M maximum	Amplitude 10% of M maximum
Latency relatively constant	Latency variable
Morphology constant	Morphology variable
Recorded from forearm and leg	Recorded from hand and foot muscles

can be recorded from many skeletal muscles (see Table 27–2).

CLINICAL APPLICATION

Proximal Conduction

The H reflex is best used as a measure of proximal conduction. It commonly is used in assessing S-1 radiculopathy. Braddom and Johnson[3] proposed that a side-to-side H-reflex latency difference greater than 1.2 milliseconds indicates a discrete lesion of the S-1 nerve root and may be the initial finding in acute radiculopathy or in cases of radiculopathy in which the only needle electromyographic findings are fibrillation poten-

tials in paraspinous muscles.[2,6,12,15] The prolongation or absence of the H reflex correlates with clinically reduced or absent ankle jerks. In using the H reflex in this way, any lesion along the pathway of the H reflex may prolong the latency; if, moreover, the largest diameter axons are not affected by a root lesion, the H reflex may not be useful. Therefore, additional corroborating electrophysiologic evidence supports the diagnosis of S-1 radiculopathy.[4b,22]

Latency of the H reflex may also be prolonged in cases of peripheral neuropathy because of slowing of peripheral or proximal conduction. Prolonged H-reflex latency has been demonstrated in diabetic, alcoholic, nutritional, paraneoplastic, cisplatin, and vasculitic neuropathies. In uremic neuropa-

Table 27–3. **NORMAL VALUES FOR H-REFLEX LATENCY**

Reference	Muscle Recorded	LATENCY, MS		Side-to-Side Comparison
		Mean	Range	
Braddom and Johnson[3]	Gastrocnemius	—	28–35	<1.2
Kimura[16]	Soleus	29.5	27–35	<1.4
Mayo Clinic	Soleus	—	25–35	<3
Shahani and Young[22]	Soleus	—	≤34	<2
Ongerboer De Visser, Schimsheimer, and Hart[20]	Flexor carpi radialis	16.8	15–20	<0.002
Mayo Clinic	Flexor carpi radialis	15.9	11–21	<2

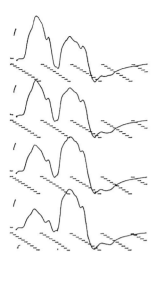

1 mV

10 ms

Figure 27–3. H reflex recorded from forearm flexors.

thy, this prolongation decreases after successful renal transplantation.[1] Prolonged latency also has been demonstrated in Guillain-Barré syndrome, chronic inflammatory polyradiculoneuropathies, and hereditary mixed sensorimotor neuropathy type 1.[21,22] Because the H reflex may be absent in persons more than 60 years old and because a large number of persons over 60 years old also have lumbar stenosis, the usefulness of the H reflex in studying neuropathy in this age group is limited.[23] The latency of the H reflex in the flexor carpi radialis muscle has proved useful in assessing slowing through the proximal median nerve fibers in radiation-induced plexopathy[20,21] and may be used to assess C-7 radiculopathy, comparable to assessment of S-1 radiculopathy with recordings from the soleus muscle (Fig. 27–3).

Central Nervous System Excitability

If the H reflex can be recorded readily from muscles other than the soleus or flexor carpi radialis, one must consider the possibility of a central nervous system disorder. The H reflex may be used to study the excitability of the motor neuron pool in two ways.[22] The first compares the ratio of maximum H amplitude with maximum M amplitude (Hmax/Mmax). This ratio may be equivalent to unity in normal subjects and may be increased in disorders causing spasticity, although it correlates poorly with the degree of spasticity noted clinically. The Hmax/Mmax ratio is normal in cases of rigidity.[12,17]

A second way is to produce H-reflex recovery curves, first described by Magladery and McDougal.[18] The H reflex is elicited before, during, and after conditioning stimuli, usually electrical, of various interstimulus intervals. It is important to keep the latency and amplitude of the M response constant during these trials. The resulting curve is divided into a series of phases for the purpose of comparison. These curves have also been used to study spasticity. However, the changes in the curves are rarely striking. The curve is greatly influenced by patient position and comfort, angle of lower extremity joints, relaxation, and head positioning. Construction of these curves is time-consuming, and reproducibility is poor. The pathophysiologic basis of these curves is poorly understood, and they are impractical.[7,12,22]

CHAPTER SUMMARY

The H reflex, a monosynaptic reflex usually evoked with tibial nerve stimulation, is simple to elicit and can be a useful adjunct to routine nerve conduction studies in assessing peripheral nerve disorders and central motor neuron excitability. Presence of the H reflex on submaximal stimulation and suppression of the reflex with supramaximal stimulation help distinguish this late response from the F wave.

REFERENCES

1. Bolton, CF: Metabolic neuropathy. In Brown, WF and Bolton, CF (eds): Clinical Electromyography. Butterworth, Boston, 1987, p 256.
2. Braddom, RL and Johnson, EW: H reflex: Review

and classification with suggested clinical uses. Arch Phys Med Rehabil 55:412–417, 1974.

3. Braddom, RL and Johnson, EW: Standardization of H reflex and diagnostic use in S1 radiculopathy. Arch Phys Med Rehabil 55:161–166, 1974.

4. Brown, WF: The Physiological and Technical Basis of Electromyography. Butterworth, Boston, 1984, (a) p 147; (b) p 477.

5. Brunia, CHM: The influence of diazepam and chlorpromazine on the Achilles tendon and H-reflexes. In Desmedt, JE (ed): New Developments in Electromyography and Clinical Neurophysiology. Vol 3. Karger, Basel, 1973, pp 367–370.

6. DeLisa, JA, Mackenzie, K, and Baran, EM: Manual of Nerve Conduction Velocity and Somatosensory Evoked Potentials, ed 2. Raven Press, New York, 1987, pp 125–127.

7. Delwaide, PJ: Contribution of human reflex studies to the understanding and management of the pyramidal syndrome. In Shahani, BT (ed): Electromyography in CNS Disorders: Central EMG. Butterworth, Boston, 1984, p 90.

8. Deschuytere, J, Rosselle, N, and De Keyser, C: Monosynaptic reflexes in the superficial forearm flexors in man and their clinical significance. J Neurol Neurosurg Psychiatry 39:555–565, 1976.

9. Falco, FJ, Hennessey, WJ, Goldberg, G, and Braddom, RL: H reflex latency in the healthy elderly. Muscle Nerve 17:161–167, 1994.

10. Fisher, MA: AAEM minimonograph #13: H reflexes and F waves: Physiology and clinical indications. Muscle Nerve 15:1223–1233, 1992.

11. Gassel, MM: An objective technique for the analysis of the clinical effectiveness and physiology of action of drugs in man. In Desmedt, JE (ed): New Developments in Electromyography and Clinical Neurophysiology, Vol 3. Karger, Basel, 1973, pp 342–359.

12. Goodgold, J and Eberstein, A: Electrodiagnosis of Neuromuscular Diseases, ed 3. Williams & Wilkins, Baltimore, 1983, pp 263–271.

13. Hoffmann, P: Ueber die Beziehungen der Schnenreflexe zur willkürlichen Bewegung und zum Tonus. Z Biol 68:351–370, 1918.

14. Hugon, M: Methodology of the Hoffmann reflex in man. In Desmedt, JE (ed): New Developments in Electromyography and Clinical Neurophysiology. Vol 3. Karger, Basel, 1973, pp 277–293.

15. Johnson, EW: Electrodiagnosis of radiculopathy. In Johnson, EW (ed): Practical Electromyography, ed 2. Williams & Wilkins, Baltimore, 1988, p 238.

16. Kimura, J: Electrodiagnosis in Diseases of Nerve and Muscle: Principles and Practice, ed 2. FA Davis, Philadelphia, 1989, pp 359–361.

17. Leonard, CT and Moritani, T: H-reflex testing to determine the neural basis of movement disorders of neurologically impaired individuals. Electromyogr Clin Neurophysiol 32:341–349, 1992.

18. Magladery, JW and McDougal, DB, Jr: Electrophysiological studies of nerve and reflex activity in normal man. I. Identification of certain reflexes in the electromyogram and the conduction velocity of peripheral nerve fibres. Bull Johns Hopkins Hosp 86:265–290, 1950.

19. Mayer, RF and Mosser, RS: Maturation of human reflexes. In Desmedt, JE (ed): New Developments in Electromyography and Clinical Neurophysiology, Vol 3. Karger, Basel, 1973, pp 294–307.

20. Ongerboer De Visser, BW, Schimsheimer, RJ, and Hart, AAM: The H-reflex of the flexor carpi radialis muscle; a study in controls and radiation-induced brachial plexus lesions. J Neurol Neurosurg Psychiatry 47:1098–1101, 1984.

21. Schimsheimer, RJ, Ongerboer De Visser, BW, Kemp, B, and Bour, LJ: The flexor carpi radialis H-reflex in polyneuropathy: Relations to conduction velocities of the median nerve and the soleus H-reflex latency. J Neurol Neurosurg Psychiatry 50:447–452, 1987.

22. Shahani, BT and Young, RR: Studies of reflex activity from a clinical viewpoint. In Aminoff, MJ (ed): Electrodiagnosis in Clinical Neurology. Churchill Livingstone, New York, 1980, pp 290–304.

23. Wilbourn, AJ: The diabetic neuropathies. In Brown, WF and Bolton, CF (eds): Clinical Electromyography. Butterworth, Boston, 1987, p 339.

CHAPTER 28

CRANIAL REFLEXES

J. Clarke Stevens, M.D.
Benn E. Smith, M.D.

Integrity of the trigeminal and facial nerves can be evaluated electrophysiologically by study of their brain stem reflexes. The reflexes discussed in this chapter are the electrically evoked blink reflex and jaw jerk.

BLINK REFLEX

Definition

The blink reflex is a two-component exteroceptive response that can be elicited by mechanical or electrical stimulation over the face with a graded threshold; the lowest is that for stimuli around the eye. Supraorbital nerve (V_1) stimulation in normal subjects produces an ipsilateral first response (R1) and bilateral second responses (R2). Infraorbital nerve (V_2) stimulation elicits the blink reflex in 20% of subjects more often with mechanical than electrical stimuli. Stimulation of the mental nerve (V_3) rarely produces a blink reflex. The R2 response correlates with orbicularis oculi contraction. The physiologic significance of the first component of the reflex is unknown.

There are other methods of provoking eye blink.[77] The corneal reflex, part of the routine neurologic examination, is a trigeminofacial reflex obtained by mechanical stimulation of the cornea. It can also be elicited by electrical stimulation, which evokes a bilateral response similar to the R2 response of the blink reflex.[9] Also, a single-component reflex occurs after a loud noise or a light flash.[77]

Normal Blink Reflex

HISTORY

The blink reflex after a tap to the forehead was first described by Overend in 1896; he considered it a cutaneous reflex.[19,57] The term "orbicularis oculi reflex" was coined by Wartenberg,[83] who suggested it was a simple stretch reflex of facial musculature. Kugelberg[38] was the first to detect two components of the reflex, ipsilateral R1 and bilateral R2 responses. He postulated that R1 represented a myotatic reflex mediated by the mesencephalic nucleus in the midbrain and that R2 was mediated by the spinal trigeminal nucleus in the pons and medulla. Tokunaga and coworkers[78] placed lesions in the brain stem of cats and concluded that R1

321

was mediated by the principal sensory nucleus of the trigeminal nerve in the pons and that R2 originated from the spinal tract and nucleus of the trigeminal nerve in the pons and medulla. Rushworth,[66] after studying patients with cranial neuropathies, theorized that R1 was a proprioceptive reflex and R2 was nociceptive. Shahani[71] evoked the reflex by stimulation of skin over a wide area of the face, including the nose, showing that the reflex was mediated by cutaneous receptors rather than by proprioceptive afferents. He succinctly and precisely described the characteristics of R1, which is usually biphasic or triphasic in configuration and is briefer in duration, simpler in configuration, and more stable in size and shape than R2. Although R1 has a more constant latency than R2, R1 latency is also somewhat variable. Finally, R1 is less likely to show habituation than R2.

NEUROANATOMY

The afferent limb of the reflex arc is the first division of the trigeminal nerve, and the efferent limb is the motor axons of the facial nerve. The early response is relayed centrally through an oligosynaptic pathway involving the principal sensory nucleus of the trigeminal nerve in the pons.[80] Afferent fibers relaying the R2 response descend into the medulla to the spinal trigeminal nucleus through polysynaptic pathways that pass ipsilaterally and contralaterally before synapsing with facial motor neurons.

METHODS

The recording electrodes are tin disks taped to the skin overlying the orbicularis oculi muscle bilaterally. The supraorbital nerve is stimulated at the eyebrow, first one side and then the other, while simultaneous recordings are made from both sides. The stimuli are given irregularly at least 5 seconds apart so that habituation is minimized. The stimulating current needed to evoke the reflex is small and not painful. With a slowly increasing stimulus, the R2 responses usually appear before R1.[68] If an early response cannot be obtained, a paired stimulus can be used with an interstimulus interval of 5 milliseconds to take advantage of a period of facilitation that may last 1 to 9 millisec-

onds.[12] Confusion may arise if the stimulating electrode is too close to the midline, because a contralateral R1 response may be obtained by stimulation of the opposite trigeminal nerve.[73] Amplitude measurements are usually not made because differences in amplitude of up to 40% occur in normal subjects.[74] A normal blink reflex study is shown in Figure 28–1 and normal values (in milliseconds) are as follows:

Ipsilateral R1   <13
Ipsilateral R2   <41
Contralateral R2   <44
Ipsilateral versus contralateral differences
     R1, <1.2; R2, <8.0

The R1 response latency in young children reaches adult values by age 2 years; however, the R2 responses are sometimes absent in children younger than 2 years and attain adult values at age 5 to 6 years. Because of their variability, the R2 responses are less helpful in children less than 6 years old.[2,13]

INTERPRETATION

Delay or absence of R1 only indicates dysfunction of either the trigeminal or facial nerve (or both) on the side of stimulation. Knowledge of clinical findings, for example, trigeminal sensory neuropathy, may indicate which nerve is responsible. If both nerves are affected, as in acoustic neuroma, it may be impossible to provide a more precise interpretation. If the R2 components are affected, the examiner can usually clarify which of the nerves is involved. Delay or absence of the ipsilateral and contralateral R2 with stimulation of the affected side indicates trigeminal nerve dysfunction.[56] Facial nerve involvement is signified by prolongation or absence of R2 on the affected side with stimulation of either supraorbital nerve. Abnormality of the ipsilateral R2 or bilateral R2 responses with a normal early reflex suggests a lesion of the spinal tract and nucleus of the trigeminal nerve.

Indications for Use

PERIPHERAL NEUROPATHY

Extremely delayed direct (latency after facial nerve stimulation at the mastoid pro-

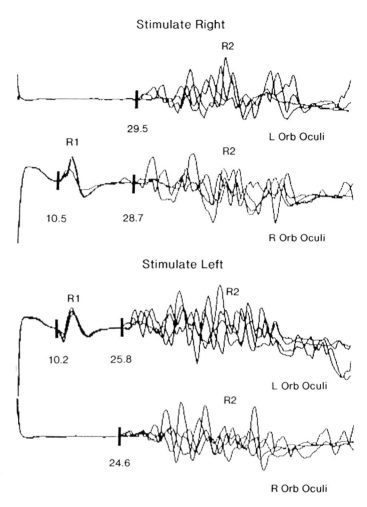

Stimulate Right

R2

29.5

L Orb Oculi

R1

R2

10.5   28.7

R Orb Oculi

Stimulate Left

R1

R2

10.2   25.8

L Orb Oculi

R2

24.6

R Orb Oculi

Figure 28–1. Normal blink reflex study. L = left; Orb oculi = orbicularis oculi muscle; R = right.

cess) and R1 responses are found in hereditary motor and sensory neuropathy type I (Charcot-Marie-Tooth disease).[29] The delay is most marked in the distal segment of the facial nerve and is present despite no detectable facial muscle weakness. In contrast, the delayed latencies in Guillain-Barré syndrome are associated with facial muscle weakness. The blink reflexes may also be abnormal in other demyelinating neuropathies (Figs. 28–2 and 28–3). Slowing of conduction in patients with diabetic neuropathy is much less frequent and milder and has little utility in assessing the presence or severity of neuropathy.[37] In most axonal sensorimotor peripheral neuropathies, the blink reflex is unaffected.

Trichloroethylene, long known to produce trigeminal neuropathy, is associated with prolonged mean R1 latency in people with occupational exposure or contaminated water supplies.[17,18]

FACIAL NERVE LESIONS

Bell's Palsy. Routine conduction studies of the facial nerve evaluate the segment of the nerve distal to the stylomastoid foramen, which is clinically helpful only when degeneration begins distally. The blink reflex, however, assesses the entire nerve, including the intraosseous portion. All patients with Bell's palsy studied by Kimura, Giron, and Young[32] showed delayed or absent R1 on the affected side of the face. On the weak side, R2 was also abnormal in all patients. In 100 of 127 patients tested serially, the previously absent R1 and R2 returned, whereas re-

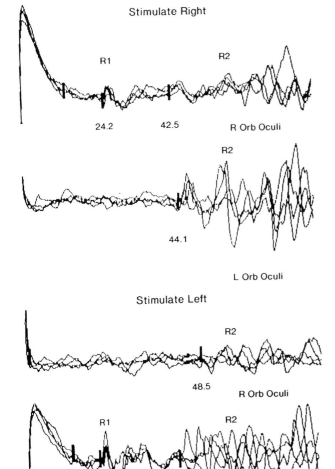

Stimulate Right

R1

R2

24.2   42.5

R Orb Oculi

R2

44.1

L Orb Oculi

Stimulate Left

R2

48.5

R Orb Oculi

R1   R2

21.2   43.3

L Orb Oculi

Figure 28–2. Bilateral, markedly prolonged R1 latencies and milder R2 prolongation in a 74-year-old woman with chronic inflammatory demyelinating polyradiculoneuropathy. The first unmarked bar indicates the normal upper limit of R1 latency. L = left; Orb oculi = orbicularis oculi muscle; R = right.

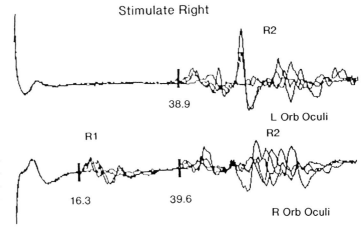

Stimulate Right

R2

38.9

L Orb Oculi

R1   R2

16.3   39.6

R Orb Oculi

Figure 28–3. Prolonged R1 latency in an elderly man with a demyelinating peripheral neuropathy associated with an IgA monoclonal gammopathy. L = left; Orb oculi = orbicularis oculi muscle; R = right.

sponses with stimulation of the facial nerve at the mastoid process remained relatively normal. The R1 latency was increased by more than 2 milliseconds initially, suggesting demyelination of facial nerve fibers. The latency of R1 decreased during the second month and returned to normal by the fourth month. These patients generally showed a good return of function in a few months. The other 27 patients showed smaller amplitudes with direct stimulation of the facial nerve, and the blink reflex responses did not return. These patients had degeneration of the facial nerve and poorer recovery.

More recent studies have reached similar conclusions—when the blink reflex is normal or R1 is only delayed, prognosis is excellent. An absent blink reflex is associated with poor prognosis in 56% of cases.[20,25] Nacimiento and associates[47] reported that 64% of patients with Bell's palsy may have an R1 response on the affected side with stimulation of the normal side as compared with 13% of controls. Synaptic reorganization of the facial nucleus unmasking preexisting crossed pathways is suggested to explain this finding. Some patients who have "postparalytic facial dysfunction," characterized by synkinesis and muscle contracture, have larger blink reflexes on the previously affected side. These and other findings suggest hyperexcitability of facial motoneurons.[82]

Aberrant Regeneration. In normal persons, R1 and R2 are recorded from the orbicularis oculi muscle and only rarely from other facial muscles. After Bell's palsy with axonal degeneration, aberrant regeneration may be present in which some fibers that originally innervated the orbicularis oculi supply other facial muscles. Clinically, this is recognized by observing contraction of lower facial muscles when the patient blinks. In these patients, when recording from the orbicularis oris and platysma on the affected side, an R1 and R2 are often recorded.

Hemifacial Spasm. Hemifacial spasm is characterized by unilateral involuntary twitching in periorbital and lower portions of the face and may result from various intraaxial and extra-axial brain stem lesions affecting the facial nerve. In most patients, the disorder is probably associated with compression of the nerve by avascular loop near the brain stem. Focal demyelination ensues, producing ectopic generation of action potentials or ephaptic transmission to other axons near the site of compression.[39,50] Recent evidence suggests that in hemifacial spasm the facial motor nucleus may be hyperactive, with spasms resulting from "backfiring" that is similar to an exaggerated F-response.[45,81] These patients also have synkinesis, which appears similar to that seen in aberrant regeneration but, unlike it, is not present between spasms (Fig. 28–4).[3]

Routine R1 and R2 reflex latencies are normal in most patients with hemifacial spasm, although the R1 amplitude may be increased. There may also be after-activity and late activity, suggesting hyperexcitability of the reflex. Abnormal communication between facial axons, termed "lateral spread,"[51] can be demonstrated by stimulating the mandibular branch of the facial nerve while recording from the orbicularis oculi or by stimulating the zygomatic nerve while recording from the mentalis muscle (Fig. 28–5). Surgical microvascular decompression of the facial nerve typically results in loss of lateral spread and synkinesis and relief of hemifacial spasm.[7,46]

Figure 28–4. Blink reflex demonstrating synkinesis of the orbicularis oculi and mentalis muscles with stimulation of the left supraorbital nerve. (From Harper, CM Jr: AAEM case report 21: Hemifacial spasm: Preoperative diagnosis and intraoperative management. Muscle Nerve 14:213–218, 1991. By permission of Mayo Foundation.)

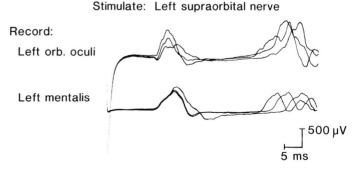

Stimulate: Left supraorbital nerve

Record:

Left orb. oculi

Left mentalis

500 µV

5 ms

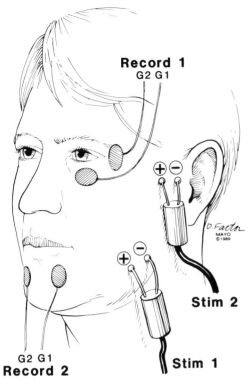

Record 1
G2 G1

(+) (−)

(+) (−)

Stim 2

G2 G1
Record 2

Stim 1

Figure 28–5. Technique for recording the lateral spread response from the orbicularis oculi muscle (Record 1) with stimulation of the mandibular branch of the facial nerve (Stim 1) or from the mentalis muscle (Record 2) with stimulation of the zygomatic branch of the facial nerve (Stim 2). (From Harper, CM Jr: AAEM case report 21: Hemifacial spasm: Preoperative diagnosis and intraoperative management. Muscle Nerve 14:213–218, 1991. By permission of Mayo Foundation.)

Cerebellopontine Angle Lesions. Delay in the R1 and R2 responses after glabellar tap in a patient with acoustic neuroma was first reported in 1964.[60] Eisen and Danon[16] studied the blink reflex in 11 patients, most of whom had large cerebellopontine angle tumors (nine acoustic neuromas and two meningiomas). All patients had delay or absence of R1. Analysis of R2 latencies indicated trigeminal nerve involvement in eight and facial nerve involvement in two others (Fig. 28–6). Brain stem auditory evoked potentials are more often abnormal than blink reflexes in evaluation of posterior fossa tumors, but both are usually abnormal with cerebellopontine angle tumors.[87] In these cases, the blink reflex is helpful both for preoperative assessment and prognosis.

TRIGEMINAL NERVE LESIONS

In lesions of the trigeminal nerve, latencies of all components are usually prolonged when stimulating the involved side, whereas they are normal when stimulating the unaffected side. No response is obtained with stimulation of the involved side when the lesion is more severe. In patients with trigeminal sensory neuropathy associated with connective tissue diseases, there may be abnormalities of R1 only or bilateral abnormalities of R1 and R2; it is uncommon that only R2 is affected (Fig. 28–7).[23] The blink reflex is also helpful in identifying lesions of the sensory root and gasserian ganglion, including tumors, vascular malformations, and postherpetic lesions. The blink reflex is usually normal in idiopathic trigeminal neuralgia and atypical facial pain (Figs. 28–8 and 28–9).[35]

MULTIPLE SCLEROSIS

Like visual, auditory, and somatosensory evoked potentials, the blink reflex can be useful diagnostically in patients suspected of having multiple sclerosis. The blink reflex can be helpful in detecting clinically silent lesions and may document objective abnormality in patients with vague facial paresthesia of uncertain significance. In 260 patients with multiple sclerosis observed over a 7-year period, Kimura[31] found R1 was delayed on one or both sides in 66% of those with definite multiple sclerosis, in 56% with probable multiple sclerosis, and in 29% with possible multiple sclerosis. R1 was abnormal in 78% of those with neurologic signs that suggested pontine involvement and in 57% of those with signs of disease of the medulla or midbrain. In patients who had no brain stem signs, R1 was abnormal in 40%. R2 was less diagnostic than R1 in detecting brain stem lesions and was most often abnormal in those with pontine signs. When R1 was normal and R2 delayed, patients had symptoms suggesting medullary involvement. As might be expected in patients with multiple sclerosis, the incidence of delayed R1 responses increases with time, although there may be improvement when patients are in remission. Similar results were found in another

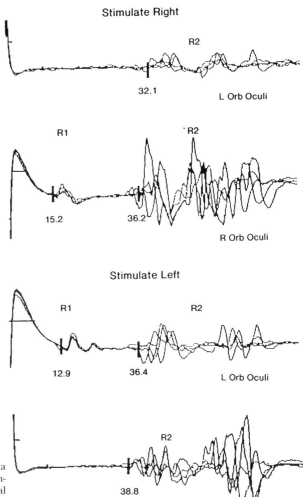

Figure 28–6. Small right acoustic neuroma in a 50-year-old man without trigeminal nerve involvement produces prolongation of ipsilateral R1 with normal R2 responses. L = left; Orb oculi = orbicularis oculi muscle; R = right.

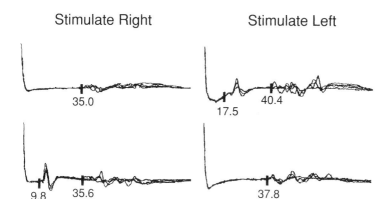

Figure 28–7. A 69-year-old woman with scleroderma and numbness and pain in the distribution of left V_2 and V_3. The left R1 is prolonged.

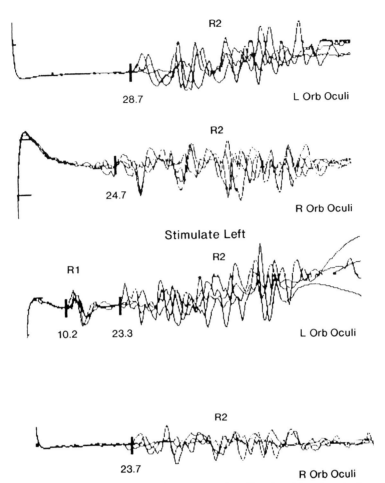

Figure 28–8. A 71-year-old man with squamous cell carcinoma invading the right trigeminal nerve. Note loss of ipsilateral R1 with normal R2 responses. L = left; Orb oculi = orbicularis oculi muscle; R = right.

large series in which 74% of patients with caudal brain stem signs had delayed latencies.[41] Most patients with internuclear ophthalmoplegia had R1 delay, suggesting that blink reflex pathways lie near the medial longitudinal fasciculus.[49] Somewhat surprisingly, induced hyperthermia does not affect blink reflex latencies in most patients with multiple sclerosis.[64]

In a study of patients with chronic progressive myelopathy (almost half of whom probably also had multiple sclerosis), blink reflex latencies were abnormal in 56%.[58] The blink reflex evoked by light stimulation is rarely used for testing, but when used, results are frequently abnormal in cases of multiple sclerosis.[85]

BRAIN STEM LESIONS

Brain Stem Glioma. The R1 response is often abnormal in children and adults with intrinsic brain stem tumors, particularly if the pons is affected.[13] The R2 components are less sensitive and may be normal when a tumor is confined to the medulla.[34]

Syrinx. The range of blink reflex abnormalities in syringobulbia has not been studied in detail, although the reflex can be abnormal when the pons is involved.

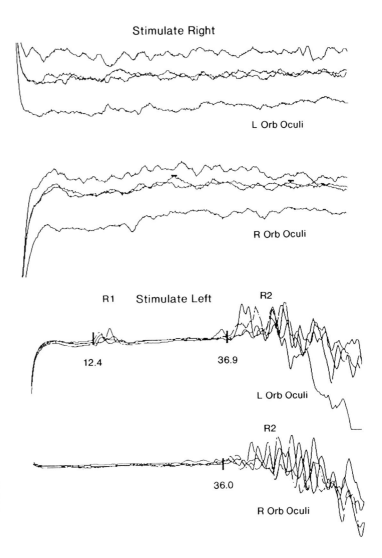

Stimulate Right

L Orb Oculi

R Orb Oculi

R1   Stimulate Left   R2

12.4   36.9

L Orb Oculi

R2

36.0

R Orb Oculi

Figure 28–9. Right V_1 postherpetic neuralgia with loss of R1 and bilateral R2 responses with right-sided stimulation. L = left; Orb oculi = orbicularis oculi muscle; R = right.

VASCULAR LESIONS

The R1 response is affected by most vascular lesions that affect the pons or the brain stem at several levels.[30,44] Blink reflex abnormalities are also found in patients who have survived episodes of vertebrobasilar insufficiency.[65] Typical findings in Wallenberg's syndrome are normal R1 and delayed ipsilateral R2 responses, suggesting involvement of the ipsilateral spinal tract and nucleus of the trigeminal nerve. Other abnormalities with stimulation of the affected side include bilateral delay or bilateral absence of R2. In large lesions, there is bilateral absence with stimulation of the *affected* side and absence of the contralateral late response with stimulation of the *unaffected* side.[55]

TRAUMATIC COMA

The blink reflex may be helpful in determining prognosis of patients in coma due to head trauma. A normal reflex excludes a severe brain stem injury, whereas bilateral absence of R1 is associated with a uniformly bad prognosis. Loss of R2 with time implies a secondary brain stem process and a poor prognosis.[70]

MISCELLANEOUS DISORDERS AFFECTING THE BLINK REFLEX

Although blink reflex pathways are confined to the brain stem, lesions rostral to the brain stem may affect latencies and, therefore, diminish their localizing value. In patients with supratentorial stroke associated with weakness on one side, R1 may be delayed in almost half of the patients for up to 1 week, and both direct and consensual R2 responses may be absent or diminished for several weeks.[8] Therefore, it is not possible to determine whether facial paresis is due to a central or a peripheral process solely on the basis of prolonged R1 latency. These effects may be the result of removal of crossed cortical facilitation on the brain stem.[36]

Habituation of the reflex does not occur in patients with Parkinson's disease, a finding sometimes helpful in neurologic evaluation. L-Dopa may restore habituation, possibly by regulating the excitability of facial nerve neurons.[59] Various changes have been noted in Huntington's disease, including early habituation.[1] In hemidystonia, there is unilateral increase in amplitude on the affected side, whereas latencies remain normal.[61] Abnormalities of the blink reflex recovery cycle have been found in idiopathic torsion dystonia and in patients with focal dystonia.[48,79]

Nontraumatic coma due to brain disease at any level may be associated with decrease or absence of late responses. Alteration of R1, however, is generally associated with primary or secondary structural changes in the pons.[42] Chronic renal failure is associated with prolongation of R1 latencies.[75] The pathophysiology of this finding is unknown. Absence of the tap-induced blink reflex has been found in diabetic patients who have increased serum osmolality.[76] Abnormalities also have been described in cluster headache, periodic ataxia, Lesch-Nyhan syndrome, Gilles de la Tourette's syndrome, spasmodic dysphonia, and Friedreich's ataxia.[14,15,24,62,72,86] General anesthetics usually abolish the reflex. Sleep increases R2 latency and duration. Although R2 is potentiated by a conditioning stimulus during sleep, R1 remains relatively unaffected.[33]

JAW JERK

Neuroanatomy

The jaw jerk, or masseter reflex, is a monosynaptic muscle stretch reflex elicited by a tap on the jaw. Afferent impulses from masseter muscle spindles are conveyed by the motor root of the trigeminal nerve to the mesencephalic nucleus.[43] Fibers from this nucleus pass to the motor nucleus to activate the efferent limb of the reflex arc. Evidence from studies on patients with neurosurgical lesions of the trigeminal root or gasserian ganglion suggests that proprioceptive affer-

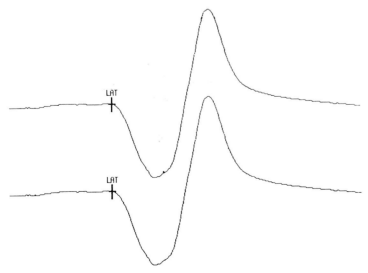

Figure 28–10. Normal jaw jerk. LAT = latency.

ent fibers travel in the sensory root.[52] In contrast to deep tendon reflexes of lower limb muscles, the masseter deep tendon reflex is potentiated rather than inhibited by vibration, suggesting different central connections at the two levels.[21] This reflex is also unique in that the afferent nerve cell bodies lie intra-axially in the mesencephalic nucleus rather than extra-axially in the craniospinal ganglia that contain the afferent nerve cell bodies subserving the limb reflexes.[4] A similar reflex can also be recorded from the medial pterygoid muscle.[27]

Methods

Recording electrodes are taped over the anterior border of the masseter muscle bilaterally just below the zygoma, with a mandibular reference electrode placed 2 to 3 cm below the recording electrode. A special reflex hammer that triggers a sweep of the oscilloscopel is used to tap the examiner's finger as it rests on the patient's chin. The latency is measured to the initial reproducible deflection from baseline. The normal range of latencies is 6 to 10.5 milliseconds, with no more than 1-millisecond difference in side-to-side latency. Wide variation in amplitude among normal subjects precludes using amplitude measurements in clinical studies. The reflex is sometimes difficult to record in normal subjects (Fig. 28–10).

Indications for Use

Ongerboer De Visser and Goor[54] studied jaw jerk and masseter electromyogram in patients with vascular or neoplastic disease of the midbrain and pons. Mesencephalic lesions were associated with abnormal jaw reflexes and normal masseter electromyograms. In the group with pontine lesions, both masseter electromyograms and jaw jerks were often abnormal. The combination of blink reflex and jaw jerk testing can help to localize medial longitudinal fasciculus lesions causing internuclear ophthalmoplegia. An abnormal jaw jerk suggests midbrain disease, whereas an abnormal R1 with or without an associated masseter reflex change suggests a rostral pontine lesion.[28] An earlier study[22] of multiple sclerosis patients without evidence of trigeminal nerve involvement showed abnormality on one or both sides in 38% of patients. The most common abnormality is absence of the reflex rather than prolongation of the latency.[84] Although the jaw jerk is less sensitive than the blink reflex, it is occasionally abnormal when the blink reflex is normal (Fig. 28–11).[67]

Masseter Inhibitory Reflex

The masseter inhibitory reflex, a rarely used test, has been employed to study various brain stem lesions. With voluntary con-

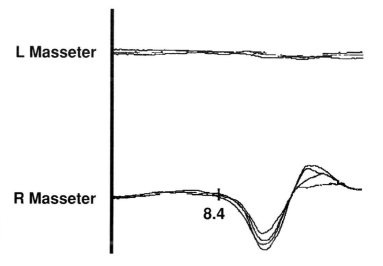

L Masseter

R Masseter

8.4

Figure 28–11. Absent left jaw jerk due to a left posterior fossa meningioma. L = left; R = right.

traction of the masseters, a brief silent period of variable duration occurs after the jaw jerk is elicited. In normal subjects, this silent period lasts from 14 to 30 milliseconds. The masseter inhibitory reflex is absent or impaired in patients with severe trigeminal sensory neuropathy who have impaired intraoral sensation. Its loss may contribute to difficulty with chewing (Fig. 28–12).[6] Similarly, the masseter inhibitory reflex is absent or shortened in tetanus, probably due to failure of Renshaw cell inhibition, leading to the clinical finding of trismus.[63] The reflex is normal in stiff-man syndrome. An abnormal masseteric silent period has also been found in patients with brain stem lesions, pseudobulbar palsy,[69] tension headaches, temporomandibular joint dysfunction,[11] jaw closing spasm,[40] blepharospasm with oromandibular dystonia, and hemimasticatory spasm.[5,10] The masseter inhibitory reflex after electrical stimulation of the mental nerve produces two silent periods, SP1 and SP2. The masseter inhibitory reflex elicited in this fasion is abnormal in lesions involving the pontine tegmentum, with some lesions selectively affecting one component or the other.[53]

CHAPTER SUMMARY

Although a certain level of expertise is necessary, electrophysiologic study of cranial reflexes is not technically demanding, time consuming, or associated with significant patient discomfort. The information obtained may document objective abnormality and assist with localization.[26] Blink reflexes are useful to study the function of the trigeminal nerve and, in combination with the jaw jerk, can suggest abnormality in the periphery as well as in several levels of the brain stem. When limb nerve conduction studies indicate a demyelinating peripheral neuropathy, the blink reflex can be performed to obtain information about involvement of proximal nerve segments. Patterns of involvement of the facial and trigeminal nerves are often helpful in suggesting the type of neuropathy under investigation. Patients with Bell's palsy can benefit from both routine nerve con-

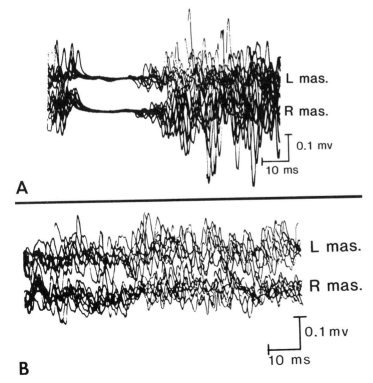

Figure 28–12. (*A*) Superimposed responses from the left and right masseter (mas) muscles to five successive taps on the chin with a reflex hammer in a normal subject. The silent period begins at 14 msec after the stimulus. (*B*) Superimposed responses from the left and right masseter muscles to five successive taps applied to the chin in a patient with trigeminal sensory neuropathy. No silent period is present. (From Auger and McManis.[6] By permission of Neurology.)

duction studies and the blink reflex to assess the entire course of the nerve. Although not discussed in this chapter, these studies are usually done in combination with careful needle electrode examination of facial muscles innervated by the trigeminal and facial cranial nerves.

REFERENCES

1. Agostino, R, Berardelli, A, Cruccu, G, et al: Correlation between facial involuntary movements and abnormalities of blink and corneal reflexes in Huntington's chorea. Mov Disord 3:281–289, 1988.
2. Anday, EK, Cohen, ME, and Hoffman, HS: The blink reflex: Maturation and modification in the neonate. Dev Med Child Neurol 32:142–150, 1990.
3. Auger, RG: Hemifacial spasm: Clinical and electrophysiologic observations. Neurology (Minneap) 29:1261–1272, 1979.
4. Auger, RG: Preservation of the masseter reflex in Friedreich's ataxia. Neurology 42:875–878, 1992.
5. Auger, RG, Litchy, WJ, Cascino, TL, and Ahlskog, JE: Hemimasticatory spasm: Clinical and electrophysiologic observations. Neurology 42:2263–2266, 1992.
6. Auger, RG and McManis, PG: Trigeminal sensory neuropathy associated with decreased oral sensation and impairment of the masseter inhibitory reflex. Neurology 40:759–763, 1990.
7. Auger, RG, Piepgras, DG, Laws, ER Jr, and Miller, RH: Microvascular decompression of the facial nerve for hemifacial spasm: Clinical and electrophysiologic observations. Neurology (N Y) 31:346–350, 1981.
8. Berardelli A, Accornero N, Cruccu G, et al: The orbicularis oculi response after hemispheral damage. J Neurol Neurosurg Psychiatry 46:837–843, 1983.
9. Berardelli, A, Cruccu, G, Manfredi, M, et al: The corneal reflex and the R2 component of the blink reflex. Neurology 35:797–801, 1985.
10. Berardelli, A, Rothwell, JC, Day, BL, and Marsden, CD: Pathophysiology of blepharospasm and oromandibular dystonia. Brain 108:593–608, 1985.
11. Bessette, R, Bishop, B, and Mohl, N: Duration of masseteric silent period in patients with TMJ syndrome. J Appl Physiol 30:864–869, 1971.
12. Broggi, G, Caraceni, T, and Negri, S: An analysis of a trigemino-facial reflex in normal humans. Confin Neurol 35:263–270, 1973.
13. Clay, SA and Ramseyer, JC: The orbicularis oculi reflex: Pathologic studies in childhood. Neurology (Minneap) 27:892–895, 1977.
14. Cohen, LG, Ludlow, CL, Warden, M, et al: Blink reflex excitability recovery curves in patients with spasmodic dysphonia. Neurology 39:572–577, 1989.
15. De Pablos, C, Berciano, J, and Calleja, J: Brain-stem auditory evoked potentials and blink reflex in Friedreich's ataxia. J Neurol 238:212–216, 1991.
16. Eisen, A and Danon, J: The orbicularis oculi reflex in acoustic neuromas: A clinical and electrodiag-

nostic evaluation. Neurology (Minneap) 24:306–311, 1974.
17. Feldman, RG, Chirico-Post, J, and Proctor, SP: Blink reflex latency after exposure to trichloroethylene in well water. Arch Environ Health 43:143–148, 1988.
18. Feldman, RG, Niles, C, Proctor, SP, and Jabre, J: Blink reflex measurement of effects of trichloroethylene exposure on the trigeminal nerve. Muscle Nerve 15:490–495, 1992.
19. Fine, EJ, Sentz, L, and Soria, E: The history of the blink reflex. Neurology 42:450–454, 1992.
20. Ghonim, MR, and Gavilan, C: Blink reflex: Prognostic value in acute peripheral facial palsy. ORL J Otorhinolaryngol Relat Spec 52:75–79, 1990.
21. Godaux, E and Desmedt, JE: Human masseter muscle: H- and tendon reflexes; their paradoxical potentiation by muscle vibration. Arch Neurol 32:229–234, 1975.
22. Goodwill, CJ and O'tuama, L: Electromyographic recording of the jaw reflex in multiple sclerosis. J Neurol Neurosurg Psychiatry 32:6–10, 1969.
23. Hagen, NA, Stevens, JC, and Michet, CJ Jr: Trigeminal sensory neuropathy associated with connective tissue diseases. Neurology 40:891–896, 1990.
24. Hatanaka T, Higashino H, Woo M, et al: Lesch-Nyhan syndrome with delayed onset of self-mutilation: Hyperactivity of interneurons at the brainstem and blink reflex. Acta Neurol Scand 81:184–187, 1990.
25. Heath, JP, Cull, RE, Smith, IM, and Murray, JA: The neurophysiological investigation of Bell's palsy and the predictive value of the blink reflex. Clin Otolaryngol 13:85–92, 1988.
26. Hopf, HC: Topodiagnostic value of brain stem reflexes. Muscle Nerve 17:475–484, 1994.
27. Hopf, HC, Ellrich, J, and Hundemer, H: The pterygoid reflex in man and its clinical application. Muscle Nerve 15:1278–1283, 1992.
28. Hopf, HC, Thömke, F, and Gutmann, L: Midbrain vs. pontine medial longitudinal fasciculus lesions: The utilization of masseter and blink reflexes. Muscle Nerve 14:326–330, 1991.
29. Kimura, J: An evaluation of the facial and trigeminal nerves in polyneuropathy: Electrodiagnostic study in Charcot-Marie-Tooth disease, Guillain-Barré syndrome, and diabetic neuropathy. Neurology (Minneap) 21:745–752, 1971.
30. Kimura, J: Electrodiagnostic study of brainstem strokes. Stroke 2:576–586, 1971.
31. Kimura, J: Electrically elicited blink reflex in diagnosis of multiple sclerosis: Review of 260 patients over a seven-year period. Brain 98:413–426, 1975.
32. Kimura, J, Giron, LT Jr, and Young, SM: Electrophysiological study of Bell palsy: Electrically elicited blink reflex in assessment of prognosis. Arch Otolaryngol 102:140–143, 1976.
33. Kimura, J and Harada, O: Recovery curves of the blink reflex during wakefulness and sleep. J Neurol 213:189–198, 1976.
34. Kimura, J and Lyon, LW: Alteration of orbicularis oculi reflex by posterior fossa tumors. J Neurosurg 38:10–16, 1973.
35. Kimura, J, Rodnitzky, RL, and Van Allen, MW: Electrodiagnostic study of trigeminal nerve: Orbicularis oculi reflex and masseter reflex in trigeminal neuralgia, paratrigeminal syndrome, and other lesions

of the trigeminal nerve. Neurology (Minneap) 20:574–583, 1970.

36. Kimura, J, Wilkinson, JT, Damasio, H, et al: Blink reflex in patients with hemispheric cerebrovascular accident (CVA): Blink reflex in CVA. J Neurol Sci 67:15–28, 1985.

37. Kirk, VH Jr, Litchy, WJ, Karnes, JL, and Dyck, PJ: Measurement of blink reflexes not useful in detection or characterization of diabetic polyneuropathy (abstr). Muscle Nerve 14:910–911, 1991.

38. Kugelberg, E: Facial reflexes. Brain 75:385–396, 1952.

39. Kumagami, H: Neuropathological findings of hemifacial spasm and trigeminal neuralgia. Arch Otolaryngol 99:160–164, 1974.

40. Lagueny, A, Deliac, MM, Julien, J, et al: Jaw closing spasm—a form of focal dystonia? An electrophysiological study. J Neurol Neurosurg Psychiatry 52:652–655, 1989.

41. Lowitzsch, K, Kuhnt, U, Sakmann, C, et al: Visual pattern evoked responses and blink reflexes in assessment of MS diagnosis: A clinical study of 135 multiple sclerosis patients. J Neurol 213:17–32, 1976.

42. Lyon, LW, Kimura, J, and McCormick, WF: Orbicularis oculi reflex in coma: Clinical, electrophysiological, and pathological correlations. J Neurol Neurosurg Psychiatry 35:582–588, 1972.

43. McIntyre, AK and Robinson, RG: Pathway for the jaw-jerk in man. Brain 82:468–474, 1959.

44. Meincke, U and Ferbert, A: Blink reflex in patients with an ischaemic lesion of the brain-stem verified by MRI. J Neurol 241:37–44, 1993.

45. Møller, AR: Interaction between the blink reflex and the abnormal muscle response in patients with hemifacial spasm: Results of intraoperative recordings. J Neurol Sci 101:114–123, 1991.

46. Møller, AR and Jannetta, PJ: Monitoring facial EMG responses during microvascular decompression operations for hemifacial spasm. J Neurosurg 66:681–685, 1987.

47. Nacimiento, W, Podoll, K, Graeber, MB, et al: Contralateral early blink reflex in patients with facial nerve palsy: Indication for synaptic reorganization in the facial nucleus during regeneration. J Neurol Sci 109:148–155, 1992.

48. Nakashima, K, Rothwell, JC, Thompson, PD, et al: The blink reflex in patients with idiopathic torsion dystonia. Arch Neurol 47:413–416, 1990.

49. Namerow, NS and Etemadi, A: The orbicularis oculi reflex in multiple sclerosis. Neurology (Minneap) 20:1200–1203, 1970.

50. Nielsen, VK: Pathophysiology of hemifacial spasm: I. Ephaptic transmission and ectopic excitation. Neurology 34:418–426, 1984.

51. Nielsen, VK: Pathophysiology of hemifacial spasm: II. Lateral spread of the supraorbital nerve reflex. Neurology 34:427–431, 1984.

52. Ongerboer De Visser, BW: Afferent limb of the human jaw reflex: Electrophysiologic and anatomic study. Neurology (N Y) 32:563–566, 1982.

53. Ongerboer De Visser, BW, Cruccu, G, Manfredi, M, and Koelman, JHTM: Effects of brainstem lesions on the masseter inhibitory reflex: Functional mechanisms of reflex pathways. Brain 113:781–792, 1990.

54. Ongerboer De Visser, BW and Goor, C: Jaw reflexes and masseter electromyograms in mesencephalic and pontine lesions: An electrodiagnostic study. J Neurol Neurosurg Psychiatry 39:90–92, 1976.

55. Ongerboer De Visser, BW and Kuypers, HGJM: Late blink reflex changes in lateral medullary lesions: An electrophysiological and neuro-anatomical study of Wallenberg's syndrome. Brain 101:285–294, 1978.

56. Ongerboer De Visser, BW, Melchelse, K, and Megens, PHA: Corneal reflex latency in trigeminal nerve lesions. Neurology (Minneap) 27:1164–1167, 1977.

57. Overend, W: Preliminary note on a new cranial reflex. Lancet 1:619, 1896.

58. Paty, DW, Blume, WT, Brown, WF, et al: Chronic progressive myelopathy: Investigation with CSF electrophoresis, evoked potentials, and CT scan. Ann Neurol 6:419–424, 1979.

59. Penders, CA and Delwaide, PJ: Blink reflex studies in patients with Parkinsonism before and during therapy. J Neurol Neurosurg Psychiatry 34:674–678, 1971.

60. Pulec, JL and House, WF: Facial nerve involvement and testing in acoustic neuromas. Arch Otolaryngol 80:685–692, 1964.

61. Raffaele, R, Palmeri, A, Ricca, G, et al: Unilateral enhancement of early and late blink reflex components in hemidystonia. Electromyogr Clin Neurophysiol 30:469–473, 1990.

62. Raudino, F: The blink reflex in cluster headache. Headache 30:584–585, 1990.

63. Risk WS, Bosch EP, Kimura J, et al: Chronic tetanus: Clinical report and histochemistry of muscle. Muscle Nerve 4:363–366, 1981.

64. Rodnitzky, RL and Kimura, J: The effect of induced hyperthermia on the blink reflex in multiple sclerosis. Neurology (Minneap) 28:431–433, 1978.

65. Ronchi O, Arnetoli G, Campostrini R, et al: Somatic brain-stem reflexes in vertebral-basilar insufficiency: An electrophysiological study. Electromyogr Clin Neurophysiol 23:577–585, 1983.

66. Rushworth, G: Observations on blink reflexes. J Neurol Neurosurg Psychiatry 25:93–108, 1962.

67. Sanders, EACM, Ongerboer De Visser, BW, Barendswaard, EC, and Arts, RJHM: Jaw, blink and corneal reflex latencies in multiple sclerosis. J Neurol Neurosurg Psychiatry 48:1284–1289, 1985.

68. Sanes, JN, Foss, JA, and Ison, JR: Conditions that affect the thresholds of the components of the eyeblink reflex in humans. J Neurol Neurosurg Psychiatry 45:543–549, 1982.

69. Schenk, E and Beck, U: Somatic brain stem reflexes in clinical neurophysiology. Electromyogr Clin Neurophysiol 15:107–116, 1975.

70. Schmalohr, D and Linke, DB: The blink reflex in cerebral coma: Correlations to clinical findings and outcome. Electromyogr Clin Neurophysiol 28:233–244, 1988.

71. Shahani, B: The human blink reflex. J Neurol Neurosurg Psychiatry 33:792–800, 1970.

72. Smith, SJM and Lees, AJ: Abnormalities of the blink reflex in Gilles de la Tourette syndrome. J Neurol Neurosurg Psychiatry 52:895–898, 1989.

73. Soliven, B, Meer, J, Uncini, A, et al: Physiologic and anatomic basis for contralateral R1 in blink reflex. Muscle Nerve 11:848–851, 1988.

74. Stoehr, M and Petruch, F: The orbicularis oculi re-

flex: Diagnostic significance of the reflex amplitude. Electromyogr Clin Neurophysiol 18:217–224, 1978.

75. Strenge, H: The blink reflex in chronic renal failure. J Neurol 222:205–214, 1980.

76. Tachibana, Y, Yasuhara, A, and Ross, M: Tap-induced blink reflex and central nervous system dysfunction in diabetics with hyperosmolality. Eur Neurol 30:145–148, 1990.

77. Tackmann, W, Ettlin, T, and Barth, R: Blink reflexes elicited by electrical, acoustic and visual stimuli. I. Normal values and possible anatomical pathways. Eur Neurol 21:210–216, 1982.

78. Tokunaga, A, Oka, M, Murao, T, et al: An experimental study on facial reflex by evoked electromyography. Med J Osaka Univ 9:397–411, 1958.

79. Tolosa, E, Montserrat, L, and Bayes, A: Blink reflex studies in patients with focal dystonias. Adv Neurol 50:517–524, 1988.

80. Trontelj, MA and Trontelj, JV: Reflex arc of the first component of the human blink reflex: A single motoneurone study. J Neurol Neurosurg Psychiatry 41:538–547, 1978.

81. Valls-Sole, J and Tolosa, ES: Blink reflex excitability cycle in hemifacial spasm. Neurology 39:1061–1066, 1989.

82. Valls-Sole, J, Tolosa, ES, and Pujol, M: Myokymic discharges and enhanced facial nerve reflex responses after recovery from idiopathic facial palsy. Muscle Nerve 15:37–42, 1992.

83. Wartenberg, R: The Examination of Reflexes: A Simplification. Year Book Publishers, Chicago, 1945.

84. Yates, SK and Brown, WF: The human jaw jerk: Electrophysiologic methods to measure the latency, normal values, and changes in multiple sclerosis. Neurology (N Y) 31:632–634, 1981.

85. Yates, SK and Brown, WF: Light-stimulus-evoked blink reflex: Methods, normal values, relation to other blink reflexes, and observations in multiple sclerosis. Neurology (N Y) 31:272–281, 1981.

86. Yokota, T, Hayashi, H, Hirose, K, and Tanabe, H: Unusual blink reflex with four components in a patient with periodic ataxia. J Neurol 237:313–315, 1990.

87. Zileli, M, Idiman, F, Hiçdönmez, T, et al: A comparative study of brain-stem auditory evoked potentials and blink reflexes in posterior fossa tumor patients. J Neurosurg 69:660–668, 1988.

CHAPTER 29

LONG LATENCY REFLEXES AND THE SILENT PERIOD

Joseph Y. Matsumoto, M.D.

LONG LATENCY REFLEXES

There is no exact division when the reflex response to a stimulus ends and voluntary reaction begins. Rather, from the onset of the monosynaptic stretch reflex to the time of the first conscious voluntary reaction, cortical input gradually increases its influence over spinal and brain stem reflex activity. Unique electromyographic (EMG) phenomena called *long latency reflexes* arise in this transition period. Long latency reflexes have predictable latencies like those of spinal reflexes, but their amplitudes are profoundly modulated by context and volition. These characteristics make them important tools in the study of normal and abnormal motor control.[12]

Hammond[10] discovered long latency reflexes while studying EMG responses evoked in the biceps by sudden stretch. EMG activity appeared at 70 milliseconds, later than the monosynaptic stretch reflex and earlier than the voluntary reaction time of 113 milliseconds. A command to "resist" the stretch augmented these long latency responses, whereas the instruction to "let go" resulted in their virtual disappearance. Marsden, Merton, and Morton[15] studied similar stretch reflexes in the long flexor of the thumb during movement and theorized that long latency reflexes served to reinforce volition or intent against unexpected perturbations. The reflexes were seen to emanate from a "transcortical loop," with one arm of the loop ascending to the sensorimotor cortex in the dorsal column–medial lemniscus system and the other descending in the corticospinal tract.

Several lines of evidence support the concept of a transcortical reflex loop. Long latency reflexes are delayed or absent in patients with lesions of the dorsal columns or contralateral sensorimotor cortex.[16,17] Cortical potentials precede long latency reflexes by 30 to 50 milliseconds, and the two events correlate in amplitude.[7] Finally, patients with cortical reflex myoclonus have hyperexcitable long latency reflexes, which are clearly cortically mediated.[9] However, the persistence of long latency reflexes in spinal animals forces one to entertain other possible explanations. Repetitive firing of muscle spindles or transmission of sensory influences by slowly conducting fibers could also explain the appearance of reflex activity at long latencies. Further, descending cortical commands could modulate segmental spinal reflexes without requiring an ascending "loop" to the cortex. It is likely that the neural circuits generating long latency reflexes depend on the type of stimulus.[22,23] Because the physiologic basis of long latency reflexes remains a subject of controversy, the use of the term "long loop reflex" is premature.

Long latency reflexes have been recorded by various techniques using stretch or different forms of electrical stimulation. The precise character of the reflexes depends on the testing paradigm, and it is unlikely that all long latency reflexes have an identical physiologic basis.[8]

Long Latency Reflexes to Stretch

All paradigms for stretch reflex testing involve a computer-controlled torque motor that can be programmed to maintain a steady load or to introduce rapid perturbations. Generally, the torque is delivered through a manipulandum that the subject holds. The subject receives visual feedback about the position of the manipulandum and attempts to hold it stationary against a low constant torque. The computer delivers random torque pulses; the surface EMG signals are recorded over the agonist and antagonist of the joint that is stretched. The EMG signal is rectified and averaged.

In a normal response, one would observe an M1 response, corresponding to the monosynaptic stretch reflex, at 30 milliseconds. The M2 response, the most frequent long latency component, appears in the wrist at 55 to 65 milliseconds. Occasionally, a later M3 component may be seen individually or it may merge with the M2 response.

In Parkinson's disease, the M2 component is enlarged in the wrist flexors. This abnormality corresponds to the degree of the patient's rigidity.[13] In contrast, a decreased or absent M2 distinguishes Huntington's disease.[21] The M2 may be prolonged in dystonia. As noted above, lesions of the ipsilateral dorsal columns or contralateral sensorimotor cortex may delay or abolish the M2 response.

Long Latency Reflexes to Mixed Nerve Stimulation

Upton, McComas, and Sica[28] discovered a late response, termed *V2*, in response to electrical stimulation. Since then, a series of long latency reflexes in the hand muscles have been differentiated in response to median nerve stimulation.

The median nerve is stimulated at the wrist, with the cathode proximal. Surface EMG electrodes are placed over the abductor pollicis brevis. The stimulus duration is set to 1 millisecond, and the intensity is increased to the level that produces the first small twitch. Initially, recordings are made with the muscle at rest and single shocks are given. Next, the patient maintains a moderate contraction and stimuli are given at a rate of 1 to 3 Hz. The signal must be rectified and then averaged. A total of 100 to 500 stimuli provide reproducible results.

In normal subjects, no late response is seen after the F wave and before the voluntary reaction time with the hand at rest. With contraction and averaging, a short latency reflex develops at around 28 milliseconds which is identical with the H reflex. A late response called the LLR II appears in all normal subjects at 50 milliseconds, and in one third of them, an additional LLR I may be recorded at 40 milliseconds or an LLR III at 75 milliseconds.

In myoclonic disorders, a late response, which corresponds to a reflex myoclonic jerk, is recorded at 40 to 60 milliseconds in the hand at rest. Sutton and Mayer[27] named this the *C reflex*, a distinct abnormality that may be seen in cortical reflex myoclonus, reticular reflex myoclonus, hyperekplexia, Alzheimer's disease, and other symptomatic myoclonic disorders. A pattern of an increased LLR I and LLR III with a normal LLR II may typify Parkinson's disease. The LLR II is absent in Huntington's disease yet is unaffected in other choreatic movement disorders.[3,21] Possibly in a subgroup of patients with essential tremor, the LLR I is increased.[2] A delayed LLR II reflects central conduction slowing in patients with multiple sclerosis.[4]

Cutaneous Reflexes

Long latency reflexes to entirely cutaneous nerve stimulation demonstrate well-defined inhibitory, as well as excitatory, periods.[1] Stimuli are delivered to the index finger by ring electrodes. Surface EMG is recorded with an electrode over the belly of the first dorsal interosseous and a reference electrode over the radial styloid. The patient maintains a steady force at 20% of maximum. Stimulus intensity is adjusted to four times sensory threshold, with a stimulus duration of 0.2 millisecond. With a stimulus rate of 3 Hz, 100 to 500 samples are collected, full-wave rectified, and then averaged.

A first excitatory period, E1, is present at around 40 milliseconds, followed by inhibition, I1, at around 51 milliseconds (Fig. 29–1). A final excitatory wave, E2, appears at around 66 milliseconds.[5,11] I1 and E2 depend on the motor cortex and are absent with contralateral cortical lesions.[11] E2 may be delayed in multiple sclerosis. The depth of I1 is decreased in patients with Parkinson's disease as compared with control subjects.[6]

The Flexor Reflex

Certain stimuli may trigger reflex withdrawal, *flexor reflexes.* Generally, the necessary stimuli must be cutaneous and noxious. Such stimuli act in a polysynaptic network of spinal neurons, termed *flexor reflex afferents,* which program patterned withdrawal behavior.

Meinck and associates[19] standardized a technique for eliciting flexor reflexes. The medial plantar nerve is stimulated in the ball of the foot with a train of five shocks by using a stimulus duration at 0.1 millisecond and an interstimulus interval of 3 to 5 milliseconds. The stimulus intensity is adjusted to the motor threshold of the flexor hallucis brevis muscle, a level that will be perceived as mildly painful. EMG is recorded from the anterior tibial muscle, rectified, and then averaged. A total of 8 trains are averaged, with stimuli repeated every 1 to 3 seconds. With this paradigm, the normal activity is triphasic, with a large F1 response at 70 milliseconds, a period of silence, then a small F2 burst at approximately 150 milliseconds.

With spinal cord lesions, exaggerated withdrawal corresponds to a large response at or beyond the latency of F2. Very high intensity stimulation may shorten the latency of this response to a value consistent with F1. In patients with spastic disease caused by hemispheric lesions, stimulation may trigger alternating clonic bursts in the anterior tibial and gastrocnemius muscles.[25]

THE SILENT PERIOD

If a strong shock is delivered to the nerve of a muscle that is tonically contracting, a period of relative or absolute silence begins immediately and persists for about 100 milliseconds (Fig. 29–2). The depth of the *silent period* depends entirely on the intensity of the shock. With supramaximal shocks, which

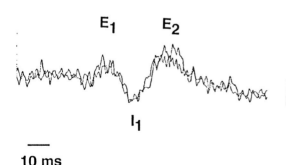

E_1 E_2

I_1

100 uV

10 ms

Figure 29–1. The cutaneous reflex. EMG activity is recorded from the flexor digitorum indicis after electrical stimulation of the index finger. Two phases of excitation, E_1 and E_2, and an intervening phase of inhibition, I_1, are recorded.

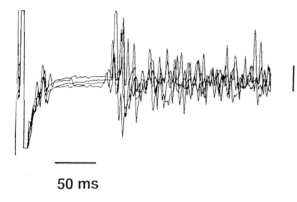

500 uV

———
50 ms

Figure 29–2. The silent period. After a supramaximal shock, thenar EMG activity is inhibited. The silence is interrupted by an M wave and H reflex.

are commonly used, the silence is generally complete except for an intervening F wave. With lower stimulation intensities, the LLR I, II, and III described previously will appear.

Initially, the silent period was thought to result from the muscle twitch and unloading of muscle spindles induced by the shock.[20] This hypothesis became untenable with the demonstration that the silent period persists with stimulation of a cutaneous nerve or a nonhomologous nerve or with stimulation proximal to a nerve block—all conditions in which twitch is absent.[14,18] The silent period should be viewed as a multifactorial phenomenon. With supramaximal stimulation, approximately the first 30 milliseconds of silence results from the collision of impulses in the nerve trunk. The next period, up to about 60 milliseconds, may reflect activation of recurrent collaterals to Renshaw cells.[26] The final period of silence should be viewed as a long latency inhibitory reflex. Likely, much of this is spinal in origin. However, silent periods after cortical stimulation raise the possibility of cortically mediated inhibitory reflexes.

Few normative data exist about the depth and duration of the silent period; thus, it is interpreted in an "all or nothing" fashion. In states of hyperexcited distal nerve or muscle, the silent period may be absent as ectopic impulses arise distal to the stimulus. In tetanus, the silent period may be abbreviated or absent.[24] An absent or shortened silent period has been reported in the case of a cervical cord tumor that produced arm rigidity.[29]

CHAPTER SUMMARY

Long latency reflexes and the silent period are electromyographic phenomena that reflect the complex interplay of spinal, brain stem, and cortical influences in motor control. These techniques have been applied to the study of disorders of motor control such as Parkinson's disease, Huntington's disease, and dystonia. Abnormalities of these reflexes may help to detect lesions of the central nervous system in multiple sclerosis.

REFERENCES

1. Caccia, MR, McComas, AJ, Upton, ARM, and Blogg, T: Cutaneous reflexes in small muscles of the hand. J Neurol Neurosurg Psychiatry 36:960–977, 1973.
2. Deuschl, G, Lücking, CH, and Schenck, E: Essential tremor: Electrophysiological and pharmacological evidence for a subdivision. J Neurol Neurosurg Psychiatry 50:1435–1441, 1987.
3. Deuschl, G, Lücking, CH, and Schenck, E: Hand muscle reflexes following electrical stimulation in choreatic movement disorders. J Neurol Neurosurg Psychiatry 52:755–762, 1989.
4. Deuschl, G, Strahl, K, Schenck, E, and Lücking, CH: The diagnostic significance of long-latency reflexes in multiple sclerosis. Electroencephalogr Clin Neurophysiol 70:56–61, 1988.
5. Fuhr, P and Friedli, WG: Electrocutaneous reflexes in upper limbs—reliability and normal values in adults. Eur Neurol 27:231–238, 1987.
6. Fuhr, P, Zeffiro, T, and Hallett, M: Cutaneous reflexes in Parkinson's disease. Muscle Nerve 15:733–739, 1992.
7. Goodin, DS, Aminoff, MJ, and Shih, PY: Evidence that the long-latency stretch responses of the human wrist extensor muscle involve a transcerebral pathway. Brain 113:1075–1091, 1990.

8. Hallett, M: Long-latency reflexes. In Quinn, NP and Jenner, PG (eds): Disorders of Movement: Clinical, Pharmacological and Physiological Aspects. Academic Press, London, 1989, pp 529–541.

9. Hallett, M, Chadwick, D, and Marsden, CD: Cortical reflex myoclonus. Neurology (Minneap) 29:1107–1125, 1979.

10. Hammond, PH: The influence of prior instruction to the subject on an apparently involuntary neuromuscular response (abstr). J Physiol (Lond) 132:17P–18P, 1956.

11. Jenner, JR and Stephens, JA: Cutaneous reflex responses and their central nervous pathways studied in man. J Physiol (Lond) 333:405–419, 1982.

12. Johnson, MT, Kipnis, AN, Lee, MC, and Ebner, TJ: Independent control of reflex and volitional EMG modulation during sinusoidal pursuit tracking in humans. Exp Brain Res 96:347–362, 1993.

13. Lee, RG and Tatton, WG: Motor responses to sudden limb displacements in primates with specific CNS lesions and in human patients with motor system disorders. Can J Neurol Sci 2:285–293, 1975.

14. Leis, AA, Ross, MA, Emori, T, et al: The silent period produced by electrical stimulation of mixed peripheral nerves. Muscle Nerve 14:1202–1208, 1991.

15. Marsden, CD, Merton, PA, and Morton, HB: Servo action in the human thumb. J Physiol (Lond) 257:1–44, 1976.

16. Marsden, CD, Merton, PA, Morton, HB, and Adam, J: The effect of lesions of the sensorimotor cortex and the capsular pathways on servo responses from the human long thumb flexor. Brain 100:503–526, 1977.

17. Marsden, CD, Merton, PA, Morton, HB, and Adam, J: The effect of posterior column lesions on servo responses from the human long thumb flexor. Brain 100:185–200, 1977.

18. McLellan, DL: The electromyographic silent period produced by supramaximal electrical stimulation in normal man. J Neurol Neurosurg Psychiatry 36:334–341, 1973.

19. Meinck, H-M, Küster, S, Benecke, R, and Conrad, B: The flexor reflex—influence of stimulus parameters on the reflex response. Electroencephalogr Clin Neurophysiol 61:287–298, 1985.

20. Merton, PA: The silent period in a muscle of the human hand. J Physiol (Lond) 114:183–198, 1951.

21. Noth, J, Podoll, K, and Friedemann, H-H: Long-loop reflexes in small hand muscles studied in normal subjects and in patients with Huntington's disease. Brain 108:65–80, 1985.

22. Palmer, E and Ashby, P: Evidence that a long latency stretch reflex in humans is transcortical. J Physiol (Lond) 449:429–440, 1992.

23. Palmer, E and Ashby, P: The transcortical nature of the late reflex responses in human small hand muscle to digital nerve stimulation. Exp Brain Res 91:320–326, 1992.

24. Risk WS, Bosch EP, Kimura J, et al: Chronic tetanus: Clinical report and histochemistry of muscle. Muscle Nerve 4:363–366, 1981.

25. Shahani, BT and Young, RR: Human flexor reflexes. J Neurol Neurosurg Psychiatry 34:616–627, 1971.

26. Shahani, BT and Young, RR: Studies of the normal human silent period. In Desmedt, JE (ed): New Developments in Electromyography and Clinical Neurophysiology, Vol 3. Karger, Basel, 1973, pp 589–602.

27. Sutton, GG and Mayer, RF: Focal reflex myoclonus. J Neurol Neurosurg Psychiatry 37:207–217, 1974.

28. Upton, ARM, McComas, AJ, and Sica, REP: Potentiation of 'late' responses evoked in muscles during effort. J Neurol Neurosurg Psychiatry 34:699–711, 1971.

29. Weinberg, DH, Logigian, EL, and Kelly, JJ Jr: Cervical astrocytoma with arm rigidity: Clinical and electrophysiologic features. Neurology 38:1635–1637, 1988.

CHAPTER 30

SURFACE ELECTROMYOGRAPHIC STUDIES OF MOVEMENT DISORDERS

Joseph Y. Matsumoto, M.D.

The study of movement disorders encompasses abnormalities of motor control that result in either too little or too much movement. At one pole are akinetic-rigid syndromes, such as Parkinson's disease, and at the other are involuntary movements, such as tremor, myoclonus, dystonia, chorea, and tics. Movement disorders stem from complex and poorly understood pathophysiologic processes that occur in the central nervous system. These processes are largely inaccessible using even the most sophisticated electrophysiologic techniques and neuroimaging procedures. Thus, the enduring tool in evaluating clinical movement disorders is the trained human eye that, together with the clinical history, provides an accurate diagnosis in most cases.

Clearly, reliance on diagnosis by visual inspection is not sufficient. Although observation is excellent for perceiving the overall pattern of movement, it is less proficient in discerning the fine details of movement, such as timing (Which body part moved first?) and regularity (Is the movement tremulous or irregular?). At times, details such as these are critical in diagnosis. Also, experimental studies in motor control clearly demonstrate that the brain, spinal cord, and musculoskeletal system are able to produce a specific movement with a large number of different motor patterns. As a practical example, rapid elbow flexion may be the result of either a brief, isolated contraction of the biceps muscle or prolonged activity of the biceps and triceps muscles. In this example, identification of the underlying motor pattern may distinguish myoclonus from dystonia.

Surface electromyographic (EMG) studies provide a method for obtaining data that

341

complements and extends clinical examination. Multichannel EMG studies map the temporal pattern of movement with a resolution measured in milliseconds. Furthermore, the surface EMG reflects not only alpha motor neuron activity but also, by inference, specific abnormal central commands that underlie a movement disorder. Surface EMG studies are noninvasive and within the capability of any EMG laboratory. We find them particularly useful in the study of involuntary movement disorders in which the patterns of surface EMG activity are best described.[18,20,22]

TECHNIQUES

Surface EMG recording can be performed with any high-quality disc electrodes. We find disposable adhesive electrodes convenient, because they can be rapidly applied when multiple muscles are recorded. After cleansing and mildly abrading the skin, the electrodes are placed 2 to 3 cm apart over the motor point of the muscle and oriented parallel to the course of the muscle fibers.[16] The iliac crest provides a relatively inactive site for the ground electrode.

A major technical limitation of surface EMG studies is their lack of selectivity. The activity of a *single muscle* is never actually recorded because adjacent muscles inevitably contribute "cross talk" to the signal. This effect is minimized by use of short interelectrode distances and by recording from relatively superficial and isolated muscles, such as the biceps, deltoid, quadriceps, tibialis anterior, or first dorsal interosseus. At times, a group of muscles, such as the forearm flexors or extensors, are intentionally recorded.

The quality of the surface EMG signal must be carefully assessed before analysis. This signal represents the interference pattern of multiple motor units with high frequencies filtered out by the intervening skin and subcutaneous tissue. Deep muscles, such as the gluteus maximus or any muscle in an obese person, may produce a signal that is too degraded for analysis. The frequency spectrum of the signal contains power throughout the range between 10 and 1000 Hz, with maximum power at approximately 100 Hz. In practice, a low-frequency

filter cutoff of at least 30 Hz must be used to eliminate the unwanted effects of movement artifact. A high-frequency filter setting of 2000 Hz passes the important high-frequency components of the signal. The amplification factor is arbitrarily set to display a maximal voluntary contraction that fills the amplifier range without blocking. After the EMG signal is collected, it may be displayed as the raw interference pattern or digitally processed to display a full-wave rectified signal or smoothed EMG envelope. For movement disorder studies, the important measurements are the onset latencies and burst durations. The amplitude of the bursts is extremely variable and rarely useful in routine clinical studies.

If a study demands highly selective recording, however, intramuscular electrodes must be used. Electrodes fashioned from fine wire are useful for this purpose. Pairs of wires are inserted into the selected muscle through a hypodermic needle that is then withdrawn. After the electrodes are in position, they remain stable for many hours and resist displacement by even vigorous body movement. When selective recording is needed for only short recording periods, standard concentric or monopolar needle recording may be suitable. In any situation, the improved selectivity of intramuscular recording must be balanced against the added discomfort to the patient.

NORMAL PATTERNS

Three normal patterns of surface EMG activity are recognized: *reflex, tonic,* and *ballistic* (Fig. 30–1).[20] Reflex activity, such as the monosynaptic tendon jerk, produces brief, synchronized discharges of alpha motor neurons. The surface EMG appearance of this discharge is a short (10 to 30 milliseconds) burst of activity in agonist and, at times, antagonist muscles. This *reflex pattern* is an involuntary response to stimulation. Indeed, most persons are unable to voluntarily generate bursts that have this short duration. On the other hand, voluntary movement produces two types of EMG patterns. When a person moves a limb slowly or holds it in a static posture, a *tonic pattern* results. This pattern consists of a continuous and steady

A   B   C

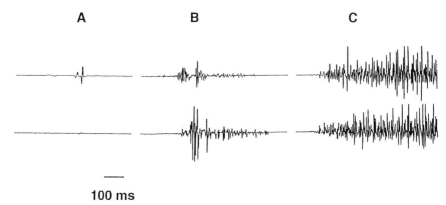

100 ms

Figure 30–1. The three normal surface EMG patterns: (*A*) reflex pattern; (*B*) triphasic pattern; and (*C*) tonic pattern. Upper trace, agonist muscle; lower trace, antagonist muscle.

EMG discharge, often with cocontraction of agonist and antagonist muscles. Cocontraction is a normal mechanism of motor control that increases the stiffness, or resistance, across a joint. By contrast, when one wills a very rapid movement of a joint, a *ballistic pattern* develops. An initial agonist burst of 50- to 100-millisecond duration leads the pattern, followed by an antagonist burst of 50- to 100-millisecond duration and a silent period in the agonist. A final agonist burst completes the EMG activity. The triphasic pattern appears to be fundamental to the motor control of ballistic limb movements.[28] The theory explaining this is that the initial agonist burst scales the size of the ballistic movement, the antagonist burst brakes the limb, and the final agonist burst determines its destination. Movement disorders reflect alterations in the excitability or control of one of these three patterns.

TREMOR

Tremor is an oscillation of a body part at a regular frequency. Not all tremor is generated by rhythmic muscle activity. Ballistocardiographic tremor, for example, is a rhythmic limb movement driven by the energy of a forceful heartbeat. However, in pathologic tremors, abnormal muscular activity drives the involuntary movements. Normally, motor unit recruitment produces a random pattern of firing, with motor unit contractions separated evenly in time. In these circumstances, twitch tensions fuse to produce a steady force. In tremor, the firing of motor unit potentials becomes synchronized, increasing the probability that two or more motor unit potentials will fire in close temporal proximity and create a jerky, unfused force. As synchronization increases, so does the amplitude of the tremor.[35] The surface EMG of pathologic tremors records the grouping of motor unit potentials as discrete bursts of activity. Analysis of these EMG bursts helps to establish whether a movement is truly tremulous. Disorders such as phasic dystonia may appear regular to the eye but may be shown as irregular when measured on an EMG recording. Conversely, a very coarse tremor may appear so jerky as to be myoclonic.

Recording Techniques

Electrodes are placed over the agonist and antagonist pairs of muscles active in driving the tremor. The forearm flexors and extensors often are most active and care must be taken to eliminate cross talk between the signals by electrode positioning, because this technical error may lead to misinterpretation. Amplitude of the tremor bursts is assessed in various limb positions. Tremors are grouped into those occurring at rest and those occurring with action. Action tremors are subdivided into intention tremors, which are maximal during or at the termination of movement, and static tremors, which persist while the limb is held in a fixed posture. Another variable used in tremor analysis is the pattern of agonist-antagonist firing. One muscle may fire while the other is silent in an *alternating pattern*. In other tremors, the pair may fire simultaneously in a *synchronous pattern*.[45] Rarely, the EMG pattern may shift from one pattern to another during the period of recording.[44]

Abnormal Patterns

EXAGGERATED PHYSIOLOGIC TREMOR

Even normal subjects have a fine, often imperceptible tremor when holding part of the body in a static posture. This *physiologic tremor* arises from the unfused twitch tensions of motor units firing from 8 to 12 Hz as well as from biomechanical properties of the limbs. This tremor often has no surface EMG correlate. Under circumstances that increase catecholaminergic drive, such as anxiety, thyrotoxicosis, or pheochromocytoma, physiologic tremor increases in amplitude and becomes visible. In this state, known as *exaggerated physiologic tremor*, underlying surface EMG activity emerges. This may consist of a low-grade interference pattern or synchronous bursts of 50- to 100-millisecond duration and a frequency of 8 to 12 Hz. This pattern is often best recorded in distal upper extremity muscles, such as finger extensors and flexors.

ESSENTIAL TREMOR

Essential, or *familial*, *tremor* is the commonest form of tremor. It is an action tremor visible when the affected body part maintains a static posture. The hands, head, or voice are commonly involved. The surface EMG shows strongly circumscribed bursts of activity that are synchronous in agonist and antagonist muscles, with a frequency of 7 to 10 Hz. When the tremor is severe, less prominent activity may be recorded, even with the limb at rest. However, Sabra and Hallett[44] have reported patients with essential tremor who clearly had an alternating EMG pattern. This group of patients appeared to have a tremor that was slower in frequency and less responsive to propranolol. Although un-

common, this phenomenon dictates caution in diagnosing tremor on the basis of a single EMG characteristic. The surface EMG findings in *neuropathic tremor* are identical with those in essential tremor.

EXTRAPYRAMIDAL TREMOR

Tremor commonly accompanies akinetic-rigid syndromes such as Parkinson's disease. Clinically, extrapyramidal tremors appear maximal at rest and attenuate with action. The dominant frequency is 4 to 7 Hz, slower than that of most essential tremors. Surface EMG studies of extrapyramidal tremors demonstrate an alternating pattern of contraction that is constant through the period of recording. Burst durations are typically in the range of 50 to 100 milliseconds. The frequency of the bursts varies little through the period of the recording. Bursts in muscles throughout the body share the same frequency. Static postures initially attenuate the burst amplitudes; however, when such postures are held for 20 seconds or longer, the tremor bursts may reappear. Because the EMG appearance of this apparently static tremor is identical to that of the rest tremor, a superimposed essential tremor should not be diagnosed.

CEREBELLAR TREMOR

Three forms of cerebellar tremor are distinguished on the basis of the differential visual and surface EMG characteristics.[39,44] *Intention tremor* occurs with target-directed limb movements and is most prominent toward the termination of the movement. The tremor frequency is often 5 to 8 Hz, and the EMG pattern is alternating. *Serial dysmetria* often appears similar to intention tremor; however, surface EMG studies indicate that the terminal movements are not the regular oscillations of tremor but rather a series of inaccurate, irregular ballistic movements. Cerebellar disease may also cause a severe postural tremor that is maximal in proximal muscles and has a frequency of 2.5 to 4 Hz. The amplitude of this tremor increases markedly the longer the posture is maintained. *Severe cerebellar postural tremor* is seen most often in multiple sclerosis and has often been labeled *rubral tremor*. This alternate term is misleading, however, because disease of the cerebellar hemispheres or deep nuclei can produce an identical pattern.

PRIMARY WRITING TREMOR

Tremor may be curiously selective and occur with certain activities but not others. *Primary writing tremor* is the commonest of these tremors. Although this tremor is predominant with writing, it often spills over into other activities, such as eating or grooming. The surface EMG correlate consists of bursts of 100-millisecond duration which occur maximally in the pronator teres, supinator, or wrist flexors and extensors. The pattern may be synchronous or alternating. Taps to the forearm, particularly in a direction that produces supination of the forearm, may stimulate bursts of tremor.[42,43]

ORTHOSTATIC TREMOR (SHAKY LEGS SYNDROME)

A distinctive tremor that occurs predominately in the elderly has been described by Heilman.[30] Patients complain of "quivering" in their legs shortly after they stand. This may cause a sense of instability so severe that walking becomes difficult. The surface EMG pattern is distinctive and displays high-amplitude 13- to 18-Hz tremor bursts (Fig. 30–2). The bursts are recorded in the legs and paraspinal muscles with the patient standing. The agonist-antagonist relationship may vary during the recording. At rest, the legs are electrically silent.[38]

FUNCTIONAL TREMOR

Occasionally, tremor may have a pattern that is atypical, and hysteria or malingering is suspected. Surface EMG can provide sup-

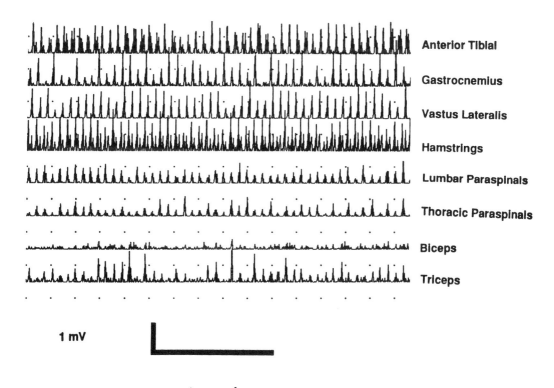

1 mV

1 second

Figure 30–2. Orthostatic tremor. Surface EMG activity recorded with the patient standing and rectified for display. Tremor activity at approximately 15 Hz is recorded maximally from the leg muscles.

portive evidence by demonstrating a pattern that does not correspond to those described above. Functional tremors tend to be confined to a single limb and seldom display a dominant frequency throughout a prolonged recording. Indeed, the tremor frequency and amplitude tend to vary widely with time, with change of position, or with distraction. However, diagnostic proof for a functional or voluntary origin for a tremor cannot be offered.

MYOCLONUS

Myoclonus is a brief, lightning-fast jerk of a body part. Of the involuntary movement disorders, it presents the greatest challenge to diagnosis and classification.[1,36] Myoclonus is best considered a general term for various motor phenomena that appear to be similar but have diverse pathologic and physiologic causes.

Myoclonus may be approached through several classification schemes. The first describes the distribution of the muscle jerks. They may be *focal* (involving only a single limb), *multifocal* (affecting more than one body part in a random, independent fashion), *generalized* (involving all body parts simultaneously), or *segmental* (involving only muscles of a given cranial or spinal segment). An etiologic classification categorizes the immense number of disease states that may have myoclonus as a prominent symptom. In one such classification, myoclonus is divided into four etiologic groups: (1) *physiologic myoclonus* includes motor phenomena such as hiccups, hypnic jerks, and the startle response which have the appearance of myoclonus but occur in normal subjects; (2) *essential myoclonus* designates disease in which abnormal muscle jerks are the sole feature of the illness; (3) *epileptic myoclonus* includes many variants of primary generalized epilepsy in which intermittent myoclonus is an accompaniment; and (4) *symptomatic myoclonus* represents all other disease states in which myoclonus occurs as a sign, often in the setting of encephalopathy. In this grouping are included the progressive myoclonic encephalopathies, such as Lafora's body disease, Creutzfeldt-Jakob disease, Alzheimer's disease, metabolic encephalopathies, anoxic encephalopathies, and other conditions.

Physiologic analysis represents yet another approach to myoclonus. By using available experimental and clinical data, Halliday[29] divided myoclonus into three categories: (1) *pyramidal myoclonus*, characterized by brief EMG bursts that follow a cortical discharge with a fixed, short latency; (2) *extrapyramidal myoclonus*, thought to arise from deeper, noncortical sites and characterized by long EMG burst durations and the absence of a clear preceding cortical discharge; and (3) *segmental myoclonus*, viewed as a local phenomenon arising from brain stem or spinal cord lesions.

Hallett[19] cited contemporary clinical neurophysiologic data to suggest a classification that divides myoclonus into epileptic and nonepileptic forms. By definition, *epileptic myoclonus* is a "fragment of epilepsy." Its supposed generator is a brief, focal, and synchronous neuronal discharge similar to the paroxysmal depolarization shift recorded in experimental epilepsy. This concept is supported by back-averaging studies that demonstrate brief and spontaneous electroencephalographic (EEG) transients preceding myoclonic jerks and by the extremely short surface EMG burst durations that resemble a single clonic burst of a seizure. The discharges generating epileptic myoclonus are usually cortical and may be focal or diffuse. However, epileptic myoclonus may also originate from spontaneous discharges in the brain stem. Experimental models implicate the nucleus reticularis gigantocellularis in the medulla as a possible brain stem generator; thus, the term *reticular myoclonus* is synonymous with *brain stem myoclonus*. Epileptic myoclonus may be reflexive in nature, indicating that the jerks are triggered by somatosensory or acoustic stimulation at short and reproducible reflex latencies.

Nonepileptic myoclonus refers to muscle jerks that are presumed to be generated by more complex mechanisms of the basal ganglia, brain stem, or spinal cord. These mechanisms involve circuits more widely distributed in space and time than those of epileptic discharges. As such, a discrete EEG correlate is not recorded in these disorders. In general, the EMG burst durations are longer than 100 milliseconds and are never as consistently short as those seen in epileptic myoclonus.

The physiologic abnormalities found in myoclonus do not always correspond to an anatomic abnormality. A similar phenomenon occurs in epilepsy, for example, when a temporal lobe sharp wave is recorded in the absence of any demonstrable pathologic condition. Nevertheless, physiologic studies offer important guidance in evaluation and treatment of myoclonus.

Recording Techniques

Our approach is first to attempt to record the surface EMG pattern associated with spontaneous muscle jerks. The critical measurement is that of the burst duration, for which we use the median value of 10 to 20 jerks. Multiple channel recording allows one to determine the order of muscle activation for each burst and also the agonist-antagonist patterns across a joint. Capturing the surface EMG bursts responsible for intermittent muscle jerks is a challenge unless the movements occur at least once every few minutes. By using a triggering device and the delay buffer available on most EMG machines, data acquisition can be set to begin with EMG activity in the most active muscle and to center the data points around the onset of the EMG trigger.

After demonstrating the areas of maximal EMG activity, we select the most active channels for C-reflex or other stimulation studies. Somatosensory evoked potentials may also demonstrate cortical neurons that are hypersensitive to peripheral stimulation. Finally, back-averaging may be used if epileptic myoclonus is suspected and no cortical accompaniments are observed on routine EEG.

Abnormal Patterns

CORTICAL MYOCLONUS

Cortical myoclonus, by far the commonest form of epileptic myoclonus, exists in many physiologic forms.[19] *Focal cortical myoclonus* is a fragment of focal epilepsy. Patients present with spontaneous and stimulus-sensitive muscle jerks confined mainly to a single limb. If the condition is severe, generalized jerks may occur. Speech may be affected by focal tongue and palatal jerks. In these patients, EEG studies reveal that a focal area of hyperexcitable cortex transiently depolarizes approximately 10 to 30 milliseconds before EMG discharges. These discharges tend to be localized to a group of contiguous muscles and are 15 to 40 milliseconds in duration. Muscles discharge synchronously, and there is little variation in the discharge pattern of the jerks. When cranial muscles are monitored, there is a cranial-to-caudal progression of activation involving the masseter muscle, facial muscles, and sternocleidomastoid muscle in succession. This pattern may be expressed over several milliseconds and requires fast recording speeds. In *cortical reflex myoclonus*, these same EMG bursts may be elicited by electrical or tapping stimuli, generally at a latency of 40 to 60 milliseconds, in the hand (C-response).[25] Somatosensory evoked potentials typically demonstrate enlarged amplitudes of the P25 and N33 peaks.[12,46] The causes of focal cortical myoclonus include anoxia, tumors, trauma, and infections. One group of patients with focal cortical dysplasia has been described.[34]

Patients with primary generalized epilepsy manifest a different form of cortical myoclonus termed *primary generalized epileptic myoclonus.* The common manifestation of this fragmentary form of epilepsy is random, spontaneous twitching of the fingers. This movement disorder, called *polyminimyoclonus,* is nonspecific,[52] was first observed in patients with neuromuscular disease, such as spinal muscular atrophy, and is the outward manifestation of frequent fasciculation. In the setting of epilepsy, however, it should be recognized as a form of cortical myoclonus, oftentimes accompanied by larger trunk and limb jerks. Analysis using EEG reveals bifrontal and diffuse cortical events that precede the jerks. At times, these events are visualized on the raw EEG as spike or polyspike discharges. Rarely, a similar movement disorder may be seen in focal cortical myoclonus.[31] The surface EMG manifestations of polyminimyoclonus are random, 10- to 30-millisecond duration bursts in the small hand muscles. Occasionally, the jerks may appear rhythmic and simulate a tremor;

however, a surface EMG study exposes the random nature of the events.

Cortical myoclonus may be present in different forms in several symptomatic myoclonias. In Alzheimer's disease[51] and Creutzfeldt-Jakob disease,[47] multifocal, short-duration EMG jerks are observed, similar to those in cortical reflex myoclonus. Typically, touch, shock, or noise triggers reflex myoclonus. A focal, contralateral EEG transient precedes the myoclonic jerks in both diseases; however, timing, duration, and distribution of this event differentiate the two diseases. *Asterixis* is a cortically driven negative myoclonus seen in various metabolic encephalopathies. The surface EMG demonstrates that voluntary tonic contraction is interrupted by periods of silence that correspond to jerk-like lapses of tone. Silent period-locked averaging clearly demonstrates that this inhibition is time-locked to a focal cortical discharge.[2,50]

RETICULAR MYOCLONUS

Reticular reflex myoclonus provides the only clear example yet described of epileptic brain stem myoclonus.[23] The jerks predominantly affect proximal flexor muscles bilaterally and occur spontaneously 5 to 10 times per minute. Joint stretch, noise, or touch may stimulate jerks. The EEG correlates are not time-locked to the myoclonus and may occur before or after the EMG bursts. This finding has been interpreted to reflect nonspecific activation of the cortex by ascending brain stem discharges. The EMG bursts are 10 to 30 milliseconds in duration and are widespread throughout the body. There is considerable variability in the activation pattern of muscles from jerk to jerk. This so-called central jitter is believed to represent polysynaptic pathways in the brain stem neuronal generator. An ascending activation pattern is recorded in the cranial muscles, suggesting a medullary generator of the activity. The EMG bursts are linked to taps or shocks at short reflex latencies of 30 to 60 milliseconds in the hand. Somatosensory evoked potentials are normal. The commonest cause of this syndrome is anoxia. Other causes include uremia and brain stem encephalitis. Identification of reticular reflex myoclonus may be important therapeutically, because Chadwick and associates[6]

found that it responded to 5-hydroxytryptophan and clonazepam better than other physiologic types of myoclonus. Cortical and reticular myoclonus may coexist.[4]

PHYSIOLOGIC MYOCLONUS

Hiccups. Although an everyday occurrence, *hiccups* may become a medical problem when prolonged and intractable. Davis[11] described the surface EMG characteristics of individual hiccups. EMG activity is recorded synchronously over the diaphragm and inspiratory muscles. The burst has a duration of 500 milliseconds; however, inspiratory airflow is interrupted abruptly by glottic closure shortly after burst onset. The bursts may show a periodicity in intractable hiccups of one every 2 to 10 seconds.

Hypnic Jerks. A few myoclonic jerks, called *hypnic jerks,* are common during the initial descent from drowsiness to sleep. A vivid sensory experience, such as the feeling of falling, often accompanies these jerks. The jerks generally are marked by a K complex visible on the EEG, but little is known about their EMG features.[41]

ESSENTIAL MYOCLONUS

Essential myoclonus denotes various disorders in which myoclonus is the predominant feature of the disease and in which encephalopathy or seizures are absent. Most kindreds demonstrate autosomal dominant inheritance with variable penetrance.[10,15] Generalized myoclonus is usually noted from early childhood. Stress, anxiety, and actions such as writing exacerbate the jerks. Many kindreds have demonstrated a dramatic response to alcohol. Some family members may display associated tremor or focal dystonias. Under this rubric, various surface EMG patterns have been described in patients who otherwise were clinically similar. The common forms are characterized by 40- to 250-millisecond burst durations that occur asynchronously between muscles. Distal upper extremity or neck muscles are affected most frequently. There is no reflex activation of the jerks by shocks or taps.

Another pattern is seen in the syndrome of *myoclonic dystonia.*[40] This syndrome implies the coexistence of myoclonic jerks with prominent dystonia. Frequently, torticollis is the accompanying dystonia. An autosomal dominant or sporadic form has been described, and improvement with alcohol is variable. The surface EMG often defines this syndrome, showing prolonged synchronous bursts with durations of 50 to 250 milliseconds blending with prolonged dystonic activity that lasts for seconds. Antagonist-agonist cocontraction is a consistent feature of the recordings.

Ballistic movement overflow myoclonus is another pattern. Hallett, Chadwick, and Marsden[24] described patients with myoclonus affecting all extremities present from birth. Voluntary movement, especially rapid movement, triggered large-amplitude jerks. The patients reported no amelioration with alcohol. With rapid voluntary movement, for example, in the thumb, surface EMG studies record "overflow" of the typical triphasic EMG pattern from the distal limb into more proximal muscles, such as the biceps and triceps.

STARTLE DISORDERS

The *startle reflex* is a whole body jerk that commonly occurs in response to sudden unexpected noise or touch. Characteristic EMG onset latencies to loud noise are well-defined, with the orbicularis oculi invariably leading activation at 30 to 40 milliseconds and the sternocleidomastoid following at 55 to 85 milliseconds. Limb muscles are less consistently active, with the biceps activated at 85 to 100 milliseconds and leg muscles at 100 to 140 milliseconds.[53] Burst durations range from 50 to 400 milliseconds. The reflex habituates rapidly. As a normal phenomenon, startle represents another form of physiologic myoclonus.

Many disorders have been characterized as "exaggerated startle," but thus far only one disease clearly results from a hyperexcitable startle reflex. *Hyperexplexia* is a hereditary or sporadic disease in which myoclonus occurs in response to loud stimuli, especially noise. The jerks often result in unprotected falls and injury.[48] The audiogenic myoclonic jerks in hyperexplexia clearly correspond to the startle pattern described previously.[3] The startle reflex, however, is increased in magnitude and is poorly habituating in this disorder.[37]

PALATAL MYOCLONUS

Palatal myoclonus is a rhythmic, involuntary movement that causes palatal elevation. At times, it may spread to involve tongue, neck, facial, and even limb muscles.[17] Because of the rhythmicity, the movement might better be regarded as tremor rather than myoclonus. Palatal EMG activity is detected with surface electrodes linked to the mastoid processes or placed on the anterior neck. Bursts recur at a regular frequency of approximately 2 Hz. Although typical palatal myoclonus reflects a structural brain stem lesion, Deuschl and colleagues[14] have described a group of patients with essential palatal myoclonus who had a benign course and no evidence of brain stem injury. This group of patients had no spread of the movement disorder outside of cranial muscles, and their movements often had a more rapid rate.

Diaphragmatic flutter may be a variant of palatal myoclonus. Patients with this disorder complain of involuntary abdominal movements caused by bilateral diaphragmatic contractions at rates of 60 to 200 Hz.[32]

SEGMENTAL MYOCLONUS

Myoclonus confined to muscles of several root distributions has long been recognized as a sign of spinal cord lesions, such as arteriovenous malformations, tumors, traumatic injury, or infections.[33] The EMG bursts have durations of 250 to 1000 milliseconds, appear rhythmic at frequencies of 1 to 3 Hz, and often persist into sleep. A possible variant of this condition is *dancing umbilicus syndrome*, or *belly-dancers' dyskinesia*. As these colorful terms suggest, the segmental myoclonic movements are confined to the abdominal wall and cause distressing, continuous undulations.[32] Because the involved muscles are deep, at times needle recordings may be needed to uncover the cause of these symptoms.

AXIAL SPASMS
OF PROPRIOSPINAL ORIGIN

In this syndrome, generalized flexion jerks centered around the trunk and neck develop in middle life.[5] The jerks are not triggered by sensory stimuli. Surface EMG shows initial activation of truncal muscles, especially the rectus abdominis, with subsequent spread up and down spinal segments. The EMG activity is bilateral and the onset latencies are consistent from jerk to jerk. Burst durations are typically long but vary widely between 40 and 4000 milliseconds. The estimated conduction velocity up and down the cord is on the order of 5 m/s, suggesting spread through slow propriospinal pathways.

PERIODIC LIMB MOVEMENTS OF SLEEP

Periodic jerks of the legs may interrupt sleep and cause insomnia or excessive daytime somnolence. Such *periodic limb movements of sleep* commonly accompany restless legs syndrome.[8] They also may appear after spinal cord trauma or vascular injury, implicating damage to descending inhibitory pathways. During the transition between drowsiness and light sleep, the movements begin their cyclic occurrence, with an average period lasting between 30 and 45 seconds. These movements resemble dystonia more than myoclonus; thus, the previous designation of "nocturnal myoclonus" has been abandoned. The surface EMG pattern varies. Most often, the burst durations are longer than 500 milliseconds. The earliest and most actively involved muscle is often the anterior tibial muscle. Although the jerks may appear unilateral, bilateral asynchronous EMG activation is the rule.

FUNCTIONAL MYOCLONUS

It is not uncommon for myoclonic-appearing jerks to be the primary manifestation of psychiatric illness such as hysteria. This diagnosis often is suspected clinically, but the neurophysiologic laboratory is called on to provide supportive evidence for the nonorganic nature of the jerks. Thompson and co-workers[49] have found several features of such jerks on surface EMG recordings. Most important, the jerks should be triggered by a measured stimulus. When this is done, the onset latencies of the EMG bursts are variable and longer than the shortest voluntary reaction time in a normal subject. The jerks often habituate rapidly to any given stimulus, and the activation pattern varies from jerk to jerk.

DYSTONIA

Dystonia is a prolonged abnormal posture maintained by involuntary muscular contraction. It may be *focal* or *generalized.* The commonest focal dystonia is cervical dystonia, or torticollis. Blepharospasm, oromandibular dystonia, and writer's or occupational cramps are other common focal dystonias. Generalized dystonia is usually a manifestation of hereditary torsion dystonia. Generally, neurophysiologic studies are most helpful in evaluating the focal dystonias, often as a prelude to therapeutic injections of botulinum toxin.

Recording Techniques

The physiologic hallmark of dystonia is intense cocontraction of agonist and antagonist muscles, producing a marked increase in stiffness across the joint and abnormal posturing. Thus, muscles acting across the postured joint should be studied to look for simultaneous interference patterns. Intramuscular electrodes often are required to ensure selective recordings. Whereas cocontraction is not specific for dystonia, it does rule out joint contractures or hysteria, in which abnormal limb posture is unaccompanied by EMG activity. The EMG discharges may be tonic or may occur in a pseudorhythmic pattern termed *phasic dystonia.* This pattern is distinguished from tremor by the lack of true rhythmicity, the variability of the burst durations, and the frequent intrusion of tonic dystonia.

Abnormal Patterns

TORTICOLLIS

Deuschl and associates[13] have described the patterns of EMG discharge in spasmodic *torticollis.* With rotational torticollis, the contralateral sternocleidomastoid and the ipsilateral splenius capitus are the muscles most often active. In retrocollis, all posterior neck muscles are active, and in laterocollis, the ipsilateral splenius capitus and sternocleidomastoid muscles are active. We have found similar patterns, but variations in a particular pattern of muscle activity are common.

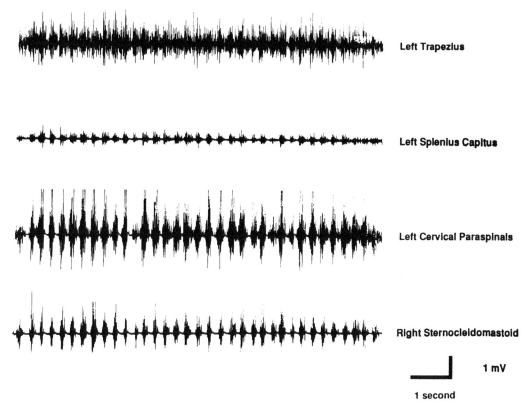

Figure 30–3. Dystonia. EMG activity recorded with intramuscular electrodes in a patient with spasmodic torticollis. Both tonic and irregular phasic EMG bursts are present.

For this reason, we perform multichannel EMG recording with intramuscular electrodes in all patients before injecting botulinum toxin (Fig. 30–3). Preinjection EMG studies may account for some degree of additional therapeutic benefit.[9]

OROMANDIBULAR DYSTONIA

In oromandibular dystonia, patients may be unable to eat or speak because of abnormal jaw posturing. *Jaw-opening dystonia* frequently reflects dystonic activity in the lateral pterygoid and digastric muscles. *Jaw-closing dystonia* reflects activity in the temporalis, masseter, and medial pterygoid muscles.

SPASTIC DYSPHONIA

Spastic dysphonia may be of the adductor or abductor type. Patients with *adductor spastic dysphonia* present with a strained or tremulous voice caused by dystonia of the thyro-arytenoid muscles. These muscles can be recorded by needle examination performed percutaneously or by direct laryngoscopy. With either method, the participation of an experienced otorhinolaryngologist is advised. In contrast, *abductor spastic dysphonia* is manifested as a whispering voice and the dystonic activity is in the posterior cricoarytenoid muscles, which are inaccessible to routine examination.

WRITER'S AND OCCUPATIONAL CRAMPS

In various circumstances, repetitive skilled motions may become complicated by painful and disabling dystonia of the hand or wrist. There is no typical posture, and a combination of flexion and extension dystonia may occur. Intramuscular EMG recordings reveal the individual pattern of phasic or tonic spasms in multiple cocontracting forearm muscles.[7]

TICS, CHOREA, AND ATHETOSIS

Surface EMG recording is of limited usefulness in the evaluation of tics, chorea, and athetosis. Although these involuntary movements are clinically distinct, the surface EMG patterns are nonspecific and may appear similar in all three. Burst durations can be 100 to 300 milliseconds and can have reflex, tonic, or ballistic patterns.[18]

SURFACE EMG STUDY OF VOLUNTARY MOVEMENT DISORDERS

Techniques are available for recording voluntary movement by surface EMG analysis. True quantitative evaluation in this realm requires additional equipment to record movement position or velocity. Qualitative evaluation is occasionally helpful but must be interpreted with caution. Such evaluation can be performed by having the patient make a short ballistic movement such as elbow flexion while recordings are made from agonist and antagonist muscles. Under most conditions when this movement is performed as fast as possible, the triphasic pattern appears and its configuration can be examined. Abnormalities have been reported in various diseases. The ballistic movements in Parkinson's disease are characterized by low amplitude of the initial agonist burst. This results in a small amplitude movement that the patient compensates for by making sequential small triphasic bursts.[26] The movements of cerebellar ataxia are characterized by long-duration first-agonist bursts or antagonist bursts that misdirect the limb.[27] With pyramidal lesions, the first agonist or antagonist burst is prolonged. Finally, in patients with athetosis due to cerebral palsy, excessive activity occurs in muscles not normally involved in the main action, and agonist and antagonist muscles often cocontract.[21]

CHAPTER SUMMARY

Surface EMG recordings provide a simple and noninvasive means of studying movement disorders. These techniques are particularly helpful in classifying involuntary movements such as tremor and myoclonus.

REFERENCES

1. Aigner, BR and Mulder, DW: Myoclonus: Clinical significance and an approach to classification. Arch Neurol 2:600–615, 1960.
2. Artieda, J, Muruzabal, J, Larumbe, R, et al: Cortical mechanisms mediating asterixis. Mov Disord 7:209–216, 1992.
3. Brown, P, Rothwell, JC, Thompson, PD, et al: The hyperekplexias and their relationship to the normal startle reflex. Brain 114:1903–1928, 1991.
4. Brown, P, Thompson, PD, Rothwell, JC, et al: A case of postanoxic encephalopathy with cortical action and brainstem reticular reflex myoclonus. Mov Disord 6:139–144, 1991.
5. Brown, P, Thompson, PD, Rothwell, JC, et al: Axial myoclonus of propriospinal origin. Brain 114:197–214, 1991.
6. Chadwick, D, Hallett, M, Harris, R, et al: Clinical, biochemical, and physiological features distinguishing myoclonus responsive to 5-hydroxytryptophan, tryptophan with a monoamine oxidase inhibitor, and clonazepam. Brain 100:455–487, 1977.
7. Cohen, LG and Hallett, M: Hand cramps: Clinical features and electromyographic patterns in a focal dystonia. Neurology 38:1005–1012, 1988.
8. Coleman, RM, Pollack, CP, and Weitzman, ED: Periodic movements in sleep (nocturnal myoclonus): Relation to sleep disorders. Ann Neurol 8:416–421, 1980.
9. Comella, CL, Buchman, AS, Tanner, CM, et al: Botulinum toxin injection for spasmodic torticollis: Increased magnitude of benefit with electromyographic assistance. Neurology 42:878–882, 1992.
10. Daube, JR and Peters, HA: Hereditary essential myoclonus. Arch Neurol 15:587–594, 1966.
11. Davis, JN: An experimental study of hiccup. Brain 93:851–872, 1970.
12. Dawson, GD: Investigations on a patient subject to myoclonic seizures after sensory stimulation. J Neurol Neurosurg Psychiatry 10:141–162, 1947.
13. Deuschl, G, Heinen, F, Kleedorfer, B, et al: Clinical and polymyographic investigation of spasmodic torticollis. J Neurol 239:9–15, 1992.
14. Deuschl, G, Mischke, G, Schenck, E, et al: Symptomatic and essential rhythmic palatal myoclonus. Brain 113:1645–1672, 1990.
15. Fahn, S and Sjaastad, O: Hereditary essential myoclonus in a large Norwegian family. Mov Disord 6:237–247, 1991.
16. Goldberg, G: Paper presented at the AAEM Workshop for the American Association of Electrodiagnostic Medicine, Charleston, South Carolina, October 14 and 15, 1992.
17. Guillain, G: The syndrome of synchronous and rhythmic palato-pharyngo-oculo-diaphragmatic myoclonus. Proc R Soc Med 31:1031–1038, 1938.
18. Hallett, M: Analysis of abnormal voluntary and involuntary movements with surface electromyography. Adv Neurol 39:907–914, 1983.

19. Hallett, M: Myoclonus: Relation to epilepsy. Epilepsia 26 Suppl 1:S67–S77, 1985.
20. Hallett, M: Electrophysiologic evaluation of movement disorders. In Aminoff, MJ (ed): Electrodiagnosis in Clinical Neurology, ed 3. Churchill Livingstone, New York, 1992, pp 403–419.
21. Hallett, M and Alvarez, N: Attempted rapid elbow flexion movements in patients with athetosis. J Neurol Neurosurg Psychiatry 46:745–750, 1983.
22. Hallett, M, Berardelli, A, Delwaide, P, et al: Central EMG and tests of motor control. Report of an IFCN committee. Electroencephalogr Clin Neurophysiol 90:404–432, 1994.
23. Hallett, M, Chadwick, D, Adam, J, and Marsden, CD: Reticular reflex myoclonus: A physiological type of human post-hypoxic myoclonus. J Neurol Neurosurg Psychiatry 40:253–264, 1977.
24. Hallett, M, Chadwick, D, and Marsden, CD: Ballistic movement overflow myoclonus: A form of essential myoclonus. Brain 100:299–312, 1977.
25. Hallett, M, Chadwick, D, and Marsden, CD: Cortical reflex myoclonus. Neurology (Minneap) 29:1107–1125, 1979.
26. Hallett, M and Khoshbin, S: A physiological mechanism of bradykinesia. Brain 103:301–314, 1980.
27. Hallett, M, Shahani, BT, and Young, RR: EMG analysis of patients with cerebellar deficits. J Neurol Neurosurg Psychiatry 38:1163–1169, 1975.
28. Hallett, M, Shahani, BT, and Young, RR: EMG analysis of stereotyped voluntary movements in man. J Neurol Neurosurg Psychiatry 38:1154–1162, 1975.
29. Halliday, AM: The electrophysiological study of myoclonus in man. Brain 90:241–284, 1967.
30. Heilman, KM: Orthostatic tremor. Arch Neurol 41:880–881, 1984.
31. Ikeda, A, Kakigi, R, Funai, N, et al: Cortical tremor: A variant of cortical reflex myoclonus. Neurology 40:1561–1565, 1990.
32. Iliceto, G, Thompson, PD, Day, BL, et al: Diaphragmatic flutter, the moving umbilicus syndrome, and "belly dancer's" dyskinesia. Mov Disord 5:15–22, 1990.
33. Jankovic, J and Pardo, R: Segmental myoclonus: Clinical and pharmacologic study. Arch Neurol 43:1025–1031, 1986.
34. Kuzniecky, R, Berkovic, S, Andermann, F, et al: Focal cortical myoclonus and rolandic cortical dysplasia: Clarification by magnetic resonance imaging. Ann Neurol 23:317–325, 1988.
35. Logigian, EL, Wierzbicka, MM, Bruyninckx, F, et al: Motor unit synchronization in physiologic, enhanced physiologic, and voluntary tremor in man. Ann Neurol 23:242–250, 1988.
36. Marsden, CD, Hallett, M, and Fahn, S: The nosology and pathophysiology of myoclonus. In Marsden, CD and Fahn, S (eds): Movement Disorders. Butterworth Scientific, London, 1982, pp 196–248.
37. Matsumoto, J, Fuhr, P, Nigro, M, and Hallett, M: Physiological abnormalities in hereditary hyperekplexia. Ann Neurol 32:41–50, 1992.
38. McManis, PG and Sharbrough, FW: Orthostatic tremor: Clinical and electrophysiologic characteristics. Muscle Nerve 16:1254–1260, 1993.
39. Nakamura, R, Kamakura, K, Tadano, Y, et al: MR imaging findings of tremors associated with lesions in cerebellar outflow tracts: Report of two cases. Mov Disord 8:209–212, 1993.
40. Obeso, JA, Rothwell, JC, Lang, AE, and Marsden, CD: Myoclonic dystonia. Neurology (N Y) 33:825–830, 1983.
41. Oswald, I: Sudden bodily jerks on falling asleep. Brain 82:92–103, 1959.
42. Ravits, J, Hallett, M, Baker, M, and Wilkins, D: Primary writing tremor and myoclonic writer's cramp. Neurology 35:1387–1391, 1985.
43. Rothwell, JC, Traub, MM, and Marsden, CD: Primary writing tremor. J Neurol Neurosurg Psychiatry 42:1106–1114, 1979.
44. Sabra, AF and Hallett, M: Action tremor with alternating activity in antagonist muscles. Neurology 34:151–156, 1984.
45. Shahani, BT and Young, RR: Physiological and pharmacological aids in the differential diagnosis of tremor. J Neurol Neurosurg Psychiatry 39:772–783, 1976.
46. Shibasaki, H, Kakigi, R, and Ikeda, A: Scalp topography of giant SEP and pre-myoclonus spike in cortical reflex myoclonus. Electroencephalogr Clin Neurophysiol 81:31–37, 1991.
47. Shibasaki, H, Motomura, S, Yamashita, Y, et al: Periodic synchronous discharge and myoclonus in Creutzfeldt-Jakob disease: Diagnostic application of jerk-locked averaging method. Ann Neurol 9:150–156, 1981.
48. Suhren, O, Bruyn, GW, and Tuynman, JA: Hyperexplexia: A hereditary startle syndrome. J Neurol Sci 3:577–605, 1966.
49. Thompson, PD, Colebatch, JG, Brown, P, et al: Voluntary stimulus-sensitive jerks and jumps mimicking myoclonus or pathological startle syndromes. Mov Disord 7:257–262, 1992.
50. Ugawa, Y, Shimpo, T, and Mannen, T: Physiological analysis of asterixis: Silent period locked averaging. J Neurol Neurosurg Psychiatry 52:89–93, 1989.
51. Wilkins, DE, Hallett, M, Berardelli, A, et al: Physiologic analysis of the myoclonus of Alzheimer's disease. Neurology 34:898–903, 1984.
52. Wilkins, DE, Hallett, M, and Erba, G: Primary generalised epileptic myoclonus: A frequent manifestation of minipolymyoclonus of central origin. J Neurol Neurosurg Psychiatry 48:506–516, 1985.
53. Wilkins, DE, Hallett, M, and Wess, MM: Audiogenic startle reflex of man and its relationship to startle syndromes: A review. Brain 109:561–573, 1986.

CHAPTER 31

VERTIGO AND BALANCE

Robert H. Brey, Ph.D.

Balance is a complex mechanism that relies primarily on input from the visual, somatosensory, and vestibular systems. Dizziness or disruption of balance occurs when the information sent to the central nervous system (CNS) by one of these systems conflicts with information provided by the other two systems. An example is the false sensation of movement experienced by a person sitting in an automobile when a large vehicle parked alongside begins slowly to pull forward. The visual information of the person is consistent with the sensation of the car rolling backward; the reflexive action is to step on the brakes. The instant it is realized that the car fails to come to a lurching stop, the CNS begins to resolve the sensory conflict by determining that it was the larger vehicle that was moving. If conflicting information is continually fed to the CNS because of vestibular malfunction, the person experiences the classic symptoms of vertigo (i.e., the false sense of motion), nausea, and vomiting. These symptoms may continue from seconds to days. The three categories of causes of vertigo are (1) peripheral causes, including the peripheral vestibular sensory mechanisms and nerve fibers leading to the brain stem; (2) CNS causes, including all the brain stem vestibular nuclei and their connections with the visual and somatosensory systems and the cerebral cortex; and (3) systemic causes, including vascular problems. Many patients report such severe symptoms that they wanted to die.

The vestibular system has two types of sensory organs: the *semicircular canals* and the *otolith organs*. These sensory organs contain hair cells that respond either to relative fluid motion in the membranous labyrinth (caused by rotating the head) or to gravitational force exerted on calcium carbonate crystals, called *otoconia* (moving the head linearly or tilting it). The anatomy and physiology of the vestibular mechanism have been reviewed recently.[16]

354

OFFICE PROCEDURES
AND HISTORY TAKING

A complete history and description of the patient's symptoms are critical for a work-up leading to the diagnosis of vertigo. The areas to be covered are (1) description of the vertiginous symptoms, (2) their onset and duration, (3) concomitant aural symptoms, (4) visual symptoms, (5) general neurologic symptoms, (6) medical history, and (7) current and previous medications, including use of drugs such as alcohol, nicotine, and caffeine.

Descriptions of Symptoms

SYMPTOMS CAUSED
BY PERIPHERAL LESIONS

Patients who have an acute peripheral vestibular attack usually can describe in detail what happened. Most patients experience a spinning sensation (either of the environment or of themselves), nausea, vomiting, disorientation or loss of control, and sweating, followed by extreme fatigue. Some patients also describe a vertical roll or tilt of the whole environment. Others describe being thrown to the floor or feeling that they are being bounced against the walls. During an acute attack, any movement of the head aggravates the symptoms. Such patients are often hospitalized for 1 to 2 days, mainly because of concern that they are having a stroke.

Visual symptoms can help define a patient's problems. With a unilateral peripheral loss of vestibular function, blurring of vision or retinal slip occurs because the vestibular ocular reflex (VOR) is unable to keep the visual image centered on the macula of the retina, where the image is sharpest. Bilateral loss of the peripheral vestibular mechanism produces *oscillopsia*, in which the entire visual environment appears to oscillate as the patient's head moves; this is because the VOR is unable to provide the necessary compensatory eye movements to keep the image stable on the retina.

Diseases that can affect the peripheral vestibular mechanism include Ménière's disease (endolymphatic hydrops), labyrinthitis (viral or bacterial infection), vestibular neuritis, perilymphatic fistula, and autoimmune disease of the inner ear. Although tumors are another peripheral lesion that can affect the vestibular branch of cranial nerve VIII, the patients often have no symptoms of dizziness, because the tumor grows so slowly that the CNS is able to compensate physiologically for slow changes caused by tumor growth. Head trauma and benign paroxysmal positional vertigo are not disease processes, but they are related to mechanical problems in the vestibular mechanism that can cause vertigo.

SYMPTOMS CAUSED
BY CENTRAL LESIONS

Patients with CNS vestibular disorders most often complain of a slow onset of symptoms that worsen over time. These patients tend to complain of unsteadiness, lightheadedness, nausea, and other neurologic symptoms. An exception to this, however, are patients who have an acute onset of symptoms because of a stroke that damages the vestibular nuclei. Examples of CNS diseases that cause vestibular symptoms are multiple sclerosis, vertebrobasilar insufficiency, migraine headache, compression of cranial nerve VIII by a vascular loop, cervical disk problems, head trauma, and Arnold-Chiari malformation.

SYSTEMIC SYMPTOMS

Symptoms found in patients with systemic vestibular problems tend to be similar to those of patients with CNS problems and can be caused by involvement of all three components—visual, somatosensory, and vestibular. Examples of systemic problems are orthostatic problems, diabetes mellitus, hypoglycemia, hypothyroidism, drug effects, stress, and allergies.

CONCOMITANT AURAL SYMPTOMS

The auditory and vestibular mechanisms share the membranous labyrinth of the inner ear and its fluid system. To determine whether a relationship may exist between aural and vestibular symptoms, it is critical to ask patients about tinnitus, aural fullness, pressure in the ears or head, fluctuating hearing loss, or changes in the ability to understand speech. For example, hydrops, or Ménière's syndrome, can cause inner ear pressure, tinnitus, vertigo, and a fluctuating low-frequency sensorineural hearing loss. Acoustic tumors of the auditory nerve often cause tinnitus, hearing loss, and problems with understanding speech. Previous ear surgery could be related to a perilymphatic fistula that could lead to hearing loss and vertigo.

Medications or Drugs

Patients must be asked about any ototoxic medications they might have taken. Aminoglycosides, salicylates, quinine, and chemotherapeutic agents such as cisplatin can directly affect and, in some cases, permanently damage the hearing and vestibular mechanisms.

PHYSICAL EXAMINATION OF THE VESTIBULAR SYSTEM

The symptoms of a patient examined during an acute attack of vertigo are different from those revealed during a later examination. However, both examinations can provide valuable information for making a diagnosis. During an acute peripheral vestibular attack, nystagmus may be observed even when the patient's eyes are open and fixed. Subsequently, however, fixation may suppress the nystagmus. In this case, electrooculographic examination (electrical measurement of the corneoretinal potential) with the patient's eyes closed is needed to observe the abnormal eye movements. Examination using Frenzel lenses on a patient in a darkened room with eyes open also allow nystagmus to be observed. With central vestibular lesions, nystagmus tends to be present or even more intense when the patient's eyes are open and fixed.

Nystagmus

The varieties of nystagmus include pendular, rotatory, jerk (horizontal and verti-

cal), and congenital nystagmus. Congenital nystagmus, which can incorporate both pendular and jerk components, exhibits a null point where the patient's nystagmus is diminished.

The types of nystagmus commonly seen during vestibular attacks are horizontal, vertical, and rotatory nystagmus. The corneoretinal potential (approximately 1 mV, with the front of the eye positive relative to the back of the eye) allows horizontal and vertical eye movements to be measured with electrodes placed around the eyes. Rotatory nystagmus can be recorded only with some type of video monitor or magnetic search coil placed over the lens of the eye, like a contact lens. The traditional method for plotting horizontal and vertical nystagmus over time is shown in Figure 31–1. For horizontal and vertical eye movements, "pen up" describes the movements that are up or to the right and "pen down" those that are down or to the left. The chart is read from left to right. Nystagmus direction is defined by the fast, or jerk, component, which is generated by the CNS. The slower component is in the opposite direction and is generated by the vestibular tracking mechanism. Generally, the slope (rise/run) of the slow component is used to calculate, in degrees per second, the speed of the eye movement.

During the fast component, vision is interrupted. Thus, rather than sensing a jerking movement, one has a sensation of smooth rotation in one direction. This means that if a person experiences a spinning sensation to the right, the fast component of the nystagmus is to the right or in the direction to which the head seems to be turning. If the room or environment seems to be spinning around the person to the right, then the fast component of the nystagmus is to the left.

OBSERVATION OF EYE MOVEMENT

Any abnormal eye movements detected during examination of a patient should be noted. The term *spontaneous nystagmus* refers to nystagmus present with the eyes open and fixed in the primary position. In this case, the examiner must look closely at the patient's eyes to observe any small rhythmic movements.

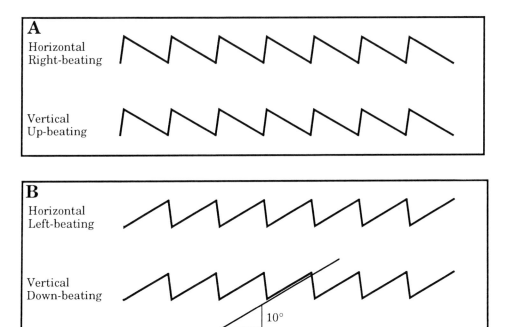

Figure 31–1. Electro-oculographic recordings of horizontal and vertical nystagmus. (*A*) Right-beating horizontal nystagmus and up-beating vertical nystagmus. (*B*) Left-beating horizontal nystagmus and down-beating vertical nystagmus. On the latter tracing is an example of how the velocity (slope = rise/run) of the slow component is measured in degrees per second.

Gaze testing is performed with the patient's eyes in the primary gaze position and ± 30 degrees from midline, both horizontally and vertically. It is more helpful to observe a small vessel on the side of the eye than to watch the pupil. This allows detection of low-amplitude (< 1 degree) horizontal, vertical, or rotatory nystagmus. Electro-oculography is capable of measuring only eye movements that are greater than 1 degree. If the center of the eye is observed, rotatory nystagmus may be overlooked. Patients with either a CNS vestibular abnormality or an acute peripheral attack show abnormal nystagmus, with their eyes open and fixed. However, if a peripheral lesion is not of recent onset, the CNS likely suppresses any nystagmus. To measure the eye movements in these cases, Frenzel lenses, which prevent patients from visually fixating, or electro-oculography must be used while the patients' eyes are closed (or while the patients are in a dark room). Nystagmus observed in this manner is referred to as *spontaneous nystagmus without fixation* (or with eyes closed). It is also re-ferred to as *latent nystagmus*, because it is present only when visual fixation is pre-vented. *The general rule is that nystagmus caused by a peripheral lesion is suppressed by vi-sual fixation* but nystagmus caused by a CNS lesion is not.

Any abnormal eye movements should then be documented on a chart similar to the one shown in Figure 31–2; lines and ar-rows are used to indicate the direction and magnitude of eye movement. The concept of Alexander's law—that nystagmus beats stronger when gaze is in the direction of the fast component of the nystagmus—is illus-trated in Figure 31–2A.[5] The response shown would be defined by Alexander as a *3rd-degree nystagmus* (i.e., nystagmus present in all positions of gaze and greater toward the unaffected ear). As the CNS physiologi-cally compensates for the deficit caused by the peripheral lesion, the nystagmus pro-gresses to the *2nd degree* (i.e., nystagmus with gaze away from the lesion and midline) and finally to the *1st degree* (i.e., nystagmus only with gaze away from the side of the lesion).

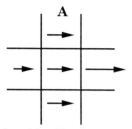

Right peripheral hypofunction; movement observed with Frenzel lenses. Nystagmus beats toward unaffected ear.

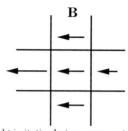

Right irritative lesion; movement observed with Frenzel lenses. Nystagmus beats toward diseased ear.

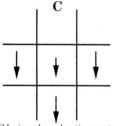

CNS lesion; down-beating nystagmus observed with patient's eyes open and fixed.

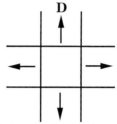

CNS lesion; direction changing nystagmus observed with patient's eyes open and fixed.

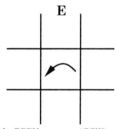

Right BPPV; rotatory (CCW) nystagmus seen with patient's eyes open and fixed during Dix-Hallpike maneuver.

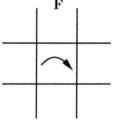

Left BPPV; rotatory (CW) nystagmus seen with patient's eyes open and fixed during Dix-Hallpike maneuver..

Figure 31–2. Method for documenting abnormal eye movements. Arrows indicate the direction of the quick eye movement, and the length of the line represents the amplitude of the movement. BPPV = benign paroxysmal positioning vertigo; CCW = counterclockwise; CNS = central nervous system; CW = clockwise.

POSITIONING-INDUCED NYSTAGMUS

A positioning maneuver to test for benign paroxysmal positioning vertigo (BPPV) is different from static positional testing (discussed later in conjunction with the electronystagmography [ENG] test battery). Dynamic positioning testing dates to Bárány,[4] Nylen,[19] and Dix and Hallpike,[8] who discovered that moving certain patients from a sitting position to a head-hanging right (or left) position generated burst rotatory nystagmus with accompanying vertigo; it lasted only a few seconds and then fatigued with repetition of the maneuver. Schuknecht[22–24] postulated that the cause was basophilic deposits on the cupula in the posterior semicircular canal; he found these deposits in pa-

tients who had BPPV at the time of death. Because he thought that gravity affected the cupula by acting on this deposit, he coined the term *cupulolithiasis*.

The *Dix-Hallpike maneuver* is illustrated in Figure 31–3. The patient is initially seated with the head turned 45 degrees to the right or the left and is then placed in the head-hanging right (or left) position with his or her eyes open so the examiner can observe any rotatory nystagmus. The classic response consists of a brief delay and then a burst of rotatory nystagmus that lasts several seconds. This maneuver duplicates the patient's symptoms. The head-hanging right position produces counterclockwise nystagmus, whereas the head-hanging left position causes clockwise nystagmus (see Fig. 31–2).

Figure 31–3. Dix-Hallpike maneuver (Nylen maneuver) for the head-hanging right position. (*A*) In the sitting position, the patient turns head 45 degrees to the right with the eyes open and fixed. Patient is then put quickly into the supine position. (*B*) Patient in the head-hanging right position with eyes open and fixed. Observe for nystagmus and dizziness for at least 30 seconds. The classic positive response is counterclockwise nystagmus with the head down and turned to the right and clockwise nystagmus with the head down and turned to the left. The nystagmus is reversed with sitting up.

When the patient sits up, the nystagmus is reversed but less intense. The response weakens with repeated trials.

Early surgical treatment for BPPV involved neurectomy of the vestibular branch of cranial nerve VIII; this innervated the ampulla of the affected posterior semicircular canal.[13] Currently, this procedure is rarely performed, because of newer, less invasive approaches. Other approaches that were used included exercises designed to habituate the CNS.[6]

Epley[9,10] postulated the existence of floating particles in the posterior semicircular canal and proposed a physical maneuver, the *canalith repositioning procedure*, for removing them. Several investigators have reported excellent success rates with this procedure.[9,10,27] In 1991, Parnes and McClure[20] discovered these free-floating particles in the posterior semicircular canals of patients with BPPV.

Semont, Freyss, and Vitte in 1988[25] reported on another procedure, the *liberatory maneuver*, designed to remove debris from the posterior semicircular canal. Herdman et al.[15] gave a good description of this maneuver and made some minor modifications. Results of the canalith repositioning procedure and the liberatory maneuver indicated that approximately 90% of the patients had either a cure or significant relief from their symptoms of BPPV. These two procedures are the treatment of choice for most patients with BPPV.

Vestibulospinal Reflexes

PAST-POINTING TEST

Bárány, in 1907,[4] described a *past-pointing test* in which a patient was asked to sit with closed eyes and an arm extended, pointing first above the head and then back to an imaginary spot. The examiner's finger served as a reference to determine how much deviation occurred. Bárány found that pointing tended to drift toward the ear with a peripheral lesion (i.e., the direction of the slow component of nystagmus). However, he indicated that this test could not be used in isolation. Baloh and Honrubia[3] pointed out that the standard finger-to-nose test did not identify a past-pointing error because of the many proprioceptive clues available to the patient.

STATIC POSTURE TESTING

In 1946, Romberg[21] noted that patients with vestibular problems had difficulty standing with their feet together and eyes closed. They tended to sway toward the affected side, that is, with the slow component of nystagmus. This was called *Romberg's test*. The sharpened *Romberg's test* was proposed in 1974 by Fregly.[11] The patient is asked to stand with feet aligned in the tandem heel-to-toe position and with closed eyes and arms folded against the chest. Most normal subjects younger than age 70 should be able to stand in this position for 30 seconds.

DYNAMIC WALKING TESTS

Unterberger[26] and Fukuda[12] developed a test in which the patient walks in place with closed eyes and arms parallel, horizontal, and extended forward. The examiner notes the angle of rotation. Again, the direction of rotation, as seen in the previous discussion of tests, is that of the slow component of nystagmus. Fregly[11] described the tandem walking test that is often used for neurologic evaluation. If patients perform this test with their eyes open, it primarily measures cerebellar function. However, patients with acute vestibular lesions may also have difficulty performing the test. For vestibular testing, the patient is blindfolded or asked to walk in tandem fashion with closed eyes and arms folded against the chest. The test begins with the patient's feet in a tandem position; the patient then takes ten tandem steps at a comfortable speed. The number of normal steps taken (without sidestepping) is scored on three trials. Most normal subjects are able to make a minimum of ten normal steps in the three trials. However, patients with acute or chronic vestibular dysfunction fail the test, but the direction of fall is not related to the fast or slow component of nystagmus.

HEAD-SHAKING NYSTAGMUS

Hain, Fetter, and Zee[14] reported that spontaneous nystagmus develops in patients with peripheral and central vestibular dysfunction after ten quick head shakes in the horizontal plane. These authors suggested that spontaneous nystagmus develops because of asymmetrical velocity storage within the central vestibulo-ocular pathways. In cases of unilateral peripheral vestibular lesions, the slow component is toward the weaker side.

DYNAMIC VISUAL ACUITY

A Snellen, or illegible E, eye chart can be used to measure the VOR. The patient is asked to sit, with head immobile, at the standard distance from the chart and to read the lowest line possible. The head is then rotated back and forth at 2 Hz. If visual acuity shifts more than one line, the result is abnormal.[3]

Bilateral vestibular weakness that causes os-cillopsia produces abnormal test results.

Visual suppression of nystagmus can be measured by placing a patient in a chair that can be rotated and asking the patient to extend a finger and fixate on it. Abnormal fixation suppression (inability to suppress) is consistent with CNS dysfunction.

LABORATORY EXAMINATION: ELECTRONYSTAGMOGRAPHY TEST BATTERY

Preparation for Testing

As mentioned previously, a carefully documented history should rule out any preexisting condition (e.g., congenital nystagmus and use of medications or drugs) that could influence test results. Many medications, such as vestibular suppressants, sedatives, tranquilizers, antidepressants, and pain relievers, have side effects related to dizziness. To avoid the adverse effects these medications can have on test results, their use must be discontinued 48 hours before the patient undergoes vestibular testing. Patients should also be counseled to refrain from smoking tobacco, because nicotine constricts blood vessels and, thus, impairs the blood supply to the vestibular mechanism. As a stimulant, caffeine can also adversely affect the vestibular system. Also, patients should avoid drinking alcohol for at least 24 hours before testing, because alcohol alters the chemical balance of the perilymph and endolymph and induces nystagmus.[18] *Geotropic* (beating toward the earth) *positional nystagmus* can be induced within one-half hour of alcohol ingestion and continue up to 4 hours. *Ageotropic* (beating away from the earth) *positional nystagmus* can be observed from 5 to 24 hours after alcohol is ingested.

Use of anticonvulsants and medications for vascular regulation should not be discontinued before testing. Patients with diabetes should eat a light meal at least 2 hours before testing and should not avoid taking their insulin. Leigh and Zee[17] provide an in-depth discussion of the effects of drugs on eye movement.

Electrodes are placed at the outer canthus of each eye to measure the horizontal component of eye movement, and electrodes placed above and below one eye are used to measure vertical movements. A common, or ground, electrode is placed on the forehead just above the nose. If a patient is blind in one eye, the electrodes must be placed around the other eye. If a patient is blind in both eyes and has no corneoretinal potential, electro-oculography cannot be used to measure eye movement. Other procedures for measuring eye movements (e.g., infrared systems, magnetic search coils, and visual image processing) are under development and will permit measurement of horizontal, vertical, and torsional nystagmus, but these systems are not yet practical for routine clinical testing.

Gaze Testing

When gaze testing (as already described) is included as part of an ENG test battery, electro-oculography can be used to measure and quantitate the direction and velocity of the eye movement. Gaze testing is performed by having the patient fixate on spots ± 20 and ± 30 degrees from center in both the horizontal and the vertical planes. Gaze greater than ± 30 degrees may cause end-point nystagmus, which is normal. Remember, however, that visually the human eye is capable of picking up eye movements less than 1 degree, but with electro-oculography the movements must be greater than 1 degree to be detected and recorded. Therefore, always inspect the patient's eyes during the gaze testing part of the protocol.

Oculomotor Testing

SACCADIC EYE MOVEMENT TESTING

During calibration for electro-oculography, the patient looks back and forth ± 10 degrees, which provides the examiner an opportunity to look for undershoot or overshoot during saccadic eye movements. If undershooting or overshooting occurs consistently, it may be related to a CNS vestibular disorder. A major improvement for evaluating the ability of the ocular motor system to produce accurate and timely saccadic eye movements is the new computerized *random saccade test*. In addition to allowing examiners to look at undershoot and overshoot, the new system produces random signals of varying degrees over a range of ± 30 degrees by turning on and off light-emitting diodes (LEDs) on a light bar. The analog signals from the eye movements are digitized, and computer algorithms are used to calculate the accuracy, latency, and velocity of the eye movements relative to the stimulating signal. These values are compared with those of normal subjects matched for age and sex. Abnormal test findings indicate a central vestibular abnormality, perhaps the most significant of which is low eye velocity. Poor cooperation by the patient must be ruled out if latency and accuracy are abnormal.

SMOOTH OCULAR PURSUIT TEST

The *smooth ocular pursuit test* can be conducted by having the patient hold the head still and then visually follow a pendulum. However, newer computerized systems produce a range of frequencies, using pendular-like signals, by turning on and off adjacent LEDs on the light bar. Ocular pursuit operates well up to a frequency of 1 Hz. The computerized pursuit tests usually cover a range from 0.2 to 0.7 Hz. The cooperation of the patient is critical, and several trials may be needed to ensure that the patient is putting forth a best effort.

Abnormal test results are consistent with CNS dysfunction. The commonest abnormality is *cogwheeling pursuit*, in which the eye is continually making saccadic movements to catch up with the target. In another type of abnormality, in which the pursuit keeps breaking up, the patient is unable to follow the target consistently.

POSITIONAL TESTING WITHOUT FIXATION

The purpose of *positional testing without fixation* is to measure eye movement with the patient's head held in various static positions, such as sitting, supine, supine head-right, supine head-left, lateral right (no neck torsion), lateral left (no neck torsion), head hanging, and supine 30 degrees (Fig. 31–4). The purpose for testing supine at 30 degrees is to have a reference for caloric irrigation (see later discussion).

Visual fixation and mental suppression must be avoided during positional testing. This is accomplished by testing in total darkness with the patient's eyes open, or by using Frenzel lenses in a darkened room, or by having the patient remain with closed eyes in a semidarkened room. The last procedure is the one used most often. The patient must be mentally alert so as not to mentally suppress the nystagmus. To do this, have the patient carry out a task that cannot be done automatically, for example, counting backward by 3s or naming states and cities. Normally, patients do not have nystagmus in any of the positions when their eyes are closed.

Abnormal findings are categorized as (1) *direction-fixed* in one or more positions and (2) *direction-changing nystagmus* in two or more positions. Direction-changing can be geotropic, ageotropic, or direction-changing in a single head position.

There is also a continuum that ranges from persistent nystagmus (always present) to intermittent nystagmus (occasional beats). During testing, the examiner must distinguish *positioning nystagmus* due to the patient's moving into a new position from *positional nystagmus*, which is present in the static head position. Recordings are made for 30 to 60 seconds.

Although the findings are nonlocalizing, there are some general rules for their interpretation. Chronic hypofunction usually results in nystagmus beating toward the unaffected or less affected ear. With an irritative lesion, nystagmus can beat toward the diseased ear. Observance of positional nystagmus is also useful in monitoring physiologic compensation during the disease process, because the nystagmus diminishes as compensation occurs.

Direction-changing nystagmus in a single head position (excluding positioning) is usually a sign of CNS dysfunction and is referred to as *periodic alternating nystagmus*. It changes direction about every 2 minutes.

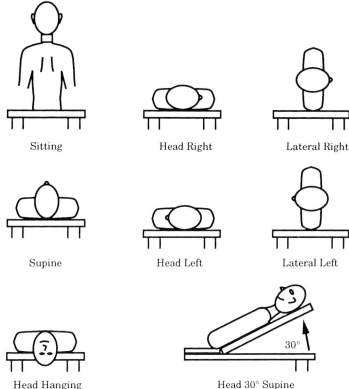

Figure 31–4. Head positions used in the static positional test. The patient's eyes remain closed, but the patient must be alert.

OPTOKINETIC NYSTAGMUS

Optokinetic nystagmus (OKN) is another test of smooth ocular pursuit or CNS function. The stimulus is usually generated as a series of light and dark vertical bars that move from right to left or left to right of the patient at 20, 40, or 60 degrees per second. Ideally, the entire visual field of the patient should be filled with these stimuli in a darkened room. Less acceptable alternatives are small hand-held rotating drums with black and white stripes or a series of LEDs that appear to move across a light bar. Abnormal findings include asymmetry between right and left beating or an inability of the patient to increase eye speed with increased stimulus speed.

Another application of OKN is the *optokinetic after nystagmus test* (OKAN). If a patient observes the OKN stimulus for 1 minute and the room is then darkened (i.e., visual fixation not possible), there should be little or no nystagmus. However, if the nystagmus persists, it is thought to indicate a CNS or peripheral abnormality.

Caloric Irrigation

Caloric irrigation is performed with the patient's head in a supine position at 30 degrees (see Fig. 31–4), which places the lateral semicircular canals vertically. The stimulus used is water ($\pm$ 7°C relative to normal body temperature) or air ($\pm$ 13°C relative to normal body temperature). The ear is irrigated with water for 30 to 40 seconds or with air for 60 seconds. The water stimulus is preferred by most examiners because it transfers heat more efficiently to the petrous bone surrounding the vestibular mechanism. However, in cases of perforated tympanic membrane, air is the better stimulus, because it is not advisable to introduce water into the middle ear.

The theory is that as the stimulus warms or cools the bone surrounding the lateral semicircular canal, a convection current is induced in the endolymph and this current causes utriculopetal or utriculofugal flow. The left lateral semicircular canal at rest and after warm (44°C) and cool (30°C) caloric stimulation is shown in Figure 31–5. Warm stimuli cause the nystagmus to beat toward the stimulated ear, and cool stimuli cause it

to beat toward the nonstimulated ear, thus providing the acronym **COWS** (cold opposite, warm same).

Caloric irrigation is the standard test for determining the laterality of the lesion. However, certain pitfalls must be avoided. Again, the patient must not be allowed to visually fixate but must remain mentally

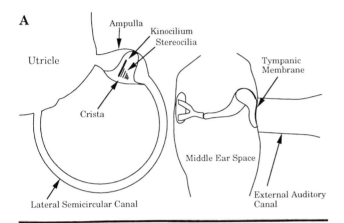

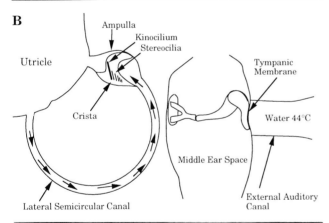

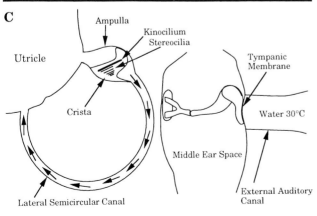

Figure 31–5. (*A*) Lateral semicircular canal at rest (i.e., no caloric stimulation or rotation). Kinocilium and stereocilia are vertical, producing a normal resting potential in vestibular nerve. (*B*) Stimulation with warm water (44°C) causes upward convection current (utriculopetal endolymph flow). Stereocilia bend toward the kinocilium, depolarizing the dendrites at base of hair cells or increasing the firing rate of the vestibular nerve. (*C*) Stimulation with cool water (30°C) induces downward convection current (utriculofugal endolymph flow), causing the kinocilium to bend toward stereocilia, which hyperpolarizes dendrites at the base of hair cells or decreases firing rate of the vestibular nerve.

alert. Also, congenital nystagmus or any drugs or medication that could influence the results must be ruled out.

Results in patients with perforated tympanic membranes may be misleading. If the middle or external ear is wet because of infection and drainage, a warm air stimulus will initially cool rather than heat the bone and the nystagmus will beat in the direction produced by a cold stimulus. This could be misinterpreted as a CNS abnormality.

Unilateral weakness (UW) is determined by comparing the nystagmus generated on each side:

$$UW = \frac{(RW + RC) - (LW + LC)}{(RW + RC + LW + LC)} \times 100$$

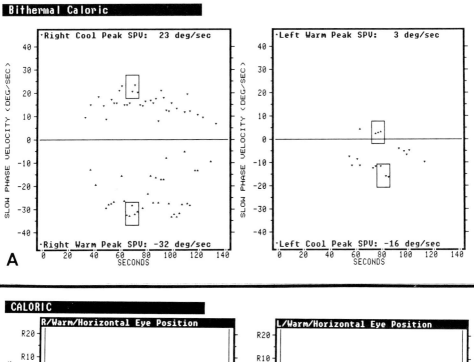

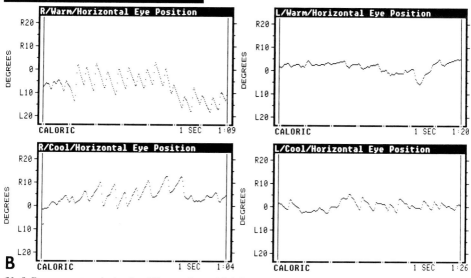

Figure 31–6. Responses to caloric stimuli in a patient with left unilateral vestibular weakness. (*A*) Calculated response of the nystagmus over time. The small boxes represent peak eye velocity values averaged for each of the four irrigations. These responses show a 49% left peripheral weakness and a 30% right-beating directional preponderance. (*B*) Raw data obtained during peak eye velocities. Note the weak responses obtained by stimulating the left ear. SPV = slow phase velocity.

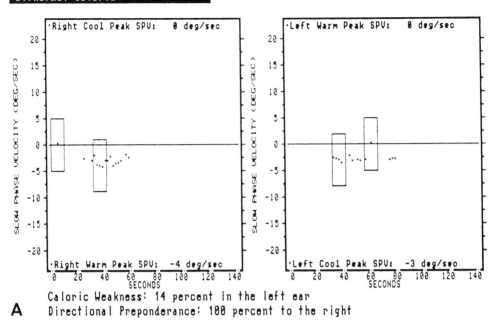

Figure 31–7. Responses to caloric stimuli in a patient with peripheral vestibular weakness bilaterally. (*A*) Calculated response of nystagmus over time. The small boxes represent peak eye velocity values averaged for each of the four irrigations. These responses show a total of 7 degrees of right-beating nystagmus, which is probably due to a positional nystagmus present whenever eyes are closed. (*B*) Raw data obtained during peak eye velocities. Note weak responses with stimulation of both ears. SPV = slow phase velocity.

where *RW* is the right warm-peak slow-component velocity, *RC* is the right cool-peak slow-component velocity, *LW* is the left warm-peak slow-component velocity, and *LC* is the left cool-peak slow-component velocity. The result is a percentage difference between the values for each ear. In most testing laboratories, the percentage difference must be 20% to 25% to be clinically significant. An example of a left peripheral weakness, 49% weaker on the left side, is shown in Figure 31–6.

Another way of analyzing the response is to measure the directional preponderance (DP):

$$DP = \frac{(RW + LC) - (LW + RC)}{(RW + RC + LW + LC)} \times 100$$

The difference between the two directions must be at least 30% to be considered significant. Directional preponderance is nonlocalizing. It usually accompanies a direction-fixed positional nystagmus, because nystagmus sums algebraically.

Bilateral weakness is another possible result. If the sum of all four irrigations is less than 25 degrees of peak nystagmus, it is considered bilaterally weak. An example of this is shown in Figure 31–7. The total nystagmus generated (in degrees) is 0 + 4 + 0 + 3 = 7 per second.

Another test is to measure fixation suppression shortly after the eyes reach their maximal velocity. If the patient fails to suppress the nystagmus by at least 30% to 40%, the result is abnormal and indicative of a CNS lesion.[1]

A summary of ENG test battery results is given in Table 31–1.[7]

Table 31–1. ELECTRONYSTAGMOGRAPHY TEST BATTERY: CORRELATION OF ABNORMAL FINDINGS AND SUSPECTED SITE OF LESION*

Test	Type of Abnormality	Suspected Site of Lesion
Saccade	Ipsilateral dysmetria	Cerebellopontine angle
	Bilateral dysmetria	Cerebellum
	Decreased velocity	Throughout the central nervous system, muscle weakness or peripheral nerve palsy
	Internuclear ophthalmoplegia	Medial longitudinal fasciculus
Pursuit	Break-up	Brain stem or cerebrum
	Saccadic	Cerebellum
Gaze	Direction-fixed and horizontal	Peripheral vestibular
	Direction-changing and vertical	Brain stem
	Up-beating	Brain stem or cerebellum
	Down-beating	Cervicomedullary junction or cerebellum
	Rotary	Vestibular nuclei/brain stem
Failure of fixation suppression	Less than 40% decrease	Brain stem or cerebellum
Positional	Direction-fixed	Nonlocalizing or peripheral
	Direction-changing	Nonlocalizing or central
Dix-Hallpike	Classic	Peripheral vestibular–undermost ear
	Nonclassic	Nonlocalizing
Caloric	Unilateral or bilateral weakness	Peripheral vestibular
	Directional preponderance	Nonlocalizing

* Exceptions to the rule may occur.
(From Cyr, DG: Vestibular system assessment. In Rintelmann, WF (ed): Hearing Assessment, ed 2. Pro-Ed, Austin, Texas, 1991, p. 777, with permission.)

COMPUTERIZED ROTARY CHAIR TEST (TORSION SWING CHAIR)

Another approach for assessing VOR is to place the patient in a rotary chair in a darkened room and to measure the nystagmus electro-oculographically as the chair rotates back and forth. This is analogous to the torsion swing chair test. The advantages of the rotary chair test are (1) eye movements can be quantitated with or without visual stimulation, (2) patients tolerate it better than caloric irrigation, (3) angular rotation is a more consistent stimulus than caloric irrigation, (4) small children can be tested without difficulty, and (5) it appears to be better for monitoring changes over time. Its major disadvantages are that it stimulates both lateral semicircular canals simultaneously and does not help in lateralizing lesions. Thus,

caloric irrigation is still the primary test for evaluating each ear independently.

The test used most often is the *low-frequency rotary chair test*, with the patient kept in total darkness. It consists of accelerating and decelerating the chair from 0 to 50 degrees per second in a sinusoidal fashion from 0.01 to 0.64 Hz. Most systems use an infrared camera to monitor the patient's eyes to ensure that they are open. As the chair rotates, the computer digitizes the analog signals and compares eye movement with chair rotation. The algorithms compare the velocity, phase, and gain of the two signals. At low frequencies (e.g., 0.01 Hz), eye velocity leads chair velocity. As the chair frequency increases and approaches 0.64 Hz, the phase difference approaches zero. With the patient rotating in the dark, the gain (ratio of eye velocity to chair velocity) is low at low frequencies and increases at higher frequencies. The relationships of phase, gain,

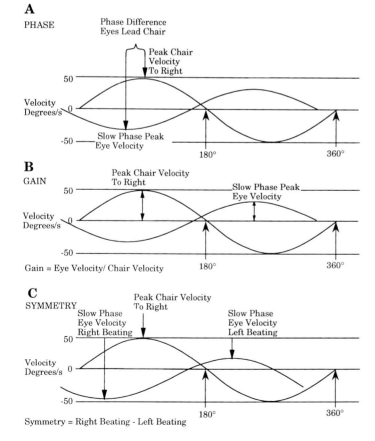

Figure 31–8. (*A*) Measurement of phase, (*B*) gain, and (*C*) symmetry by using a computerized rotary chair. Sine waves represent velocity of the chair and eyes, as indicated.

and symmetry of chair velocity and eye velocity are shown in Figure 31–8. Normal phase gain and symmetry are shown for a patient in Figure 31–9. The data from a patient with left peripheral vestibular weakness, as indicated by a 59% caloric difference between the two ears and a 22% right-beating directional preponderance, are shown in Figure 31–10. Gain is normal from 0.01 to 0.32 Hz (0.64 Hz was not tested). Phase is abnormal or borderline abnormal from 0.01 to 0.16 Hz. Asymmetry, although within normal range, is slightly below the line, indicating that right-beating is greater than left-beating. Often in such patients, the gain recovers but phase can remain abnormal, particularly with complete loss of function on one side.

Patients with total bilateral vestibular weakness have poor gain at all frequencies and no response to caloric irrigation. When gain is below normal from 0.01 to 0.64 Hz, phase and symmetry are meaningless because there is no eye velocity to use for comparison.

Another approach is the *vestibular visual ocular reflex* (VVOR) test. The patient, with eyes open, is rotated in a lighted room. Thus, visual and vestibular clues are available. In normal subjects, the test produces gain measurements that approach 1.0 and phase measurements that approach 0 degrees.

The *step test* provides the information necessary to assess the time constant (i.e., the time it takes for nystagmus to decay to 37% of its maximum after stimulation has stopped). The patient is quickly accelerated from 0 to 50 or 100 degrees per second in 0.5 second. The chair then continues to rotate at 50 or 100 degrees per second for 1 minute. As the flow of the endolymphatic fluid in the semicircular canals approaches the velocity of the head, the nystagmus decays. The response time for the cupula to bend and to return to its resting state is about 6 seconds. However, the nystagmus continues for 10 to 30 seconds, which is attributed to CNS vestibular integration. When the chair is stopped suddenly (0.5 sec-

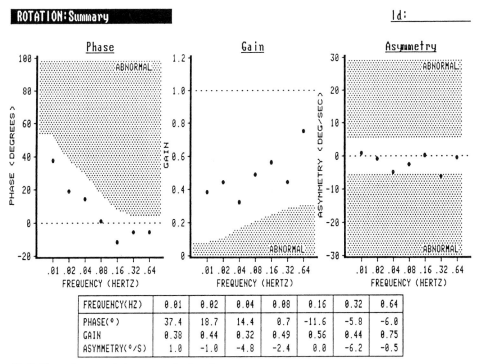

Figure 31–9. Normal rotary chair test results for phase, gain, and symmetry obtained with patient rotating in the dark. Results in shaded areas are abnormal.

FREQUENCY(HZ)	0.01	0.02	0.04	0.08	0.16	0.32	0.64
PHASE(°)	37.4	18.7	14.4	0.7	-11.6	-5.8	-6.0
GAIN	0.38	0.44	0.32	0.49	0.56	0.44	0.75
ASYMMETRY(°/S)	1.0	-1.0	-4.8	-2.4	0.0	-6.2	-0.5

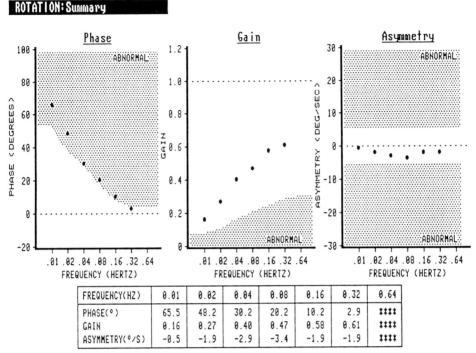

Figure 31–10. Results of rotary chair test (conducted in darkness) in patient with right peripheral weakness. Phase is abnormal but gain and symmetry are normal. Results in shaded areas are abnormal.

ond), the fluid continues to move relative to the head but in the opposite direction, thus generating nystagmus. The time constant during this period is also measured. Abnormal findings can be caused by either CNS dysfunction or the inability of the peripheral system to send the appropriate information for integration.

The consistent stimulus and reliable nature of rotary chair testing make it the best choice for monitoring changes in the VOR over time. Rotary chair testing is valuable when monitoring physiologic compensation or change in the vestibular mechanism induced by ototoxic medications.

COMPUTERIZED DYNAMIC POSTUROGRAPHY

Balance is a complex function that requires input from three major sensory systems. Somatosensory information is the dominant input, followed by visual and then vestibular inputs. The inputs from these

three systems are integrated, analyzed, and incorporated into a complex network by the CNS for maintenance of balance. For many years, physicians have used subjective methods for assessing a person's ability to maneuver and to maintain balance, with and without vision. Tests such as Romberg's and the tandem gait tests are two examples.

Computerized dynamic posturography (CDP) is a recently developed test for assessing balance. The test provides quantitative information that can be correlated with the types of problems patients have with balance deficits. The CDP is designed to provide real-life experiences in a controlled laboratory environment so that the examiner can evaluate the patient objectively and subjectively.

This test consists of two major components, each containing subtests. The first tests motor control to maintain balance. The second measures the patient's use of sensory information as it relates to maintaining balance. On July 20, 1990, the American Academy of Otolaryngology–Head and Neck Surgery, Inc.[2] ruled that dynamic posturography is an acceptable test procedure for assessing the balance system.

The patient's anterior-posterior sway is monitored by measuring vertical force with the four strain gauges that are mounted underneath the two platforms on which the patient stands (i.e., one foot on each platform). A fifth strain gauge is mounted in another direction to measure the shear force that is generated as the hips are thrown forward or backward to maintain balance instead of rotating about the ankles. The platform can move forward or backward to produce perturbations of small, medium, or large magnitude. It can also tilt up or down, rotating at the axis of the patient's ankle. The visual surround can be made to sway in the same anterior-posterior direction as the patient. In addition to measuring the patient's anterior-posterior sway, the data can be used to produce sway-referenced signals to drive the visual surround or platform (or both) to sway according to the input from the patient. This forces the patient to ignore or to compensate for the adverse stimulation.

Data are compared with those from control subjects matched for age. Much of the analysis is based on research that indicates that humans have a *cone of stability* of 12.5 degrees for anterior-posterior sway (Fig. 31–11).[9] Thus, a person who sways about 12.5 degrees is at the limits of stability and is likely to fall. A fall is scored any time a patient reaches out and touches the visual surround to keep from falling or any time the patient moves the feet to keep the center of gravity over the base of support.

Motor Control Test

With the *motor control test* (MCT), the patient is presented with forward and backward perturbations (varying from small to medium to large) of the platform. The computer algorithms then calculate the latency of the response to each perturbation. This value is the latency of the long motor arc loop and serves as a screening test for problems in that system. The equipment is also capable of making electromyographic recordings with surface electrodes to measure the actual responses of the gastrocnemius and anterior tibialis muscles. The symmetry

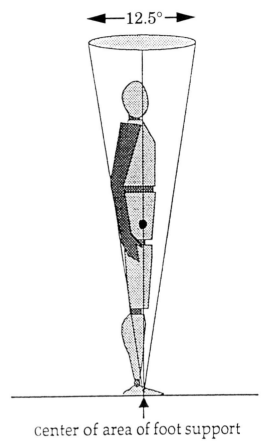

center of area of foot support

Figure 31–11. The cone of stability is 12.5 degrees and is the amount of sway against which anterior-posterior sway is judged. (From EquiTest Systems. Installing EquiTest version 4.03. NeuroCom International, Clackamas, Oregon, with permission.)

and amount of force and weight distribution are also measured and displayed with the MCT.

In the *test for adaptation,* "toes up/down," the platform tilts up or down to generate a stimulus analogous to that of walking on uneven surfaces. It is expected that patients will perform poorly on the first trial, but on the second through fifth trials, they should adapt and perform more normally.

Sensory Organization Test

The sensory organization test consists of six different conditions, with three trials possible for each condition. These conditions, including the support surface condition

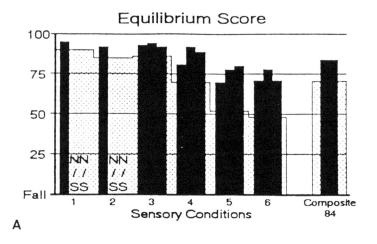

Figure 31–12. The six sensory conditions of the sensory organization test. (From EquiTest Systems. Installing EquiTest version 4.03. Neuro-Com International, Clackamas, Oregon, with permission.)

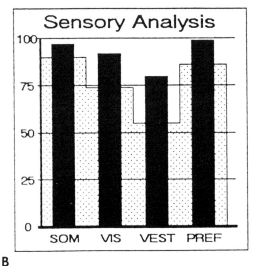

Figure 31–13. (*A*) Sensory organization equilibrium scores of normal subject for the six test conditions and the composite score. The latter is the mean of 14 scores (the mean of conditions 1 and 2, and all 12 trials of conditions 3 to 6). N/S = no score (trial was not run). (*B*) Normal sensory analysis scores. PREF = visual preference; SOM = somatosensory; VEST = vestibular; VIS = visual. Results in shaded areas are abnormal. (From EquiTest Systems. Installing EquiTest version 4.03. NeuroCom International, Clackamas, Oregon, with permission.)

(fixed or sway-referenced) and the visual condition (fixed, eyes closed, or sway-referenced) are summarized in Figure 31–12. Recall that sway-referenced means that the visual surround or platform is driven by the patient's anterior-posterior sway and the patient must either ignore or compensate for the inappropriate information. Test results for conditions 1 through 6 for a normal subject are shown in Figure 31–13. Scores approaching 100 indicate little sway, whereas those near zero represent a large amount of sway (relative to the 12.5 degrees of the cone of stability). If a patient touches the wall or moves the feet, the trial is scored as a fall (zero points). Scores falling into the shaded area are abnormal (see Fig. 31–13A). From these scores, a sensory analysis is performed (see Fig. 31–13B). The scores (ratios) are derived from comparisons of the various conditions, as shown in Figure 31–14. This information provides a fairly complete picture of a person's ability to maintain stability.

Abnormal results typical of patients with acute vestibular disorders are low scores on conditions 5 and 6, with an abnormal vestibular ratio. Patients with bilateral vestibular hypofunction generally fall "like a stick" on conditions 5 and 6 because they receive no information from the vestibular system and their vision and somatosensation are compromised. Patients who are unable to suppress inappropriate visual information function poorly on conditions 3 and 6.

One of the most effective uses for CDP is testing patients who have functional abnormalities or at least functional overlays. Because patients must cooperate to complete the test, they have ample opportunity to exaggerate their responses. Interestingly, many patients with functional problems perform relatively better on the difficult tests, such as conditions 5 and 6, and more poorly on the easier tests, such as conditions 1 and 2. Results such as these are thus physiologically inconsistent.

SENSORY ANALYSIS			
RATIO NAME	TEST CONDITIONS	RATIO PAIR	SIGNIFICANCE
SOM Somatosensory	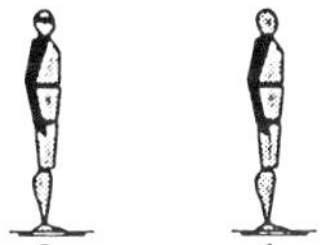 2 1	Condition 2 / Condition 1	Question: Does sway increase when visual cues are removed? Low scores: Patient makes poor use of somatosensory references.
VIS Visual	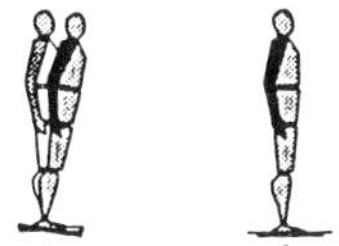 4 1	Condition 4 / Condition 1	Question: Does sway increase when somatosensory cues are inaccurate? Low scores: Patient makes poor use of visual references.
VEST Vestibular	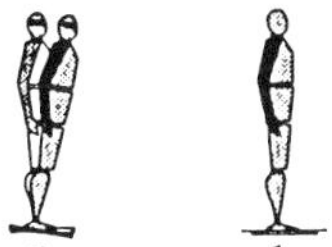 5 1	Condition 5 / Condition 1	Question: Does sway increase when visual cues are removed and somatosensory cues are inaccurate? Low scores: Patient makes poor use of vestibular cues, or vestibular cues unavailable.
PREF Visual Preference	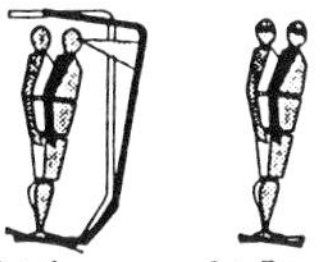 3 + 6 2 + 5	Condition 3 + 6 / Condition 2 + 5	Question: Do inaccurate visual cues result in increased sway compared to no visual cues? Low scores: Patient relies on visual cues even when they are inaccurate.

Figure 31–14. Summary of sensory analysis for six sensory conditions and the significance of their outcomes. (From EquiTest Systems. Installing EquiTest version 4.03. NeuroCom International, Clackamas, Oregon, with permission.)

VESTIBULAR REHABILITATION

CDP also provides quantitative information to help document the need for, and the results of, vestibular rehabilitation. This relatively new area has begun to provide a mechanism for helping to treat the condition of patients who previously were told they had to live with their balance problem. We have learned that the vestibular mechanism is extremely plastic. Appropriate exercise, prescribed by a physical therapist trained in vestibular rehabilitation, can help speed the recovery of many patients.

CHAPTER SUMMARY

Vertigo and imbalance are becoming more common and more serious problems, particularly as the population increases in age. The patient's history is critical in determining whether the problem is peripheral, central, systemic, or some combination of the three. Aspects of a bedside or office examination for evaluating nystagmus or the status of the vestibular ocular reflex are discussed. The physician's information can then be supplemented by laboratory examination, including the electronystagmography test, computerized rotary chair test, a modern torsion swing chair test, and computerized dynamic posturography. On the basis of information gathered from all these sources, physicians can then make a reasonable diagnosis and develop a plan for providing relief from the symptoms. In addition to surgical or medical treatment or only to monitor the problem, vestibular rehabilitation helps patients recover from their symptoms.

REFERENCES

1. Alpert, JN: Failure of fixation suppression: A pathologic effect of vision on caloric nystagmus. Neurology 24:891–896, 1974.
2. American Academy of Otolaryngology–Head and Neck Surgery: Personal communication, July 20, 1990.
3. Baloh, RW and Honrubia, V: Clinical Neurophysiology of the Vestibular System, ed. 2. FA Davis, Philadelphia, 1990.
4. Bárány, R: Physiologie und Pathologie (Funktions-Prüfung) des Bogengang-Apparatus beim Menschen. Franz Deuticke, Leipzig, 1907.
5. Barber, HO and Stockwell, CW: Manual of Electronystagmography, ed 2. CV Mosby, St. Louis, 1980, p 87.
6. Cawthorne, T: The physiologic basis for head exercises. Journal of Chartered Society of Physiotherapy 30:106–107, 1944.
7. Cyr, DG: Vestibular system assessment. In Rintelmann, WF (ed): Hearing Assessment, ed 2. Pro-ed, Austin, Texas, 1991, p 777.
8. Dix, MR and Hallpike, CS: The pathology, symptomatology and diagnosis of certain common disorders of the vestibular system. Ann Otol Rhinol Laryngol 61:987–1016, 1952.
9. Epley, JM: New dimensions of benign paroxysmal positional vertigo. Otolaryngol Head Neck Surg 88:599–605, 1980.
10. Epley, JM: The canalith repositioning procedure: For treatment of benign paroxysmal positional vertigo. Otolaryngol Head Neck Surg 107:399–404, 1992.
11. Fregly, AR: Vestibular ataxia and its measurement in man. In Kornhuber, HH (ed): Handbook of Sensory Physiology. Vol 6, part 2. Springer-Verlag, Berlin, 1974, pp 321–360.
12. Fukuda, T: The stepping test: Two phases of the labyrinthine reflex. Acta Otolaryngol (Stockh) 50:95–108, 1959.
13. Gacek, R: Transection of the posterior ampullary nerve for the relief of benign paroxysmal positional nystagmus. Ann Otol Rhinol Laryngol 83:596–605, 1974.
14. Hain, TC, Fetter, M, and Zee, DS: Head-shaking nystagmus in patients with unilateral peripheral vestibular lesions. Am J Otolaryngol 8:36–47, 1987.
15. Herdman SJ, Tusa RJ, Zee, DS, et al: Single treatment approaches to benign paroxysmal positional vertigo. Arch Otolaryngol Head Neck Surg 119:450–454, 1993.
16. Jacobson, GP, Newman, CW, and Kartush, JM: Handbook of Balance Function Testing. Mosby-Year Book, St. Louis, 1993, p 24.
17. Leigh, RJ and Zee, DS: The Neurology of Eye Movements, ed. 2. FA Davis, Philadelphia, 1991.
18. Money, KE, Myles, WS, and Hoffert, BM: The mechanism of positional alcohol nystagmus. Can J Otolaryngol 3:302–313, 1974.
19. Nylen, CO: Positional nystagmus: A review and future prospects. J Laryngol Otol 64:295–318, 1950.
20. Parnes, LS and McClure, JA: Free-floating endolymph particles: A new operative finding during posterior canal occlusion. Presented at the Meeting of the Eastern Section of the American Laryngological, Rhinological, and Otological Society, Philadelphia, February 1991.
21. Romberg, MH: Lehrbuch der Nervenkrankheiten des Menschen. Dunker, Berlin, 1946.
22. Schuknecht, HF: Positional vertigo: Clinical and experimental observations. Trans Am Acad Ophthalmol Otolaryngol 66:319–332, 1962.
23. Schuknecht, HF: Pathology of the Ear. Harvard University Press, Cambridge, 1974.
24. Schuknecht, HF and Ruby, RRF: Cupulolithiasis. Adv Otorhinolaryngol 20:434–443, 1973.

25. Semont, A, Freyss, G, and Vitte, E: Curing the BPPV with a liberatory maneuver. Adv Otorhinolaryngol 42:290–293, 1988.

26. Unterberger, S: Neue objectiv registrierbare Vestibularis-Körperdrehreaktion, erhalten durch Treten auf der Stelle: Der ''Tretversuch.'' Archiv für Ohren-, Nasen- und Kehlkopfheilkunde sowie die angrenzenden Gebiete 145:478–492, 1938.

27. Weider, DJ, Ryder, CJ, and Stram, JR: Benign paroxysmal positional vertigo: Analysis of 44 cases treated by the canalith repositioning procedure of Epley. Am J Otol 15:321–326, 1994.

SECTION 2
Electro-physiologic Assessment of Neural Function
PART F
Autonomic Function

The autonomic nervous system regulates visceral function and the internal environment of the body through its effect on the heart, gut and other internal organs, peripheral blood vessels, and sweat glands. Neural activity in the autonomic nervous system is difficult to record directly. Although sympathetic nerve function in peripheral nerves can be recorded with fine-tipped tungsten electrodes, this technique generally is not clinically useful. Therefore, assessment of autonomic function depends primarily on measuring the response of the autonomic nervous system to external stimuli. The measurement of sweating, cardiovascular activity, peripheral blood flow, and other autonomically mediated reflexes provides insight into the broad range of disorders that affect the central and peripheral components of the autonomic nervous system—from the hypothalamus to the autonomic axons in the trunk and the limbs. These measurements of autonomic function are not made as frequently as they should be. With better understanding of their importance, they will be used more often to benefit the patient.

CHAPTER 32

CLINICAL PHYSIOLOGY OF THE AUTONOMIC NERVOUS SYSTEM

Eduardo E. Benarroch, M.D.

perform, but interpretation of their results may be difficult. The effector organs react slowly to variations in neural input, and their response is affected by hormonal, local chemical, and mechanical influences. Because of complex interactions of sympathetic and parasympathetic outputs at a single target level, it is often difficult to characterize the function of a particular test as being purely sympathetic or parasympathetic. This chapter provides an overview of some aspects of autonomic function that may help in interpreting the results of noninvasive autonomic tests commonly used clinically.

The autonomic nervous system consists of *sympathetic,* or thoracolumbar, and *parasympathetic,* or craniosacral, *divisions.* Both divisions carry general visceral afferent and efferent fibers. The *enteric nervous system* is located in the wall of the gut and is considered a third division of the autonomic nervous system. Unlike the function of somatic motor or sensory nerves, autonomic nerve function is difficult to evaluate precisely in humans. In general, evaluation of autonomic function has been restricted to noninvasive recordings of heart rate, blood pressure, blood flow, or sweat production.[13,17,25,31,33,51] These tests are simple to

GENERAL ORGANIZATION OF THE AUTONOMIC SYSTEM

Visceral Afferents

Visceral receptors generally are slowly adapting mechanoreceptors or chemoreceptors that have a low level of spontaneous activity. They are innervated by small myelinated and unmyelinated fibers. Visceral afferent signals may mediate local reflexes in peripheral organs or provide collateral input to autonomic ganglion cells. However, most visceral afferent fibers enter the central ner-

379

vous system at a spinal or medullary level to initiate segmental or suprasegmental viscerovisceral reflexes or to relay visceral information to higher centers. Spinal sympathetic afferents mediate visceral pain and initiate segmental and suprasegmental viscerovisceral reflexes. Visceral afferents that enter at the medullary level are carried by the vagus (cranial nerve X) and glossopharyngeal (cranial nerve IX) nerves and relay in the nucleus of the tractus solitarius (NTS). This nucleus is critically involved in respiratory and cardiovascular reflexes through its extensive connections with neurons in the so-called intermediate reticular zone of the medulla.[30]

Visceral Efferents

Sympathetic and parasympathetic autonomic outflow systems initiate responses that are longer in latency and duration and more continuous and generalized than those mediated by the somatic motor system. This reflects important differences in the functional organization of autonomic and somatic efferents. Autonomic output involves a two-neuron pathway that has at least one synapse in the autonomic ganglia. The preganglionic axons are small myelinated (type B) cholinergic fibers that emerge from neurons of the general visceral efferent column in the brain stem or spinal cord and pass through peripheral nerves to the autonomic ganglia.

Postganglionic autonomic neurons send varicose, unmyelinated (type C) axons to innervate peripheral organs. Autonomic terminals contain varicosities. *The primary postganglionic neurotransmitter at most sympathetic neuroeffector junctions is norepinephrine*, which acts using different subtypes of adrenergic receptors. In the sweat glands, sympathetic effects are mediated by acetylcholine. *The primary postganglionic neurotransmitter at all parasympathetic neuroeffector junctions is acetylcholine*, which acts through different subtypes of muscarinic receptors. Whereas the main consequence of denervation in striated muscle is paralysis and atrophy, most autonomic effectors develop denervation supersensitivity in response to postganglionic efferent denervation. Thus, *denervation super-*

sensitivity is evidence of a lesion involving postganglionic neurons.

Activity of most autonomic effectors is modulated by dual, continuous sympathetic and parasympathetic controls. Sympathetic control originates from preganglionic neurons in segments T-2 to L-1 of the cord and predominates in blood vessels, sweat glands, and cardiac muscle. Parasympathetic outflow originates from preganglionic neurons in nuclei of cranial nerves III, VII, IX, and X and in segments S-2 to S-4 of the spinal cord. It predominates in control of the salivary glands, sinoatrial node, gastrointestinal tract, and bladder. Sympathetic-parasympathetic interactions are not simply antagonistic but are functionally complementary. The sympathetic and parasympathetic systems may interact at several levels, including that of the central nervous system, autonomic ganglia, neuroeffector junction, and target organ.

SYMPATHETIC FUNCTION

Functional Anatomy of the Sympathetic Outflow

Preganglionic sympathetic neurons have a slow, irregular, tonic activity that depends mainly on multiple segmental afferent and descending inputs. Spinal sympathetic activity is not diffuse; therefore, there is no generalized sympathetic tone. On the contrary, preganglionic sympathetic neurons are organized into different *spinal sympathetic functional units* that control specific targets and are differentially influenced by input from the hypothalamus and brain stem.[11,23] Sympathetic functional units include skin vasomotor, muscle vasomotor, visceromotor, pilomotor, and sudomotor units.[24] Thus, there are clear differences between the sympathetic outflows to the skin and muscle, but similarities in sympathetic activity recorded simultaneously from different muscles at rest or from skin sympathetic nerves innervating the palms and the feet are remarkable.[43,46]

Preganglionic sympathetic output has a segmental organization. Segmental distribution of preganglionic fibers does not follow the dermatomal pattern of somatic nerves. For example, sudomotor functional units in segments T-1 and T-2 innervate the head and neck, units in T-3 to T-6 innervate the upper extremities and thoracic viscera, those in T-7 to T-11 innervate the abdominal viscera, and others in T-12 to L-2 innervate the lower extremities and pelvic and perineal organs. Preganglionic sympathetic axons exit through ventral roots and pass through the white ramus communicans of the corresponding spinal nerve to reach the paravertebral sympathetic chain.

Ganglia termed sympathetic include the paravertebral (sympathetic trunk) and prevertebral (autonomic plexus). Paravertebral ganglia innervate all tissues and organs except those in the abdomen, pelvis, and perineum. Their postganglionic fibers that are destined for the trunk and limbs follow the course of spinal nerves, blood vessels, or both. Spinal fibers join the peripheral spinal (somatic) nerve through the gray ramus communicans. These fibers provide vasomotor, sudomotor, and pilomotor input to the extremities and trunk. Sympathetic fibers are intermingled with somatic motor and sensory fibers, and their distribution is similar to that of the corresponding somatic nerve. Most sympathetic fibers are destined for the hand and foot and are carried mainly by the median, peroneal, tibial, and, to a lesser extent, ulnar nerves.

Sympathetic Function in Humans

Microneurographic technique allows recording of postganglionic sympathetic nerve activity in humans.[43,46] Nerve recordings are made with tungsten microelectrodes inserted percutaneously into an underlying nerve, especially the median, peroneal, and tibial nerves. This technique allows multiunit recordings of two different types of outflow: muscle sympathetic nerve activity and skin sympathetic nerve activity.[46]

SYMPATHETIC INNERVATION OF SKIN

Sympathetic vasomotor, pilomotor, and sudomotor innervation of skin effectors has a primarily thermoregulatory function.[44,50] Skin sympathetic activity is a mixture of su-

domotor and vasoconstrictor impulses and may sometimes also include pilomotor and vasodilator impulses. The average conduction velocity for skin sudomotor and vasomotor fibers is 1.3 m/s and 0.8 m/s, respectively.[18] The intensity of skin sympathetic activity is determined mainly by environmental temperature and the emotional state of the subject. Decreased or increased environmental temperature can produce selective activation of the vasoconstrictor or sudomotor system, respectively, with suppression of activity in the other system.[24,46] Emotional stimuli or inspiratory gasp also increases spontaneous skin sympathetic activity, but in this case, the bursts are due to simultaneous activation of sudomotor and vasomotor impulses.[46]

There is a prominent noradrenergic innervation to cutaneous arteries and veins, which regulates both nutritive and arteriovenous skin blood flow.[21] Nutritive skin flow is carried by capillaries and is regulated by sympathetic (alpha- and beta-adrenergic) and local nonadrenergic mediators. Arteriovenous skin blood flow is carried by low-resistance arteriovenous shunts, which receive abundant sympathetic vasoconstrictor input and play a key role in thermoregulation.[21] Skin vasoconstrictor neurons may coordinate their activity with vasomotor neurons in other vascular beds to maintain cardiac output, but they are not sensitive to baroreflex input. Skin blood flow is also controlled by somatosympathetic reflexes[39,40] and three local axon reflexes: (1) axon flare response, (2) sudomotor axon reflex, and (3) venoarteriolar reflex. The axon flare response is mediated by nociceptive C-fiber terminals. Activation by noxious chemical or mechanical stimuli produces antidromic release of neuropeptides (substance P and others) that cause skin vasodilatation directly and through stimulation of histamine release by mast cells. The sudomotor axon reflex is mediated by sympathetic sudomotor C fibers. The venoarteriolar reflex is mediated by sympathetic vasomotor axons innervating small veins and arterioles. Skin vasomotor activity has been studied clinically by using several noninvasive methods for measuring skin blood flow, including plethysmography[21] and laser Doppler flowmetry.[33]

Eccrine sweat glands in humans have a

major role in thermoregulation. The segmental pattern of distribution of sudomotor fibers to the trunk and limbs is irregular and varies substantially among individuals and even between the right and left sides of an individual.

MUSCLE SYMPATHETIC ACTIVITY

Muscle sympathetic activity is composed of vasoconstrictor impulses that are strongly modulated by arterial baroreceptors.[16,46,47] Conduction velocity of the postganglionic C fibers has been estimated to be 0.7 m/s in the median nerve and 1.1 m/s in the peroneal nerve.[18] At rest, there is a striking similarity between muscle sympathetic activity recorded in different extremities.[18,43,46] However, this activity in the arm and leg can be dissociated during mental stress[5] and during forearm ischemia after isometric exercise.[49]

Muscle sympathetic activity is important for buffering acute changes of blood pressure and decreases in response to baroreceptor influence. However, it has much less importance for long-term control of blood pressure; the mean level of muscle sympathetic activity does not correlate with mean arterial pressure.[43,46] At rest, muscle sympathetic activity correlates positively with antecubital venous plasma norepinephrine levels.[48]

Muscle sympathetic activity is also inhibited by cardiopulmonary receptors. Respiratory cycle, changes of posture, or the Valsalva maneuver may affect muscle sympathetic activity secondary to changes in arterial pressure. On the other hand, hypercapnia, hypoxia, isometric handgrip, emotional stress, or the cold pressor test increases muscle sympathetic activity despite unchanged or increased arterial pressure.[46]

AUTONOMIC CONTROL OF HEART RATE

Heart rate depends on the effects of parasympathetic and sympathetic modulation of the intrinsic firing rate of the sinus node.[10,28,29] Parasympathetic control is provided by cardiovagal neurons in the nucleus ambiguus and dorsal motor nucleus in the medulla.[4,9] Sympathetic noradrenergic control derives mainly from the cervical and upper thoracic ganglia. In humans at rest, parasympathetic tone predominates over the excitatory sympathetic beta-adrenergic influence.[27–29] Effects mediated by the vagus have a shorter latency and duration than those mediated by the sympathetic nerves.[29]

Heart rate has spontaneous fluctuations, which reflect changing levels of autonomic activity modulating sinus-node discharge. Use of power spectral analysis of heart rate fluctuations allows a noninvasive quantitative assessment of beat-to-beat modulation of neuronal activity affecting the heart.[2,3,7,35–37] Fluctuations of heart rate at respiratory frequencies (approximately 0.15 Hz) are mediated almost exclusively by the vagus.[2,3,7,36,37] Vagal influences on the heart are associated directly with respiratory activity and are minimal during inspiration and maximal at the end of inspiration and in early expiration.[20] This is the basis of the *respiratory sinus arrhythmia.*[6,13–15,22,45] Spontaneous and baroreflex-induced firing of central cardiovagal neurons is inhibited during inspiration and is maximal during early expiration.[13,15] Increased tidal volume increases respiratory sinus arrhythmia, whereas increased respiratory frequency decreases it.[22] Heart rate variability is inversely correlated with age in resting normal subjects.[17]

Spontaneous lower frequency fluctuations (those less than 0.15 Hz) of heart rate are mediated by both the vagus and the sympathetic nerves and may be related to baroreflex activity, temperature regulation, and other factors.[7,36,37] Upright posture in humans dramatically increases sympathetic nerve activity as well as a large increase in low-frequency heart rate power.[36,37,51]

CARDIOVASCULAR REFLEXES

Arterial Baroreflexes

In normal humans, control of arterial pressure depends primarily on the sympathetic innervation of the blood vessels, particularly in the splanchnic bed.[38] The number of splanchnic sympathetic preganglionic neurons progressively decreases with age, and orthostatic hypotension occurs after more than 50% of them have been lost.[32] Changes in sympathetic outflow are regulated by arterial baroreceptors and chemoreceptors, cardiopulmonary receptors, and receptors in skeletal muscles (i.e., ergoreceptors).[42] In addition, these reflexes are modulated by central commands, particularly from the amygdala and hypothalamus. In humans, arterial pressure is regulated primarily by the carotid sinus and aortic baroreceptors innervated by branches of cranial nerves IX and X, respectively, and by cardiopulmonary mechanoreceptors innervated by vagal and sympathetic afferents.

The primary role of arterial baroreflexes is the rapid adjustment of arterial pressure around the existing mean arterial pressure.[1,34,42] They provide negative feedback that buffers the magnitude of arterial pressure oscillations throughout the day. Baroreflexes induce short-term changes in heart rate opposite in direction to the changes in arterial pressure, thus increasing heart rate variability.

The carotid baroreflex has been studied by applying negative and positive pressures to the neck, which increase and decrease, respectively, carotid-sinus transmural pressure.[12,13,34,47] Combined influence of carotid and aortic baroreceptors in control of heart rate has been studied by measuring heart rate responses to changes in arterial pressure induced by intravenous infusion of vasoconstrictor or vasodilator agents. Despite their major influence on heart rate, the buffering effects of the carotid baroreflex depend predominantly on changes in total peripheral resistance. Baroreflex control of regional circulation is heterogeneous and largely affects resistance vessels in the splanchnic area.[26] Sympathetic vasoconstriction in skeletal muscle is also strongly modulated by the baroreflex; but this control is dynamic and more suitable for buffering short-term than long-term variations of arterial pressure.[46,47] During exercise, carotid baroreceptor activity is rapidly adjusted to a higher level; this allows increased arterial pressure to meet metabolic demands of the contracting muscles.[8]

Cardiopulmonary Reflexes

Cardiopulmonary receptors are innervated by vagal and sympathetic myelinated and unmyelinated afferent fibers.[34,41,42] Atrial receptors innervated by vagal myelinated fibers are activated by atrial distention or contraction and initiate reflex tachycardia caused by selective increase of sympathetic outflow to the sinus node. Cardiopulmonary receptors with unmyelinated vagal afferents, similar to arterial baroreceptors, tonically inhibit vasomotor activity. Unlike baroreceptors, cardiopulmonary receptors provide sustained rather than phasic control in sympathetic activity and vasomotor tone in the muscle and have no major effect in controlling heart rate.[34]

Venoarteriolar Reflexes

The venoarteriolar reflex is a sympathetic postganglionic C-fiber axon reflex, with receptors in small veins and effectors in muscle arterioles. Venous pooling activates receptors in small veins of the skin, muscle, and adipose tissue; the result is vasoconstriction in the arterioles supplying these tissues. During limb dependency, this local reflex vasoconstriction may decrease blood flow by 50%. The main function of the venoarteriolar reflexes is to increase total peripheral resistance.

Ergoreflexes

Static muscle contraction increases heart rate and blood pressure. The mechanisms underlying these responses are thought to involve (1) reflexes initiated by activation of chemosensitive endings of small myelinated and unmyelinated afferent fibers by local metabolites in the contracting muscle and (2) a central command that influences descending autonomic pathways. Cardiovascular responses to moderately intense static contraction may be produced primarily by the motor command, which is solely responsible for increased heart rate. At higher intensity, responses depend on both the motor command and muscle chemoreflexes.[19]

CHAPTER SUMMARY

The autonomic nervous system consists of three divisions: sympathetic or thoracolumbar, parasympathetic or craniosacral, and the enteric nervous system. The sympathetic and parasympathetic autonomic outflow involves a two-neuron pathway with a synapse in an autonomic ganglion. Preganglionic sympathetic neurons are organized into different functional units that control specific targets and include skin vasomotor, muscle vasomotor, visceromotor, pilomotor, and sudomotor units. Microneurographic technique allows recording of postganglionic sympathetic nerve activity in humans. Skin sympathetic activity is a mixture of sudomotor and vasoconstrictor impulses and is regulated mainly by environmental temperature and emotional influences. Muscle sympathetic activity is composed of vasoconstrictor impulses that are strongly modulated by arterial baroreceptors. Heart rate is controlled by vagal parasympathetic and thoracic sympathetic inputs. Vagal influence on the heart rate is strongly modulated by respiration; it is more marked during expiration and is absent during inspiration. This is the basis for the so-called *respiratory sinus arrhythmia*, which is an important index of vagal innervation of the heart. Power spectral analysis of heart rate fluctuations allows a noninvasive assessment of beat-to-beat modulation of neuronal activity affecting the heart. Arterial baroreflex, cardiopulmonary reflexes, venoarteriolar reflex, and ergoreflexes control sympathetic and parasympathetic influences on cardiovascular effectors.

REFERENCES

1. Abboud, FM: Interaction of cardiovascular reflexes in humans. In Lown, B, Malliani, A, and Prosdocimi, M (eds): Neural Mechanisms and Cardiovascular Disease. Liviana Press, Padova, Italy, 1986, pp 73–84.
2. Akselrod, S, Gordon, D, Madwed, JB, et al: Hemodynamic regulation: Investigation by spectral analysis. Am J Physiol 249:H867–H875, 1985.
3. Akselrod, S, Gordon, D, Ubel, FA, et al: Power spectrum analysis of heart rate fluctuation: A quantitative probe of beat-to-beat cardiovascular control. Science 213:220–222, 1981.
4. Ammons, WS, Blair, RW, and Foreman, RD: Vagal afferent inhibition of primate thoracic spinothalamic neurons. J Neurophysiol 50:926–940, 1983.
5. Anderson, EA, Wallin, BG, and Mark, AL: Dissociation of sympathetic nerve activity in arm and leg muscle during mental stress. Hypertension 9 Suppl 3:III-114–III-119, 1987.
6. Angelone, A and Coulter, NA Jr: Respiratory sinus arrhythmia: A frequency dependent phenomenon. J Appl Physiol 19:479–482, 1964.
7. Berger, RD, Akselrod, S, Gordon, D, and Cohen, RJ: An efficient algorithm for spectral analysis of heart rate variability. IEEE Trans Biomed Eng BME-33:900–904, 1986.
8. Christensen, NJ and Galbo, H: Sympathetic nervous activity during exercise. Ann Rev Physiol 45:139–153, 1983.
9. Ciriello, J and Calaresu, FR: Medullary origin of vagal preganglionic axons to the heart of the cat. J Auton Nerv Syst 5:9–22, 1982.
10. Clarke, JM, Hamer, J, Shelton, JR, et al: The rhythm of the normal human heart. Lancet 2:508–512, 1976.
11. Dampney, RA: Functional organization of central pathways regulating the cardiovascular system. Physiol Rev 74:323–364, 1994.
12. Eckberg, DL: Baroreflex inhibition of the human sinus node: Importance of stimulus intensity, duration, and rate of pressure change. J Physiol (Lond) 269:561–577, 1977.
13. Eckberg, DL: Parasympathetic cardiovascular control in human disease: A critical review of methods and results. Am J Physiol 239:H581–H593, 1980.
14. Eckberg, DL: Human sinus arrhythmia as an index of vagal cardiac outflow. J Appl Physiol 54:961–966, 1983.
15. Eckberg, DL, Nerhed, C, and Wallin, BG: Respiratory modulation of muscle sympathetic and vagal cardiac outflow in man. J Physiol (Lond) 365:181–196, 1985.
16. Eckberg, DL, Rea, RF, Andersson, OK, et al: Baroreflex modulation of sympathetic activity and sympathetic neurotransmitters in humans. Acta Physiol Scand 133:221–231, 1988.
17. Ewing, DJ: Analysis of heart rate variability and other non-invasive tests with special reference to diabetes mellitus. In Bannister, R and Mathias, CJ (eds): Autonomic Failure: A Textbook of Clinical Disorders of the Autonomic Nervous System, ed 3. Oxford University Press, Oxford, 1992, pp 312–333.
18. Fagius, J and Wallin, BG: Sympathetic reflex latencies and conduction velocities in normal man. J Neurol Sci 47:433–448, 1980.
19. Gandevia, SC and Hobbs, SF: Cardiovascular responses to static exercise in man: Central and reflex contributions. J Physiol (Lond) 430:105–117, 1990.
20. Gilbey, MP, Jordan, D, Richter, DW, and Spyer, KM: Synaptic mechanisms involved in the inspiratory modulation of vagal cardio-inhibitory neurones in the cat. J Physiol (Lond) 356:65–78, 1984.
21. Henriksen, O: Local sympathetic reflex mechanism in regulation of blood flow in human subcutaneous adipose tissue. Acta Physiol Scand Suppl 450:7–48, 1977.
22. Hirsch, JA and Bishop, B: Respiratory sinus arrhythmia in humans: How breathing pattern modulates heart rate. Am J Physiol 241:H620–H629, 1981.

23. Jänig, W: Pre- and postganglionic vasoconstrictor neurons: Differentiation, types, and discharge properties. Ann Rev Physiol 50:525–539, 1988.

24. Jänig, W, Sundlöf, G, and Wallin, BG: Discharge patterns of sympathetic neurons supplying skeletal muscle and skin in man and cat. J Auton Nerv Syst 7:239–256, 1983.

25. Johnson, JM: Nonthermoregulatory control of human skin blood flow. J Appl Physiol 61:1613–1622, 1986.

26. Johnson, JM, Rowell, LB, Niederberger, M, and Eisman, MM: Human splanchnic and forearm vasoconstrictor responses to reductions of right atrial and aortic pressures. Circ Res 34:515–524, 1974.

27. Jose, AD and Collison, D: The normal range and determinants of the intrinsic heart rate in man. Cardiovasc Res 4:160–167, 1970.

28. Levy, MN: Cardiac sympathetic-parasympathetic interactions. Fed Proc 43:2598–2602, 1984.

29. Levy, MN and Martin, PJ: Neural control of the heart. In Geiger, SR (ed): Handbook of Physiology. Section 2: The Cardiovascular System. Vol 1. American Physiological Society, Bethesda, Maryland, 1979, pp 581–620.

30. Loewy, AD: Anatomy of the autonomic nervous system: An overview. In Loewy, AD and Spyer, KM (eds): Central Regulation of Autonomic Functions. Oxford University Press, New York, 1990, pp 3–16.

31. Low, PA, Caskey, PE, Tuck, RR, et al: Quantitative sudomotor axon reflex test in normal and neuropathic subjects. Ann Neurol 14:573–580, 1983.

32. Low, PA and Dyck, PJ: Splanchnic preganglionic neurons in man. III. Morphometry of myelinated fibers of rami communicantes. J Neuropathol Exp Neurol 37:734–740, 1978.

33. Low, PA, Newmann, C, Dyck, PJ, et al: Evaluation of skin vasomotor reflexes by using laser Doppler velocimetry. Mayo Clin Proc 58:583–592, 1983.

34. Mancia, G, Grassi, G, Ferrari, A, and Zanchetti, A: Reflex cardiovascular regulation in humans. J Cardiovasc Pharmacol 7 Suppl 3:S152–S159, 1985.

35. Novak, P, Novak, V, Li, Z, and Remillard, G: Time-frequency analysis of slow cortical activity and cardiovascular fluctuations in a case of Alzheimer's disease. J Clin Auton Res 4:141–148, 1994.

36. Pagani, M, Lombardi, F, Guzzetti, S, et al: Power spectral analysis of heart rate and arterial pressure variabilities as a marker of sympatho-vagal interaction in man and conscious dog. Circ Res 59:178–193, 1986.

37. Pomeranz, B, Macaulay, RJB, Caudill, MA, et al: Assessment of autonomic function in humans in heart rate spectral analysis. Am J Physiol 248:H151–H153, 1985.

38. Rowell, LB, Detry, J-MR, Blackmon, JR, and Wyss, C: Importance of the splanchnic vascular bed in human blood pressure regulation. J Appl Physiol 32:213–220, 1972.

39. Sato, A and Schmidt, RF: Somatosympathetic reflexes: Afferent fibers, central pathways, discharge characteristics. Physiol Rev 53:916–947, 1973.

40. Shahani, BT, Halperin, JJ, Boulu, P, and Cohen, J: Sympathetic skin response—a method of assessing unmyelinated axon dysfunction in peripheral neuropathies. J Neurol Neurosurg Psychiatry 47:536–542, 1984.

41. Shepherd, JT and Mancia, G: Reflex control of the human cardiovascular system. Rev Physiol Biochem Pharmacol 105:1–99, 1986.

42. Shepherd, RFJ and Shepherd, JT: Control of blood pressure and the circulation in man. In Bannister, R and Mathias, CJ (eds): Autonomic Failure: A Textbook of Clinical Disorders of the Autonomic Nervous System, ed 3. Oxford University Press, Oxford, 1992, pp 78–93.

43. Somers, VK, Dyken, ME, Mark, AL, and Abboud, FM: Sympathetic-nerve activity during sleep in normal subjects. N Engl J Med 328:303–307, 1993.

44. Uno, H: Sympathetic innervation of the sweat glands and piloarrector muscles of macaques and human beings. J Invest Dermatol 69:112–120, 1977.

45. van Ravenswaaij-Arts, CM, Kollee, LA, Hopman, JC, et al: Heart rate variability. Ann Intern Med 118:436–447, 1993.

46. Wallin, BG and Fagius, J: Peripheral sympathetic neural activity in conscious humans. Ann Rev Physiol 50:565–576, 1988.

47. Wallin, BG, Sundlöf, G, and Delius, W: The effect of carotid sinus nerve stimulation on muscle and skin nerve sympathetic activity in man. Pflugers Arch 358:101–110, 1975.

48. Wallin, BG, Sundlöf, G, Eriksson, B-M, et al: Plasma noradrenaline correlates to sympathetic muscle nerve activity in normotensive man. Acta Physiol Scand 111:69–73, 1981.

49. Wallin, BG, Victor, RG, and Mark, AL: Sympathetic outflow to resting muscles during static handgrip and postcontraction muscle ischemia. Am J Physiol 256:H105–H110, 1989.

50. Wyss, CR, Brengelmann, GL, Johnson, JM, et al: Control of skin blood flow, sweating, and heart rate: Role of skin vs. core temperature. J Appl Physiol 36:726–733, 1974.

51. Yutaka, A, Saul, JP, Albrecht, P, et al: Modulation of cardiac autonomic activity during and immediately after exercise. Am J Physiol 256:H132–H141, 1989.

CHAPTER 33

QUANTITATIVE SUDOMOTOR AXON REFLEX TEST AND RELATED TESTS

Phillip A. Low, M.D.

LABORATORY EVALUATION OF AUTONOMIC FUNCTION

The *quantitative sudomotor axon reflex test* (QSART) is a routine test of autonomic function (Table 33–1). A detailed description of tests of autonomic function is beyond the scope of this chapter but is considered elsewhere.[7] Because autonomic tests are affected significantly by many compounding variables, standardization,[13] recognition of pitfalls,[8] and patient preparation are critically important (Table 33–2). The autonomic reflex screen noninvasively and quantitatively evaluates sudomotor, adrenergic, and cardiovagal functions. Sudomotor function is discussed in this chapter, and adrenergic and cardiovagal functions are considered in subsequent chapters.

Indications for autonomic testing are summarized in Table 33–3. The presence of generalized autonomic failure, a serious medical problem, adversely affects prognosis. It is important to quantitate the deficits by system and by autonomic level. Certain benign disorders, such as chronic idiopathic anhidrosis[10] and benign syncopes, need to be differentiated from the more serious dysautonomias. Distal small fiber neuropathy is diagnosable by autonomic studies in 80% of patients.[20] Sympathetically maintained pain is associated with unilateral vasomotor and sudomotor abnormalities[11] and is described in detail elsewhere.[7]

QUANTITATIVE SUDOMOTOR AXON REFLEX TEST (QSART)

Many tests of sudomotor function of historical and limited clinical and research interest exist. For modern clinical neurophysiology laboratories, only the thermoregulatory sweat test, QSART, pe-

Table 33–1. **ROUTINE TESTS OF AUTONOMIC FUNCTION**

Autonomic reflex screen
   QSART distribution
   Orthostatic blood pressure and heart rate
      response to tilt
   Heart rate response to deep breathing
   Valsalva ratio
   Beat-to-beat blood pressure response to
      Valsalva maneuver, tilt, and deep
      breathing
Reflex sympathetic dystrophy screen
   Telethermographic skin temperature
      distribution
   Comparative resting sweat-output study
   Comparative QSART study

QSART = Quantitative sudomotor axon reflex test.

ripheral autonomic surface potential, and sweat imprint test (SIT) require consideration. The thermoregulatory sweat test is described in Chapter 35. QSART and SIT are described later in this chapter.

Normal Response

The neural pathway evaluated by QSART consists of an axon reflex mediated by the postganglionic sympathetic sudomotor axon. The axon terminal is activated by iontophoresed acetylcholine. The impulse travels first antidromically to a branch point, and then orthodromically to release acetylcholine from the nerve terminal. Acetylcholine traverses the neuroglandular junction

Table 33–2. **PATIENT PREPARATION FOR AUTONOMIC STUDIES**

- No food for 3 hours before testing. The antecedent meal should be a light breakfast or lunch without coffee or tea.
- Treatment with anticholinergic, diuretic, and, where practicable, vasoactive agents should be discontinued at least 48 hours and preferably 4 days before the study.
- The patient should be comfortable and pain-free (e.g., bladder recently emptied).
- The room should be warm and quiet.

Table 33–3. **INDICATIONS FOR AUTONOMIC LABORATORY EVALUATION**

Strong
   Suspicion of generalized autonomic failure
   Diagnosis of benign autonomic disorders
      that mimic life-threatening disorders
   Detection of distal small-fiber neuropathy
   Diagnosis of autonomic neuropathy
   Sympathetically maintained pain
   Postural orthostatic tachycardia syndrome
Moderate
   Monitoring course of autonomic failure
   Evaluation of response to therapy
   Peripheral neuropathies
   Syncope
   Amyotrophic lateral sclerosis
   Extrapyramidal and cerebellar
      degenerations
   Research questions

and binds to muscarinic receptors on eccrine sweat glands to evoke the sweat response.

Equipment needed to perform QSART includes the sudorometer, a multicompartmental sweat cell, a constant current generator, and some method of displaying the sweat response. We build our own sudorometer consoles; each consists of four sudorometers, thus permitting dynamic recording of sweat output from four sites simultaneously. Single or dual water-loss units are available commercially (Abrams Instruments, Okemos, Michigan).

The multicompartmental sweat cell (Fig. 33–1) is attached to the skin and permits iontophoresis of acetylcholine through the stimulus compartment (C) and a constant-current generator. Axon-reflex-mediated sweat response is recorded from compartment A.

QSART responses are recorded from standard sites, which include the distal forearm, proximal leg, distal leg, and proximal foot. The tests are sensitive and reproducible in control subjects[9] and in patients with diabetic neuropathy. The coefficient of variation is between 10% and 25%.

Extensive control data are available. We studied QSART responses in 139 normal subjects (74 females and 65 males) between

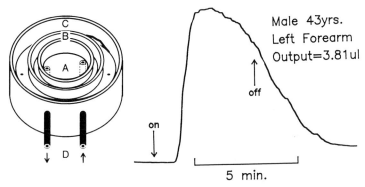

Figure 33–1. Quantitative sudomotor axon reflex test (QSART) sweat cell (*left*) and QSART response (*right*). The sweat response in compartment *A* is evoked in response to iontophoresis of acetylcholine in compartment *C*. Compartment *B* is an air gap. (*Right*) Sweating causes a change in thermal mass of the nitrogen stream (*D*) that is sensed by the sudorometer and displayed. (From Low, PA: Sudomotor function and dysfunction. In Asbury, AK, McKhann, GM, and McDonald, WI [eds]: Diseases of the Nervous System: Clinical Neurobiology, Vol 1. WB Saunders, Philadelphia, 1986, pp 596–605, with permission of Butterworth-Heinemann.)

10 and 83 years old.[11,12] Mean sweat output was 3.01 μL/cm^2 and 1.15 μL/cm^2 for the forearm of males and females, respectively. This difference was significant ($p < 0.001$). Values for the foot were 2.65 μL/cm^2 and 1.15 μL/cm^2 for males and females, respectively. Again, the difference was statistically significant ($p < 0.001$). Thus, the two groups were considered separately. For the forearm, QSART response did not regress with age. For the foot response, however, there was a consistent, albeit small, negative slope of QSART responses with increasing age.[7,12]

Abnormal Patterns

Several abnormal QSART patterns are recognized (Fig. 33–2). The response patterns may be (1) normal, (2) reduced, (3) absent, (4) excessive, (5) persistent, or (6) associated with ultrashort latencies. Short latencies are commonly seen with patterns 4 and 5. Pattern 5 consists of a "hung-up" sweat response that fails to turn off when the stimulus ceases; it is often seen in painful diabetic and other neuropathies and in florid reflex sympathetic dystrophy.

Significance of the Test

Normal findings on QSART indicate integrity of the postganglionic sympathetic sudomotor axon.

Absent response indicates failure of this axon, providing that iontophoresis is successful and that eccrine sweat glands are present. Because the axonal segment mediating the axon is likely to be short, the test probably evaluates distal axonal function.

The presence of persistent sweat activity occurs commonly in mild or painful neuropathies. Because it is unlikely that continuous activity occurs at the level of the sweat gland, the mechanism of persistent sweat activity may be due to repetitive firing of the sympathetic axon. Damaged axons can fire spontaneously, and on stimulation they may have more persistent activity than normal axons.

The latency is sometimes markedly decreased in the context of painful neuropathies with persistent sweat activity. The mechanism of short latency (10 to 30 seconds) is likely due to augmented somatosympathetic reflexes with a reduced threshold of polymodal C nociceptors—it often occurs with allodynia.

Applications of QSART

QSART has been used to detect postganglionic sudomotor failure in neuropathies,[9,14] aging,[12] Lambert-Eaton myasthenic syndrome,[15] amyotrophic lateral sclerosis,[6] and preganglionic neuropathies with pre-

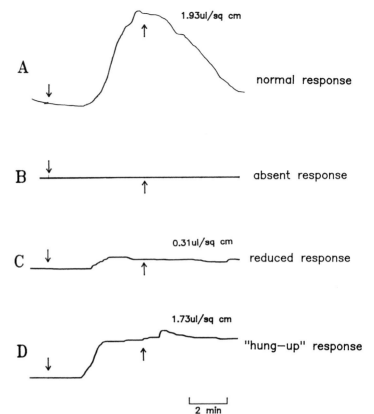

1.93ul/sq cm

A normal response

B absent response

0.31ul/sq cm

C reduced response

1.73ul/sq cm

D "hung—up" response

2 min

Figure 33–2. (*A to D*) Types of responses on quantitative sudomotor axon reflex tests. The response may be normal, absent, reduced, or show persistent sweat activity ("hung-up" response). (From Low,[7] p 175, with permission.)

sumed transsynaptic degeneration.[1,18] In patients with distal small fiber neuropathy, it is the most sensitive diagnostic test.[20] It is an important part of the autonomic reflex screen, when QSART distribution is taken in conjunction with cardiovascular heart rate tests and Finapres recordings of the Valsalva maneuver and tilt.[7] In studies of sympathetically maintained pain, QSART recordings are performed bilaterally and simultaneously for evidence of asymmetry of latency, volume, and morphology of responses.[11]

Another application of QSART is detection of the site of the lesion. QSART in combination with the thermoregulatory sweat test can be used to determine whether a lesion is preganglionic or postganglionic. A preganglionic site is deduced when anhidrosis on the thermoregulatory sweat test is associated with normal findings on QSART, and the lesion is postganglionic when both tests show anhidrosis.

Imprint Methods of Sweat Measurement

Several imprint methods have been used. The subject has been reviewed recently.[4]

STARCH-IODINE METHOD

The starch-iodine method was introduced by Minor[16] in 1927 and consists of the application of an iodine solution (2 g iodine, 10 mL castor oil, and alcohol to 100 mL). Sweat droplets appear violet-black. Wada[22] applied a 3% solution of iodine and potassium iodide in 95% ethanol and, after it was dry, applied a thin film of a corn starch and castor oil mixture (50 g corn starch and 100 mL castor oil).

BROMOPHENOL-BLUE METHOD

The pH indicator bromophenol-blue (BPB) is applied in a powder base. It is simple to use but sweat droplets run together so that the test is good only for detecting the onset of sweating.[2] Harris, Polk, and Willis[2] improved the method by using a 5% mixture of BPB in stopcock grease mixed with a dispersing agent.

PLASTIC AND SILICONE IMPRINTS

A colloidal graphite plastic method was introduced in 1952.[21] The plastic is initially dissolved in a solvent, which dries in about 30 seconds. The sweat droplet forms an imprint. Alternatives are silicone rubber monomers[19] and colloidal graphite plastic.[2] Harris, Polk, and Willis[2] compared the BPB, colloidal graphite plastic, and silicone monomer imprint methods and determined the last was the most sensitive and reliable.

Kennedy and associates[5] used a combination of pilocarpine iontophoresis, the Silastic imprint method, and imaging techniques to study the sweat response in humans and rodents systematically. For human studies, a 1% solution of pilocarpine is applied iontophoretically for 5 minutes over a 1-cm^2 area of skin on the dorsum of the hand and foot. The imprint material is spread thinly over the skin. The material sets within 3 minutes.

Counts of sweat gland density are made under a dissecting microscope and counting grid. To estimate volume, the size of the droplet can be determined using computerized imaging analysis techniques. The mean density is 311 and 281 active glands/cm^2 for the hand and the foot, respectively, without differences for age or sex.[3]

This technique has been applied systematically to patients with diabetic neuropathy.[3,5] Counts of active sweat glands were abnormal in 24% for the hand and 56% for the foot of patients with diabetic neuropathy. Of interest is that approximately one third of patients with normal findings on electrophysiologic tests have abnormal results on SIT.[4,17]

CHAPTER SUMMARY

In conclusion, application of noninvasive, sensitive, quantitative, and dynamic tests of sudomotor function significantly enhances our ability to quantitate one aspect of an au-

tonomic deficit. Tests offer considerable promise in applications such as the better definition of the course of neuropathy, its response to treatment, and in further exploration of sudomotor physiology.

REFERENCES

1. Cohen, J, Low, P, Fealey, R, et al: Somatic and autonomic function in progressive autonomic failure multiple system atrophy. Ann Neurol 22:692–699, 1987.
2. Harris, DR, Polk, BF, and Willis, I: Evaluating sweat gland activity with imprint techniques. J Invest Dermatol 58:78–84, 1972.
3. Kennedy, WR and Navarro, X: Sympathetic sudomotor function in diabetic neuropathy. Arch Neurol 46:1182–1186, 1989.
4. Kennedy, WR and Navarro, X: Evaluation of sudomotor function by sweat imprint methods. In Low, PA (ed): Clinical Autonomic Disorders: Evaluation and Management. Little, Brown & Company, Boston, 1993, pp 253–261.
5. Kennedy, WR, Sakuta, M, Sutherland, D, and Goetz, FC: Quantitation of the sweating deficiency in diabetes mellitus. Ann Neurol 15:482–488, 1984.
6. Litchy, WJ, Low, PA, Daube, JR, and Windebank, AJ: Autonomic abnormalities in amyotrophic lateral sclerosis. Neurology 37 Suppl 1:162, 1987.
7. Low, PA: Laboratory evaluation of autonomic failure. In Low, PA (ed): Clinical Autonomic Disorders: Evaluation and Management. Little, Brown & Company, Boston, 1993, pp 169–195.
8. Low, PA: Pitfalls in autonomic testing. In Low, PA (ed): Clinical Autonomic Disorders: Evaluation and Management. Little, Brown & Company, Boston, 1993, pp 355–365.
9. Low, PA, Caskey, PE, Tuck, RR, et al: Quantitative sudomotor axon reflex test in normal and neuropathic subjects. Ann Neurol 14:573–580, 1983.
10. Low, PA, Fealey, RD, Sheps, SG, et al: Chronic idiopathic anhidrosis. Ann Neurol 18:344–348, 1985.
11. Low, PA and McManis, PG: Autonomic reflex testing in sympathetic reflex dystrophy. Neurology 35 Suppl 1:148, 1985.
12. Low, PA, Opfer-Gehrking, TL, Proper, CJ, and Zimmerman, I: The effect of aging on cardiac autonomic and postganglionic sudomotor function. Muscle Nerve 13:152–157, 1990.
13. Low, PA and Pfeifer, MA: Standardization of clinical tests for practice and clinical trials. In Low, PA (ed): Clinical Autonomic Disorders: Evaluation and Management. Little, Brown & Company, Boston, 1993, pp 287–296.
14. Low, PA, Zimmerman, BR, and Dyck, PJ: Comparison of distal sympathetic with vagal function in diabetic neuropathy. Muscle Nerve 9:592–596, 1986.
15. McEvoy, KM, Windebank, AJ, Daube, JR, and Low, PA: 3,4-Diaminopyridine in the treatment of Lambert-Eaton myasthenic syndrome. N Engl J Med 321:1567–1571, 1989.
16. Minor, V: Ein neues Verfahren zu der klinischen

Untersuchung der Schweissabsonderung. Deutsche Zeitschrift für Nervenheilkunde 101:302–308, 1928.

17. Navarro, X and Kennedy, WR: Sweat gland reinnervation by sudomotor regeneration after different types of lesions and graft repairs. Exp Neurol 104:229–234, 1989.

18. Sandroni, P, Ahlskog, JE, Fealey, RD, and Low, PA: Autonomic involvement in extrapyramidal and cerebellar disorders. Clin Auton Res 1:147–155, 1991.

19. Sarkany, I and Gaylarde, P: A method for demonstration of sweat gland activity. Br J Dermatol 80:601–605, 1968.

20. Stewart, JD, Low, PA, and Fealey, RD: Distal small fiber neuropathy: Results of tests of sweating and autonomic cardiovascular reflexes. Muscle Nerve 15:661–665, 1992.

21. Sutarman, and Thomson, ML: A new technique for enumerating active sweat glands in man. J Physiol (Lond) 117:51P–52P, 1952.

22. Wada, M: Sudorific action of adrenalin on the human sweat glands and determination of their excitability. Science 111:376–377, 1950.

CHAPTER 34

PERIPHERAL ADRENERGIC FUNCTION

Phillip A. Low, M.D.

SKIN VASOMOTOR REFLEXES
Venoarteriolar Reflex
Denervation Supersensitivity
Plasma Norepinephrine
Sustained Handgrip
Baroreflex Indices

Peripheral adrenergic function is important in maintaining postural normotension. It may be impaired in patients with peripheral neuropathy, as manifested by alterations in acral temperature, color, or volume. Despite its importance, simple, accurate, and reproducible tests of peripheral adrenergic function are largely unavailable. In this chapter, I describe methods that are used to determine peripheral adrenergic function and their value and shortcomings. Microneurography is sometimes used to evaluate adrenergic function, but it is invasive and not discussed in this chapter.

SKIN VASOMOTOR REFLEXES

The integrity of postganglionic sympathetic adrenergic fibers can be evaluated by testing skin vasoconstrictor reflexes.[8] Studies are usually performed on the toe or finger pads, because sympathetic innervation in these sites is purely vasoconstrictor, whereas other sites, such as the forearm, have both

vasoconstrictor and vasodilator fibers.[17]

Skin blood flow (SBF) is measured by a laser Doppler flowmeter or by plethysmography, and the vasoconstrictor response to an autonomic maneuver is thus determined. Vasoconstriction can be induced by maneuvers such as inspiratory gasp, response to standing (for the finger), contralateral cold stimulus, or the Valsalva maneuver. The response can be expressed as percentage of vasoconstriction:

Percentage of vasoconstriction

$$= \frac{\text{resting SBF} - \text{minimal SBF}}{\text{resting SBF}} \times 100$$

The pathways of these reflexes are complex. For instance, the response to standing is mediated by the venoarteriolar reflex,[6] low-pressure and high-pressure baroreceptors,[19,23] and, to a lesser extent, by increased levels of epinephrine,[2] norepinephrine,[20] and renin.[12] The advantage of these reflexes is that they have different afferents but also have an identical final efferent pathway. Thus, the relevant afferent pathway can be evaluated. However, this advantage is offset by a major shortcoming of tests of skin adrenergic function, that is, the marked sensitivity of skin sympathetic fibers to emotional and temperature changes,[1] which means there is considerable ambient fluctuation[8] of SBF.

Skin vasomotor reflexes have a coefficient of variation in excess of 20% and are best regarded as semiquantitative tests. The tests are useful in comparing vasoconstrictor reflexes from identical sites simultaneously, however.

Venoarteriolar Reflex

When venous transmural pressure is increased by 25 mm Hg (e.g., by lowering a limb by 40 cm), reflex arteriolar vasoconstriction occurs, decreasing blood flow by 50%.[6] This reflex, termed the *venoarteriolar reflex*, has its receptor in small veins, and its neural pathway appears to be that of the sympathetic C-fiber local "axon reflex" (Fig. 34–1).[6] Function of the "reflex" is to increase total peripheral resistance, compensating by up to 45% of the orthostatic decrease in cardiac output.[6,7] It may also reduce the orthostatic increase in tissue fluid by adjusting precapillary-to-postcapillary resistance ratio.

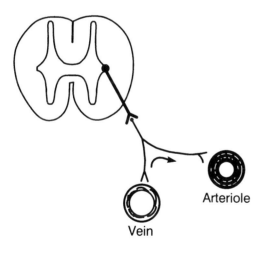

Venoarteriolar
reflex

Figure 34–1. Presumed pathway of venoarteriolar reflex. The receptor is the small vein and the effector is the arteriole. The impulse passes antidromically along the postganglionic sympathetic sudomotor axon to the branch point and then orthodromically to activate the effector. Preganglionic axons emanating from the spinal cord are shown as thick structures, and postganglionic axons as thin structures. (From Moy, et al,[11] p 1491, with permission.)

In one study, the venoarteriolar reflex was measured in the feet of patients with diabetes and was reported to be reduced in those with diabetic neuropathy.[16] The potential value of this test is in its presumed ability to examine the status of sympathetic vasoconstrictor fibers at the postganglionic level.

We studied the venoarteriolar reflex, quantitative sudomotor axon reflex test, and heart-rate responses to deep breathing and the Valsalva maneuver in 40 control subjects, 49 patients with diabetes, and 29 patients with other neuropathies. The mean vasoconstrictor response was greater in control subjects than in patients with diabetes or other neuropathies, but the overlap among groups was marked (Fig. 34–2). The venoarteriolar reflex appeared to have less specificity and much less sensitivity than other tests of autonomic function. This test lacks the necessary sensitivity and specificity to be used as a clinical test of autonomic function.[11]

Denervation Supersensitivity

An exaggerated pressor response to intra-arterial or intravenous application of direct-acting alpha-agonists (i.e., phenylephrine or norepinephrine) indicates denervation supersensitivity. It may be seen when widespread denervation of postganglionic sympathetic fibers exists.[9,10,18] The mechanism of denervation supersensitivity is an increase in receptor density, affinity, efficacy of receptor-effector coupling, or other postreceptor events. These blood pressure tests of adrenergic denervation supersensitivity are too invasive for routine use and are insensitive.

Acral adrenergically mediated vasoconstriction should be measured in response to the above-mentioned infusion. Unfortunately, recordings of vasoconstriction of the muscle bed are indirect. Plethysmography can be performed, but recorded flow is contaminated by SBF.

Plasma Norepinephrine

Plasma norepinephrine results from a spillover of norepinephrine from adrenergic postganglionic nerve terminals. The supine

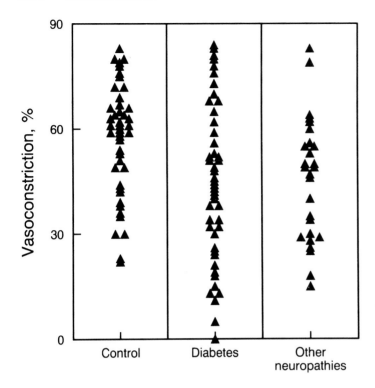

Figure 34–2. Venoarteriolar reflex, expressed as a percentage of vasoconstriction, in control subjects and in patients with diabetes or other neuropathies. (From Moy, et al,[11] p 1491, with permission.)

value is an index of net sympathetic activity[15,21,22] and is affected by the rate of norepinephrine secretion and clearance.[14] Plasma norepinephrine has been used to separate postganglionic from preganglionic failure. In a disorder in which the lesion is preganglionic, resting supine norepinephrine values are normal but the response to standing would be absent because of failure of activation. In cases of postganglionic lesion, the supine value is decreased if the lesion is widespread. The disadvantage of this method is its low sensitivity.

Sustained Handgrip

Sustained muscle contraction increases systolic and diastolic blood pressure and heart rate. The stimulus derives from exercising muscle, is reflexive in nature, and the increase in blood pressure is mediated by an increase in cardiac output and peripheral resistance.[3] The test has been adapted as a simple test of sympathetic autonomic failure.[5] Ewing and colleagues[5] recommend 30% maximal contraction for up to 5 minutes. Many patients have difficulty sustaining the handgrip for 5 minutes; however, 3 minutes is sufficient. The test evaluates generalized rather than peripheral adrenergic function.

Baroreflex Indices

Studying baroreflex indices must evaluate the heart-period (reciprocal of heart rate) responses to induced increases and decreases in blood pressure. Phenylephrine or norepinephrine can be used to increase blood pressure, and tilt or trinitroglycerin, to decrease it. One approach, adapted from the method of Korner, is to determine the heart-period range and mean gain.[9] These two indices express the range of heart period to a moderate pressor-hypotensive stress and the mean rate of change in heart period in response to sudden changes in blood pressure.[9] Another related approach, described by Pickering and associates,[13] relates beat-to-beat systolic blood pressure to heart rate. An alternative approach to stimulating baroreceptors is to use a neck chamber whose pressure can be increased or decreased.[4] Finally, the heart-period responses to the decrease and increase in blood pressure during the Valsalva maneuver can be related to blood pressure changes.

CHAPTER SUMMARY

For noninvasive evaluation of autonomic function, tests of peripheral adrenergic function are neither as accurate nor as reproducible as tests of sudomotor and cardiovagal function. A noninvasive or minimally invasive method for evaluation of muscle blood flow needs to be developed.

REFERENCES

1. Burton, AC and Taylor, RM: A study of the adjustment of peripheral vascular tone to the requirements of the regulation of body temperature. Am J Physiol 129:565–577, 1940.
2. Celander, O: The range of control exercised by the 'sympathico-adrenal system': A quantitative study on blood vessels and other smooth muscle effectors in the cat. Acta Physiol Scand Suppl 116:1–132, 1954.
3. Coote, JH, Hilton, SM, and Perez-Gonzalez, JF: The reflex nature of the pressor response to muscular exercise. J Physiol (Lond) 215:789–804, 1971.
4. Eckberg, DL, Cavanaugh, MS, Mark, AL, and Abboud, FM: A simplified neck suction device for activation of carotid baroreceptors. J Lab Clin Med 85:167–173, 1975.
5. Ewing, DJ, Irving, JB, Kerr, F, et al: Cardiovascular responses to sustained handgrip in normal subjects and in patients with diabetes mellitus: A test of autonomic function. Clin Sci Mol Med 46:295–306, 1974.
6. Henriksen, O: Local sympathetic reflex mechanism in regulation of blood flow in human subcutaneous adipose tissue. Acta Physiol Scand Suppl 450:1–48, 1977.
7. Henriksen, O, Skagen, K, Haxholdt, O, and Dyrberg, V: Contribution of local blood flow regulation mechanisms to the maintenance of arterial pressure in upright position during epidural blockade. Acta Physiol Scand 118:271–280, 1983.
8. Low, PA, Caskey, PE, Tuck, RR, et al: Quantitative sudomotor axon reflex test in normal and neuropathic subjects. Ann Neurol 14:573–580, 1983.
9. Low, PA, Walsh, JC, Huang, CY, and McLeod, JG: The sympathetic nervous system in diabetic neuropathy: A clinical and pathological study. Brain 98:341–356, 1975.
10. Moorhouse, JA, Carter, SA, and Doupe, J: Vascular responses in diabetic peripheral neuropathy. BMJ 1:883–888, 1966.
11. Moy, S, Opfer-Gehrking, TL, Proper, CJ, and Low, PA: The venoarteriolar reflex in diabetic and other neuropathies. Neurology 39:1490–1492, 1989.
12. Oparil, S, Vassaux, C, Sanders, CA, and Haber, E: Role of renin in acute postural homeostasis. Circulation 41:89–95, 1970.
13. Pickering TG, Gribbin B, Petersen ES, et al: Effects of autonomic blockade on the baroreflex in man at rest and during exercise. Circ Res 30:177–185, 1972.
14. Polinsky, RJ, Goldstein, DS, Brown, RT, et al: Decreased sympathetic neuronal uptake in idiopathic orthostatic hypotension. Ann Neurol 18:48–53, 1985.
15. Polinsky, RJ, Kopin, IJ, Ebert, MH, and Weise, V: Pharmacologic distinction of different orthostatic hypotension syndromes. Neurology (N Y) 31:1–7, 1981.
16. Rayman, G, Williams, SA, Spencer, PD, et al: Impaired microvascular hyperaemic response to minor skin trauma in type I diabetes. BMJ 292:1295–1298, 1986.
17. Roddie, IC, Shepherd, JT, and Whelan, RF: A comparison of the heat elimination from the normal and nerve-blocked finger during body heating. J Physiol (Lond) 138:445–448, 1957.
18. Smith, AA and Dancis, J: Exaggerated response to infused norepinephrine in familial dysautonomia. N Engl J Med 270:704–707, 1964.
19. Sundlöf, G and Wallin, BG: Human muscle nerve sympathetic activity at rest: Relationship to blood pressure and age. J Physiol (Lond) 274:621–637, 1978.
20. Vendsalu, A: Studies on adrenaline and noradrenaline in human plasma. Acta Physiol Scand Suppl 173:1–123, 1960.
21. Wallin, BG, Sundlöf, G, Eriksson, B-M, et al: Plasma noradrenaline correlates to sympathetic muscle nerve activity in normotensive man. Acta Physiol Scand 111:69–73, 1981.
22. Ziegler, MG, Lake, CR, and Kopin, IJ: The sympathetic-nervous-system defect in primary orthostatic hypotension. N Engl J Med 296:293–297, 1977.
23. Zoller, RP, Mark, AL, Abboud, FM, et al: The role of low pressure baroreceptors in reflex vasoconstrictor responses in man. J Clin Invest 51:2967–2972, 1972.

CHAPTER 35

THERMOREGULATORY SWEAT TEST

Robert D. Fealey, M.D.

METHOD
NORMAL THERMOREGULATORY SWEAT
   DISTRIBUTIONS
REPORTING RESULTS
THE TST IN CLINICAL DISORDERS OF
   THE AUTONOMIC NERVOUS SYSTEM
DIFFICULTIES AND PITFALLS IN
   INTERPRETATION

From a physiologic viewpoint, the thermoregulatory sweat test (TST) consists of giving a controlled heat stimulus at tolerable levels to the body to produce a generalized sweat response (i.e., recruiting all areas of skin capable of sweating). The TST assesses the integrity of efferent (central and peripheral) sympathetic sudomotor pathways. Central (preganglionic) structures tested include the hypothalamus, bulbospinal projections, intermediolateral cell column in the thoracic and upper lumbar spinal cord, and white rami. Peripheral (postganglionic) autonomic structures tested include the sympathetic chain and postganglionic sudomotor axons and the sweat glands. Because the entire anterior body surface is tested simultaneously for both preganglionic and postganglionic lesions, the TST is well suited for screening patients for autonomic involvement in many disorders.

Thermoregulatory sweating is influenced by the mean and local skin temperature as well as by the central (blood/core) temperature.[13] A maximal sweat response occurs when both central (oral) and mean skin temperatures are increased in a moderately humid (about 35% relative humidity) environment in which some degree of sweat evaporation can occur.[15,16] Therefore, proper technique includes controlling the ambient air temperature and humidity as well as the patient's core and skin temperatures.

Several techniques, including hot baths and infrared and incandescent heat lamps, have been used for the last 50 years to produce sweating, but the most satisfactory method is to use a cabinet in which the environment is controlled and the entire body (including the head) is heated. Guttmann[8] described such a cabinet and demonstrated the usefulness of the TST in diagnosis and monitoring of spinal cord and peripheral nerve lesions. The TST conducted in the Mayo Clinic Thermoregulatory Laboratory is a modification of Guttmann's quinizarin sweat test[5,12] and is described in the next section.

METHOD

The patient, covered with a towel to maintain modesty, is placed in a supine position on a cart and enclosed in the cabinet (Fig. 35–1A). The "head end" of the cabinet is a

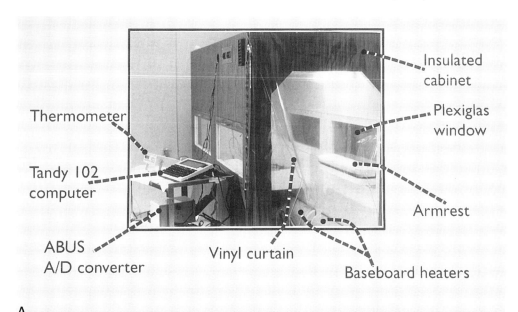

A

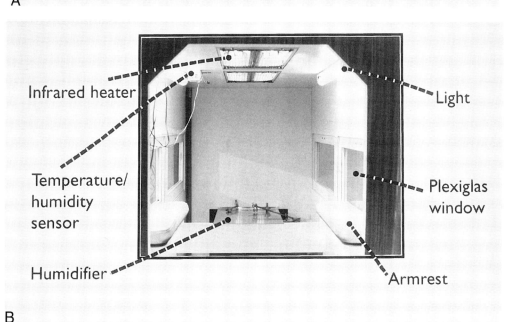

B

Figure 35–1. Thermoregulatory sweat cabinet. (*A*) Exterior, with temperature monitoring equipment. (*B*) Interior.

clear vinyl curtain tented over the cart; this arrangement allows the patient's head to be in a heated environment that provides a clear view of the technician and surroundings to minimize the claustrophobia. Suitable parameters for achieving a generalized sweat response within 50 minutes include an ambient temperature of 45°C to 50°C, a rel-

ative humidity of 30% to 40%, and a skin temperature between 39°C and 40°C. The interior sweat cabinet is shown in Figure 35–1*B*. The major components are the insulated cabinet walls containing four baseboard heaters and a humidifier, and overhead infrared heaters that heat the skin and are carefully regulated by skin temperature feedback control. Armrests comfort patients as they rest supine with palms down. Clear Plexiglas windows that can be opened allow the technician to adjust the skin temperature probes if necessary and to observe the developing sweat distribution during the test.

Thermistor probes continuously measure skin and oral temperatures during the test. Sweating on the skin surface is best visualized by an indicator powder that is applied to the body before heating. A mixture of alizarin red, cornstarch, and sodium carbonate in a 50:100:50-g ratio, respectively, is suitable.[12] It appears as a light orange color on nonsweating skin and turns dark purple on sweating skin. Other indicators currently used include iodinated cornstarch[19] and combined starch and iodine in solution.[14]

The average response of the oral temperature in 30 healthy control subjects (20 to 60 years old) who achieved full-body sweating during the TST was an increase of 1.2°C in 35 to 40 minutes. Because all subjects had sweated profusely at an oral temperature of 38°C, we use this temperature as a test end point. During 1991, mean oral temperature increase in 467 patients was 1.4°C, with 38°C or a 1°C increase above baseline (whichever yielded the higher temperature) as an end point. These observations indicate that the often-quoted 1°C increased temperature criterion for an adequate TST[1] is inadequate in patients with low (<37°C) initial temperatures. A 38°C end point is a far more reliable standard than some arbitrary amount of heating time. Of course, if generalized sweating occurs at a lower body temperature, the test can be ended. For reasons of patient comfort and safety, we do not allow oral temperature to increase above 38.5°C nor extend the heating period beyond 60 minutes.

NORMAL THERMOREGULATORY SWEAT DISTRIBUTIONS

The normal variants in sweat distribution observed in our laboratory in more than 50 healthy control subjects (20 men, 30 women) from 20 to 75 years old are shown in Figure 35–2. Areas of "normal" anhidro-

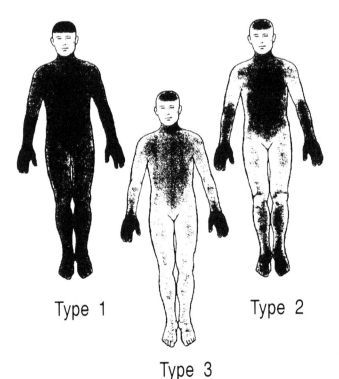

Type 1   Type 2   Type 3

Figure 35–2. Normal thermoregulatory sweating patterns, types 1 to 3. Black = areas of sweating. (From Fealey, et al,[5] p 619, with permission.)

sis may occur over bony prominences (e.g., the patella and clavicle) and, with the subject supine, in the lateral calves and inner thighs. The proximal extremities frequently show less sweating than the distal, but left-right symmetry is always the rule. Males tend to show the type 1 (heavy generalized sweating) pattern, whereas females usually show the type 2 (heavy generalized sweating but less in proximal extremities) or type 3 (generalized sweating but less in proximal arms and lower body) pattern. Elderly men and women tend to have types 2 and 3, and the lighter sweating areas have a higher threshold of activation.[7]

We have noted anhidrosis in elderly women (70 years and older) whose results on neurologic examination and autonomic inventory are normal except for their reporting having had dry skin for many years.[4] Whether this represents a variant of chronic idiopathic anhidrosis[10] or a loss of functional sweat glands or reflects the loss of preganglionic autonomic neurons known to occur with aging[11] is unclear. Severely dehydrated persons and those taking anticholinergic medications have diffusely diminished sweating; thus, we encourage increased fluid intake and discontinuation of anticholinergic medications at least 48 hours before the TST.

We have described seven types of thermoregulatory sweat patterns or distributions that are used to report test results.[5]

1. *Distal anhidrosis* is characterized by sweat loss greatest in the fingers, the legs below the knees, and the lower anterior abdomen (Fig. 35–3*a*).
2. *Segmental anhidrosis* involves large contiguous zones of the body surface bordered by areas of normal sweating; these usually respect sympathetic dermatomal borders (Fig. 35–3*b*).
3. *Focal sweat loss* is confined to isolated dermatomes, peripheral nerve territories, or small localized areas of skin (Fig. 35–3*c*).
4. *Global anhidrosis*, by definition, occurs when more than 80% of the body surface is involved (Fig. 35–3*d*).
5. *Regional anhidrosis* refers to large anhidrotic areas (but less than 80% of body surface) that blend gradually into sweating areas and that may or may not

be contiguous; anhidrosis of the trunk alone and anhidrosis of the proximal parts of all four extremities are examples of this pattern.
6. *Mixed patterns* combine any of the above in the same patient, for example, focal and distal patterns of anhidrosis (Fig. 35–3*c*).
7. A *normal sweat distribution* has no areas of anhidrosis or minor areas of sweat loss as observed in control subjects, as described above (Fig. 35–3*e*).

REPORTING RESULTS

Data about the age, sex, identity number, clinical problem of the patient, and the date of the TST are indicated on the report (Fig. 35–4).

The temperature and humidity ranges, the amount of time the patient was exposed to the heat stress, and the initial and final oral temperatures are indicated.

The main part of the report includes a brief anatomic description of the results and states the clinical significance of the findings.

The anatomic figure is shaded in by the technician and is checked by the reporter before the patient showers, or else the areas of anhidrosis are photographed using a video camera and these are reviewed by the reporter on a video monitor.

In our laboratory, the final report also contains the percentage of anterior body surface anhidrosis (TST%) along with the sweat distribution pattern.[5,20] We also draw a digitized body image that is shaded in by a computer graphics program and suitable for output on a dot-matrix printer (Fig. 35–4) for a permanent record. This image is also used to determine TST% using a pixel counting program. Another accurate method for deriving the TST% is to use planimetry. The drawn body image is scrutinized and the region(s) where no sweating has occurred outlined and measured; individual areas are summed to produce a total anterior body surface estimate. The procedure we use is as follows:

1. Sketch the sweat distribution accurately, and mark in pencil the perimeter of the area(s) of anhidrosis.

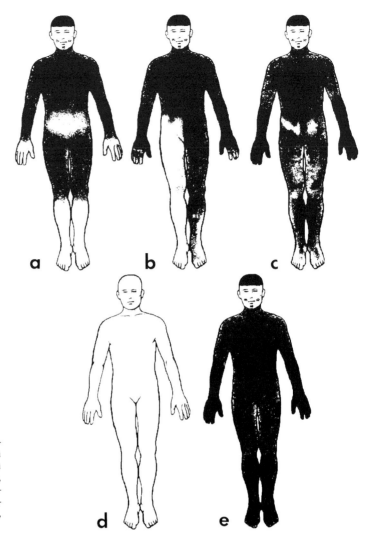

Figure 35–3. Sweat distribution patterns: examples from patients with diabetes mellitus. Sweating in shaded areas. (Modified from Low, PA and Fealey, RD: Sudomotor neuropathy. In Dyck, PJ, et al [eds]: Diabetic Neuropathy. WB Saunders, Philadelphia, 1987, p 141.)

2. Place the body figure on its side in parallel with the tracer arm of the planimeter.
3. Turn on the processor and check the unit's calibration.
4. Place the rolling disc planimeter so the magnifier lens is over the body figure and the tracer arm is at right angles to the wheel axle.
5. Make counterclockwise traces around the perimeter of each anhidrotic area.
6. Read the total area (cm²) from the processor's display; divide this value by 17.57 cm² (area of entire body figure) and multiply by 100 for the percentage of anhidrosis.

THE TST IN CLINICAL DISORDERS OF THE AUTONOMIC NERVOUS SYSTEM

The TST is helpful in identifying autonomic involvement in many neurologic disorders, including primary autonomic failure,[2,18] secondary autonomic failure due to neuropathy (diabetic, primary systemic amyloid, lepromatous, idiopathic small fiber,[20,21] paraneoplastic), myelopathy (syrinx, spinal cord injury, multiple sclerosis), surgical sympathectomy, and chronic idiopathic anhidrosis.[3,9,17] The TST can monitor disease progression or recovery. Body areas that are not readily accessible to measure-

THERMOREGULATORY SWEAT TEST

Clinic # _____

Date 10/3/88

Name _____

Clinical Problem: Reports loss of sweating from the right arm and face.

Diagnosis 7030

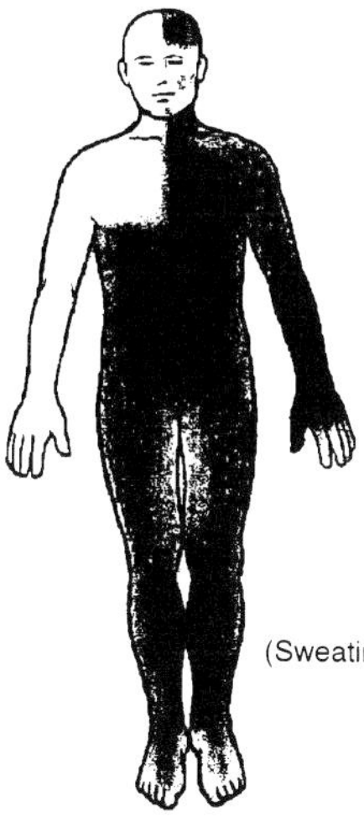

Cabinet temperature range: 43.0 to 43.5 °C

Cabinet humidity range: 40 to 45 %

Time . . . . 28 min.

Oral temperature:
before test. . . 36.8 °C
after test . . . . 38.0 °C

Body surface anhidrosis %. . . 24

Pattern of anhidrosis . . . . MIXED

RESULTS:

Sweating was completely absent in the T_2 through T_4 sympathetic dermatomes on the right. There was distal sweat loss in the left hand and both feet as well.

Results indicate a pre or postganglionic segmental lesion of the upper sympathetic chain on the right. The distal loss suggests a mild peripheral neuropathy.

(Sweating in shaded areas)

Physician   Robert D. Fealey M.D.

Figure 35–4. Thermoregulatory sweat test report of a patient with a right sympathetic chain lesion (Pancoast's tumor) and a peripheral neuropathy. Black = areas of sweating.

ment using other techniques are easily examined using the TST, and several patterns are diagnostic. TST results in specific diseases are discussed elsewhere.[4]

DIFFICULTIES AND PITFALLS IN INTERPRETATION

The interpreter must be aware of the normal patterns of thermoregulatory sweating, including areas where anhidrosis may be seen in a normal subject, such as over bony prominences or the lateral calves.

Patients who have worn pressure wraps (i.e., ace bandages and abdominal binders) within 12 hours before the test many show anhidrosis in the areas that had been covered. This usually can be recognized by the straight edges of the deficit.

Severely dehydrated patients may sweat less overall,[6] but they generally do not have focal defects. Anticholinergic drugs, including most tricyclic antidepressants, may inhibit thermoregulatory sweating and should be stopped 48 hours before the TST is performed.

The application of skin lotions such as moisturizing creams may discolor alizarin-covered skin, making it difficult to discern areas of anhidrosis. Use of lotions is thus prohibited on the day of the TST.

Anhidrosis in elderly patients may present an interpretative challenge, because the effect of aging on the autonomic nervous system may be responsible for the regional anhidrosis (most often affecting the lower half of the body and proximal arms) that is seen in some women older than age 65. Foster and coworkers[7] found evidence of an age-related decrease in sweat gland activity caused by thermal stimulation. Although there is evidence for attrition of neurons in the intermediolateral cell columns of the spinal cord,[11] the evidence for failure of sweating with age is inconclusive.[3] Senile atrophy of the skin and differences in sweat-gland training between young and old may also be a factor. A certain percentage of healthy elderly persons who report that they have been unable to sweat much for years may have a variety of chronic idiopathic anhidrosis.[10]

Drawbacks of the TST include the untidiness and duration of the test, the possibility of injury to skin by heat, or skin irritation caused by alizarin. Contact dermatitis occurs only rarely (observed frequency, 1:1000 subjects) and is readily treated with oral and topical agents. A more common but not harmful problem is the persistence of purple discoloration of small skin areas; it may take several days for the color to wash off. Because of repeated exposure to the indicator powder, technicians in our laboratory wear masks, gloves, and goggles when applying the powder, so as to minimize inhalation, oral ingestion, or contact with the eyes. Patients who are extremely claustrophobic, who indicate a history of severe contact dermatitis, or who are less than 16 years old are not tested.

CHAPTER SUMMARY

The thermoregulatory sweat test assesses the integrity of central and peripheral efferent sympathetic sudomotor neural pathways. A controlled heat and humidity stimulus is given to produce a generalized sweating response in all skin areas capable of sweating. Sweating is visualized by placing an indicator powder on the skin beforehand. The entire anterior body surface can be examined and abnormalities can usually be detected at a glance. Characteristic normal and abnormal sweat distributions are described herein, as are methods to prepare a report of the test results, including a technique to quantify the response, known as the "percentage of anhidrosis." Important parameters of the heat stimulus, the patient's oral and skin temperature response, and pitfalls in the interpretation of the sweat test results are described.

REFERENCES

1. Bannister, R, Ardill, L, and Fentem, P: Defective autonomic control of blood vessels in idiopathic orthostatic hypotension. Brain 90:725–746, 1967.
2. Cohen, J, Low, P, Fealey, R, et al: Somatic and autonomic function in progressive autonomic failure and multiple system atrophy. Ann Neurol 22:692–699, 1987.
3. Collins, KJ: Autonomic control of sweat glands and disorders of sweating. In Bannister, R (ed): Auto-

nomic Failure: A Textbook of Clinical Disorders of the Autonomic Nervous System, ed 2. Oxford University Press, Oxford, 1988, pp 748–765.
4. Fealey, RD: The thermoregulatory sweat test. In Low, PA (ed): Clinical Autonomic Disorders: Evaluation and Management. Little, Brown & Company, Boston, 1993, pp 217–229.
5. Fealey, RD, Low, PA, and Thomas, JE: Thermoregulatory sweating abnormalities in diabetes mellitus. Mayo Clin Proc 64:617–628, 1989.
6. Fortney, SM, Nadel, ER, Wenger, CB, and Bove, JR: Effect of blood volume on sweating rate and body fluids in exercising humans. J Appl Physiol 51:1594–1600, 1981.
7. Foster, KG, Ellis, FP, Doré, C, et al: Sweat responses in the aged. Age Ageing 5:91–101, 1976.
8. Guttmann, L: The management of the Quinizarin Sweat Test (Q.S.T.). Postgrad Med J 23:353–366, 1947.
9. Low, PA and Fealey, RD: Testing of sweating. In Bannister, R and Mathias, CJ (eds): Autonomic Failure: A Textbook of Clinical Disorders of the Autonomic Nervous System, ed 3. Oxford University Press, Oxford, 1992, pp 413–420.
10. Low, PA, Fealey, RD, Sheps, SG, et al: Chronic idiopathic anhidrosis (CIA). Ann Neurol 18:344–348, 1985.
11. Low, PA, Okazaki, H, and Dyck, PJ: Splanchnic preganglionic neurons in man. 1. Morphometry of preganglionic cytons. Acta Neuropathol (Berl) 40:55–61, 1977.
12. Low, PA, Walsh, JC, Huang, CY, and McLeod, JG: The sympathetic nervous system in diabetic neuropathy: A clinical and pathological study. Brain 98:341–356, 1975.
13. McCaffrey, TV, Wurster, RD, Jacobs, HK, et al: Role of skin temperature in the control of sweating. J Appl Physiol 47:591–597, 1979.
14. Minor, V: Ein neues Verfahren zu der klinischen Untersuchung der Schweißabsonderung. Dtsch Z Nervenheilkunde 101:302–308, 1928.
15. Pandolf, KB, Griffin, TB, Munro, EH, and Goldman, RF: Persistence of impaired heat tolerance from artificially induced miliaria rubra. Am J Physiol 239:R226–R232, 1980.
16. Pinnagoda, J, Tupker, RA, Coenraads, PJ, and Nater, JP: Transepidermal water loss with and without sweat gland inactivation. Contact Dermatitis 21:16–22, 1989.
17. Quinton, PM: Sweating and its disorders. Annu Rev Med 34:429–452, 1983.
18. Sandroni, P, Ahlskog, JE, Fealey, RD, and Low, PA: Autonomic involvement in extrapyramidal and cerebellar disorders. Clin Auton Res 1:147–155, 1991.
19. Sato, KT, Richardson, A, Timm, DE, and Sato, K: One-step iodine starch method for direct visualization of sweating. Am J Med Sci 295:528–531, 1988.
20. Stewart, JD, Low, PA, and Fealey, RD: Distal small fiber neuropathy: Results of tests of sweating and autonomic cardiovascular reflexes. Muscle Nerve 15:661–665, 1992.
21. Suarez, GA, Fealey, RD, Camilleri, M, and Low, PA: Idiopathic autonomic neuropathy: Clinical, neurophysiologic, and follow-up studies on 27 patients. Neurology 44:1675–1682, 1994.

CHAPTER 36

CARDIOVASCULAR AND OTHER REFLEXES

Eduardo E. Benarroch, M.D.
Phillip A. Low, M.D.

Cardiovascular heart rate tests are a non-invasive, reliable, reproducible, and widely used way to measure autonomic function in human subjects. Although the physiologic basis of these autonomic maneuvers has been known for a long time, it is mainly in the last two decades, because of the enthusiastic promotion by Ewing and associates,[31] that these tests have been applied clinically. Controversy exists about the definition of an adequate battery.[31,77] The pitfalls of cardiovascular heart rate tests remain underappreciated.[56,90]

In the following description, the focus is on the most useful tests. Their underlying physiology, clinical utility, and shortcomings are described, followed by a brief description of newer approaches.

HEART RATE RESPONSE TO DEEP BREATHING

The heart rate response to deep breathing is probably the most reliable of the cardiovascular heart rate tests, because the major afferent and efferent pathways are both vagal.[45] That vagal efferents are important has been demonstrated experimentally in animals and in humans. In dogs, heart rate variations correlate well with recorded vagal efferent activity[46] and can be blocked by atropine or by freezing the vagus nerve.[2] In humans, atropine (a muscarinic antagonist) totally abolishes respiratory sinus arrhythmia.[97]

The basis of respiratory sinus arrhythmia appears to be the connection between the respiratory centers and cardioinhibitory centers in the medulla. Evidence for this is cessation of vagal efferent activity during the inspiratory phase of natural, but not artificial, ventilation[43,46,69] and the loss of respiratory sinus arrhythmia in some patients with brain stem infarction.[72] Respiratory sinus arrhyth-

403

mia is modulated by input from the lungs, heart, and baroreceptors. Pulmonary stretch receptors that mediate the Hering-Breuer respiratory reflex[36] modulate respiratory sinus arrhythmia, although their role may be less important in humans than in experimental animals.[33] Receptors from the right atrium initiate the vagally mediated Bainbridge reflex[3] and a venoatrial mechanoreceptor sympathetic reflex.[16,48] Finally, baroreflex sensitivity changes throughout deep breathing, thus modulating respiratory sinus arrhythmia.[28,63,65]

Factors That Affect the Heart Rate Response to Deep Breathing

Numerous factors affect the heart rate response to deep breathing. From the standpoint of clinical autonomic testing, the most important of these are the effects of age and rate of forced respiration.

A progressive reduction in the response with increasing age has been reported by all large studies.* We studied 137 control subjects and reached a similar conclusion. The heart rate response to deep breathing did not significantly differ between the sexes; therefore, the data were combined. A consistently significant regression with age was found. For response to deep breathing (y = range of heart rate in beats per minute and x = age in years; $y = 37.15 - 0.36x$; $P < .001$).

Maximal heart rate response to deep breathing occurs at a breathing frequency of 5 to 6 respirations per minute in normal subjects;[1,7,73,92] this observation forms the basis for the standard test of deep breathing.[97] However, the maximal heart rate response to deep breathing is at considerably lower frequencies in patients with vagal neuropathy.

There is a sympathetic modulation of the heart rate response to deep breathing which is inhibited by stress and enhanced by beta-adrenergic blockade.[22] Also, the response is impaired during severe tachycardia,[73] in heart failure, and in deeply unconscious patients.[88]

---

*References 9, 31, 35, 42, 44, 57, 62, 68, 72, 75, 84, 97, 99.

The position of the subject has some effect on respiratory sinus arrhythmia. The response is larger when the subject is supine than when sitting or erect.[8,24,60]

A standardized deep-breathing protocol can be used, because depth of breathing above a tidal volume of about 1.2 liters causes insignificant changes in the heart rate response to deep breathing.[33] Bennett and associates[7] and Eckberg[27] found little or no difference in the response for different depths of respiration.

The sex of the subject does not greatly affect the heart rate response to deep breathing. No sex differences have been observed.[38,59] The amount of antecedent rest is not important in relaxed subjects. After 5 minutes of rest, another 25 minutes of supine rest will not alter the response.[38] No significant differences have been found in the response whether the test is performed in the same subjects in the morning or in the afternoon.[7]

One indirect effect of prolonged hyperventilation is the reduction of P_{CO_2}, resulting in a depression in respiratory sinus arrhythmia.[13]

Basically, there have been two methods of respiratory cycling. The more commonly used is to have the breathing of a subject follow a pattern, usually a sine wave or an oscillating bar, generated by a computer. The alternative is to instruct the subject to breathe in and out when instructed. The differing effect of the two approaches has not been studied systematically, but is it likely to be minor.

Methods of Analysis

The three methods of analysis generally used are the heart rate range, the heart period range, and the E:I ratio. The E:I ratio was described by Sundkvist, Almér, and Lilja[85] who took the ratio of the shortest R-R interval during inspiration to the longest R-R interval during expiration and derived the E:I ratio. We prefer the heart rate range, because the effect of resting heart rate on the range is smaller than its effect on heart period. Weinberg and Pfeifer[95] recommended calculation of the circular mean resultant, a method based on vector analysis, to eliminate the effects of trends in heart rate over time and attenuating the effect of basal heart rate and ectopic beats in the calculated variability of heart rate.

Reproducibility

The tests have been reproducible. Typical coefficients of variation have been 11%[7] and 8.9%.[84]

Problems and Controversies

The heart rate response to deep breathing is an indirect measure of cardiovagal function. Reduced response indicates a lesion anywhere along the complicated autonomic central nervous system, that is, in the afferent, central processing unit, efferent, synapse, or effector apparatus.

To further complicate interpretation, reduced response does not unequivocally indicate cardiovagal failure. Heart rate usually increases during inspiration and decreases during expiration,[2] but even this observation is not entirely correct. Both inspiration and expiration are followed by an increase, then a decrease, in heart rate but at a different rate of change, amplitude, time of appearance, and duration.[64] Mehlsen and colleagues[64] suggested that the reason the maximal heart rate range in many individuals is 6 beats per minute is because they have well-defined heart rate maxima with positive "interference" of phases. The reason subjects have reduced heart rate range less than 7 beats per minute is because of negative "interference."

HEART RATE RESPONSE TO STANDING

The immediate heart rate response to standing can be recorded using an electrocardiogram (ECG) machine. In normal subjects, tachycardia is maximal about the 15th beat and relative bradycardia around the 30th beat.[30] The 30:15 ratio (R-R interval at beat 30)/(R-R interval at beat 15) has been recommended as an index. Reflex tachycardia is thought to be mediated by the vagus nerve, because the response is abolished by atropine but not by propranolol.[30]

A detailed evaluation of the phases and mechanisms of the response to standing has been reported.[14] The initial heart rate responses to standing consist of a tachycardia at 3 seconds and then at 12 seconds, followed by a bradycardia at 20 seconds. The initial cardioacceleration is an exercise reflex, whereas the subsequent tachycardia and bradycardia are baroreflex-mediated.

OTHER TESTS OF VAGAL FUNCTION

Potential tests of cardiovagal function are numerous; they include heart rate response to coughing and to facial immersion in cold water. Coughing results in inspiration and expiratory effort against a closed glottis, followed by an explosive expiration as the glottis suddenly opens.[98] The heart rate response consists of a cardioacceleration, which is maximal about 2 to 3 seconds after the last cough, and a return to resting values in about 12 to 14 seconds.[93,94] The mechanism is thought to be cholinergic, due initially to muscular contraction followed by baroreflex response to a decrease in blood pressure.[18]

The diving response has also been adapted as a test of cardiovagal function.[47] The application of a cold stimulus to both halves of the facial area innervated by the first division of the trigeminal nerve results in reflex bradycardia.

Certain gastrointestinal regulatory peptides are involved in autonomic regulation.[83] The responses of these peptides to stress have been suggested to be reliable indices of parasympathetic function. Suggested tests have included responses of pancreatic polypeptide,[37,51,55] somatostatin, and glucagon to hypoglycemia.[37,55]

THE VALSALVA MANEUVER

The Valsalva maneuver consists of an abrupt transient elevation of intrathoracic and intra-abdominal pressures induced by blowing against a pneumatic resistance while maintaining a predetermined pressure ("straining").*

---

*References 12, 15, 17, 32, 34, 81, 96, 101.

Normal Response

Intra-arterial recordings of arterial pressure and, more recently, noninvasive monitoring of arterial pressure with a photoplethysmographic (Finapres) technique* have provided important information about the hemodynamic changes during the Valsalva maneuver in normal and pathologic conditions.

The responses to the Valsalva maneuver have been divided into four phases[81,82] (Fig. 36–1). Phase I consists of a brisk increase in systolic and diastolic arterial pressure and a decrease in heart rate immediately after the onset of the Valsalva strain and lasts about 4

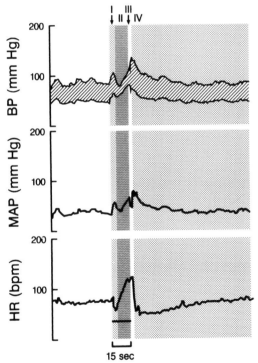

Figure 36–1. The changes in heart rate (*HR*), mean arterial pressure (*MAP*), and blood pressure (*BP*) during the four phases (*I, II, III,* and *IV*) of the Valsalva maneuver in a 34-year-old woman. The bar at the bottom of the figure denotes the duration of the maneuver performed at an expiratory pressure of 40 mm Hg. (From Benarroch, et al.,[6] p 1167, Copyright © 1991 John Wiley & Sons. Reprinted by permission of John Wiley & Sons.)

*References 6, 41, 70, 71, 79–82, 86, 89, 100.

seconds. Elevation of arterial pressure during phase I reflects mechanical factors and is not sympathetically mediated. It persists in patients with transections of the high cervical spinal cord and in normal subjects after administration of alpha$_1$-adrenergic blocking drugs.[50] The slowing of heart rate is reflexive and mediated by increased parasympathetic efferent activity.[26]

Phase II consists of a decrease (early phase II, II$_e$) and subsequent partial recovery (late phase II, II$_l$) of arterial pressure and continuous increase of heart rate during straining. Continuous straining impedes venous return to the heart and results in displacement of large amounts of blood from the thorax and abdomen to the limbs. The decrease in venous return produces a reduction in left atrial and left ventricular dimensions, left ventricular stroke volume, and cardiac output. This triggers reflex compensatory tachycardia and vasoconstriction. The tachycardia during phase II results from a prominent early component of inhibition of cardiovagal output and is abolished with muscarinic blockade with atropine.[52,54] There is also a late contribution of increased sympathetic cardioacceleratory output that is blocked with propranolol.[52]

The progressive recovery of arterial pressure during phase II reflects a similarly progressive increase in total peripheral resistance[50] due to increased sympathetic vasoconstrictor activity.[25,78,91] Increased arterial pressure during late phase II is abolished by alpha$_1$-adrenergic blockade with phentolamine.[79] Blood pressure during phase II decreases more if compensatory tachycardia is prevented by atropine and propranolol and if vasoconstriction is prevented by alpha$_1$-adrenergic blockade.[26,79]

Phase III consists of a sudden, brief (1 to 2 seconds) further decrease of arterial pressure and increase of heart rate immediately after the release of the straining. It is essentially mechanical in nature.

Phase IV is characterized by increased systolic and diastolic arterial pressure above control levels, called "overshoot," accompanied by bradycardia relative to the control level of heart rate. In phase IV, venous return to the heart, left ventricular stroke volume, and cardiac output return toward normal, whereas the arteriolar bed remains vasoconstricted, perhaps because of the long time constant of sympathetic responses.[49] This combination results in an overshoot of arterial pressure above control values. The increase in arterial pressure during phase IV can be prevented by beta-adrenergic blockade.[79] Increase in both cardiac output and total peripheral resistance is important in producing the elevation of arterial pressure in phase IV. Recent pharmacologic evidence indicates that increase in cardiac-output-mediated cardioacceleration is more important than vasoconstriction in producing arterial pressure overshoot in phase IV. This overshoot is abolished by beta-blockade with propranolol but is maintained or even exaggerated during alpha-adrenergic blockade with phentolamine.[79] The increase in arterial pressure during phase IV stimulates the baroreceptors and results in reflex bradycardia due to increased parasympathetic activity, which is abolished with atropine.[26,52,54]

Clinical Use of the Valsalva Maneuver as a Test of Autonomic Function

TECHNIQUE

For testing the responses to the Valsalva maneuver, care should be taken, as in any other autonomic test, to ensure that the patient is well hydrated and is not taking medications known to affect blood volume, cholinergic function, or vasoreactivity. At our institution, subjects are tested in supine position and are asked to maintain a column of mercury of 40 mm Hg for 15 seconds through a bugle with an air leak (to ensure an open glottis). The responses are obtained in triplicate and the largest response is accepted.[21] In most laboratories, only heart rate is continuously monitored. We[6,79] and others[41,70,86,90] also monitor beat-to-beat arterial pressure with a noninvasive photoplethysmographic technique. The "normal" Valsalva response should be defined according to the technique used in each laboratory, because several technical variables affect the magnitude of the response.

TECHNICAL VARIABLES AFFECTING THE VALSALVA MANEUVER

The cardiovascular changes during the Valsalva maneuver are determined mainly by the magnitude of hemodynamic change during the forced expiratory effort, the time course and efficiency of reflex cardiovagal and sympathetic vasomotor and cardiomotor responses, and the modification of these responses by interactions with respiratory mechanisms at both central and peripheral levels.[28,29] Accordingly, the responses to the Valsalva maneuver may be affected by (1) the position of the subject during the maneuver, (2) the magnitude and duration of the straining, (3) the breathing pattern before and after the maneuver, including depth and phase of respiration preceding the straining, and (4) the control of respiration after release of the straining.

Effects of Posture. In subjects in the supine position, changes in arterial pressure during phases II and IV may be modest, because the large intrathoracic blood volume may buffer the reduced venous return during phase II. In the supine position, some normal subjects may show a square wave response similar to that of patients with congestive heart failure. The magnitude of arterial pressure decrease during phase II, subsequent systolic blood pressure overshoot in phase IV, and the Valsalva ratio increase with change to sitting and particularly standing position.[86]

Effects of Test Duration. The duration of straining during the Valsalva maneuver has different effects in vagally and sympathetically mediated responses, because of their different latencies and time constants. When the Valsalva maneuver is performed at low expiratory pressures (20 mm Hg), the magnitude of the tachycardia in phase II is independent of the duration of the test, consistent with the short latency of vagal responses.[6] The maximal increase in arterial pressure in late phases II and IV correlates with the duration of the Valsalva maneuver, however, which may reflect the longer latency of sympathetic vasoconstrictor and cardioacceleratory responses.[50] A test duration of 10 seconds is effective, and 15 seconds is a practical optimum and may be sufficient to assess sympathetically mediated responses in a clinical setting.[20]

Effects of Expiratory Pressure. The magnitude of most heart rate and arterial pressure responses during phases II and IV correlates with the magnitude of expiratory pressure used during the Valsalva maneuver.

Maximal arterial pressure and heart rate responses are obtained with expiratory pressures of 40 to 50 mm Hg.[6,7,23,31,39,73,86]

Phase of Respiration. In normal subjects, magnitude of heart rate responses during the Valsalva maneuver is significantly lower if the expiratory strain is preceded by maximal inspiration instead of tidal inspiration.[26,28]

THE VALSALVA RATIO

The Valsalva ratio is defined as the longest R-R interval (pulse interval, in milliseconds) after the maneuver (phase IV) to the shortest R-R interval during the maneuver (phase II).[4,5,8,52,54,60,67,76] In clinical settings, the Valsalva maneuver has commonly been used to calculate the Valsalva ratio. The best of three responses is accepted. In more than 96% of control subjects, this ratio exceeds 1.5.[54,58] It is not affected by sex, but it decreases significantly with age.[57]

PITFALLS OF THE VALSALVA RATIO

There is some evidence that the Valsalva ratio in normal subjects depends mainly on cardiovagal function.[52,54] However, interpretation of this ratio as a test of cardiovagal function in the absence of simultaneous recordings of arterial pressure may be misleading.

1. The Valsalva ratio correlates better with the heart rate response in phase II than in phase IV.[6] Therefore, if sufficient tachycardia in phase II is present, the Valsalva ratio may be "normal" even in the absence of significant bradycardia in phase IV. This may occur in patients with cardiovagal impairment but intact sympathetic innervation.[58]
2. The Valsalva ratio also correlates with the magnitude of arterial pressure overshoot in phase IV.[6] The absence of bradycardia in phase IV, and thus an abnormal Valsalva ratio, may be caused not only by vagal dysfunction but also by the inability to increase arterial pressure in phase IV sufficiently to stimulate the baroreflex.
3. Both the heart rate increase in phase II and the heart rate decrease in phase IV

are critically affected by the magnitude of the decrease in venous return during the Valsalva maneuver, which depends on the position of the subject and the pooling and buffering effect of thoracic vessels.[86]

4. Assessing the integrity of the total baroreflex arc during the Valsalva maneuver by only testing the Valsalva ratio is unreliable, because the magnitude and time course of the heart rate response may be normal despite a response of arterial pressure typical of sympathetic failure.[90]

INDICES OF SYMPTHETIC FUNCTION

The integrity of reflex sympathetic responses cannot be inferred on the basis of the Valsalva ratio. Both the magnitude of the decrease in mean arterial pressure in phase II and the overshoot of arterial pressure in phase IV have been considered indices of vasomotor function.[23,77] However, the magnitude of the decrease in arterial pressure in early phase II is also affected by the heart rate responses, and the overshoot of arterial pressure in phase IV may be more dependent on cardiac output.[79] In our laboratory, the changes in arterial pressure during early and late phases II and IV are used to assess sympathetic vasomotor function.[79] Late phase II is impaired in patients with alpha-adrenergic failure caused by dopamine-beta-hydroxylase deficiency.[10]

Pitfalls of the Valsalva Maneuver

1. The test requires patient cooperation, and therefore cannot be performed in patients who are seriously ill or who have weak respiratory, facial, or oropharyngeal muscles.
2. The maneuver should be avoided in patients with proliferative retinopathy, because of the risk of intraocular hemorrhage.
3. Theoretically, the Valsalva maneuver can precipitate arrhythmias and angina and may cause syncope, particularly in elderly patients with impaired reflex mechanisms that respond to the decrease in venous return.[67]

4. Patients with congestive heart failure, mitral stenosis, aortic stenosis, constrictive pericarditis, and atrial septal defect may have an abnormal square wave response of arterial pressure to the Valsalva maneuver,[67] because of the increase in pulmonary blood volume which is capable of maintaining ventricular filling during the Valsalva strain.

OTHER TESTS OF AUTONOMIC FUNCTION

Pupil Cycle Time

Infrared recording of dark-adapted pupil diameter has been suggested as a good quantitative measure of sympathetic function[40,74] and pupil cycle time as an index of parasympathetic function.[61] Pupil cycle time refers to the frequency of oscillations of the pupil in response to a light stimulus.

Thompson,[87] who introduced the test, has advised caution with performing it. Pupil cycle time is also affected by lesions of the optic nerve,[66] by sympathetic lesions,[11] by lesions of the myoneural junction,[53] by narrow-angle glaucoma,[19] and by the level of retinal illumination.[87] It is also less accurate when pupillary movements are weak.[87]

CHAPTER SUMMARY

Noninvasive cardiovascular tests are reliable and reproducible and are widely used to evaluate autonomic function in human subjects. The heart rate response to deep breathing is probably the most reliable test for assessing the integrity of the vagal afferent and efferent pathways to the heart. This response is usually tested at a breathing frequency of five or six respirations per minute and decreases linearly with age. The Valsalva maneuver consists of a forced expiratory effort against resistance and produces mechanical (phases I and III) and reflex (phases II and IV) changes in arterial pressure and heart rate. When performed under continuous arterial pressure monitoring with a noninvasive (Finapres) technique, the

Valsalva maneuver provides valuable information about the integrity of the cardiac parasympathetic, cardiac sympathetic, and sympathetic vasomotor outputs. The responses to the Valsalva maneuver are affected by the position of the subject and the magnitude and duration of the expiratory effort. In general, it is performed at an expiratory pressure of 40 mm Hg sustained for 15 seconds. The Valsalva ratio, the relationship between the maximal heart rate response during phase II (straining) and phase IV (after release of the straining), has been considered a test of cardiac parasympathetic function. However, without simultaneous recording of arterial pressure, this may be misleading. An exaggerated decrease in arterial pressure during phase II suggests sympathetic vasomotor failure, whereas an absent overshoot during phase IV indicates inability to increase the cardiac output and cardiac adrenergic failure.

REFERENCES

1. Angelone, A and Coulter, NA Jr: Respiratory sinus arrhythmia: A frequency dependent phenomenon. J Appl Physiol 19:479–482, 1964.
2. Anrep, GV, Pascual, W, and Rössler, R: Respiratory variations of the heart rate. I. The reflex mechanism of the respiratory arrhythmia. Proc R Soc Lond [Biol] 119:191–230, 1936.
3. Bainbridge, FA: The influence of venous filling upon the rate of the heart. J Physiol (Lond) 50:65–84, 1915.
4. Baldwa, VS and Ewing, DJ: Heart rate response to Valsalva manoeuvre: Reproducibility in normals, and relation to variation in resting heart rate in diabetics. Br Heart J 39:641–644, 1977.
5. Bannister, R, Sever, P, and Gross, M: Cardiovascular reflexes and biochemical responses in progressive autonomic failure. Brain 100:327–344, 1977.
6. Benarroch, EE, Opfer-Gehrking, TL, and Low, PA: Use of the photoplethysmographic technique to analyze the Valsalva maneuver in normal man. Muscle Nerve 14:1165–1172, 1991.
7. Bennett, T, Farquhar, IK, Hosking, DJ, and Hampton, JR: Assessment of methods for estimating autonomic nervous control of the heart in patients with diabetes mellitus. Diabetes 27:1167–1174, 1978.
8. Bennett, T, Fentem, PH, Fitton, D, et al: Assessment of vagal control of the heart in diabetes: Measures of R-R interval variation under different conditions. Br Heart J 39:25–28, 1977.
9. Bergström, B, Lilja, B, Rosberg, K, and Sundkvist, G: Autonomic nerve function tests: Reference val-

ues in healthy subjects. Clin Physiol 6:523–528, 1986.

10. Biaggioni, I, Goldstein, DS, Atkinson, T, and Robertson, D: Dopamine-β-hydroxylase deficiency in humans. Neurology 40:370–373, 1990.

11. Blumen, SC, Feiler-Ofry, V, and Korczyn, AD: The pupil cycle time in Horner's syndrome. J Clin Neuroophthalmol 6:232–235, 1986.

12. Booth, RW, Ryan, JM, Mellett, HC, et al: Hemodynamic changes associated with the Valsalva maneuver in normal men and women. J Lab Clin Med 59:275–285, 1962.

13. Borgdorff, P: Respiratory fluctuations in pupil size. Am J Physiol 228:1094–1102, 1975.

14. Borst, C, Wieling, W, van Brederode, JFM, et al: Mechanisms of initial heart rate response to postural change. Am J Physiol 243:H676–681, 1982.

15. Brooker, JZ, Alderman, EL, and Harrison, DC: Alterations in left ventricular volumes induced by Valsalva manoeuvre. Br Heart J 36:713–718, 1974.

16. Brooks, CM, Lu, H-H, Lange, G, et al: Effects of localized stretch of the sinoatrial node region of the dog heart. Am J Physiol 211:1197–1202, 1966.

17. Buda, AJ, Pinsky, MR, Ingels, NB Jr, et al: Effect of intrathoracic pressure on left ventricular performance. N Engl J Med 301:453–459, 1979.

18. Cardone, C, Bellavere, F, Ferri, M, and Fedele, D: Autonomic mechanisms in the heart rate response to coughing. Clin Sci 72:55–60, 1987.

19. Clark, CV and Mapstone, R: Autonomic neuropathy in ocular hypertension. Lancet 2:185–186, 1985.

20. Clark, CV and Mapstone, R: Age-adjusted normal tolerance limits for cardiovascular autonomic function assessment in the elderly. Age Ageing 15:221–229, 1986.

21. Cohen, J, Low, P, Fealey, R, et al: Somatic and autonomic function in progressive autonomic failure and multiple system atrophy. Ann Neurol 22:692–699, 1986.

22. Coker, R, Koziell, A, Oliver, C, and Smith, SE: Does the sympathetic nervous system influence sinus arrhythmia in man? Evidence from combined autonomic blockade. J Physiol (Lond) 356:459–464, 1984.

23. Corbett, JL: Some Aspects of the Autonomic Nervous System in Normal and Abnormal Man. Thesis, University of Oxford, 1969.

24. Davies, CTM and Neilson, JMM: Sinus arrhythmia in man at rest. J Appl Physiol 22:947–955, 1967.

25. Delius, W, Hagbarth, K-E, Hongell, A, and Wallin, BG: Manoeuvres affecting sympathetic outflow in human muscle nerves. Acta Physiol Scand 84:82–94, 1972.

26. Eckberg, DL: Parasympathetic cardiovascular control in human disease: A critical review of methods and results. Am J Physiol 239:H581–H593, 1980.

27. Eckberg, DL: Human sinus arrhythmias as an index of vagal cardiac outflow. J Appl Physiol 54:961–966, 1983.

28. Eckberg, DL, Kifle, YT, and Roberts, VL: Phase relationship between normal human respiration and baroreflex responsiveness. J Physiol (Lond) 304:489–502, 1980.

29. Eckberg, DL and Orshan, CR: Respiratory and baroreceptor reflex interactions in man. J Clin Invest 59:780–785, 1977.

30. Ewing, DJ, Campbell, IW, Murray, A, et al: Immediate heart-rate response to standing: Simple test for autonomic neuropathy in diabetes. BMJ 1:145–147, 1978.

31. Ewing, DJ, Martyn, CN, Young, RJ, and Clarke, BF: The value of cardiovascular autonomic function tests: 10 years experience in diabetes. Diabetes Care 8:491–498, 1985.

32. Fox, IJ, Crowley, WP Jr, Grace, JB, and Wood, EH: Effects of the Valsalva maneuver on blood flow in the thoracic aorta in man. J Appl Physiol 21:1553–1560, 1963.

33. Freyschuss, U and Melcher, A: Sinus arrhythmia in man: Influence of tidal volume and oesophageal pressure. Scand J Clin Lab Invest 35:487–496, 1975.

34. Hamilton, WF, Woodbury, RA, and Harper, HT Jr: Physiological relationships between intrathoracic, intraspinal and arterial pressures. JAMA 107:853–856, 1936.

35. Hellman, JB and Stacy, RW: Variation of respiratory sinus arrhythmia with age. J Appl Physiol 41:734–738, 1976.

36. Hering, E: Uber eine reflectorische Beziehung zwischen Lunge und Herz Sitzungsb Akad Wissensch (Wien) 64:333–353, 1871.

37. Hilsted, J: Pathophysiology in diabetic autonomic neuropathy: Cardiovascular, hormonal, and metabolic studies. Diabetes 31:730–737, 1982.

38. Hilsted, J and Jensen, SB: A simple test for autonomic neuropathy in juvenile diabetics. Acta Med Scand 205:385–387, 1979.

39. Hosking, DJ, Bennett, T, and Hampton, JR: Diabetic autonomic neuropathy. Diabetes 27:1043–1054, 1978.

40. Hreidarsson, AB: Pupil size in insulin-dependent diabetes: Relationship to duration, metabolic control, and long-term manifestations. Diabetes 31:442–448, 1982.

41. Imholz, BPM, van Montfrans, GA, Settels, JJ, et al: Continuous non-invasive blood pressure monitoring: Reliability of Finapres device during Valsalva manoeuvre. Cardiovasc Res 22:390–397, 1988.

42. Ingall, TJ: Autonomic nervous system function in ageing and in diseases of the peripheral nervous system. Thesis, University of Sydney, 1986.

43. Joels, N and Samueloff, M: The activity of the medullary centres in diffusion respiration. J Physiol (Lond) 133:360–372, 1956.

44. Kaijser, L and Sachs, C: Autonomic cardiovascular responses in old age. Clin Physiol 5:347–357, 1985.

45. Katona, PG and Jih, F: Respiratory sinus arrhythmia: Noninvasive measure of parasympathetic cardiac control. J Appl Physiol 39:801–805, 1975.

46. Katona, PG, Poitras, JW, Barnett, GO, and Terry, BS: Cardiac vagal efferent activity and heart period in the carotid sinus reflex. Am J Physiol 218:1030–1037, 1970.

47. Khurana RK, Watabiki S, Hebel JR, et al: Cold face test in the assessment of trigeminal-brainstem-vagal function in humans. Ann Neurol 7:144–149, 1980.

48. Koizumi, K, Ishikawa, T, Nishino, H, and Brooks, CM: Cardiac and autonomic system reactions to stretch of the atria. Brain Res 87:247–261, 1975.

49. Korner, PI: Integrative neural cardiovascular control. Physiol Rev 51:312–367, 1971.

50. Korner, PI, Tonkin, AM, and Uther, JB: Reflex and mechanical circulatory effects of graded Valsalva maneuver in normal man. J Appl Physiol 40:434–440, 1976.

51. Krarup, T, Schwartz, TW, Hilsted, J, et al: Impaired response of pancreatic polypeptide to hypoglycaemia: An early sign of autonomic neuropathy in diabetics. BMJ 2:1544–1546, 1979.

52. Leon, DF, Shaver, JA, and Leonard, JJ: Reflex heart rate control in man. Am Heart J 80:729–739, 1970.

53. Lepore, FE, Sanborn, GE, and Slevin, JT: Pupillary dysfunction in myasthenia gravis. Ann Neurol 6:29–33, 1979.

54. Levin, AB: A simple test of cardiac function based upon the heart rate changes induced by the Valsalva maneuver. Am J Cardiol 18:90–99, 1966.

55. Levitt, NS, Vinik, AI, Sive, AA, et al: Studies on plasma glucagon concentration in maturity-onset diabetics with autonomic neuropathy. Diabetes 28:1015–1021, 1979.

56. Low, PA: Pitfalls in autonomic testing. In Low, PA (ed): Clinical Autonomic Disorders: Evaluation and Management. Little, Brown & Company, Boston, 1993, pp 355–366.

57. Low, PA, Opfer-Gehrking, TL, Proper, CJ, and Zimmerman, I: The effect of aging on cardiac autonomic and postganglionic sudomotor function. Muscle Nerve 13:152–157, 1990.

58. Low, PA, Walsh, JC, Huang, CY, and McLeod, JG: The sympathetic nervous system in diabetic neuropathy: A clinical and pathological study. Brain 98:341–356, 1975.

59. Low, PA, Zimmerman, BR, and Dyck, PJ: Comparison of distal sympathetic with vagal function in diabetic neuropathy. Muscle Nerve 9:592–596, 1986.

60. Mackay, JD, Page, MMcB, Cambridge, J, and Watkins, PJ: Diabetic autonomic neuropathy: The diagnostic value of heart rate monitoring. Diabetologia 18:471–478, 1980.

61. Martyn, CN and Ewing, DJ: Pupil cycle time: A simple way of measuring an autonomic reflex. J Neurol Neurosurg Psychiatry 49:771–774, 1986.

62. Masaoka, S, Lev-Ran, A, Hill, LR, et al: Heart rate variability in diabetes: Relationship to age and duration of the disease. Diabetes Care 8:64–68, 1985.

63. Matthes, K and Ebeling, J: Untersuchungen über die Atemschwankungen des Blutdrucks und der Pulsfrequenz beim Menschen. Pflugers Arch Gesamte Physiol 250:747–768, 1948.

64. Mehlsen, J, Pagh, K, Nielsen, JS, et al: Heart rate response to breathing: Dependency upon breathing pattern. Clin Neurophysiol 7:115–124, 1987.

65. Melcher, A: Respiratory sinus arrhythmia in man: A study in heart rate regulating mechanisms. Acta Physiol Scand Suppl 435:1–31, 1976.

66. Miller, SD and Thompson, HS: Edge-light pupil cycle time. Br J Ophthalmol 62:495–500, 1978.

67. Nishimura, RA and Tajik, AJ: The Valsalva maneuver and response revisited. Mayo Clin Proc 61:211–217, 1986.

68. O'Brien, IAD, O'Hare, P, and Corrall, RJM: Heart rate variability in healthy subjects: Effect of age and the derivation of normal ranges for tests of autonomic function. Br Heart J 55:348–354, 1986.

69. Ott, NT, Tarhan, S, and McGoon, DC: Circulatory effects of vagal inflation reflex in man. Z Kardiol 64:1066–1070, 1975.

70. Parati, G, Casadei, R, Groppelli, A, et al: Comparison of finger and intra-arterial blood pressure monitoring at rest and during laboratory testing. Hypertension 13:647–655, 1989.

71. Penaz, J: Photoelectric measurement of blood pressure, volume and flow in the finger. In Albert, R, Vogt, WS, and Helbig, W (eds): Digest of the International Conference on Medicine and Biological Engineering. Conference Committee of the 10th International Conference on Medicine and Biological Engineering, Dresden, 1973, p 104.

72. Persson, A and Solders, G: R-R variations, a test of autonomic dysfunction. Acta Neurol Scand 67:285–293, 1983.

73. Pfeifer, MA, Cook, D, Brodsky, J, et al: Quantitative evaluation of cardiac parasympathetic activity in normal and diabetic man. Diabetes 31:339–345, 1982.

74. Pfeifer, MA, Cook, D, Brodsky, J, et al: Quantitative evaluation of sympathetic and parasympathetic control of iris function. Diabetes Care 5:518–528, 1982.

75. Pfeifer, MA, Weinberg, CR, Cook, D, et al: Differential changes of autonomic nervous system function with age in man. Am J Med 75:249–258, 1983.

76. Pfeifer, MA, Weinberg, C, Cook, D, et al: Sensitivity of RR-variation and the Valsalva ratio in the assessment of diabetic autonomic neuropathy (abstr). Clin Res 31:394A, 1983.

77. Pfeifer, MA, Weinberg, CR, Cook, DL, et al: Autonomic neural dysfunction in recently diagnosed diabetic subjects. Diabetes Care 7:447–453, 1984.

78. Robertson, D, Johnson, GA, Robertson, RM, et al: Comparative assessment of stimuli that release neuronal and adrenomedullary catecholamines in man. Circulation 59:637–643, 1979.

79. Sandroni, P, Benarroch, EE, and Low, PA: Pharmacological dissection of components of the Valsalva maneuver in adrenergic failure. J Appl Physiol 71:1563–1567, 1991.

80. Settels, JJ and Wesseling, KH: Finapres: Noninvasive finger arterial pressure waveform registration. In Orlebeke, JF, Mulder, G, and Van Doornen, LJP (eds): Psychophysiology of Cardiovascular Control: Models, Methods, and Data. Plenum Press, New York, 1985, pp 267–283.

81. Sharpey-Schafer, EP: Effects of Valsalva's manoeuvre on the normal and failing circulation. BMJ 1:693–695, 1955.

82. Sharpey-Schafer, EP and Taylor, PJ: Absent circulatory reflexes in diabetic neuritis. Lancet 1:559–562, 1960.

83. Smith, PH and Madson, KL: Interactions between autonomic nerves and endocrine cells of the gastroenteropancreatic system. Diabetologia 20:314–324, 1981.

84. Smith, SA: Reduced sinus arrhythmia in diabetic autonomic neuropathy: Diagnostic value of an age-related normal range. BMJ 285:1599–1601, 1982.

85. Sundkvist, G, Almér, L-O, and Lilja, B: Respiratory influence on heart rate in diabetes mellitus. BMJ 1:924–925, 1979.

86. ten Harkel, ADJ, van Lieshout, JJ, van Lieshout, EJ, and Wieling, W: The assessment of cardiovascular

reflexes: Influence of posture and period of preceding rest. J Appl Physiol 68:147–153, 1990.

87. Thompson, HS: The pupil cycle time. J Clin Neuroophthalmol 7:38–39, 1987.

88. Vallbona, C, Cardus, D, Spencer, WA, and Hoff, HE: Patterns of sinus arrhythmia in patients with lesions of the central nervous system. Am J Cardiol 16:379–389, 1965.

89. Van Egmond, J, Hasenbos, M, and Crul, JF: Invasive v. non-invasive measurement of arterial pressure: Comparison of two automatic methods and simultaneously measured direct intra-arterial pressure. Br J Anaesth 57:434–444, 1985.

90. Van Lieshout, JJ, Wieling, W, Wesseling, KH, and Karemaker, JM: Pitfalls in the assessment of cardiovascular reflexes in patients with sympathetic failure but intact vagal control. Clin Sci 76:523–528, 1989.

91. Wallin, BG and Eckberg, DL: Sympathetic transients caused by abrupt alterations of carotid baroreceptor activity in humans. Am J Physiol 242:H185–H190, 1982.

92. Watkins, PJ and Mackay, JD: Cardiac denervation in diabetic neuropathy. Ann Intern Med 92:304–307, 1980.

93. Wei, JY and Harris, WS: Heart rate response to cough. J Appl Physiol 53:1039–1043, 1982.

94. Wei, JY, Rowe, JW, Kestenbaum, AD, and Ben-Haim, S: Post-cough heart rate response: Influence of age, sex and basal blood pressure. Am J Physiol 245:R18–R24, 1983.

95. Weinberg, CR and Pfeifer, MA: An improved method for measuring heart-rate variability: Assessment of cardiac autonomic function. Biometrics 40:855–861, 1984.

96. Wexler, L, Bergel, DH, Gabe, IT, et al: Velocity of blood flow in normal human venae cavae. Circ Res 23:349–359, 1968.

97. Wheeler, T and Watkins, PJ: Cardiac denervation in diabetes. BMJ 4:584–586, 1973.

98. Widdicombe, JG: Mechanism of cough and its regulation. Eur J Respir Dis 61 Suppl 110:11–15, 1980.

99. Wieling, W, van Brederode, JFM, de Rijk, LG, et al: Reflex control of heart rate in normal subjects in relation to age: A data base for cardiac vagal neuropathy. Diabetologia 22:163–166, 1982.

100. Yamakoshi, K-I, Shimazu, H, and Togawa, T: Indirect measurement of instantaneous arterial blood pressure in the human finger by the vascular unloading technique. IEEE Trans Biomed Eng 27:150–155, 1980.

101. Zema, MJ, Restivo, B, Sos, T, et al: Left ventricular dysfunction—bedside Valsalva manoeuvre. Br Heart J 44:560–569, 1980.

SECTION 2
Electro-physiologic Assessment of Neural Function
PART G
Sleep and Consciousness

Neural and respiratory impairments are frequent causes of sleep disorders and only recently have been adequately assessed. The various disorders include those of inadequate, excessive, and disordered sleep. The last includes excessive or abnormal movements during sleep. Recording the surface muscle electromyogram in association with these movements while monitoring blood pressure, pulse, and respiration, as described in this section, can help physicians identify, characterize, and define the type and severity of a sleep disorder.

Electrophysiologic assessment of sleep disorders is a superb example of the importance of combining the methods of clinical neurophysiology in assessing the condition of a patient. In sleep disorders, the surface recordings of electromyography, the electrical activity of electroencephalography, and measurements of autonomic function are each a critical part of the total assessment.

413

CHAPTER 37

PHYSIOLOGIC ASSESSMENT OF SLEEP

Peter J. Hauri, Ph.D.
Cameron D. Harris, B.S., R.PSG.T.

The clinical evaluation of sleep and its disorders began in the early 1970s. The goal of this chapter is to review recording and interpretation of sleep studies and to discuss how specific clinical issues determine what data need to be obtained. (The purpose of the chapter is *not* to teach the details of diagnosis and treatment of sleep disorders.)

DEFINITIONS

Polysomnography

Polysomnograms are recorded during the normal sleeping hours of a patient. Patients come to the sleep laboratory 1 to 2 hours before their usual bedtime. After the electrodes and sensors are applied, patients may then watch television until they are ready to go to sleep. They sleep 6 to 8 hours before either awakening spontaneously or being awakened by a technician.

The goal of polysomnography is to quantitate the amount of time spent in various stages of sleep during the night and to document clinically relevant events such as cardiopulmonary abnormalities or sleep-related abnormal motor activity.

The basic format of a polysomnogram is standardized. The required minimum includes (1) continuous monitoring of at least one channel of an electroencephalogram (EEG) (C_3-A_2 or C_4-A_1), (2) one or two channels of an electro-oculogram (EOG), (3) three channels of respiratory data (airflow, effort, and oxygen saturation), (4) electrocardiographic (ECG) recording, and (5) electromyography (EMG) of mental/sub-

415

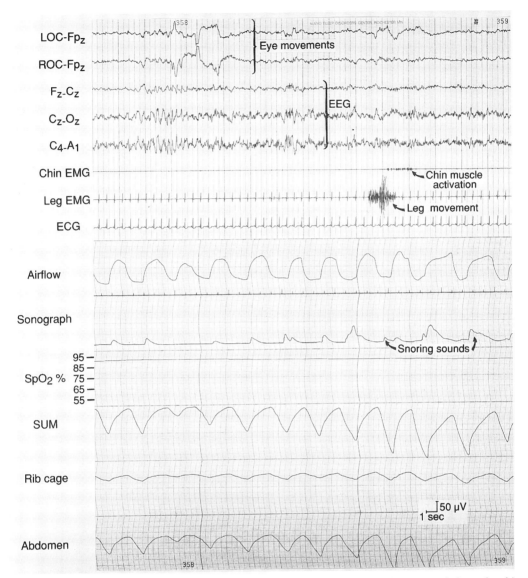

Figure 37–1. Typical polysomnogram montage, including channels for eye movement, electroencephalography, chin and leg EMG, airflow, snoring sounds, oxyhemoglobin saturation, and respiratory effort. Abdomen = abdominal breathing effort by inductive plethysmograph; Airflow = combined signal from nose and mouth; ECG = electrocardiogram; EMG = electromyogram; LOC = left outer canthus; Rib cage = thoracic breathing effort by inductive plethysmograph; ROC = right outer canthus; Sonograph = sound recording; SpO$_2$ = oxyhemoglobin saturation by pulse oximetry; SUM = electronic summation of rib cage and abdominal movements.

mental and tibial areas. Other variables may be monitored as clinically indicated, including snoring, position of the sleeper, additional EEG channels, penile circumference measurements, surface EMG recordings from all four limbs, and esophageal pH.

Polysomnograms are typically recorded at a paper speed of 10 mm/second on 12-22 channel polygraphs. The patient's nighttime behavior should also be observed and recorded during polysomnography.

Analysis of polysomnograms for neonates and infants requires special experience and skill because, for these groups, EEG stages

MC #:   0-000-000

NAME: Smith, Joe

| date of study: | 3-04-89 | age: | 59 | sex: | M | scored by: | CH |

Variables recorded: EEG,EOG, tibial and submental EMG, ECG, airflow by thermo- | **Special circumstances:**   Standard Study
couple, chest wall motion by inductive plethysmograph, sonograph, pulse oximetry

SLEEP ARCHITECTURE

Start time:	23:13	End time:	6:40
			minutes
Time in bed (TIB)			451
Time out of bed			4
Sleep Efficiency (TST/TIB)			81%
Initial sleep latency			7
Initial REM latency			65
Wake after sleep onset			81	%TST
Stage 1			30	8%
Stage2			163	45%
Stage 3/4			80	22%
Stage REM			91	25%
Total sleep time (TST)			364

DISORDERED BREATHING (DB) PROFILE

Event Type	Frequency/sleep hr			Mean Duration (sec)		
	NREM	REM	TST	NREM	REM	TST
Central apnea	0	0	0	0	0	0
Obstructive/mixed	0	0	0	0	0	0
Hypopnea	5	27	11	15	20	18
Total DB events	5	27	11	15	20	18

Effect of body position

	Sleep time		DB index		Snore rating	
		off		off		off
	back	back	back	back	back	back
NREM	47	225	NREM	19	2	%sleep time	30	20
REM	0	91	REM	0	27	grade (0-4)	2	1

AROUSAL PROFILE

duration	arousal index (#/hr)	movement related (%)	breathing related (%)
< 15 seconds	16	3%	67%
>=15 second	3	0%	6%
Total	19	3%	58%

PERIODIC MOVEMENT INDEX

| Movements/Sleep Hour | 2 |
| Percent with Arousals | 30% |

OXYGEN SATURATION (SpO2, %)

Awake baseline   94
Range: NREM 90-95   REM 80-95

| Saturation range, % |
>=90	80-89	70-79	60-69	< 60
%TIB	95.0	5.0	0.0	0.0	0.0

Mean saturation: 92

Cardiac rhythm abnormalities:
NONE

EEG abnormalities:
NONE

Technical comments:

IMPRESSION: The overnight sleep study showed a normal initial sleep latency and REM latency. Sleep architecture is also within normal limits. The patient has mild obstructive hyponea while sleeping on his back and during REM sleep. He has a significant amount of wakefulness after sleep onset, totaling more than one hour.

Figure 37–2. A typical polysomnogram report. This patient later was diagnosed to have narcolepsy (based partly on results of a multiple sleep latency test). The summary statistics are prepared by a polysomnographic technologist, and the impression is provided by the supervising physician after reviewing the raw data.

are defined differently and respiratory and other behaviors during sleep are unique.

Polysomnograms usually are interpreted visually. If scoring of sleep stages and other polysomnographic variables is done by computer,[8] visual inspection of the raw data is still necessary because interpretation of many nocturnal events requires clinical skill (e.g., other sleep disorders may masquerade as sleep-disordered breathing or some patients may have indeterminate sleep that is not easily scored by machine).

An excerpt from a typical polysomnogram is shown in Figure 37–1. Figure 37–2 shows a typical summary of a polysomnographic recording (in this case, from a patient who complained of excessive sleepiness). Table 37–1 summarizes the usual placement of electrodes and sensors and amplifications typically used for a polysomnogram at our clinic.

Multiple Sleep Latency Test

A *multiple sleep latency test (MSLT)* consists of four or five 20-minute rest periods in bed spaced 2 hours apart and almost always recorded on the day following a night with polysomnography. *The goal of the test is to quantitate physiologic sleepiness during waking hours and to determine the occurrence of REM sleep near sleep onset.* For this test, patients are asked to remain in the laboratory for the entire day. They may read, watch television, or engage in other quiet activities, but they must not sleep between scheduled naps. Recordings performed during the MSLT are simpler than those performed during polysomnography. Typically, only six channels are used: two EOG channels, one or two EEG channels (C_3-A_2 required and O_1-A_2 strongly suggested), one mental/submental EMG channel, and the ECG channel. Occa-

Table 37–1. **TYPICAL SETTINGS FOR A POLYSOMNOGRAM**

Signal	Sensor Type and Placement	Sensitivity	LFF*	HFF*
EOG	Electrodes placed 1 cm lateral and 1 cm inferior to outer canthi referenced to F_{pz}	5–7.5 μV/mm	0.3 Hz	90 Hz
EEG	Electrodes placed at F_z, C_z, O_z, C_3, C_4, A_1, and A_2[†]	5–7.5 μV/mm	0.3 Hz	90 Hz
Chin EMG	Electrodes placed over mentalis, mylohyoid, and digastric muscles	2 μV/mm	10 Hz	90 Hz
Leg EMG	Electrodes placed over belly of anterior tibialis muscle of each leg	2 μV/mm	10 Hz	90 Hz
ECG	Electrodes placed on each shoulder	50 μV/mm	0.3 Hz	90 Hz
Oxygen saturation	Light absorbance sensor placed on earlobe or finger	100 MV/cm[‡]	DC	3 Hz
Airflow	Thermocouple placed in front of both nares and the mouth, connected in series to provide one signal	Transducer dependent[‡]	0.05 Hz	3 Hz
Breathing effort	Inductors (elastic bands with embedded wires) placed around the chest, just under the axillae, and around the abdomen at the level of the navel	Patient dependent[‡]	DC–0.05 Hz	3 Hz
Snoring	Microphone placed in contact with the skin slightly superior and lateral to thyroid cartilage	Transducer dependent[‡]	10 Hz	90 Hz

*Low (LFF) or high (HFF) ½ amplitude frequency cut-off, may vary depending on manufacturer of polygraph.
[†]F_z, C_z, O_z are not required by Rechtschaffen and Kales.[13]
[‡]Optimum sensitivity may depend on manufacturer or equipment used or recording conditions. ECG = electrocardiogram; EEG = electroencephalogram; EMG = electromyogram; EOG = electro-oculogram.

sionally, airflow channels are also recorded. Detailed guidelines for the MSLT are given by Carskadon and associates.[4]

For each nap, sleep latency and the occurrence of REM sleep are noted. For the MSLT, *sleep latency* is defined as the time between "lights out" and the beginning of 15 consecutive seconds of stage I sleep. If no sleep occurs during the first 20 minutes in bed, the test is discontinued and the sleep latency is recorded as 20 minutes. After sleep onset has occurred, patients are allowed to sleep for 15 minutes to determine whether they will enter REM sleep.

A rarely used variant of the MSLT is the *maintenance of wakefulness test (MWT)*. Conditions are the same except that the patient sits in a comfortable chair in a dimly lit room and is asked to try to stay awake rather than to try to sleep.

EVALUATION OF SLEEP

An 8-hour polysomnogram recorded at the usual speed of 10 mm/second yields 960 pages, each 30 cm wide. On review of these pages, one sees that periods of wakefulness, non-REM (NREM) sleep, and REM sleep *gradually* wax and wane. For example, falling asleep usually takes a few seconds, often minutes, during which time the typical alpha waves of relaxed wakefulness gradually disappear while the slower waves of stage 1 sleep become prominent. Similarly, during the first hour of sleep, delta waves very gradually increase as sleep progressively deepens. There is no abrupt switching from one sleep stage to the next except possibly when the sleeper is suddenly awakened by an intense stimulus. Nevertheless, to make the data more manageable, each page of the polysomnogram is classified into one of six possible stages: wakefulness, REM sleep, and NREM sleep stages 1 to 4. The scoring for these six stages is done according to rules that have been modified only slightly since they were first published by Rechtschaffen and Kales[13] in 1968.

According to the Rechtschaffen and Kales system, to score wakefulness and the four stages of NREM sleep, one observes only a one-channel EEG (C_3-A_2 or C_4-A_1). Because stage 1 of NREM sleep and REM sleep are quite similar on the EEG, three additional channels are consulted to score REM sleep: two channels of EOG (left and right outer canthus) and a mental/submental EMG channel.

Traditional Sleep Stages

Wakefulness is scored if there is low-voltage random fast activity or activity in the 8- to 12-Hz (alpha) range over the majority of a 30-second epoch (Fig. 37–3). Rapid eye movements, eye blinks, and high tonus in the chin EMG are also typical of wakefulness but are not defining criteria.

Stage 1 is characterized by a relatively low-amplitude mixed-frequency EEG predominantly in the 3- to 7-Hz range (Fig. 37–4). For each page, alpha activity has to be less than 50% of the recording. Typically, there are slow eye movements and decreased tonus on the chin EMG, but the latter two findings are not defining characteristics. Amplitude and predominance of 3- to 7-Hz activity increase as stage 1 deepens. There may be numerous vertex sharp waves toward the end of stage 1.

Stage 2 is characterized by the appearance of sleep spindles (bursts of 12- to 14-Hz activity lasting 0.5 to 1.5 seconds) and by K complexes (well-delineated, negative EEG deflections that are followed by a positive component, the entire complex lasting a minimum of 0.5 second) (Fig. 37–5). The limited EEG coverage typical for sleep staging does not allow localization, so that both F waves and V waves with an initial negative component followed by a positive one and with a total duration of at least 0.5 second

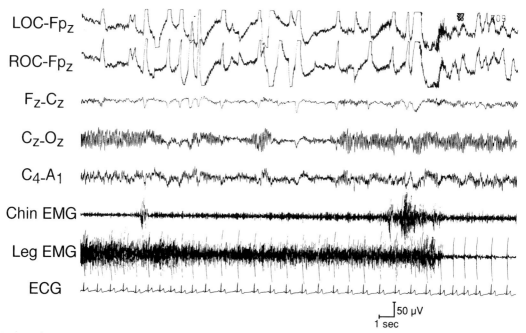

Figure 37–3. Wakefulness. Note the prominent alpha rhythm, rapid eye movements, and high EMG activity in the chin and legs.

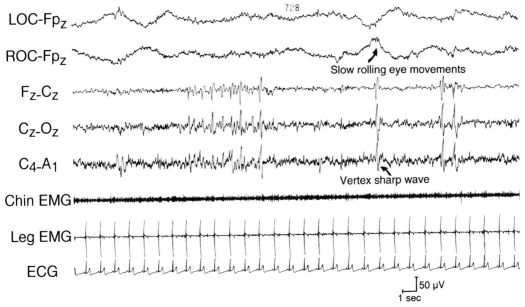

Figure 37–4. Stage 1 sleep. Note the vertex sharp waves, the slow rolling eye movements, and, also, the absence of alpha rhythm.

are considered K complexes when scoring sleep. Because sleep spindles and K complexes are discrete and intermittent, intervals as long as 3 minutes of stage 1–appearing EEG between K complexes or spindles are still scored as stage 2 unless there is evidence to the contrary (e.g., large body movements followed by alpha waves or disappearance of EMG activity and rapid eye movements).

Stages 3 and 4 are frequently combined and called *slow wave sleep* (Figs. 37–6 and 37–7). The defining criteria are large-amplitude slow waves (minimally 75 μV peak-to-peak, 0.5 to 2 Hz). Stage 3 is scored if an epoch contains 20% to 50% of 0.5- to 2-Hz waves, whereas stage 4 is scored if more than 50% of the epoch consists of 0.5- to 2-Hz waves.

With the patients' increasing age, the overall amplitude of the EEG diminishes. There still may be considerable activity in the 0.5- to 2-Hz range, but it no longer produces amplitudes exceeding 75 μV. Most sleep clinicians still score stages 3 and 4 in the elderly if the amplitudes of these waves reach at least 50 μV.

Stage REM is defined by a relatively low-amplitude mixed-frequency EEG, similar to that seen in stage 1, in combination with a markedly decreased tonus in chin EMG activity and with episodic bursts of rapid eye movements (Fig. 37–8). Sawtooth waves (sharp waves in the theta range with a small sawtooth about halfway up or down the main wave) are seen only in REM sleep, but not in all patients. Twitches and transient increases in chin EMG activity are often seen in REM sleep, but a sustained increase of chin EMG tonus is incompatible with the scoring of normal REM sleep.

REM sleep may be phasic or tonic. During phasic REM, eye movements are dense, many chin EMG twitches occur, and there is increased variability in ECG and respiration. Tonic REM epochs are relatively devoid of these phasic features.

Movement time is not a sleep stage but is scored whenever movement artifacts obscure EEG channels for at least 50% of the epoch and if preceding and subsequent epochs are scored as sleep. In such a case, one does not know whether the patient actually awakened during the movement or whether the movement occurred exclusively during sleep.

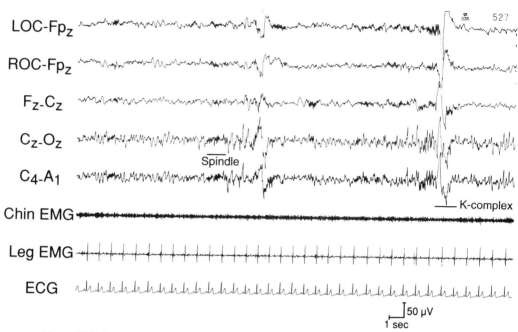

Figure 37–5. Stage 2 sleep. Note sleep spindles and K complexes that characterize stage 2 sleep.

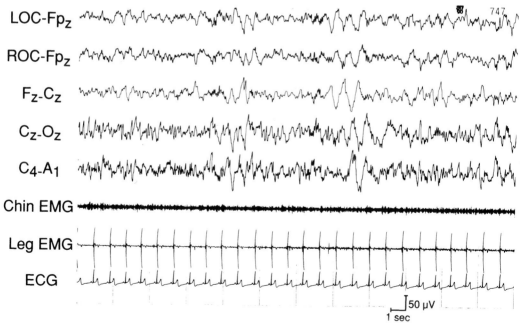

Figure 37–6. Stage 3 sleep. Between 20% and 50% of the epoch contains 0.5- to 2-Hz activity of amplitude greater than 75 μV.

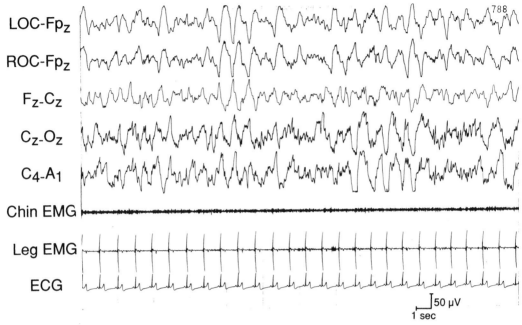

Figure 37–7. Stage 4 sleep. Of the epoch, 50% or more contains 0.5- to 2-Hz activity of amplitude greater than 75 μV.

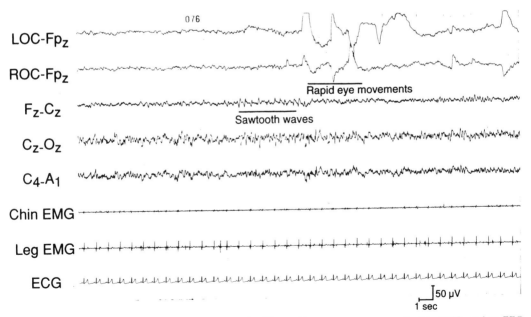

Figure 37–8. Stage REM sleep. REM sleep is characterized by rapid eye movements, atonic chin EMG, and an EEG pattern similar to that seen in stage 1 sleep. Note the sawtooth waves also.

Nontraditional Measures of Sleep

The four sleep stage phenomena that are clinically significant but not included in the original Rechtschaffen and Kales scoring system are arousals, alpha intrusions, REM sleep without atonia, and body position.

Arousals are defined as abrupt shifts toward faster EEG frequencies which last for only a few seconds and, therefore, cannot be scored as wakefulness because they do not dominate an entire epoch. A subcommittee of the American Sleep Disorders Association[14] has defined EEG arousal scoring rules, which require that (1) the subject must be asleep for at least 10 seconds before and after the arousal, (2) the EEG frequency shift must last for a minimum of 3 seconds, and (3) arousals scored in REM sleep must have both increased EEG frequency and increased chin EMG activity. Scoring these arousals is important for an assessment of sleep quality: the more arousals there are, the worse one's perception is of how well one has slept and the sleepier one is during the subsequent day.[3] Of course, if an arousal lasts longer than 15 seconds, wakefulness predominates during the epoch in question and a full awakening is scored.

Alpha intrusions are scored if alpha activity (8 to 12 Hz) that is easily visible to the eye permeates all NREM stages (Fig. 37–9). This phenomenon, also called *alpha-delta sleep*[6] or *nonrestorative sleep*,[11] is often associated with chronic pain (e.g., in fibrositis), thyroid problems, or chronic stimulant abuse, but it may also occur in patients who have no known medical disorder.

There is no formal way to score alpha intrusions. Typically, the scorer simply estimates how much of the NREM sleep record is permeated by alpha activity. The phenomenon seems to become more significant clinically if at least three-quarters of NREM sleep is permeated by alpha waves.

REM sleep without atonia is scored when both EEG and EOG suggest REM sleep but the chin EMG does not show the expected muscle atonia. The technician often observes vigorous twitching and apparently purposeful movements during this type of sleep, such as hitting a pillow or appearing to engage in a fist fight (Fig. 37–10). This is seen predominantly in middle-aged or older

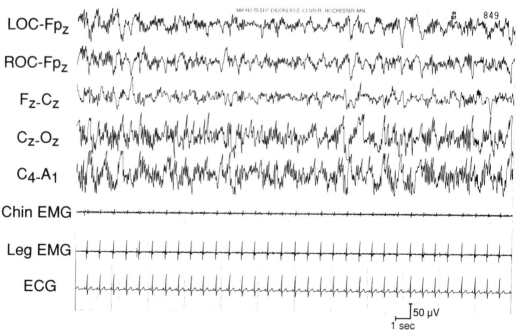

Figure 37–9. Alpha intrusions. This polysomnogram of a 50-year-old woman with a complaint of chronic fatigue illustrates intrusion of diffuse alpha activity into slow wave sleep.

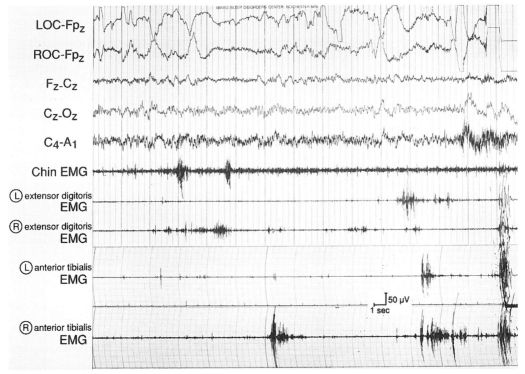

Figure 37–10. Atypical REM sleep without muscle atonia. The montage is modified to record EMG activity from all four limbs from a 67-year-old man with Parkinson's disease. There are frequent bursts of activity in the limb EMG leads. Typically, the movements associated with this activity are related to dream content.

men or in patients with neurologic disorders (e.g., Parkinson's disease) who have a REM behavior disorder (i.e., acting out of their dreams).[9]

Body position may affect the severity of disordered breathing. It is always scored in clinical polysomnograms. This is done either with a special position indicator worn by the patient or by observing the patient through a closed circuit video monitor. If the patient does not spontaneously sleep part of the time on the back and part of the time on the side during the recording, he or she is usually awakened and asked to sleep in the other position to assess respiration in both positions.

Summary Statistics for Sleep Variables

When all epochs have been scored, summary statistics are computed (see Fig. 37–2). They include the following:

Time in bed is from "lights out" to getting out of bed in the morning for the last time, minus time spent out of bed during the night (e.g., for trips to the bathroom). To have a valid polysomnogram, a minimum of 6 hours in bed is typically required.

Total sleep time includes all epochs scored as stages 1, 2, 3, and 4 or as REM sleep.

Sleep efficiency is computed as the percentage of the time in bed that is spent asleep.

The norms for sleep stages and percentages are reported in Table 37–2. It is ironic that in most sleep laboratories sleep is scored according to the Rechtschaffen and Kales[13] rules, but only weak norms exist for that system. The norms detailed in Table 37–2 are from the Williams, Karacan, and Hursch[18] system, which scores sleep from a frontal-occipital derivation and, therefore, reports more stage 3 and 4 sleep and slightly more wakefulness than the Rechtschaffen and Kales system.

Initial sleep latency is the time from "lights out" to the first epoch of scored sleep. Some

Table 37–2. **NORMATIVE DATA FOR SLEEP STAGES RELATED TO AGE**

	Age, in Years										
	4	7.5	11	14	17.5	25	35	45	55	65	75
Time in bed (min)	618.7	599.7	587.3	506.35	477.41	443.94	439.1	435.3	444.65	458.62	500.07
Total sleep time (min)	593.5	580.8	559.9	484.6	451.71	424.61	423.58	407.1	410.28	406.12	393.20
Stage 1											
Percentage	2.12	2.30	2.96	3.63	3.88	4.31	4.94	6.60	6.20	8.71	8.03
Minutes	12.7	13.4	16.8	17.8	17.8	18.5	21.3	28.0	26.7	38.6	36.9
Stage 2											
Percentage	45.05	47.92	47.76	46.33	49.24	48.96	55.3	54.38	59.76	55.78	53.86
Minutes	270.8	279.7	271.3	226.5	225.9	209.8	238.1	230.4	257.5	246.9	247.6
Delta											
Percentage	20.55	20.98	20.93	22.82	23.24	19.22	13.2	10.30	7.78	4.92	5.70
Minutes	123.5	122.4	118.9	111.8	106.6	82.4	56.8	43.6	33.5	21.8	26.2
REM											
Percentage	30.01	28.32	27.21	26.16	22.07	26.62	24.84	24.76	21.62	22.26	18.57
Minutes	180.4	165.3	154.6	128.2	101.3	114.1	107.0	104.9	93.2	98.5	85.4

Source: data from Williams et al.[18] REM = rapid eye movement sleep.

compute latency to stage 1 and others to stage 2 sleep. Because insomniacs often fall asleep quite early for a minute or two and then remain awake for a considerable time after that, some laboratories score sleep latency to the first sleep epoch that is followed by a minimum of 10 minutes of sleep. Among the three measures of sleep latency, latency to stage 2 is preferred by clinicians.

Wake after sleep onset is the total amount of wakefulness scored between the time a patient falls asleep and the time he or she gets up in the morning. Others score it only between first and last sleep, scoring as *early morning wakefulness* the time between the patient's last sleep episode and finally getting up. Although the latter scoring method is preferred because it gives more pertinent information, it is used less frequently by sleep disorders centers.

Initial REM latency is scored by adding all the time spent sleeping (stages 1, 2, 3, 4) before the first epoch scored as REM. Initial REM latencies of shorter than 15 minutes suggest sleep-onset REM periods, important in the diagnosis of narcolepsy. Initial REM latency decreases with age (Table 37–3). If the initial REM latency is considerably shorter than expected for a patient's age, it is often seen as a diagnostic marker for a psychiatric disorder, including, but not definitively, a major affective disorder,[17] unless it can be explained by medication effects, previous REM deprivation, or a severely disrupted sleep-wake schedule.

Arousal index counts the number of awakenings and arousals per sleep hour. (As discussed previously, arousals are short awakenings of a few seconds; awakenings last at least 15 seconds.) Arousal indices of 10 and fewer (per hour) are considered normal. To interpret arousal indices, it is important to know how many of the arousals and awakenings are caused by disordered breathing or by periodic limb movements. Arousals may be associated with subtle increases in upper airway resistance which are not easily recognized as hypopnea.[5] Markedly increased arousal indices unassociated with periodic limb movements or disordered breathing are often seen in patients with pain or other medical disorders or in those with psychologic distress, especially anxiety.

Interpretation of the various sleep scores and indices requires clinical judgment and should be made conservatively. Many studies[1,10] have shown that the patient's sleep in the laboratory on the first night may be atypically poor (*first-night effect*). Alternatively, some insomniacs sleep especially well on the first night in the laboratory (*reverse first-night effect*).[7] Because the first night spent in the laboratory is frequently atypical, research studies often use a minimum of three laboratory nights to assess sleep and discard the first as adaptation. Economically, however, this is not feasible for clinical studies. Therefore, clinical studies are rarely performed to assess sleep in itself but usually to assess what happens *during* sleep to breathing, leg movements, brain waves, or heart function.

A comparison between the objectively recorded sleep variables as discussed in this chapter and the patients' self-reports of their sleep (obtained by their answers to a questionnaire the following morning) may yield clinical insight, especially if there is a large discrepancy. Small discrepancies of up to about 90 minutes in total sleep time are within normal limits. If the discrepancy is much larger, it is important to consider whether it can be explained by a high arousal index or alpha intrusions into sleep or whether it may be a *sleep-stage misperception*, that is, totally normal sleep lasting a minimum of 6.5 hours in a patient who reports sleeping little or not at all.

Development of sleep over an entire night is best represented by results of a *sleep histogram* (Fig. 37–11). These histograms, often plotted by computer, may have additional data not plotted in Figure 37–11. Such data can indicate whether the sleep disturbances (e.g., disordered breathing, periodic limb

Table 37–3. MAYO SLEEP DISORDERS CENTER GUIDELINES FOR EVALUATION OF NORMAL LOWER BOUNDS FOR REM LATENCY

Age, y	Initial REM Latency, min
15–24	70
25–34	60
35–45	45
45–60	35
60+	30

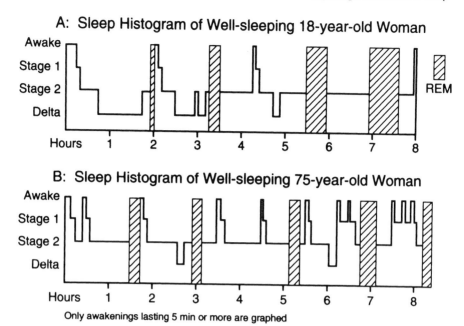

Figure 37–11. Sleep histogram of (*A*) an 18-year-old woman and (*B*) a 75-year-old woman . Note the preponderance of slow wave sleep early in the night in the younger woman and its relative absence in the older woman. The older woman has many more awakenings and earlier onset of REM sleep.

movements) are evenly distributed over the entire sleep period, associated with specific sleep stages, or amassed (e.g., in the first 2 or 3 hours of sleep). They also show whether excessive wakefulness occurs early or late (possibly suggesting influences of the circadian rhythm or of psychiatric factors).

ASSESSING RESPIRATION DURING SLEEP

The majority of polysomnographic studies are performed to assess disordered breathing during sleep.

Definitions

Apnea is defined as cessation of all airflow for longer than 10 seconds. This definition seems to be accepted in most sleep laboratories. There is considerably more disagreement among laboratories about the definition of *hypopnea*. The preferred definition for hypopnea is a temporary decrease in ventilation, as observed by measuring airflow, in combination with a greater than 2% decrease in oxygen saturation. Some define hypopnea as a decrease of 50% or more in the amplitude of the airflow channel, but this is questionable. The thermistors in front of the nose and mouth measure air temperature, which is not linearly related to airflow.

Obstructive apnea is defined as a cessation of airflow in the presence of measurable, often gradually increasing respiratory effort. During obstructive apnea, one often observes paradoxical breathing, that is, the chest expands as the stomach contracts (Fig. 37–12*A*). Obstructive hypopnea is often marked by crescendo inspiratory snoring. *Central apnea* shows a cessation of airflow coupled with a *lack* of respiratory effort (Fig. 37–12*B*). Sleep-onset central apnea may be relatively benign; it often simply indicates anxious overbreathing during wakefulness, with a normalization during sleep. Periodic central apnea *during* sleep is more worrisome, indicating either poor cardiac output (longer circulation time between lung and blood-gas sensors in the carotid body) or problems with neuronal control of respiration. *Mixed apnea* shows an initial central component followed by obstructive apnea (Fig. 37–12*C*).

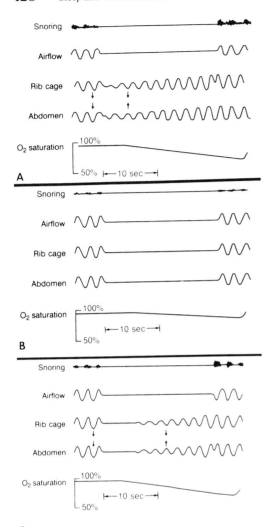

Figure 37–12. Types of sleep apnea. (*A*) Obstructive apnea is characterized by loud intermittent snoring, complete cessation of airflow, paradoxical movement of the rib cage and abdomen, and moderate-to-severe oxygen desaturation. (*B*) Central apnea has absence of snoring, simultaneous cessation of airflow and respiratory effort, and usually only mild-to-moderate oxygen desaturation. (*C*) Mixed apnea. An initial central apnea followed by obstructive apnea that usually produces moderate-to-severe oxygen desaturation. (From Kaplan, J: Diagnosis and therapy of sleep-disordered breathing. In Burton, GG, Hodgkin, JE, and Ward, JJ [eds]: Respiratory Care: A Guide to Clinical Practice, ed 3. JB Lippincott, Philadelphia, 1991, p 282, with permission.)

Questions to be answered by polysomnography include the number and duration of sleep apneas and hypopneas and whether they occur in all sleep stages, predominantly at sleep onset, or predominantly during REM sleep. The study needs to differentiate among obstructive, central, and mixed apneas and hypopneas. Does disordered breathing occur in all body positions or predominantly in only one (e.g., when a patient sleeps on the back)? How many disordered breathing events cause arousals? How much does oxygen saturation decrease? Is there a relationship between these disordered breathing events and cardiac arrhythmias?

Although scoring of sleep stages has been standardized since 1968, evaluation of respiration during sleep still varies among laboratories. To answer the questions posed earlier, five channels of data are needed. Of primary importance are measurement of airflow through the nose and mouth and measurement of breathing effort. Snoring intensity also needs to be assessed, usually with a decibel meter or by a technician's judgment. A continuous measure of blood gas (either O_2 saturation by oximetry or end tidal CO_2) is needed to score hypopneas (see previous discussion). As already mentioned, the body position of the sleeper also has to be assessed.

If continuous positive airway pressure (CPAP) is considered as a treatment for disordered breathing during sleep, polysomnography is used to determine the minimal CPAP pressure that eliminates all disordered breathing (including snoring) in all sleep stages and in all positions.[16] Do other sleep disorders remain after breathing is normalized by CPAP? For example, there may still be serious periodic limb movements or spontaneous awakenings associated with anxiety or there may be central sleep apneas not treatable with CPAP. A polysomnographic study also determines the initial reaction of the nasal membranes to CPAP. For example, in some patients, this treatment causes excessive nasal secretions or excessive swelling. Therapeutic trials in the laboratory also help the patient adapt to wearing the mask while sleeping under constant professional supervision.

Airflow

In the sleep disorders center, airflow is usually detected by thermocouples or thermistors placed in the airstream in front of the nose and mouth. Temperature changes indicate that air is flowing, because expired air is warmer than inspired air. Although this does not quantitatively measure how much air is flowing, relative changes (over a few seconds) in the amplitude of this channel's output indicate increases or decreases in breath volume. Temperature sensors placed in front of each nostril and the mouth are connected in series to form one signal, because there may be changes from nose breathing to mouth breathing or cyclic alternations between one nostril and the other during sleep.

Obviously, an absolute measure of airflow would involve putting a mask over the nose and mouth and attaching it to a pneumotachometer. This is not considered necessary or desirable for clinical studies, because it unnecessarily disturbs sleep. It is, however, often done for research.

Effort

Whether effort is expended by patients for breathing can be assessed by inductive plethysmography, various strain-gauge methods, intercostal EMG, or esophageal balloon. This is important in differentiating central apnea from obstructive apnea.

The most elegant technique for monitoring respiratory effort during sleep is *inductive plethysmography*.[15] In brief, the human breathing pump consists of two major components: the diaphragm and the rib cage. Motion of the diaphragm changes the volume of the abdominal cavity, and motion of the rib cage changes the volume of the thoracic cavity. The inductive plethysmograph measures the cross-sectional area of each of these two chambers. Changes in the cross-sectional area of the two cavities are relatively proportional to changes in the volume of each. An electrical summation of the signals from rib cage and abdomen provides a rough estimate of overall tidal volume.

The inductive plethysmograph has wide elastic bands that are placed around the chest and abdomen. Small wires sewn into these elastic bands in a zigzag pattern form an induction coil connected to an oscillator circuit. As the breathing motion changes the cross-sectional areas encircled by the bands, the frequency of the oscillator changes.

The recording of paradoxical breathing—out-of-phase movements of the rib cage and abdomen (Fig. 37–12*A*)—is crucial in marginal patients in whom it is difficult to distinguish between obstructive and central events. Well-calibrated induction plethysmography can show paradoxical breathing even during hypopnea, when upper airway resistance is increased but a considerable amount of air is still flowing.

Another device commonly used for detecting respiratory effort is the strain gauge, which takes advantage of the same chest wall motion. However, rather than measuring cross-sectional area, a strain gauge is sensitive to changes in thoracic or abdominal circumference. This device typically consists of mercury-filled capillary tubing that can be taped over a portion of the chest and abdomen. As body circumference increases, the capillary tubing stretches, elongating the mercury column. This causes measurable resistance changes, mainly because the mercury column becomes thinner.

Another device used to sense changes in circumference is a *piezoelectric transducer*. It consists of a thin crystal that is placed under physical stress, thus creating an electrical potential difference between the surfaces of the crystal. By connecting this crystal to a set of straps encircling the body, the stress level on the crystal is varied in proportion to change in body circumference.

A shortcoming of strain gauges and piezoelectric transducers is that they measure circumference rather than cross-sectional area. Particularly in obese patients, circumference changes are not necessarily proportional to cross-sectional changes. Obese patients may exert considerable respiratory effort, causing their cross-sectional area to change significantly (from a near circle to an ellipse) without much change in circumference.

Some laboratories record breathing effort by monitoring EMG activity in intercostal muscles. This is occasionally problematic because intercostal activity is inhibited during REM sleep, during which time mainly the diaphragm functions. Because of this, electrodes are placed over the sixth or seventh intercostal space, where diaphragmatic and intercostal EMG activity can be recorded.

Technically, the best measure of respiratory effort is *esophageal pressure*. This is recorded by placing a transnasopharyngeal catheter in the upper esophagus. As the diaphragm contracts and the rib cage is extended, negative pressure is created in the intrapleural space. This pressure is detected by a balloon-tipped catheter attached to an external pressure transducer or by a transducer-tipped catheter. However, this procedure is unpleasant for most patients, although patients with excessive daytime somnolence often sleep quite well with a balloon in place. We reserve its use for patients in whom it is difficult to distinguish between central apneas and obstructive apneas with respiratory plethysmography.

Blood Gases

Continuous monitoring of a blood gas (O_2 or CO_2) is mandatory, because it provides information about the severity of a respiratory dysfunction (especially during hypopneas). Monitoring of oxygen saturation is performed easily using a *pulse oximeter*, a device that measures the light absorption of two wavelengths of red light passed through a capillary bed, such as that of the earlobe or the nailbed of a finger. The wavelengths used match the peak absorption factors of oxyhemoglobin and deoxyhemoglobin. The oximeter then calculates the ratio of oxyhemoglobin to total hemoglobin and translates it into a digital display and an analog voltage that is written out on the polygraph.

In addition to using one channel of the polysomnogram for ear oximetry, it is often useful to record oximetry on an independent strip-chart recorder, running this strip at a slow speed of about 2 to 5 inches/hour. This provides a quick overview of oxygen-

ation for the entire night and allows easy identification of trouble spots, that is, when breathing was compromised (Fig. 37–13).

Blood oxygen tension can be measured by *transcutaneous PO_2* recordings. A specifically designed electrode is placed directly on the skin and heated to about 44°C. It measures the transpired PO_2, a relatively accurate measure of arterial PO_2 if tissue perfusion is good. Although this method is widely used in neonatal intensive care units, the skin of most adults is too thick to permit accurate assessment. Also, the time constant in this method is inherently slower and does not accurately reflect the fluctuations in PO_2 associated with individual episodes of apnea.

Another measurement of blood gases used in sleep laboratories is assessment of *end-expired CO_2 (capnography)*. During inspiration, very little CO_2 is present in the air. During expiration, PCO_2 increases until it approximates alveolar PCO_2 at the end of expiration. This end-expiration PCO_2 is high on the first breath after apnea and returns to baseline levels after a few breaths. However, the technical problems with this method are much greater than those with pulse oximetry.

Snoring Sounds

In many sleep laboratories, a small microphone is attached to the throat of the patient or a sound-pressure measuring device is mounted on the bedboard. Crescendo snoring and gasping are often observed during disordered breathing. Recording these sounds (a *sonogram*) helps with the final

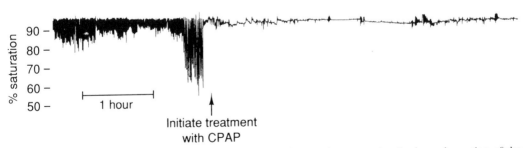

Figure 37–13. Oximetry strip chart recording. The repetitive desaturations occurring in the early portion of the tracing are caused by obstructive apneas. Initiation of continuous positive airway pressure (CPAP) eliminates the apnea during the later portion of the recording.

characterization of the type of disordered breathing. In addition, many laboratories use the technician's personal judgment, grading loudness of snoring in four steps: 1, barely audible; 2, audible at the bedside; 3, audible from the open door to the bedroom; and 4, audible through the closed door.

Summary Statistics for Respiratory Variables

The following respiratory variables are typically reported on the polysomnographic summary sheet:

Number of central, obstructive, or mixed apneas and hypopneas. Usually, this is given as the average number of these events per hour of sleep. Occasionally, this average is computed for each stage of sleep.

Mean duration of episodes of apnea and hypopnea during NREM, REM, and total sleep time is often reported, as well as the longest duration of an apneic episode in each stage of sleep.

Apnea index indicates the number of episodes of apnea per sleep hour. An index of 15 or higher is clinically significant. Others report an apnea plus hypopnea index for total sleep or report the apnea index for each sleep stage.

Mean oxygen saturation values during wakefulness, REM sleep, and NREM sleep are usually given or a range is indicated. Also important is the percentage of time during sleep that the patient spends with oxygen saturation above 90%, 80%, or 70%, and so forth.

Effect of body position typically is indicated as the number of disordered breathing events (apneas and hypopneas) per sleep hour spent on the back and off the back, both for NREM sleep and REM sleep.

Snoring is often reported on a scale from 1 to 4 (see above). Clinically, one looks for crescendo snoring that leads to arousal, as the sleeper struggles harder and harder for air.

The number of arousals caused by disordered breathing is especially important, because occasionally breathing is not disordered enough to be scored as apnea or hypopnea but nevertheless awakens the sleeper (upper airway resistance syndrome).

ASSESSING PERIODIC LIMB MOVEMENTS

Periodic movements of the legs (more rarely the arms) at intervals of 10 to 90 seconds are another cause of disturbed sleep. Periodic limb movements during sleep are recorded with surface electrodes applied over the anterior tibialis muscle of both legs. Either one channel of EMG is recorded from both legs or two channels are recorded, one each per leg. The latter arrangement eliminates much of the ECG artifact that is typical in a one-channel recording from both legs. Periodic leg movements are only scored if at least four, about equally spaced, occur in sequence[2] (Fig. 37–14). Periodic limb movements need to be distinguished from gross body movements, which also show EMG artifact in the ECG and EEG leads, and from body movements associated with sleep apnea.

Episodes of periodic limb movements occur in many normal sleepers as well as in those with sleep disturbances. When limb movements are not associated with arousal, these movements seem to have almost no effect on subsequent waking alertness. However, if the movements cause the sleeper to awaken, they can become a major reason for excessive daytime somnolence.

Videotaping the movements is often instructive. Periodic limb movements show flexions of hip, knee, and ankle joints, together with an extension of the toes, similar to that seen in Babinski's sign.

In a few laboratories, accelerometers that better indicate the force of the movement are used. These accelerometers may also be useful in patients with various tremors, if there is a question about the degree to which these tremors interfere with sleep.

Summary Statistics for Periodic Limb Movements

First, a period limb movement index is reported (the number of periodic limb movements per sleep hour), followed by a periodic limb movement-arousal index. This is the number of periodic limb movements per sleep hour that led to an arousal or an awakening.

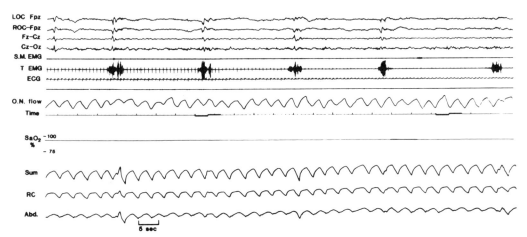

Figure 37–14. Periodic limb movements. Bursts of tibial surface EMG (T EMG) corresponding to leg movements occurring at approximately 20-second intervals. Abd. = abdomen; O.N. flow = oronasal airflow; RC = rib cage; SaO_2 = oxygen saturation; S.M. EMG = submental (chin) EMG; T EMG = tibial surface EMG. (From Fredrickson, PA, and Krueger, BR: Insomnia associated with specific polysomnographic findings. In Kryger, MH, Roth, T, and Dement, WC [eds]: Principles and Practice of Sleep Medicine. WB Saunders, Philadelphia, 1989, p 486, with permission.)

ASSESSING OTHER PHYSIOLOGIC VARIABLES

Nocturnal Penile Tumescence

Males normally have erections associated with REM sleep, irrespective of dream content. This phenomenon is used to help determine whether impotence has a psychogenic or organic cause. An organic cause is less likely if there is normal erectile activity during REM sleep. Mercury strain gauges of appropriate size are applied to the base and the tip of the penis to measure circumference changes during periods of REM sleep. Penile circumference reflects vascular filling of the penis but may not reflect penile rigidity. Therefore, at least once or twice per night, the patient is awakened at a time of maximum circumference change and penile rigidity is assessed by measuring the amount of pressure that must be applied to the tip of the penis to cause it to buckle.[12] Both the frequency and duration of nocturnal penile tumescence decrease with age. Table 37–4 gives the normative values used as guidelines at the Mayo Clinic for assessing nocturnal penile tumescence.

Core Temperature

Monitoring core temperature has become important as an index of the phase and am-plitude of the circadian cycle; this may be important in some insomniacs. Core temperature may be measured with a rectal probe or, less commonly, with a swallowed capsule from which temperature is reported using radio waves. Core temperature depends on both activity and on the sleep-wake cycle. With a special technique called *constant routine*, the patient is maintained in a semirecumbent position in bed for 40 hours, always remaining awake and eating small aliquots of food every hour. This technique allows exact assessment of the timing of a patient's temperature peak and trough, which is not possible when temperature is affected by activity and sleep. This technique is used mainly for circadian rhythm research and not for clinical work.

Esophageal pH Measurements

Many patients experience reflux of stomach contents into the esophagus during sleep. Clearing of this material from the esophagus is delayed markedly during sleep. Occasionally, the reflux may be related to apnea or be associated with insomnia. Measurement of esophageal pH is rarely performed in sleep laboratories. Each laboratory that does make this measurement has developed its own technique.

Table 37–4. GUIDELINES CURRENTLY USED AT THE MAYO CLINIC SLEEP DISORDERS CENTER FOR THE EVALUATION OF NOCTURNAL PENILE TUMESCENCE*

Circumference change, cm	Tip	Base
Marginal	>1.0	>1.5
Adequate	>1.5	>2.5
Duration of NPT, min		
Adequate	>20	
Frequency of NPT episodes		
Linked to frequency of REM episodes; 3–4 are adequate for interpretation		
Buckle pressure, mm Hg		
Marginal	>100	
Adequate	>150	
Typical normal values		
NPT-to-REM ratio, %		
Age 19		160
Age 80		90
NPT percentage of total sleep		
Age 19		33
Age 80		20

NPT = nocturnal penile tumescence.

*Age affects the measurements. The values listed are for middle-aged men.

CHAPTER SUMMARY

The tests and recording techniques used for assessing sleep and physiologic variables during sleep are discussed, including a review of polysomnography (all-night sleep studies) and the multiple sleep latency test (daytime nap studies to assess excessive daytime fatigue and sleep-onset REM periods). The recording and scoring of sleep stages are explained, as is the assessment of respiration and periodic limb movements during sleep. The chapter also includes discussion of recording nontraditional sleep variables, such as the evaluation of nocturnal penile tumescence, a testing procedure often performed in sleep disorders centers even though it does not assess a disorder of sleep.

REFERENCES

1. Agnew, HW Jr, Webb, WB, and Williams, RL: The first night effect: An EEG study of sleep. Psychophysiology 2:263–266, 1966.

2. ASDA Atlas Task Force: Recording and scoring leg movements. Sleep 16:749–759, 1993.

3. Bonnet, MH: Performance and sleepiness as a function of frequency and placement of sleep disruption. Psychophysiology 23:263–271, 1986.

4. Carskadon MA, Dement WC, Mitler MM, et al: Guidelines for the multiple sleep latency test (MSLT): A standard measure of sleepiness. Sleep 9:519–524, 1986.

5. Guilleminault, C, Stoohs, R, Clerk, A, et al: A cause of excessive daytime sleepiness. The upper airway resistance syndrome. Chest 104:781–787, 1993.

6. Hauri, P and Hawkins, DR: Alpha-delta sleep. Electroencephalogr Clin Neurophysiol 34:233–237, 1973.

7. Hauri, PJ and Olmstead, EM: Reverse first night effect in insomnia. Sleep 12:97–105, 1989.

8. Hirshkowitz, M and Moore, CA: Issues in computerized polysomnography. Sleep 17:105–112, 1994.

9. Mahowald, MW and Schenck, CH: REM sleep behavior disorder. In Kryger, MH, Roth, T, and Dement, WC (eds): Principles and Practice of Sleep Medicine. WB Saunders, Philadelphia, 1989, pp 389–401.

10. Mendels, J and Hawkins, DR: Sleep laboratory adaptation in normal subjects and depressed patients ("first night effect"). Electroencephalogr Clin Neurophysiol 22:556–558, 1967.

11. Moldofsky, H and Lue, FA: The relationship of alpha and delta EEG frequencies to pain and mood in 'fibrositis' patients treated with chlorpromazine and L-tryptophan. Electroencephalogr Clin Neurophysiol 50:71–80, 1980.

12. Nofzinger, EA, Fasiczka, AL, Thase, ME, et al: Are buckling force measurements reliable in nocturnal penile tumescence studies? Sleep 16:156–162, 1993.

13. Rechtschaffen, A and Kales, A: A manual of standardized terminology, techniques and scoring system for sleep stages of human subjects. Brain Information Service/Brain Research Institute, Los Angeles, California, 1968.

14. Sleep Disorders Atlas Task Force of the American Sleep Disorders Association: EEG arousals: Scoring rules and examples. Sleep 15:174–184, 1992.

15. Staats, BA, Bonekat, HW, Harris, CD, and Offord, KP: Chest wall motion in sleep apnea. Am Rev Respir Dis 130:59–63, 1984.

16. Sullivan, CE and Crunstein, RR: Continuous positive airway pressure in sleep disordered breathing. In Kryger, MH, Roth, T, and Dement, WC (eds): Principles and Practice of Sleep Medicine. WB Saunders, Philadelphia, 1994, pp 694–705.

17. Vogel, G, Reynolds, CF III, Akiskal, HS, et al: Psychiatric disorders. In Kryger, MH, Roth, T, and Dement, WC (eds): Principles and Practice of Sleep Medicine. WB Saunders, Philadelphia, 1989, pp 413–430.

18. Williams, RL, Karacan, I, and Hursch, CJ: Electroencephalography (EEG) of Human Sleep: Clinical Applications. John Wiley & Sons, New York, 1974.

CHAPTER 38

ASSESSING SLEEP DISORDERS

Peter J. Hauri, Ph.D.
Cameron D. Harris, B.S., R.PSG.T.

According to the Therapeutics and Technology Assessment Subcommittee of the American Academy of Neurology,[6] the American Thoracic Society,[2] and current clinical practice, polysomnographic studies are indicated in the diagnosis of the following disorders:

1. Assessment of sleep-disordered breathing, including (a) patients with symptoms of excessive daytime sleepiness or sleep maintenance insomnia who have been observed to snore loudly or to have stopped-breathing events; (b) patients with chronic obstructive pulmonary disease whose PaO_2 level when awake is greater than 55 mm Hg but whose illness is complicated by pulmonary hypertension, right-side heart failure, or polycythemia; (c) patients with restrictive ventilatory impairment whose illness is complicated by chronic hypoventilation, polycythemia, pulmonary hypertension, disturbed sleep, morning headaches, or daytime somnolence or fatigue; and (d) patients with cardiovascular manifestations of sleep apnea—nocturnal cyclic alternations between bradycardia and tachycardia, nocturnal abnormalities of atrioventricular conduction, and ventricular ectopy during sleep that appears increased relative to wakefulness. The mere presence of snoring, obesity, systemic hypertension, or nocturnal nonspecific cardiac arrhythmias without other symptoms is not an indication for polysomnography.

2. Assessment of sleep-related violent or injurious behavior. An extended polysomnogram, using 8-12 EEG channels and increased paper speed (at least 15 mm/second), is necessary to identify the cause of this behavior. For example, such behavior may be associated with nocturnal seizures, REM behavior disorder, or confused wakefulness.

3. Assessment of periodic limb movements of sleep (nocturnal myoclonus). One needs to know how many of these

occur per sleep hour and how many lead to an arousal. There are two types of periodic limb movements: one type occurs throughout the night, and the other type occurs mainly in the first third of the night.

4. Assessment of excessive daytime somnolence (narcolepsy, idiopathic hypersomnia). This assessment may require the use of polysomnography and a subsequent multiple sleep latency test (MSLT).

For cases of insomnia, no criterion for polysomnography has been officially accepted. In our judgment, polysomnography is performed in cases of insomnia only if (1) the insomnia complaint is very severe and has been present for a minimum of 6 months, (2) insomnia has not yielded to behavioral or sleep hygiene therapy or to a short course of treatment with hypnotic agents, and (3) a medical or psychiatric reason for the insomnia has been excluded or treatment for it has been unsuccessful.

PARTIAL EVALUATIONS

Because complete polysomnographic studies are expensive, screening with partial recordings is often performed instead. The American Sleep Disorders Association has recently reviewed this practice in the case of assessing obstructive sleep apnea syndrome and has found that under certain, carefully spelled-out conditions, unattended (portable) recording for the assessment of obstructive sleep apnea is acceptable.[5] This includes the following:

1. *Nocturnal oximetry*—Although the presence of repetitive desaturations in an oximetric tracing strongly suggests a sleep apnea syndrome, normal findings on oximetry do not rule out disordered breathing. Patients, especially younger ones, may have sleep apneas serious enough to cause repeated arousals from sleep without causing significant oxygen desaturations or the patient may not have slept much during the night when oximetry was performed and therefore shows no desaturations or also the patient may have positional

sleep apnea and happened to sleep only on the side during that night.

2. *Holter monitoring*—Frequent bradycardia alternating with tachycardia, demonstrated with Holter monitoring, during the night suggests a sleep apnea syndrome. However, many patients with a serious obstructive sleep apnea syndrome show entirely normal cardiac rhythms throughout the period of sleep.

3. *Wrist actigraphy*—A small, watch-like device is worn by the patient on the wrist of the nondominant hand, usually for an entire week. This device counts and stores in its memory the number of wrist movements that occur for each 1-minute epoch. Periods of relative absence of such movements are interpreted as sleep and periods of high activity as wakefulness. Although the validity of this method is adequate for estimating sleep in normal subjects,[4] its accuracy is significantly less in people with sleep disorders such as apnea or insomnia.[3] Patients with periodic limb movements or obstructive sleep apnea show an excessive number of movements and on the actigraph look as though they are awake when they are actually asleep. Because patients with depression or psychophysiologic insomnia may move very little while lying in bed awake, their total sleep may be overestimated by wrist actigraphy.

4. *Continuous portable EEG monitoring*—This may be adequate to assess sleep architecture, but it cannot determine the cause of frequent arousals, such as disordered breathing or periodic limb movements.

5. *Full ambulatory monitoring*—This may be adequate if the system contains a minimum of four channels to evaluate sleep (two EOG channels, one EEG channel, and one chin EMG channel) and other channels to evaluate airflow, oxygen saturation, respiratory effort, anterior tibialis EMG, and ECG. Also, the home environment has to be adequately monitored to ensure the possibility of normal sleep (no telephone calls, children crying, and so forth).

ORGANIZATION OF A SLEEP DISORDERS CENTER

Before evaluation, patients are usually sent extensive questionnaires about sleep and a sleep log to be filled out for a minimum of 1 week. The patients are then interviewed by a sleep disorder specialist, ideally by a person certified by the American Board of Sleep Medicine. If indicated, polysomnography with or without a subsequent MSLT is then scheduled.

On the appointed night, patients come to the sleep center at about 8 PM. First, they fill out a questionnaire about their activity during the past day; their intake of medicine, coffee, and alcohol; and emotional issues that might disrupt sleep. A technician applies the electrodes and sensors necessary for the polysomnographic study. Patients then watch television or read until they are ready to go to bed. After the patients are connected to the polysomnograph, biocalibrations are performed (patients are asked to move and blink their eyes, move their legs, grit their teeth, breathe exclusively through either the nose or mouth, and perform an isovolume maneuver to maximize the recording of paradoxical breathing). After lights out, patients stay in bed for a minimum of 6 hours (often longer) except for short trips to the bathroom. A technician monitors the recording throughout the night, observes the patients through video monitoring, and notes on the record a patient's position and any potentially significant events (e.g., patient vocalization, environmental noise that arouses the patient, and loudness of snoring).

If a diagnosis of obstructive sleep apnea syndrome is made early in the night, the second half of the night may be used for a therapeutic trial of continuous positive airway pressure (CPAP). On other occasions, medications (such as Sinemet [DuPont] for periodic movements) are administered during the second part of the night. Although the American Sleep Disorders Association strongly encourages a second night in the laboratory for evaluating treatment, economic pressures have forced an increasing number of sleep laboratories to do "split nights," that is, to perform both diagnostic work and therapeutic trials in the same night.

In the morning, patients fill out another questionnaire regarding their perception of the quality of their sleep in the laboratory. The recordings are then scored by a technician and various indices are computed (see Figure 37–2). After the sleep specialist has reviewed the record, a follow-up conference with the patient is scheduled to discuss the results, make a final diagnosis, and initiate treatment if indicated.

DISORDERS OF EXCESSIVE DAYTIME SOMNOLENCE

Excessive daytime somnolence is defined as the actual tendency to fall asleep very easily. The ability to fall asleep quickly differentiates excessive daytime somnolence from chronic fatigue, malaise, or low-grade depression, in which patients may be exhausted and lie in bed for most of the day but are usually unable to fall asleep quickly when asked to do so.

Excessive daytime somnolence may be caused by *sleep deprivation*, either voluntary or involuntary. Voluntary sleep deprivation is often found in people who work two jobs or have other reasons for not allowing themselves enough time in bed. Normal sleep needs vary from person to person, ranging from 3 to 4 hours to 9 to 10 hours per night. It is often useful to ask patients with excessive daytime somnolence to stay in bed 1 hour longer for an entire week, regardless of how long they have stayed in bed until that time, and observe whether excessive daytime somnolence wanes.

Excessive daytime somnolence caused by involuntary sleep deprivation may be found in cases of insomnia or it may be a consequence of certain medications, endocrine dysfunction, or other medical disorders.

If these causes of excessive daytime somnolence can be ruled out in a patient's complaint of excessive somnolence, polysomnography coupled with an MSLT is indicated. During the polysomnographic study, one looks for disordered breathing, periodic limb movements, or other events that might cause excessive daytime somnolence. If such a cause is found, the MSLT scheduled for the following day is cancelled and the problem is treated before the assessment of excessive daytime somnolence continues. If nighttime sleep is adequate (a minimum of 7 hours is often cited), one proceeds with the test. In the absence of other factors that might increase rapid eye movement (REM) pressure (e.g., withdrawal from an antidepressant medication or a stimulant) and in the presence of a history that is suggestive of narcolepsy, a finding of excessive daytime somnolence (mean MSLT sleep latency of 5 minutes or less) in combination with at least two sleep-onset REM periods during the five naps in the multiple sleep latency test suggest *narcolepsy.* If the results of polysomnography and MSLT are normal except for a pathologically short sleep latency on the test (less than 5 minutes), *idiopathic hypersomnia* is the likely diagnosis.

Occasionally, a problem in assessing excessive daytime somnolence is caused by *delayed sleep phase syndrome,* the inability to fall asleep until the early morning hours coupled with routine sleep until noon or later. In such cases, polysomnography is performed during routine sleeping hours (e.g., 4 AM until 1 PM) and the MSLT is conducted during routine wakefulness (e.g., naps at 3 PM, 5 PM, 7 PM, 9 PM, and 11 PM).

PARASOMNIAS

Parasomnias are undesirable phenomena that occur during sleep, such as nightmares, sleep terrors, bruxism, and sleepwalking. As mentioned above, polysomnography is indicated if there is sleep-related violent behavior or potentially injurious behavior or if the parasomnias cause serious distress in patients or caretakers (e.g., nocturnal sleep terrors that arouse the family) and have not yielded to commonly effective treatment.

Sleep-related seizure disorders can masquerade as almost any parasomnia. Therefore, when polysomnography is performed for the evaluation of parasomnias, an extended EEG montage that uses a minimum of 12 EEG channels is required. A minimal paper speed of 15 mm/second is used for most of the night, with 5-minute episodes every hour in which the speed is increased to 25 or 30 mm/second (Fig. 38–1). In addition, split-screen video recordings are very useful in such cases, showing simultaneously the patient's behavior and the polysomnographic recording. Patients with nocturnal seizure disorders sometimes show entirely normal sleep for the first few hours, whereas epileptiform EEG activity occurs only toward the morning.

The diagnosis of *REM behavior disorder* requires special consideration. In normal subjects, REM sleep is accompanied by a marked decrease in electromyographic (EMG) tonus, rendering the sleeper nearly paralyzed. In some patients, especially in middle-aged and older men and in those with certain brain stem disorders such as parkinsonism, the typical REM sleep atonia is only partially present and patients are able to perform various semipurposeful movements related to their dream content. To document this, EMG recordings are obtained from all four limbs. One looks for frequent EMG activity in the limbs during REM sleep. Video recording of the patient's behavior is an essential adjunct to the polysomnogram when evaluating REM behavior disorder.

PERIODIC LIMB MOVEMENT DISORDER AND RESTLESS LEGS

Periodic limb movement disorder is characterized by episodes of repetitive and highly stereotyped limb movements. Of interest in the evaluation is not only the number of such movements (as defined in the preceding chapter) but also whether they arouse or awaken the sleeper and for how long. A few periodic limb movements that do not lead to evidence of arousal are considered normal.

Periodic limb movements often occur in association with disordered breathing. It needs to be determined whether true periodic limb movements (stereotyped limb movements without accompanying gross body movements) are being recorded or whether the movements are the leg components of gross body movements accompanying repetitive termination of obstructive apneas.

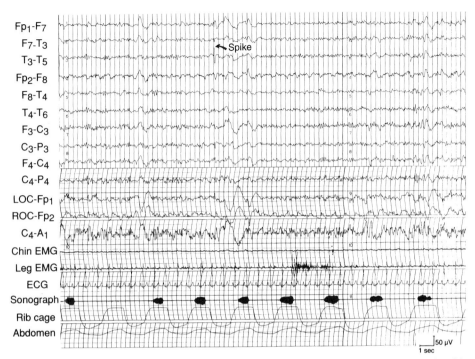

Figure 38–1. Parasomnia recording, using an expanded EEG montage, from a 25-year-old man with a history of nocturnal "spells." A standard wake and sleep EEG was essentially normal; however, note the left temporal spike as observed approximately 3 hours after sleep onset. Abdomen = abdominal breathing effort by inductive plethysmograph; ECG = electrocardiogram; EMG = electromyogram; LOC = left outer canthus; Rib cage = thoracic breathing effort by inductive plethysmograph; ROC = right outer canthus; Sonograph = snoring sound.

Periodic leg movements are often associated with restless leg syndrome. In such cases, the repetitive stereotyped limb movements may occur even when the patient is awake or drowsy.

In patients with restless legs, a "voluntary immobilization test" is often performed. Patients are asked to lie quietly in bed and to signal when they perceive a strong urge to move their legs without actually doing so. If the urge to move the legs comes in regularly spaced intervals (similar to periodic leg movements during sleep), a diagnosis of restless legs is more likely than if the urge occurs randomly.

INSOMNIA

Currently, there are no accepted indications for assessing insomnia in a sleep disorders center. Obviously, if there is a strong suspicion that either disordered breathing or periodic limb movements disturb sleep, polysomnography is carried out for those reasons. In rare cases, polysomnography is also performed if a sleep-state misperception syndrome is suspected. Such a syndrome is defined as a complaint of insomnia or excessive sleepiness that occurs without any objective evidence of sleep disturbance. The polysomnogram would demonstrate normal sleep latency (less than 15 to 20 minutes), a normal number of arousals and awakenings (fewer than 10 per hour), and a normal sleep duration (longer than 6.5 hours). The MSLT may or may not show abnormal sleep latencies.

STANDARDS OF PRACTICE

There are two levels of practice: the laboratory interpretation of polysomnographic and MSLT results and the clinical diagnosis and treatment of patients with sleep disorder.

According to the American EEG Society,[1] to interpret polysomnograms a person should meet the minimal qualifications of a clinical electroencephalographer. Such a person should also be able to document a minimum of 6 months of supervised experience in clinical polysomnography before serving without supervision. A technician who conducts polysomnographic studies and MSLTs should be familiar with the recording instruments and techniques discussed in Chapter 37. Such a person also should be proficient in basic EEG technology, classification of sleep disorders and normal attendant EEG characteristics of adults and children, and scoring of sleep stages. Such a technologist should also know basic cardiopulmonary resuscitation.

Although the standards given are adequate for laboratory work in sleep, sleep disorders clinicians need to possess extensive clinical knowledge and skills in the diagnosis and treatment of patients with sleep disorders. Proof that such knowledge and skills have been obtained is provided by a 2-day examination that leads to board certification in Sleep Medicine. This examination is given yearly by the American Board of Sleep Medicine (1610 14th Street, N.W., Rochester, MN 55901).

CHAPTER SUMMARY

This chapter discusses indications for polysomnographic studies and organization of a sleep disorders center. We then review some special techniques for dealing with excessive daytime somnolence, parasomnias, periodic limb movement disorders, and insomnias. Current standards of practice are reviewed. Laboratory interpretation of polysomnography and MSLT is an important part of the clinical practice of sleep disorders medicine, which, in addition, encompasses clinical interviewing and assessment skills, the understanding of relationships between sleep disorders and other medical and psychologic aspects of the person, and treatment techniques—medical, psychiatric, and behavioral.

REFERENCES

1. American Electroencephalographic Society: Guideline fifteen: Guidelines for polygraphic assessment of sleep-related disorders (polysomnography). J Clin Neurophysiol 11:116–124, 1994.
2. American Thoracic Society: Indications and standards for cardiopulmonary sleep studies. Am Rev Respir Dis 139:559–568, 1989.
3. Hauri, PJ and Wisbey, J: Wrist actigraphy in insomnia. Sleep 15:293–301, 1992.
4. Sadeh, A, Sharkey, KM, and Carskadon, MA: Activity-based sleep-wake identification: An empirical test of methodological issues. Sleep 17:201–207, 1994.
5. Standards of Practice Committee of the American Sleep Disorders Association: Practice parameters for the use of portable recording in the assessment of obstructive sleep apnea. Sleep 17:372–377, 1994.
6. Therapeutics and Technology Assessment Subcommittee of the American Academy of Neurology: Assessment: Techniques associated with the diagnosis and management of sleep disorders. Neurology 42:269–275, 1992.

SECTION 2
Electro-physiologic Assessment of Neural Function

PART H
Intraoperative Monitoring

The central and peripheral nervous systems are at risk for damage during surgical procedures, particularly of an orthopedic and neurosurgical nature. Although some damage may be expected because of the nature of the procedure, other types can occur unexpectedly. In either case, the damage may be reversible if the surgeon takes appropriate action. However, neural function cannot be assessed in the standard clinical fashion during surgical procedures. Therefore, surgeons have relatively little information on which to base decisions about modifying the procedure in light of potential damage to neural tissue.

Most electrophysiologic measurements described thus far in this book can be made intraoperatively to monitor neural function. Electroencephalography can be used to monitor the status of cortical function; somatosensory evoked potentials, to monitor sensory pathways in the periphery, spinal cord, and brain; auditory evoked potentials, to monitor peripheral and central auditory pathways; nerve conduction studies and electromyography, to monitor peripheral nerve damage; and motor evoked potentials, to monitor descending motor pathways in the brain stem and spinal cord. Each of these techniques has been modified so it can be used in the operating room for a wide variety of surgical procedures. Such monitoring techniques provide helpful guidance to the surgeon during the procedure and have reduced morbidity associated with certain procedures.

Similar continuous monitoring can be helpful in the intensive care unit. Patients with diseases severe enough to warrant intensive care either have a major neurologic disease or are at risk for one as a complication of their disorder. Electrophysiologic techniques can be used to monitor neural function in this setting, just as they can be used intraoperatively to identify early or otherwise unrecognizable neural damage. The chapters in this section illustrate the applications of various techniques both intraoperatively and in the intensive care unit.

CHAPTER 39

CEREBRAL FUNCTION MONITORING

Frank W. Sharbrough, M.D.

Among the first applications of intraoperative recording of cerebral electrical activity were corticography[1,5,6,10,11] and, later, acute depth studies during the course of surgery for epilepsy. More recently, scalp electroencephalographic (EEG) recordings (with or without special computer processing) have been used routinely in many medical centers to monitor cerebral electrical activity during endarterectomy[2,14,17] and cardiac bypass operations. However, EEG monitoring during the latter procedure is limited by the effect of hypothermia, which often suppresses EEG activity,[12] making it ineffective as a monitoring tool for ischemia. EEG monitoring for cardiac surgery is further limited because easily correctable

causes of ischemia occur less commonly than during operations on the carotid artery.

TECHNICAL FACTORS IN INTRAOPERATIVE EEG MONITORING

To ensure reliable EEG recordings during intraoperative monitoring, it is important to pay special attention to technical factors:[19] stable applications of electrodes, an adequate number of electrodes (at least 8 and preferably 16 channels), filtering of background noise, use of adequate sensitivity, and proper grounding. Paper speed is a particularly important technical factor, because a large amount of data is generated during intraoperative EEG monitoring and it must be compressed without losing essential data. Computer processing with spectral analysis presented as a compressed spectral array has been used for this purpose.[2,15] However, data can be compressed without using computers simply by slowing the paper speed, initially to a rate of 15 mm/second (i.e., half the speed used in standard EEG recordings),[14,17,19] (Figs. 39–1 through 39–4). An even slower paper speed of 5 mm/second (one-sixth of that used in standard recordings) is adequate for detecting important

443

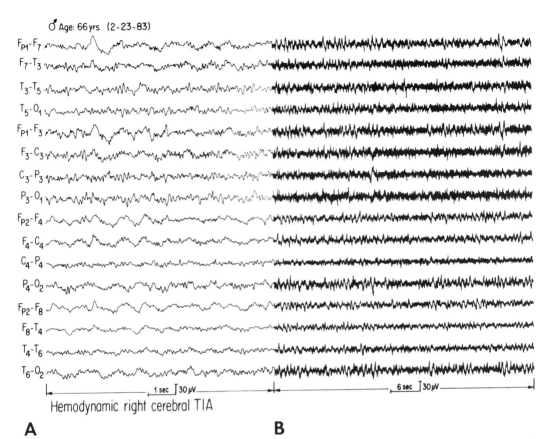

Figure 39–1. Electroencephalogram (EEG) with isoflurane anesthesia. (*A*) EEG recorded with routine paper speed (30 mm/sec). Common patterns of anesthetic—ARF alpha pattern, ATS pattern, and PAS pattern—are seen in the left hemisphere. In the right hemisphere, there is an increase in the amount of irregular slowing, especially in temporal distribution, and reduction of the normal ARF alpha pattern. (*B*) The same changes can be qualitatively appreciated even at a reduced paper speed of 5 mm/sec (one sixth the usual paper speed). Note that this patient has a focal abnormality under anesthesia, even though the patient had had only right cerebral transient ischemic attack (TIA) without residual deficit. (From Daube, JR, et al: Intraoperative monitoring. In Daly, DD and Pedley, TA [eds]: Current Practice of Clinical Electroencephalography, ed 2. Raven Press, New York, 1990, p. 743, with permission of Mayo Foundation.)

EEG changes to ischemia during intraoperative clamping (Fig. 39–5). Moreover, if unusual EEG changes develop that cannot be immediately recognized at slower paper speeds (5 mm/second), simply returning for a brief time to the regular paper speed (30 mm/second) allows rapid identification of the pattern (see Fig. 39–1).

SYMMETRICAL EEG PATTERNS DURING ANESTHESIA

Although much has been written about the different effects of anesthetic agents on the EEG during anesthesia, most of these agents produce similar EEG patterns when used at concentrations below their minimal alveolar concentration (MAC) level (i.e., that necessary to prevent movement in response to a painful stimulus in about 50% of patients). The common anesthetic agents (thiopental, halothane, enflurane, isoflurane, nitrous oxide) produce certain EEG changes.

At subanesthetic concentrations, thiopental characteristically produces maximum beta activity in the anterior midline. Halothane, enflurane, isoflurane, and 50% nitrous oxide, when administered at subanesthetic concentrations, produce a similar but less prominent pattern of beta activity.[15]

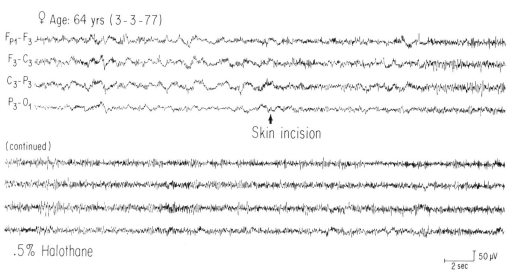

Figure 39–2. The effect of painful stimulation (skin incision) on the electroencephalogram during levels of anesthesia below minimal alveolar concentration; such stimulation tends to reduce the amount of slow activity and to accentuate the amount of fast activity seen with a given concentration of anesthetic agent. Paper speed is 15 mm/sec (half the usual speed). (From Daube, JR, et al: Intraoperative monitoring. In Daly, DD and Pedley, TA [eds]: Current Practice of Clinical Electroencephalography, ed 2. Raven Press, New York, 1990, p 742, with permission of Mayo Foundation.)

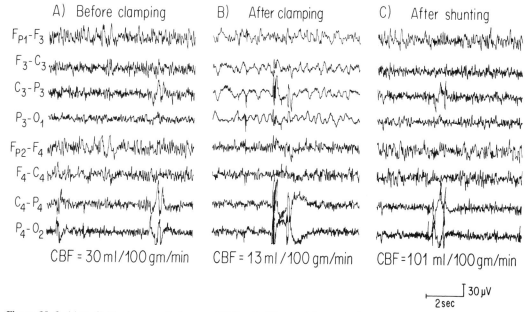

Figure 39–3. (*A* to *C*) Electroencephalogram (EEG) of a 54-year-old man undergoing left carotid endarterectomy. The EEG result demonstrates the type of reduction in the faster anesthetic components and retention of rhythmic slowing that occurs with less severe reduction in cerebral blood flow (CBF) below the critical level. This change is easy to identify despite significant artifact affecting the posterior electrodes. Paper speed is 15 mm/sec (half the usual paper speed). (From Daube, JR, et al: Intraoperative monitoring. In Daly, DD and Pedley, TA [eds]: Current Practice of Clinical Electroencephalography, ed 2. Raven Press, New York, 1990, p 742, with permission of Mayo Foundation.)

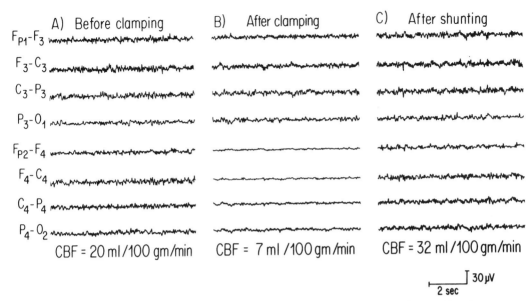

Figure 39–4. (*A* to *C*) Electroencephalogram (EEG) of a 74-year-old man undergoing right carotid endarterectomy. The EEG result demonstrates the more severe type of EEG "suppression" that occurs in association with more severe reduction of cerebral blood flow (CBF) below the critical level. Paper speed is 15 mm/sec (half the usual paper speed). (From Daube, JR, et al: Intraoperative monitoring. In Daly, DD and Pedley, TA [eds]: Current Practice of Clinical Electroencephalography, ed 2. Raven Press, New York, 1990, p 743, with permission of Mayo Foundation.)

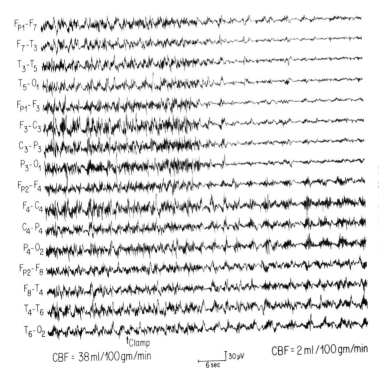

Figure 39–5. Electroencephalogram (EEG) of a 68-year-old man undergoing left carotid endarterectomy. The EEG shows the type of dramatic, rapid attenuation of all components often associated with cerebral blood flow (CBF) of less than 6 to 7 mL/100 g per minute. Paper speed is 5 mm/sec (one sixth the usual paper speed). (From Daube, JR, et al: Intraoperative monitoring. In Daly, DD and Pedley, TA [eds]: Current Practice of Clinical Electroencephalography, ed 2. Raven Press, New York, 1990, p 744, with permission of Mayo Foundation.)

Most surgical inductions are performed relatively rapidly with thiopental and are accompanied by the following characteristic sequence of changes. Beta activity, described under the subanesthetic pattern, rapidly becomes more widespread, increases in amplitude, and slows in frequency toward the alpha range.[13] During this phase, the faster rhythm may be intermixed with bursts of high amplitude, **f**rontal **i**ntermittent, **r**hythmic **d**elta **a**ctivity (the FIRDA pattern). With a slower rate of induction when nonbarbiturate inhalation agents are used, there is less tendency toward intermittent rhythmic bursting.

When a light level of steady-state anesthesia is reached, a characteristic **a**nteriorly maximum, **r**hythmic **f**ast (ARF) pattern, usually in the lower beta- or alpha-frequency range, is seen with all agents, including halothane, enflurane, isoflurane, and even thiopental (see Fig. 39–1). The frequency of this pattern tends to slow with increasing concentrations of the agent. This anesthetic ARF pattern is strikingly similar to postanoxic, widespread, anteriorly maximum alpha-coma pattern. The origin of the postanoxic ARF pattern is uncertain. However, evidence suggests that the anesthetic ARF pattern, usually in the alpha-frequency range, is a drug-induced beta-variant pattern[13] that with anesthesia becomes more widespread, higher in amplitude, and slower in frequency—from the upper to the lower beta range and finally into the alpha or even theta range. In humans, the anesthetic ARF pattern tends to be more continuous than normal sleep spindles.

In addition to the alpha-frequency ARF anesthetic pattern, there are often **a**nteriorly maximum, **t**riangular, **s**low (ATS) waves, which are commonly diphasic and have a sharply contoured initially negative phase that is followed by a more rounded positive phase (see Fig. 39–1). These ATS waves characteristically have a duration of less than 1 second and may occur either as a single transient or in brief trains. When preceded by a waxing ARF alpha pattern, they may produce a "mitten-like" pattern.[15] In addition to the ATS waves, more **p**osterior, **a**rrhythmic, **s**low (PAS) waves, which are often of lower amplitude, may become prominent. Duration of these individual polymorphic slow waves is usually longer than 1 second.

The maximum of this activity is often difficult to identify clearly, but it is always more posterior and may be more prominent over the temporal regions (see Fig. 39–1). This pattern is least obvious with halothane, more so with enflurane, and most prominent with isoflurane. Strictly, nitrous oxide alone is not potent enough to be an anesthetic agent at atmospheric pressure, but it nonetheless potentiates the effects of other inhalation agents. During "balanced anesthesia" of 50% nitrous oxide combined with another agent, the PAS pattern tends to be more prominent than when only a single agent is used.

PREOPERATIVE FOCAL ABNORMALITIES

Most patients undergoing endarterectomies show a symmetrical baseline pattern of the type described above. However, depending on patient selection, 30% to 40% of patients may have a focal abnormality of varying degrees of severity.[18] This focal abnormality consists of unilateral reduction of the amplitude of the ARF anesthetic pattern by more than 30% to 40%. This is commonly associated with increased wave length and amplitude of persistent polymorphic slowing on the side of the reduced amplitude. The latter can generally be distinguished from the usual PAS pattern during anesthesia because it has a longer wave length and is more irregular and higher in amplitude on the pathologic side. In most patients, focal baseline EEG abnormality correlates with preoperative deficits. However, a small percentage of patients with such baseline abnormalities have experienced only transient ischemic attacks, presumably due to a hemodynamic mechanism. These patients usually have normal findings on computed tomography (CT), with ipsilateral decrease in retinal artery pressures and a low baseline cerebral blood flow (see Fig. 39–1). These patients are thought to have a low baseline cerebral blood flow sufficient to cause EEG abnormalities in the absence of residual neurologic signs or CT abnormalities.[20]

In general, a preoperative focal anesthetic EEG abnormality correlates with a preoperative waking EEG abnormality. There are occasional exceptional patients in whom anesthesia will activate an abnormality that was either not apparent or less apparent during the waking trace.[7] However, anesthesia may obscure an abnormality that was present during the waking trace and may also activate an abnormality that is minimal or not apparent during the waking state, as in the case of an anterior hemispheric insult that has left the alpha pattern normal and symmetrical. Despite such a normal symmetrical alpha pattern, the anesthetic record may show major reduction in the ARF pattern and an increase in the irregular slowing in the anterior distribution. In these patients, the drug-induced beta activity seen during the preanesthetic state is nearly always decreased on the side of a reduced anesthetic ARF pattern, which is further evidence that the anesthetic alpha-frequency ARF pattern is more directly related to drug-induced beta activity than to the normal waking alpha rhythm.

One additional situation in which the anesthetic EEG tends to activate an abnormality occurs when intermittent rhythmic slow waves are present in the temporal region on the side of ischemia. This intermittent abnormality is often converted into a more obvious and persistent focal slowing along with reduction in the ARF pattern during anesthesia. A more posterior lesion, which produces significant abnormality in the alpha pattern, may leave the anesthetic ARF pattern symmetrical, without obvious focal slowing. In such a case, preanesthetic beta activity is usually also symmetrical.

Finally, the nonlocalizing FIRDA pattern or more persistent generalized slowing, which is easily recognized as abnormal during the waking state, cannot be identified as abnormal during anesthesia because these patterns commonly occur in most patients at some stage of anesthesia, regardless of whether they have had symptoms of central nervous system lesions.

INTRAOPERATIVE FOCAL EEG CHANGES

EEG changes commonly occur with carotid artery clamping (some change occurs in approximately 25% of patients).[18] These changes almost always occur within 20 to 30 seconds after clamping and are associated with decreased cerebral blood flow below a critical level, which varies somewhat with anesthetic agent. With halothane, the critical level is between 15 and 18 mL/100 g per minute; it is slightly lower with enflurane and isoflurane.[18] The reasons for these differences with various anesthetic agents are uncertain. The severity and rapidity of the onset of change vary roughly in direct proportion to the degree of lowering of blood flow below the critical level. Minor changes consist of 25% to 50% decrease in the faster components and increase in amplitude and wave length of slower components (see Fig. 39–3). Changes seen with more severe decreases in blood flow (in the range of 6 to 7 mL/100 g per minute or less) are associated with an even greater reduction of anesthetic faster components, along with reduced amplitude of the slower components, producing a lower amplitude and featureless EEG on the side of clamping (see Figs. 39–4 and 39–5). Although up to 25% of EEGs may show some change, only 1% to 3% show the more severe degree of change.

Focal transient changes occurring other than during clamping can be seen in up to 10% of patients. In most, this is due to transient asymmetrical effects of changing levels of anesthesia on a preexisting focal abnormality and is of no consequence.[18] Embolization rarely occurs in association with shunting (in less than 1% of our shunted patients).[18] Some EEG changes are likely due to the reversible effects of embolization from the operative site. However, in about 1% of patients undergoing endarterectomy, a focal EEG change develops intraoperatively that is unassociated with carotid artery clamping and persists throughout the procedure; it ultimately is associated with a new neurologic deficit in the immediate postoperative period. These changes almost always prove to be due to embolization.[18] An effective way of identifying embolization is to measure cerebral blood flow when a new persistent focal EEG change develops.

EEG IN RELATION TO OTHER MONITORING TECHNIQUES DURING ENDARTERECTOMY

There are several ways of processing primary EEG data into a compressed spectral array[2,15] and thus integrating EEG data.[3] Somatosensory evoked potential monitoring has been used by some investigators to supplement or to replace EEG monitoring.[8] However, all these methods require reliable primary EEG recordings and are basically an attempt at refinement of EEG monitoring. A non-EEG monitoring method is measurement of blood flow, usually with intracarotid injection of xenon,[19] or measurement of the back pressure in the distal stump of the carotid artery, called stump pressure.[9] Blood flow is a more complicated but a more reliable measure than stump pressure.[9] Blood flow is clearly influenced by the $PaCO_2$ level and requires that xenon be delivered directly to the internal carotid artery while the external carotid artery is clamped. Blood flow techniques are usually performed only three or four times intraoperatively: before clamping, immediately after clamping, immediately after placing a shunt (if one is used), and at the end of the procedure. Blood flow measurement is complementary to EEG findings. Focal EEG changes due to ischemia caused by decreased perfusion pressure distal to a clamped carotid artery are always associated with low blood flow, as measured by the xenon technique. An unexpected EEG change that persists and is associated with "normal blood flow" is essentially pathognomonic of embolization. The presence of normal blood flow after embolization is explained on the basis of the so-called look-through phenomenon.[4] A simplified interpretation of this phenomenon is as follows: If the ischemia is a result of embolic occlusion of half the blood vessels to a region (the other half of the vessels are patent), injection of xenon produces a normal or at times an increased flow and washout of xenon through the patent blood vessels. Totally occluded vessels receive no xenon and thus do not contribute to the overall measurement of flow.

EEG MONITORING DURING CARDIAC SURGERY

EEG monitoring during cardiac surgery is severely limited in some institutions, including the Mayo Clinic, by the profound hypothermia that is induced intraoperatively and that suppresses most of the EEG features needed for monitoring. Even in the absence of profound hypothermia, hypothermia and changing levels of PCO_2 may produce diffuse EEG changes difficult to distinguish from global ischemia.

To better quantitate changes and improve detection of long-term trends, the efforts to try to use processed EEG have intensified. Many devices for quantitating and displaying trends are available. When evaluating such devices, their performance in the following six areas must be considered: (1) adequate number of channels and montage selection, (2) unprocessed EEG display, (3) reliable method of processing, (4) simple and useful display, (5) automatic use of quantitative data, and (6) retrievable storage of data, preferably before processing or at least after processing. A comparative evaluation of a number of commercial instruments has been published.[8]

Results of a study using processed EEG for cardiac monitoring are intriguing.[16] Strobl's study quantitated the percentage of power reduction below a critical percentage (usually 60%) of a baseline level weighted by the duration of time during which this reduction persisted. This reduction, of course, is influenced by the degree of hypothermia, anesthesia, and change in perfusion pressure. When anesthesia and hypothermia were held constant, the higher score of integrated power reduction below a critical level was associated with increased gross neurologic dysfunction in the postoperative period. Further severe unilateral reduction in power was not associated with lateralized neurologic deficit. Thus, this technique is promising.

CHAPTER SUMMARY

Electrophysiologic monitoring of cerebral function intraoperatively has found its greatest application during carotid endarterectomy surgery. EEG and somatosensory evoked potential monitoring can identify the occurrence of ischemia that immediately follows occlusion of the blood supply as part of the operation. Different types of changes are able to predict with a reasonable degree of accuracy the occurrence of postoperative deficits. This information often allows the surgeon to modify the procedure to minimize the likelihood of a deficit.

Success of the monitoring depends heavily on the technical aspects of recording, such as paper speed and anesthesia. Although there are variations among anesthetic agents and their effects on the EEG, most of them produce similar changes that can be recognized and distinguished from the effects of ischemia.

Preoperative EEG abnormalities are often seen in patients undergoing carotid endarterectomy, most commonly in relation to a measurable clinical deficit. These abnormalities may have some particular value for outcome, but the changes generally are not of significance under anesthesia and, thus, do not preclude monitoring.

Intraoperative electrophysiologic monitoring is used primarily to identify a focal EEG change. Such changes can occur abruptly and usually are seen with blood flow levels of less than 15 mL/100 g per minute. The extent of EEG change can vary from mild to severe and may occur either with clamping or, in some cases, secondary to embolization, which is associated with little measurable change in total blood flow to the cerebral hemisphere.

Recognition and identification of the range of changes enable electrophysiologists to provide useful information to the surgeon to help determine surgical decisions.

REFERENCES

1. Ajmone-Marsan, C and Baldwin, M: Electrocorticography. In Baldwin, M and Bailey, P (eds): Temporal Lobe Epilepsy. Charles C Thomas, Springfield, Illinois, 1958, pp 368–395.
2. Chiappa, KH, Burke, SR, and Young, RR: Results of electroencephalographic monitoring during 367 carotid endarterectomies: Use of a dedicated minicomputer. Stroke 10:381–388, 1979.
3. Cucchiara, RF, Sharbrough, FW, Messick, JM, and Tinker, JH: An electroencephalographic filter-processor as an indicator of cerebral ischemia during carotid endarterectomy. Anesthesiology 51:77–79, 1979.
4. Donley, RF, Sundt, TM Jr, Anderson, RE, and Sharbrough, FW: Blood flow measurements and the "look through" artifact in focal cerebral ischemia. Stroke 6:121–131, 1975.
5. Gloor, P: Contributions of electroencephalography and electrocorticography to the neurosurgical treatment of the epilepsies. Adv Neurol 8:59–105, 1975.
6. Green, JR, Duisberg, REH, and McGrath, WB: Electrocorticography in psychomotor epilepsy. Electroencephalogr Clin Neurophysiol 3:293–299, 1951.
7. Hansotia, PL, Sharbrough, FW, and Berendes, J: Activation of focal delta abnormality with methohexital and other anesthetic agents (abstr). Electroencephalogr Clin Neurophysiol 38:554, 1975.
8. Health Devices 15(4):71–107, April 1986.
9. McKay, RD, Sundt, TM, Michenfelder, JD, et al: Internal carotid artery stump pressure and cerebral blood flow during carotid endarterectomy: Modification by halothane, enflurane, and innovar. Anesthesiology 45:390–399, 1976.
10. Meyers, R, Knott, JR, Hayne, RA, and Sweeney, DB: The surgery of epilepsy: Limitations of the concept of the cortico-electrographic "spike" as an index of the epileptogenic focus. J Neurosurg 7:337–346, 1950.
11. Penfield, W and Jasper, H: Epilepsy and the Functional Anatomy of the Human Brain. Little, Brown & Company, Boston, 1954, pp 692–736.
12. Quasha, AL, Sharbrough, FW, Schweller, TA, and Tinker, JH: Hypothermia plus thiopental: Synergistic EEG suppression (abstr). Anesthesiology 51 Suppl:S20, 1979.
13. Sharbrough, FW: Nonspecific abnormal EEG patterns. In Niedermeyer, E and Lopes da Silva, F (eds): Electroencephalography: Basic Principles, Clinical Applications and Related Fields. Urban & Schwarzenberg, Baltimore-Munich, 1982, pp 135–154.
14. Sharbrough, FW, Messick, JM Jr, and Sundt, TM Jr: Correlation of continuous electroencephalograms with cerebral blood flow measurements during carotid endarterectomy. Stroke 4:674–683, 1973.
15. Stockard, J and Bickford, R: The neurophysiology of anaesthesia. Monogr Anesthesiol 2:3–46, 1975.
16. Arom, KV, Cohen, DE, and Strobl, FT: Effect of intraoperative intervention on neurological outcome based on electroencephalographic monitoring during cardiopulmonary bypass. Ann Thorac Surg 48:476–483, 1989.
17. Sundt, TM Jr, Sharbrough, FW, Anderson, RE, and Michenfelder, JD: Cerebral blood flow measurements and electroencephalograms during carotid endarterectomy. J Neurosurg 41:310–320, 1974.
18. Sundt, TM Jr, Sharbrough, FW, Piepgras, DG, et al: Correlation of cerebral blood flow and electroencephalographic changes during carotid endarterectomy: With results of surgery and hemodynamics of cerebral ischemia. Mayo Clin Proc 56:533–543, 1981.
19. Sundt, TM Jr, Sharbrough, FW, Trautmann, JC, and Gronert, GA: Monitoring techniques for carotid endarterectomy. Clin Neurosurg 22:199–213, 1975.
20. Yanagihara, T and Klass, DW: Discrepancy between CT scan and EEG in hemodynamic stroke of the carotid system. Trans Am Neurol Assoc 104:141–144, 1979.

CHAPTER 40

BRAIN STEM AND CRANIAL NERVE MONITORING

C. Michel Harper, Jr., M.D.

METHODOLOGY
Electromyography
Nerve Conduction Studies
Evoked Potentials
APPLICATIONS
Middle Cranial Fossa
Posterior Cranial Fossa
Head and Neck Surgery

Cranial nerves can be injured during surgical procedures performed in the middle and posterior cranial fossae, as well as in the head and neck region. Damage can result from compression, stretch, abrasion, or ischemia of the nerve. If axonal disruption occurs, recovery is limited. Cranial nerve function can be monitored with the patient under anesthesia by recording spontaneous or stimulus-evoked electrical activity directly from the nerve or the cranial muscles. These methods can detect damage to either the intra-axial or extra-axial portion of cranial nerves. Activity in other pathways in the brain stem can be monitored by noting changes in blood pressure, pulse, respiration, temperature, and evoked potentials in various sensory and motor pathways.

METHODOLOGY

Electromyography

Special small-diameter flexible wire electrodes are used to record electromyograms (EMGs). Such electrodes are less traumatic to local tissue and are more easily secured than monopolar or concentric needle electrodes. Electrodes placed in muscle record a variety of spontaneous and stimulus-evoked activities that arise from individual muscle fibers or motor units. Movement and electrical artifacts, fibrillation and fasciculation potentials, and random motor unit potential activity related to inadequate anesthesia are regularly recorded from muscle intraoperatively. Electrical stimulation of the innervating nerve produces a response by activation of motor unit potentials in the area immediately surrounding the electrode. Mechanical irritation (abrasion, stretch, compression), saline irrigation, and ischemia produce high-frequency bursts of motor unit potentials termed "neurotonic discharges" that have a characteristic sound and appearance.[2] Neurotonic discharges provide surgeons immediate knowledge about location and potential injury to nerves in the surgical

451

field. Neurotonic discharges can be recorded in situations that require neuromuscular blockade by titrating the dose of neuromuscular blocking agent with the motor response obtained with peripheral nerve stimulation intraoperatively.[1]

Nerve Conduction Studies

Two types of nerve conduction studies can be performed on cranial nerves intraoperatively. Compound muscle action potentials (CMAPs) represent activity in motor axons and muscle fibers. Whenever possible, CMAPs recorded from the skin surface overlying the motor point are used because they give more quantifiable information about the total number of functioning motor axons in the nerve than CMAPs recorded from intramuscular electrodes. Nerve action potentials are recorded directly from mixed or sensory nerves in the surgical field or subcutaneously. Nerve action potentials are lower in amplitude and more difficult to record than CMAPs but may provide useful information when sensory nerves are involved or when CMAPs cannot be recorded.

Evoked Potentials

VISUAL EVOKED POTENTIALS

Visual evoked potentials reflect electrical activity in optic pathways in response to photic stimulation of the retina. The most reproducible response is recorded as a broad positivity over the occipital head region approximately 100 milliseconds after the stimulus. This "P100" waveform represents electrical activity in the occipital cortex. It is well-defined in all patients who are awake when a pattern-reversal illuminated stimulus is given to the retina. Visual evoked potentials can also be recorded in anesthetized patients with a strobe flash stimulus applied through the closed eyelid or by way of specially designed contact lenses. Unfortunately, flash-evoked visual evoked potentials are greatly attenuated by general anesthesia and so have limited use intraoperatively[5] (Fig. 40–1).

AUDITORY EVOKED POTENTIALS

Auditory evoked potentials reflect electrical activity in the auditory nerve and brain stem in response to cochlear stimulation.

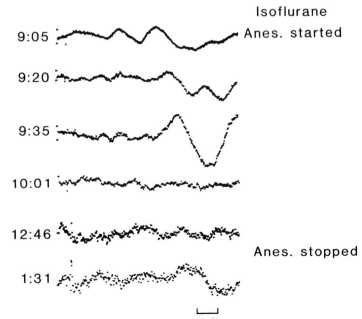

Isoflurane

9:05 — Anes. started

9:20

9:35

10:01

12:46

Anes. stopped

1:31

20 ms

Figure 40–1. Flash-evoked visual evoked potential recorded intraoperatively. The signal is often unreliable when recorded with the patient under general anesthesia.

Brain stem auditory evoked potentials (BAEPs) are recorded from surface electrodes placed either in the external ear canal or on the ear pinna and referenced to the vertex. Five distinct waveforms can be recorded: waves I and II originate from the auditory nerve and waves III to V reflect activity in brain stem structures. The electrocochleogram can be recorded using a needle electrode placed through the tympanic membrane into the wall of the middle ear cavity. The first major negative waveform (N1) of the electrocochleogram represents activity in the lateral portion of the auditory nerve. Auditory nerve action potentials can also be recorded from a small cotton-wick electrode placed directly on the nerve in the cerebellopontine angle at surgery. Auditory evoked potentials are not affected significantly by general anesthetics.

Recording BAEPs appears to be the most comprehensive technique because it monitors the entire auditory system, has few technical recording problems, can be used during the entire surgical procedure, and correlates well with postoperative hearing status.[4] The electrocochleogram is better defined in patients with acoustic neuromas, but it has many technical difficulties and adequately monitors only the lateral portion of the auditory nerve located in the auditory canal of the temporal bone. Recording auditory nerve action potentials provides immediate feedback, but it has technical difficulties and can be used only when the auditory nerve is exposed in the surgical field. Sudden changes in latency and amplitude of BAEPs occur when the nerve is avulsed or if the internal auditory artery is damaged. Recovery after sudden loss of BAEPs is unusual but correlation of the change with intraoperative events may help prevent hearing loss in future cases. Gradual changes in BAEPs may reflect excessive traction on or manipulation of the nerve and often resolve with adjustment of retractors or modification of the surgical approach.

SOMATOSENSORY EVOKED POTENTIALS AND MOTOR EVOKED POTENTIALS

Median, ulnar, and tibial somatosensory evoked potentials can be used to monitor activity in the medial lemniscus of the brain stem. Changes may occur only when sensory pathways are involved or when damage to the brain stem is extensive. Monitoring of motor evoked potentials may enhance the sensitivity of brain stem monitoring by detecting early compromise of pyramidal tract neurons. Use of motor evoked potentials to monitor brain stem function is still investigational and is limited by the exquisite sensitivity of these potentials to inhalation anesthetics.[7]

APPLICATIONS

Middle Cranial Fossa

PITUITARY REGION

Visual evoked potentials have been monitored during removal of tumors in the region of the pituitary and hypothalamus and during operations on vascular lesions that may affect the optic nerves, chiasm, or tracts. Because of the variability in latency and amplitude of the response and the poor correlation with postoperative visual function, visual evoked responses are not a reliable monitor of visual pathway function during surgery.[5]

CAVERNOUS SINUS REGION

Tumors and vascular lesions of the orbital, sphenoidal, or cavernous sinus regions can damage cranial nerves directly or distort normal anatomic relationships, making it difficult to distinguish nervous system structures from abnormalities. The oculomotor, trochlear, and abducens nerves can be monitored with EMG or CMAPs. Wire electrodes are placed in the extraocular muscles after the patient is anesthetized. Neurotonic discharges are recorded in the appropriate muscle when its nerve is mechanically stimulated in the surgical field. The surgeon may also use a hand-held stimulator to identify selected cranial nerves by recording a CMAP in the appropriate target muscle (Fig. 40–2). A nerve action potential can be recorded directly from the ophthalmic division of the trigeminal nerve with a small cotton-wick electrode.

Stimulate: Oculomotor nerve

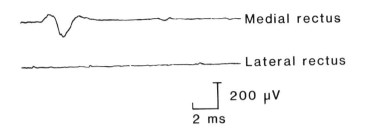

Figure 40–2. Compound muscle action potentials from extraocular muscles obtained by direct electrical stimulation of the oculomotor nerve in the surgical field in the region of the cavernous sinus.

Posterior Cranial Fossa

CEREBELLOPONTINE ANGLE

The trigeminal, facial, auditory, and vestibular nerves can be injured during operations for acoustic neuromas and other cerebellopontine angle tumors as well as during microvascular decompression or neurectomy for trigeminal neuralgia, hemifacial spasm, or vertigo. Also, the brain stem may be involved with mass lesions larger than 3 cm in diameter. The facial nerve and the motor division of the trigeminal nerve are monitored with intramuscular electrodes in muscles of facial expression and mastication. Mechanical stimulation produces neurotonic discharges in the respective muscles. Electrical stimulation can be used to identify various nerves in the cerebellopontine angle. The amplitude of CMAPs recorded over the nasalis or mentalis muscles correlates with the number of functioning axons in the nerve (Fig. 40–3). Preservation of facial CMAP at the end of the operation predicts good recovery of facial nerve function within 1 year postoperatively.[3] BAEPs can be monitored simultaneously with EMG and CMAP.[6] Changes in BAEPs correlate well with postoperative level of hearing.[4] Gradual changes are often reversible by altering the surgical approach or by moving retractors. Sudden loss of BAEPs is usually irreversible and represents either ischemia or avulsion of the auditory nerve (Fig. 40–4). Changes that correlate with postoperative function help determine the mechanism of nerve injury, thereby improving future surgical results. When brain stem compression is present, monitoring somatosensory evoked potentials or motor evoked potentials may also be useful.

JUGULAR FORAMEN REGION

Glomus tumors, meningiomas, metastatic cancer, and other lesions in the region of the jugular foramen may involve the facial, auditory, glossopharyngeal, vagal, and spinal accessory nerves. EMG activity is monitored by placing intramuscular wire electrodes in facial, laryngeal, and trapezius muscles. Electrical stimulation can distinguish between rootlets of the glossopharyngeal and vagus nerves in patients undergoing neurectomy for glossopharyngeal neuralgia. EMG monitoring of the hypoglossal nerve (by using electrodes in the tongue), in addition to monitoring of the vagus and spinal accessory nerves, is useful during operations to remove chordomas, meningiomas, and other lesions in the region of the clivus and foramen magnum.

Head and Neck Surgery

Operations for neoplasms of the parotid gland may injure one or more branches of the facial nerve that course through the gland. Each branch can be monitored selectively by placing wire electrodes in the frontalis, orbicularis oculi, orbicularis oris, and mentalis muscles. Mechanical irritation of the nerve produces neurotonic discharges in the target muscle. Electrical stimulation in the surgical field can locate and prevent damage to branches of the facial nerve.

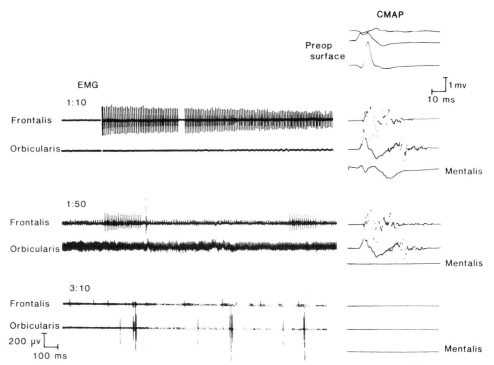

Figure 40–3. Monitoring of electromyographic (*EMG*) potentials and facial compound muscle action potentials (*CMAP*) intraoperatively for acoustic neuroma in a 55-year-old woman. Examples of neurotonic discharges observed at various times during surgery and gradual loss of facial CMAPs indicate iatrogenic injury of the facial nerve. (From Daube, JR and Harper, CM: Surgical monitoring of cranial and peripheral nerves. In Desmedt, JE [ed]: Neuromonitoring in Surgery. Elsevier Science Publishers, Amsterdam, 1989, p 118, with permission.)

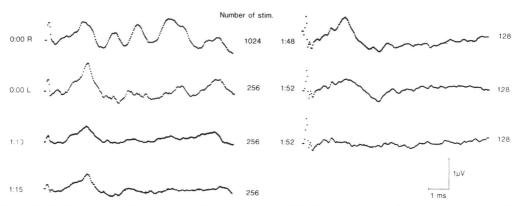

Figure 40–4. Monitoring of brain stem auditory evoked potentials during operation for acoustic neuroma. The sudden loss of the response correlated with inadvertent coagulation of the internal auditory artery. (From Harper, CM and Daube, JR: Surgical monitoring with evoked potentials: The Mayo Clinic experience. In Desmedt, JE [ed]: Neuromonitoring in Surgery. Elsevier Science Publishers, Amsterdam, 1989, p 281, with permission.)

The recurrent branch of the laryngeal nerve can be injured during carotid endarterectomy or resection of thyroid tumors. Wire electrodes can be placed by direct laryngoscopy into the vocalis muscle for monitoring the recurrent laryngeal nerve.

CHAPTER SUMMARY

Various modalities are available for monitoring the function of the cranial nerves and the brain stem during intracranial or extracranial head and neck operations. After consideration of the surgical risks, a multimodal approach can be tailored to the needs of each patient. Intraoperative monitoring has been shown to decrease cranial nerve injury during posterior fossa surgery.

REFERENCES

1. Blair, EA, Teeple, E, Jr, Sutherland, RM, et al: Effect of neuromuscular blockade on facial nerve monitoring. Am J Otol 15:161–167, 1994.
2. Harner, SG, Daube, JR, and Ebersold, MJ: Electrophysiologic monitoring of facial nerve during temporal bone surgery. Laryngoscope 96:65–69, 1986.
3. Harner, SG, Daube, JR, Ebersold, MJ, and Beatty, CW: Improved preservation of facial nerve function with use of electrical monitoring during removal of acoustic neuromas. Mayo Clin Proc 62:92–102, 1987.
4. Harper, CM, Harner, SG, Slavit, DH, et al: Effect of BAEP monitoring on hearing preservation during acoustic neuroma resection. Neurology 42:1551–1553, 1992.
5. Raudzens, PA: Intraoperative monitoring of evoked potentials. Ann N Y Acad Sci 388:308–325, 1982.
6. Wazen, JJ: Intraoperative monitoring of auditory function: Experimental observations and new applications. Laryngoscope 104:446–455, 1994.
7. Zentner, J: Noninvasive motor evoked potential monitoring during neurosurgical operations on the spinal cord. Neurosurgery 24:709–712, 1989.

CHAPTER 41

SPINAL CORD MONITORING

Jasper R. Daube, M.D.

APPLICATIONS
SOMATOSENSORY EVOKED POTENTIALS
Technical Factors
Extraoperative Recordings
Intraoperative Recordings
MOTOR EVOKED POTENTIALS
NERVE ROOT AND SPINAL NERVE
   MONITORING

APPLICATIONS

Surgical monitoring is of benefit in many surgical procedures on the spine and spinal contents. In each case, methods of surgical monitoring are selected on the basis of the structures at risk and optimal methods for monitoring their function. This often requires simultaneous use of multiple modalities of monitoring but depends heavily on evoked potentials. Applications for spinal monitoring include:

1. Orthopedic spinal operations at the lumbosacral, thoracic, and cervical levels, in particular for scoliosis, spondylosis, degenerative disk disease, and bone tumors.
2. Cauda equina operations, especially tumors and congenital deformities.
3. Intramedullary and extramedullary spinal cord and nerve root tumors.
4. Selective dorsal rhizotomy.

5. Arteriovenous malformations of the spinal cord.
6. Aortic aneurysms, especially thoracic ones.

A persistent neurologic deficit develops immediately postoperatively in a small proportion of patients, usually less than 0.5%, who undergo corrective operations for scoliosis and other surgical procedures on the spine.[10,15] Careful surgical techniques and stabilization of the spine intraoperatively have helped to keep this complication rate low. Nonetheless, the procedure is devastating for the person who awakens after a spinal operation and discovers being paraplegic. To further decrease this possible complication, the "wake-up test" was devised.[9] The patient is awakened during the surgical procedure to ensure continuity of the spinal cord pathways after correction of a spinal deformity by having the patient voluntarily move the feet. Although this test is helpful, it is difficult to perform and is associated with the problems of changing the level of anesthesia. Moreover, it is not applicable in patients undergoing surgical procedures for spine tumors in which there is no well-defined time of major hazard.

Therefore, somatosensory evoked potentials (SEPs) have been applied by many physicians as an additional method for monitoring spinal cord function intraoperatively.[1,8,16,17,19] Animal studies have demon-

457

strated that the somatosensory evoked potential is lost with spinal cord damage.[2,5] It has been shown experimentally that compressive lesions that compromise motor function also alter SEP.

SOMATOSENSORY EVOKED POTENTIALS

SEP monitoring of patients during spinal surgery has been developed by several investigators using different methods. These methods can be divided into those in which SEP recordings are made within or without the operative field and those in which motor evoked potentials (MEPs) are recorded. In addition to monitoring evoked potentials to assess spinal cord function, many spinal surgical procedures also warrant simultaneous monitoring of electromyographic (EMG) activity in the muscles innervated by the nerve roots or motor neuron pools at risk during the procedure.[10] The particular combination of SEP, MEP, and EMG monitoring performed must be individualized for each patient on the basis of the preoperative clinical deficits and the neural structures at risk for damage during the operation.

Determination of the optimal combination of recordings for each patient is often assisted by preoperative testing with SEP and EMG and nerve conduction studies (NCS) or some combination of them. When SEPs have a low amplitude or are absent, bilateral peripheral nerve stimulation, stimulation of the sciatic nerves, or stimulation of the cauda equina may be needed to elicit SEPs. Recordings may have to be made from the epidural space rather than from an extraspinal location. If the EMG/NCS show abnormalities in the distribution of particular nerve roots, these roots are more susceptible to further damage and warrant intramuscular recordings to monitor neurotonic discharges.

Technical Factors

Several technical factors must be considered with intraoperative recordings.[11] The rate of stimulation cannot be as fast as when testing a patient who is awake. Anesthetics make the sensory system more susceptible to fatigue. Although SEPs can be recorded with stimulation rates of 5 Hz or even 10 Hz in most patients who are awake, under anesthesia the scalp SEPs fatigue at rates greater than 3 Hz, and it may be necessary to stimulate at rates as low as 0.5 or 1.0 Hz, especially at deeper levels of anesthesia.

The number of stimuli that are averaged varies with magnitude of the response and level of background noise. Because many of the patients are paralyzed, muscle activity is minimal. Under ideal conditions, responses can be obtained clearly with only 64 or 128 stimuli. However, in most cases, sources of artifact other than muscle exist; thus, 500 stimuli are often needed to obtain reproducible traces.

Technical problems are common in the operation room. For example, 60-Hz artifact occurs with gas warmers/humidifiers, blood warmers, and some electric drills. Movement of the recording wires must be eliminated. Wires can be disconnected, cut, dislodged, or damaged during the course of surgical procedures (especially procedures performed near the head or neck) and with movement of the torso. Because stimulating electrodes can also be displaced, some type of peripheral monitor is required to ensure they are functioning.

The level of anesthesia and blood pressure both change latency and amplitude of the SEPs. Rarely, the scalp response is enhanced after induction of anesthesia; generally, however, it is decreased. In a small proportion of cases, the response is lost immediately after induction of anesthesia. This loss occurs more frequently in children and adolescents. Similarly, there is a gradual reduction in amplitude and an increase in latency with continued anesthesia. SEP changes due to anesthesia are much less prominent at the neck. Response varies depending on the anesthetic agent. Of the induction agents used, propofol has the least effect. Enflurane and isoflurane reduce SEPs in more than half of the patients monitored. Alterations in blood pressure can also reduce the amplitude of the evoked response, especially with a mean blood pressure below 70 mm Hg.

Extraoperative Recordings

Recordings made outside the operating field are the simplest to perform and do not require the direct assistance of the surgeon, leaving the physician free for the surgical procedure.[6] Figure 41–1 shows a typical intraoperative recording from the neck and scalp. Stimulation is applied to a peripheral nerve, usually either the peroneal or tibial nerve in the leg or to the median or ulnar nerve in the arm. Recordings are commonly made from standard scalp derivations, usually C_z–F_z with leg stimulation and C_3'/ (C_4')–F_z with arm stimulation. Recordings can also be made from the spinal cord and peripheral nerves. Scalp recordings generally give a well-defined, though unstable, response that may be altered by blood pressure and anesthetic agent used.

Patients, from infants to the elderly, are best selected for monitoring by surgeons on the basis of the risk of neural damage. The largest group of patients monitored intraoperatively are teenagers undergoing corrective surgery for scoliosis. Another large group are elderly persons operated on for cervical spondylosis (Fig. 41–2). Patients with bony spinal tumors, thoracic aneurysms, traumatic spinal damage, or spondylitis can also be monitored. Two-thirds of the monitoring performed is for surgical procedures at the thoracic level.

A frequent problem during surgical monitoring is the variability of the evoked response on sequential recordings because of artifact, blood pressure changes, anesthetic level, and other factors. Therefore, determining a change in SEP as significant requires that there be a consistent alteration in the latency (2 milliseconds more than baseline changes) and the amplitude (50% less than baseline values) at both the neck and scalp sites, with an intact peripheral response.[7] It must be shown that this change is not due to technical factors. The change

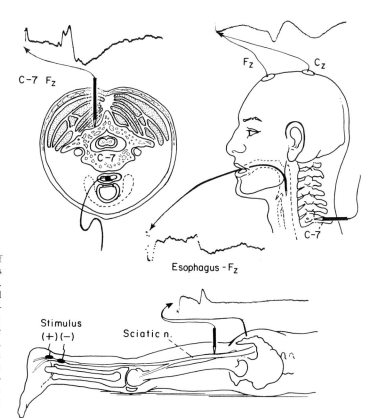

Figure 41–1. Standard locations of placement of recording electrodes for thoracolumbar spine surgery. The cervical response is recorded with either an esophageal electrode or a deep cervical (e.g., C-7) paraspinal needle electrode on the vertebral lamina. (From Daube, JR, et al: Intraoperative monitoring. In Daly, DD and Pedley, TA [eds]: Current Practice of Clinical Electroencephalography, ed 2. Raven Press, New York, 1990, p 747, with permission.)

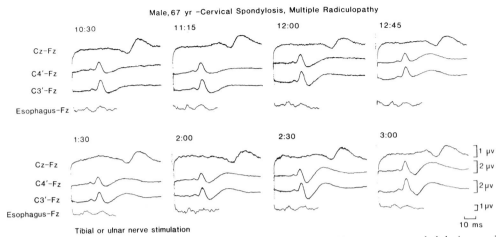

Figure 41–2. Sequential scalp and neck somatosensory evoked potential (SEP) responses recorded during cervical spondylosis surgery.

must also be greater than the baseline variation that occurred during the initial period of the operation. In a few patients, recordings can be obtained only at one of the two cephalad sites, usually the neck. In such cases, changes need to be seen with stimulation of more than one nerve before they can be considered significant. No absolute change in amplitude can be considered evidence of spinal cord damage, because in some patients in whom there is no damage, a scalp response appears to be lost transiently while other responses are intact. In other patients in whom there is damage,

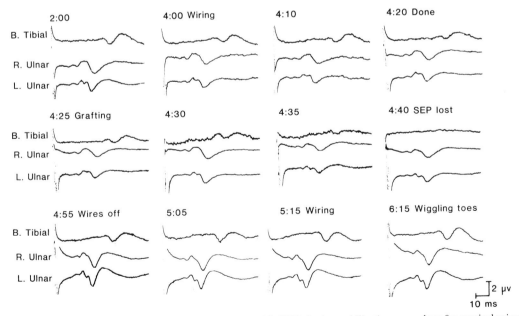

Figure 41–3. Gradual loss of somatosensory evoked potential (SEP) during stabilization procedure for cervical spine fracture in a 60-year-old man. Responses were lost within a few minutes after wiring C-5 or C-7 and returned quickly after the wires were removed. Patient awoke with no deficit. (From Daube, JR, et al: Intraoperative monitoring. In Daly, DD and Pedley, TA [eds]: Current Practice of Clinical Electroencephalography, ed 2. Raven Press, New York, 1990, p 765, with permission.)

smaller but consistent alteration at two sites of stimulation can be evidence of compression before the major changes occur.

SEP monitoring of operations on the spine is sometimes difficult because of the inability to record reliable potentials over the scalp in patients with or without a neurologic deficit preoperatively. SEPs may be recorded at the neck in many of these patients. The addition of a cervical spine and a peripheral recording location to the usual scalp recording allows reliable monitoring to be performed in more patients.

SEP monitoring during spinal surgery has proved valuable in warning surgeons of potential damage to the spinal cord. Because such damage is uncommon, few patients are available for defining the change in SEPs in relationship to cord damage. In surgical monitoring of spinal cord sensory pathways, approximately 20% of patients have significant changes in SEP intraoperatively. Up to half of these changes may be reversed surgically (removal of hematoma, rods, spine wires, and so forth) (Fig. 41–3). Patients with

no change in SEPs may have postoperative deficits consistent with anterior spinal artery syndrome, although this is rare. SEP changes are similar in most patients. The amplitude reduction begins after 10 to 30 minutes at the neck and scalp, with an increase in latency of up to 3 milliseconds. SEPs can recover to baseline value within 5 to 10 minutes after correction. The abrupt loss of SEPs is less likely to be reversible. SEP amplitude shows improvement intraoperatively in a few patients (Fig. 41–4).

Intraoperative Recordings

Several methods of recording in the operating field have been developed to allow recordings to be made closer to the neural tissue. These include subarachnoid, epidural, spinous process, and intraspinous ligament recordings.[5] Lueders and associates[14] used needle electrodes placed between the spinous processes. Epidural electrodes inserted between the spine and the dural sac

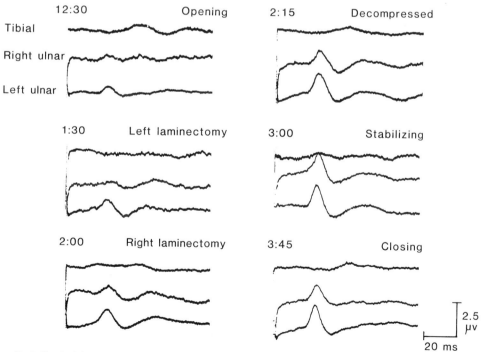

Figure 41–4. Gradual improvement of somatosensory evoked potential (SEP) amplitude and latency during cord decompression for rheumatoid arthritis. Patient's neurologic deficit also showed improvement postoperatively. (From Daube, JR, et al: Intraoperative monitoring. In Daly, DD and Pedley, TA [eds]: Current Practice of Clinical Electroencephalography, ed 2. Raven Press, New York, 1990, p 766, with permission.)

obtain large readily recorded potentials. Similar recordings have been made of descending activity by stimulating and recording from the spinal cord directly.[20]

Whereas recordings made in the surgical field give larger responses, they are associated with the technical problems of the surgical procedure. They are thus subject to mechanical artifact and are limited to surgical procedures in which the spine is opened to expose the dura mater. Such recordings generally also require more technical expertise for obtaining satisfactory results and require that the surgeon be familiar and cooperate with the recording procedure. Recordings performed in the surgical field are most useful for operations on the spinal cord (e.g., tumors or arteriovenous malformations); directly recorded potentials can localize the area of damage or record responses that are too small to be obtained with other methods.

In summary, several patterns of change in SEPs and their correlation with postoperative neurologic function have been observed.[4] Most changes evolve gradually, occur in relation to distraction or other specific operative manipulation, and gradually resolve without neurologic sequelae. The change may be late or gradual, emphasizing the importance of continuing the monitoring until the patient is awake. Gradual changes in SEP may also be caused by ischemia of the spinal cord or peripheral nerves, as during operations on the thoracic or abdominal aorta.[18] Less frequently, the SEPs change abruptly, usually in relation to contusion of the cord or to compression by an epidural or subdural hematoma. When the loss of SEPs is abrupt, an attempt should be made to localize the site of injury by recording with epidural or direct spinal electrodes, followed by careful inspection of the involved level for hematomas or other potentially reversible causes of spinal cord injury. Unfortunately, when the loss of SEPs is abrupt and nonreversible, paraplegia is likely to occur. Improvement in the amplitude of the SEP intraoperatively is usually associated with improved neurologic status postoperatively. Finally, postoperative motor deficits without associated changes in SEPs intraoperatively are infrequent but well documented.[12] The motor deficits may occur as part of the anterior spinal artery syndrome or they may be caused by direct injury to the ventral spinal cord, ventral horn cells, or nerve roots.

MOTOR EVOKED POTENTIALS

MEPs are easily and reliably obtained with direct stimulation of the spinal cord or cerebral hemispheres with the patient under nitrous/narcotic anesthesia.[3,13] MEPs can be evoked using laminar or epidural electrodes in the surgical field, but it is easier to use laminar needle cathodes and an esophageal anode, which have been used successfully in thoracic and lumbar surgery but not cervical spine surgery.[21] With laminar needle cathodes and an esophageal anode, MEPs can be obtained even with partial neuromuscular block. MEPs are markedly reduced in amplitude when halogenated anesthetics are used.

Cervical spine surgery requires cranial stimulation to evoke MEPs. Two methods of directly stimulating the cerebral cortex through the intact skull have been developed in the last few years. A small number of medical centers in Europe have begun to use these methods for surgical monitoring, but in the United States they are still considered experimental.

NERVE ROOT AND SPINAL NERVE MONITORING

Injury can occur to cervical or lumbosacral nerve roots or motor neurons during surgical procedures on the spine. Radiculopathies are an occasional complication of scoliosis surgery that may be mimimized by monitoring. Manipulation, traction, or ischemia of nerve roots produces neurotonic discharges in limb muscles innervated by the affected spinal cord segment. Ventral horn cells can be damaged during dissection of intraspinal tumors. EMG activity is monitored with fine wire electrodes inserted in each of the muscles at risk. Monitoring of muscles such as the anal sphincter is helpful in operations for myelomeningocele.

Direct stimulation of nerves in the surgical field can provide information about the location and integrity of the nerves. Stimulation can help to distinguish between different nerve roots and to differentiate them from nonneural structures in situations where the normal anatomy is distorted. Intraoperative monitoring of nerve roots and spinal nerves in combination with SEP monitoring is performed for assessing the integrity of spinal cord function.

CHAPTER SUMMARY

Potential damage that may occur to the spinal cord or spinal nerves during spine surgery can be minimized by continuous monitoring of spinal cord or spinal nerve (or both) function intraoperatively. Somatosensory evoked potentials are the easiest to use for monitoring function and have had the widest application. Although they may not detect purely motor damage (unless it is caused by a vascular insult), somatosensory evoked potentials are able to identify most spinal cord damage early enough to alert the surgeon. Recently, motor evoked potentials have been shown to be useful as well, but technical problems in some settings limit their application. Recording neurotonic discharges from a peripheral muscle is surprisingly sensitive to minor root irritation and, thus, can help surgeons recognize when and where damage may be occurring.

REFERENCES

1. Allen, A, Starr, A, and Nudleman, K: Assessment of sensory function in the operating room utilizing cerebral evoked potentials: A study of fifty-six surgically anesthetized patients. Clin Neurosurg 28:457–481, 1981.
2. Bennett, MH: Effects of compression and ischemia on spinal cord evoked potentials. Exp Neurol 80:508–519, 1983.
3. Boyd, SG, Rothwell, JC, Cowan, JMA, et al: A method of monitoring function in corticospinal pathways during scoliosis surgery with a note on motor conduction velocities. J Neurol Neurosurg Psychiatry 49:251–257, 1986.
4. Dawson, EG, Sherman, JE, Kanim, LE, and Nuwer, MR: Spinal cord monitoring. Results of the Scoliosis Research Society and the European Spinal Deformity Society survey. Spine 16 Suppl 8:S316–S361, 1991.
5. Dinner, DS, Lüders, H, Lesser, RP, and Morris, HH: Invasive methods of somatosensory evoked potential monitoring. J Clin Neurophysiol 3:113–130, 1986.
6. Dunne, JW and Field, CM: The value of non-invasive spinal cord monitoring during spinal surgery and interventional angiography. Clin Exp Neurol 28:199–209, 1991.
7. Forbes, HJ, Allen, PW, Waller, CS, et al: Spinal cord monitoring in scoliosis surgery. Experience with 1168 cases. J Bone Joint Surg Br 73:487–491, 1991.
8. Grundy, BL, Nelson, PB, Doyle, E, and Procopio, T: Intraoperative loss of somatosensory-evoked potentials predict loss of spinal cord function. Anesthesiology 57:321–322, 1982.
9. Hall, JE, Levine, CR, and Sudhir, KG: Intraoperative awakening to monitor spinal cord function during instrumentation and spine fusion. J Bone Joint Surg Am 60:533–536, 1978.
10. Harper, CM Jr, Daube, JR, Litchy, WJ, and Klassen, RA: Lumbar radiculopathy after spinal fusion for scoliosis. Muscle Nerve 11:386–391, 1988.
11. Lesser, RP, Lueders, H, Dinner, DS, et al: Technical aspects of surgical monitoring using evoked potentials. In Struppler, A and Weindl, A (eds): Electromyography and Evoked Potentials: Theories and Applications. Springer-Verlag, Berlin, 1985, pp 177–180.
12. Lesser, RP, Raudzens, P, Lüders, H, et al: Postoperative neurological deficits may occur despite unchanged intraoperative somatosensory evoked potentials. Ann Neurol 19:22–25, 1986.
13. Levy, WJ, York, DH, McCaffrey, M, and Tanzer, F: Motor evoked potentials from transcranial stimulation of the motor cortex in humans. Neurosurgery 15:287–302, 1984.
14. Lueders, H, Gurd, A, Hahn, J, et al: New technique for intraoperative monitoring of spinal cord function: Multichannel recording of spinal cord and subcortical evoked potentials. Spine 7:110–115, 1982.
15. MacEwen, GD, Bunnell, WP, and Sriram, K: Acute neurological complications in the treatment of scoliosis: A report of the Scoliosis Research Society. J Bone Joint Surg Am 57:404–408, 1975.
16. Macon, JB and Poletti, CE: Conducted somatosensory evoked potentials during spinal surgery. Part 1: Control conduction velocity measurements. J Neurosurg 57:349–353, 1982.
17. Macon, JB, Poletti, CE, Sweet, WH, et al: Conducted somatosensory evoked potentials during spinal surgery. Part 2: Clinical applications. J Neurosurg 57:354–359, 1982.
18. Okamoto, Y, Murakami, M, Nakagawa, T, et al: Intraoperative spinal cord monitoring during surgery for aortic aneurysm: Application of spinal cord evoked potential. Electroencephalogr Clin Neurophysiol 84:315–320, 1992.
19. Schramm, J and Kurthen, M: Recent developments in neurosurgical spinal cord monitoring. Paraplegia 30:609–616, 1992.
20. Tamaki, T, Tsuji, H, Inoue, S, and Kobayashi, H: The prevention of iatrogenic spinal cord injury utilizing the evoked spinal cord potential. Int Orthop 4:313–317, 1981.
21. Taylor, BA, Fennelly, ME, Taylor, A, and Farrell, J: Temporal summation—the key to motor evoked potential spinal cord monitoring in humans. J Neurol Neurosurg Psychiatry 56:104–106, 1993.

CHAPTER 42

PERIPHERAL NERVOUS SYSTEM MONITORING

C. Michel Harper, Jr., M.D.

METHODS
Nerve Conduction Studies
Electromyography
Somatosensory Evoked Potentials
APPLICATIONS
Mononeuropathy
Brachial Plexopathy

Electrophysiologic studies are performed during operations for nerve trauma, entrapment neuropathies, and primary or metastatic neoplasms that involve peripheral nerves. These studies provide information about the number, location, type, and severity of nerve lesions.[2] This information can be used to answer important questions left unresolved by preoperative electrodiagnostic studies and also to help surgeons make important therapeutic decisions regarding decompression, neurolysis, grafting, or neurotization of nerves. With minor modifications, standard techniques of electrodiagnosis such as nerve conduction studies, electromyography (EMG), and somatosensory evoked potentials are used to monitor peripheral nerve function intraoperatively. Appropriate monitoring protocols can be designed for each patient after the findings of the preoperative neurologic examination, nerve conduction studies, EMG, and surgical goals are reviewed.

METHODS

Nerve Conduction Studies

Mixed motor and sensory nerves or sensory cutaneous nerves are stimulated using a hand-held bipolar electrical stimulator placed directly on the nerve in the surgical field. Compound muscle action potentials (CMAPs) are recorded from surface or subcutaneous needle electrodes placed over distal muscles. Compound nerve action potentials (NAP) are recorded either from large mixed nerves using a hand-held bipolar recorder placed directly on the nerve or from small cutaneous nerves with needle electrodes placed next to the nerve. CMAP recordings are performed whenever possible because they have high amplitude, are less contaminated with artifact, and because they exclusively monitor the function of motor axons. NAP recordings are technically more difficult to produce but more sensitive than CMAPs in localizing abnormality along nerves. These recordings may also be able to detect early axonal regeneration through an area of nerve injury long before regeneration is reflected in CMAP recordings. The presence of CMAPs or NAPs indicates that some axons are in continuity; amplitude and area of the response are proportional to the number of functioning axons. Focal slowing of conduction velocity, conduction block, or increased threshold of stimulation can be used to localize abnormality along the nerve with 1 to 2 cm of accuracy.

Electromyography

EMG potentials are best recorded using small wire electrodes placed directly in the muscle. When intramuscular electrodes are used, electrical stimulation of the nerve produces a polyphasic EMG response that, although difficult to quantitate, is less likely than surface electrodes to record nonspecific activity from adjacent muscles. Mechanical irritation of the nerve produces a high-frequency discharge of motor unit potentials (neurotonic discharge) that can easily be distinguished from artifact and other motor unit potential activity in EMG recordings. The occurrence of neurotonic discharges is used to locate nerves in the surgical field and to warn of the potential for nerve injury should the irritation continue.

Somatosensory Evoked Potentials

Somatosensory evoked potentials are recorded from surface electrodes over the contralateral parietal scalp using direct electrical stimulation of peripheral nerve elements in the surgical field. A response indicates axonal continuity between the spinal cord and the site of peripheral nerve stimulation. This technique has several advantages over standard recordings of somatosensory evoked potentials made in the laboratory, including increased selectivity of stimulation and enhanced sensitivity due to amplification of the response by the central nervous system.

APPLICATIONS

Mononeuropathy

Intraoperative recordings of EMG, CMAPs, and NAPs can be applied to the study of any major peripheral nerve. Among surgical procedures monitored are exploration and repair of traumatic lesions, decompression for entrapment syndromes, and resection of peripheral nerve neoplasms.

Lesions of the ulnar nerve at the elbow, the median nerve in the forearm, the radial nerve in the arm or forearm, the sciatic and femoral nerves in the thigh, and the peroneal nerve at the knee are particularly amenable to intraoperative monitoring. Monitoring nerves in the leg is becoming more frequent during major hip operations.

Ulnar neuropathy at the elbow is one of the most common mononeuropathies. It can be used as an example of how nerve conduction studies are applied intraoperatively. Ulnar neuropathy can be caused by repeated trauma to the nerve in the region of the medial epicondyle, compression by bony or soft tissue deformities about the elbow joint, recurrent subluxation over the medial epicondyle, or entrapment between the heads of the flexor carpi ulnaris (cubital tunnel syndrome). There has been disagreement about the relative frequency of, and proper surgical treatment for, the various causes of ulnar neuropathy at the elbow.[1,6,11] Although ulnar transposition has been the most traditional surgical procedure, it has a higher incidence of postoperative morbidity than more conservative procedures, such as cubital tunnel release.[6] Precise localization of the lesion using preoperative and intraoperative electrophysiologic studies can help choose the proper surgical procedure. A well-localized lesion in the region of the two heads of the flexor carpi ulnaris may respond to simple cubital tunnel release, whereas a more diffuse lesion or one localized to the medial epicondyle requires anterior transposition. Preoperative electrophysiologic studies often help localize the site of entrapment. Useful findings include a localized area of conduction slowing or conduction block on short segmental stimulation ("inching") and the distribution of fibrillation potentials and changes in motor unit potential on needle examination.

Precise localization is not always possible, and results of preoperative studies may even be misleading.[8] Factors that contribute to this inaccuracy include (1) variability in the location of the cubital tunnel in relation to the medial epicondyle, (2) selective involvement of certain fascicles in the ulnar nerve, (3) technical difficulties with the recording (e.g., overstimulation that causes current spread along the nerve), and (4) the occurrence of lesions in unusual locations. For example, Campbell Pridgeon, and Sahni[3,4] have detected conduction block due to ulnar nerve compression at the most distal aspect of the cubital tunnel by using intraoperative nerve conduction studies.

Figure 42–1 illustrates results of intraoperative conduction studies performed during exploration of an ulnar nerve. Preoperative nerve conduction studies revealed an area of focal slowing with increased CMAP dispersion approximately 3 cm distal to the medial epicondyle. The CMAP and NAP were recorded intraoperatively over the abductor digiti minimi and flexor carpi ulnaris muscles and from the proximal ulnar nerve, respectively. Changes in amplitude and latency of both the CMAP and NAP were noted over a 3-cm segment at the origin of the cubital tunnel. Because no area of slowing was detected either proximal or distal to this point, cubital tunnel release was performed.

Although entrapment and traumatic neuropathies are more frequent, EMG and nerve conduction studies are useful during operations to remove primary or metastatic tumors that infiltrate or compress peripheral nerves. Mechanical manipulation of nerves during dissection may produce neurotonic discharges in distal muscles. This can assist in locating nerves or in detecting viable fascicles in nerves. Nerve action potential recordings from individual fascicles can help identify and preserve functioning axons. An attempt is made to resect only those fascicles that do not have electrophysiologic evidence of axonal continuity.

Brachial Plexopathy

The brachial plexus is frequently contused, lacerated, or stretched by violent trauma. Evaluation and treatment of brachial plexus injury are difficult because the

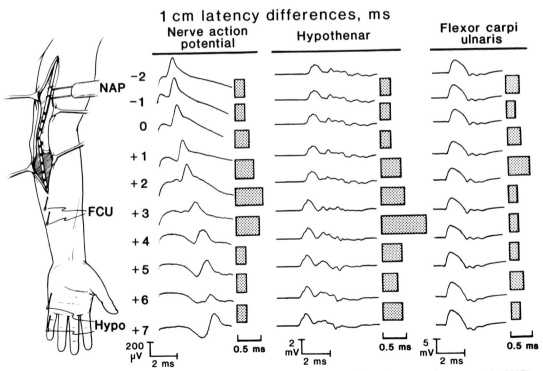

Figure 42–1. Intraoperative recording of compound muscle action potentials and nerve action potentials (NAP) during ulnar nerve exploration and stimulation at 1-cm intervals. The "0" point indicates the location of the medial epicondyle. The greatest change in latency and amplitude occurred over a 3-cm segment spanning the origin of the cubital tunnel. FCU = flexor carpi ulnaris; Hypo = hypothenar. (From Daube, JR and Harper, CM: Surgical monitoring of cranial and peripheral nerves. In Desmedt, JE [ed]: Neuromonitoring in Surgery. Elsevier Science Publishers, Amsterdam, 1989, p 133, with permission.)

damage is often severe and may involve multiple elements, including roots, trunks, cords, and peripheral nerves.[10]

The primary consideration in surgical repair of brachial plexus injury is determination of the status of the nerve roots. Avulsion of a root ("preganglionic" injury) eliminates any possibility of recovery in nerves that receive axons from that root. On the other hand, if the root is intact, then neurolysis or grafting of lesions affecting the trunks, cords, or distal elements ("postganglionic" injury) may be of great benefit to the patient. The findings of the clinical examination, EMG, and myelography may detect nerve root avulsion in some patients[5] but most require intraoperative recording of somatosensory evoked potentials to determine with certainty the functional status of the root.[9,12] A somatosensory evoked potential study is performed by selective stimulation of consecutive nerve roots using a hand-held bipolar stimulator (Fig. 42–2). A well-defined potential recorded over the contralateral parietal scalp confirms root continuity, whereas absence of a response confirms root avulsion (Fig. 42–3).

After the status of the cervical roots has been determined, intraoperative nerve conduction studies are used to help localize lesions affecting postganglionic elements of the plexus and to detect the presence or absence of axonal regeneration. A lesion is localized by detecting conduction block or focal slowing of conduction velocity over short segments in either CMAP or NAP recordings (Fig. 42–4). Nerve action potential recordings made over short segments may detect early axonal regeneration across a lesion long before a distal CMAP appears. If NAP is absent, resection and grafting of the lesion are indicated. If NAP is present, the lesion is left undisturbed or neurolysis is performed in an attempt to enhance further regenera-

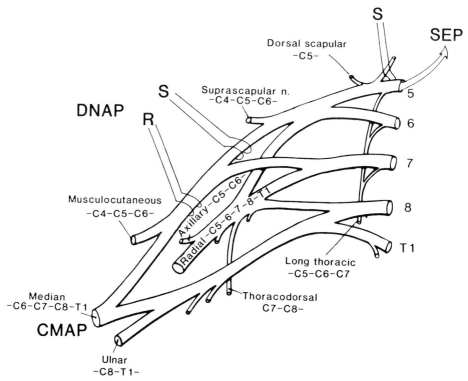

Figure 42–2. Electrophysiologic techniques for monitoring brachial plexopathy. (*Upper right*) Somatosensory evoked potentials (SEP) recorded over scalp during root stimulation. (*Center*) Nerve action potentials recorded directly (DNAP) from short segments of plexus. (*Lower left*) Compound muscle action potentials (CMAP) recorded from distal muscles during selective stimulation of plexus elements. S = stimulating electrodes; R = recording electrodes. (From Daube, JR and Harper, CM: Surgical monitoring of cranial and peripheral nerves. In Desmedt, JE [ed]: Neuromonitoring in Surgery. Elsevier Science Publishers, Amsterdam, 1989, p135, with permission.)

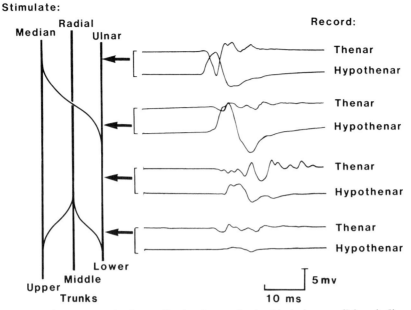

Figure 42–3. Intraoperative nerve conduction studies showing conduction block along medial cord of brachial plexus. (From Daube, JR and Harper, CM: Surgical monitoring of cranial and peripheral nerves. In Desmedt, JE [ed]: Neuromonitoring in Surgery. Elsevier Science Publishers, Amsterdam, 1989, p 136, with permission.)

tion. With the use of specially designed electrodes, the NAP can be recorded from individual nerve fascicles and used as a guide for partial resection and fascicular repair of incomplete lesions.[7,13]

Brachial plexus injuries also occur as a complication of surgery. They result from either direct trauma to nerves or from traction on the plexus caused by retraction of the ribs during thoracic surgery or positioning of the upper limb during general or orthopedic surgery. Musculocutaneous and axillary nerves are particularly susceptible to injury during operations for recurrent anterior shoulder dislocation. EMG monitoring of selected muscles for neurotonic discharges is

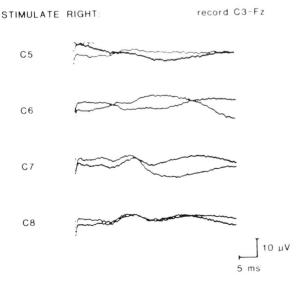

Figure 42–4. Intraoperative recording of somatosensory evoked potentials over scalp with stimulation of cervical nerve roots directly in surgical field. Well-defined responses were seen with stimulation of the C-7 and C-8 roots. No response was obtained with stimulation of C-5 or C-6 roots, indicating avulsion of the root at these levels. (From Daube, JR and Harper, CM: Surgical monitoring of cranial and peripheral nerves. In Desmedt, JE [ed]: Neuromonitoring in Surgery. Elsevier Science Publishers, Amsterdam, 1989, p 136, with permission.)

the most reliable electrophysiologic method available for preventing iatrogenic injury of the brachial plexus intraoperatively.

CHAPTER SUMMARY

Intraoperative monitoring of the peripheral nervous system can be performed using EMG and recordings of CMAPs, NAP, and somatosensory evoked potentials. Application of these techniques requires adequate preoperative electrophysiologic evaluation and consultation with surgeons about surgical goals. These techniques can assist surgeons in determining the number, location, type, and severity of peripheral nerve lesions, while simultaneously preventing iatrogenic injury to peripheral nerve elements.

REFERENCES

1. Brown, WF, Ferguson, GG, Jones, MW, and Yates, SK: The location of conduction abnormalities in human entrapment neuropathies. Can J Neurol Sci 3:111–122, 1976.
2. Brown, WF and Veitch, J: AAEM minimonograph #42: Intraoperative monitoring of peripheral and cranial nerves. Muscle Nerve 17:371–377, 1994.
3. Campbell, WW, Pridgeon, RM, and Sahni, KS: Entrapment neuropathy of the ulnar nerve at its point of exit from the flexor carpi ulnaris muscle (abstr). Muscle Nerve 9:662, 1986.
4. Campbell, WW, Sahni, SK, Pridgeon, RM, et al: Intraoperative electroneurography: Management of ulnar neuropathy at the elbow. Muscle Nerve 11:75–81, 1988.
5. Davis, DH, Onofrio, BM, and MacCarty, CS: Brachial plexus injuries. Mayo Clin Proc 53:799–807, 1978.
6. Dawson, DM, Hallett, M, and Millender, LH: Entrapment Neuropathies. Little, Brown & Company, Boston, 1983, pp 87–122.
7. Kline, DG, and DeJonge, BR: Evoked potentials to evaluate peripheral nerve injuries. Surg Gynecol Obstet 127:1239–1248, 1968.
8. Kline, DG, Hudson, AR, and Zager, E: Selection and preoperative work-up for peripheral nerve surgery. Clin Neurosurg 39:8–35, 1992.
9. Landi, A, Copeland, SA, Wynn Parry, CB, and Jones, SJ: The role of somatosensory evoked potentials and nerve conduction studies in the surgical management of brachial plexus injuries. J Bone Joint Surg Br 62:492–496, 1980.
10. Mahla, ME, Long, DM, McKennett, J, et al: Detection of brachial plexus dysfunction by somatosensory evoked potential monitoring–a report of two cases. Anesthesiology 60:248–252, 1984.
11. Miller, RG: The cubital tunnel syndrome: Diagnosis and precise localization. Ann Neurol 6:56–59, 1979.
12. Sugioka, H, Tsuyama, N, Hara, T, et al: Investigation of brachial plexus injuries by intraoperative cortical somatosensory evoked potentials. Arch Orthop Trauma Surg 99:143–151, 1982.
13. Terzis, JK, Dykes, RW, and Hakstian, RW: Electrophysiological recordings in peripheral nerve surgery: A review. J Hand Surg 1:52–66, 1976.

SECTION 3

Applications of Clinical Neurophysiology: Assessing Symptom Complexes and Disease Entities

Electrophysiologic assessments described in preceding chapters outline the wide range of measurements that can be made in patients with suspected disease of the central or peripheral nervous system. Each technique has advantages, but also shortcomings. The clinical neurophysiologic testing technique that is most appropriate for use on a patient depends on the clinical problem of that patient. Usually, it is some combination of techniques that best provides the necessary data. Selection of a technique requires, first, taking a medical history and examining the patient, and second, formulating a complete differential diagnosis. This differential diagnosis should include many possible disorders; for practical purposes, those that are most likely are considered first in selecting a diagnostic technique. Clinical guidance is also necessary in selecting individual components of the techniques, for example, which nerves to test using nerve conduction studies in a patient with suspected peripheral nerve disease. Similar decisions need to be made during clinical neurophysiologic testing to optimize data collection and to minimize the time, discomfort, and cost to the patient. The approach to particular groups of clinical problems, with suggestions on the approach to patients, is reviewed in Chapter 43. These suggestions may not be entirely correct for any individual patient, because assessment depends on history and physical examination results of the patient.

CHAPTER 43

APPLICATION OF CLINICAL NEUROPHYSIOLOGY: ASSESSING SYMPTOM COMPLEXES

Jasper R. Daube, M.D.

The many clinical neurophysiologic techniques that have been described in earlier chapters are applied either to assist clinicians in assessing disease of the central or peripheral nervous system or, less commonly, in monitoring changes in neural function. Clinical neurophysiologic techniques can be used to monitor neural function in observing progression of a disease or improvement in a patient's condition with specific treatment. These techniques are also used in the intensive care unit and operating room to identify progressive neural damage. The focus of this chapter is the application of clinical neurophysiology in assessing clinical problems.

CLINICAL NEUROPHYSIOLOGY IN THE ASSESSMENT OF DISEASE

A patient comes to a physician with specific complaints that require explanation and treatment. A physician who assesses a potential neurologic disease first develops a hypothesis about the *location* of the disease on the basis of the patient's symptoms and the results of neurologic examination. The second step is to develop a hypothesis about the *disease*. Clinical neurophysiology can be of assistance in assessing symptom complexes, localizing disease, and defining the likelihood of different diseases.

To define the appropriate treatment of a disease after the disease has been localized and identified, the physician must make a determination about the *stage* of the disease. The stage of a disease includes severity of involvement, rate of evolution of the disorder, and prognosis. Clinical neurophysiologic techniques can assist in defining the stage of a disease.

The combination of the patient's symptoms and the signs found on examination is used to localize the disease process to a neural system and to an anatomic level. Neurologic diseases may involve any combination of the following: motor system, sensory system, internal regulation (autonomic nervous) system, systems regulating consciousness and cognition, and vascular system. Each of these systems can be tested by one or more of the techniques of clinical neurophysiology.

Selection of a clinical neurophysiologic technique to assist in evaluating a patient can be based in part on the patient's presenting symptoms. The significance of individual clinical symptoms can be assessed by applying the clinical neurophysiologic technique that tests that portion of the nervous system usually responsible for those symptoms. For example, sensory symptoms can be assessed with tests such as sensory nerve conduction studies, somatosensory evoked potentials, brain stem auditory evoked potentials, and visual evoked potentials.

Anatomic localization is first approached by major level: supratentorial, posterior fossa, spinal cord, and peripheral nervous system. Within each of these major levels, more specific localization is possible. At the supratentorial level, the disease may involve the cerebral cortex, subcortical white matter, thalamus, hypothalamus, or basal ganglia. At the posterior fossa level, involvement may be at the level of the midbrain, pons, medulla, or cerebellum. Spinal cord levels from C-2 to S-2 may be distinguished clinically. In the periphery, disease can be localized to the spinal nerve, plexus, peripheral nerve, neuromuscular junction, or muscle. Specific clinical neurophysiologic techniques can help to localize to each of these levels.

Selecting a treatment for the cause of symptoms requires identifying the underlying disease entity. For example, it is not sufficient to know that a patient's symptoms are due to a localized lesion in the midthoracic spinal cord without knowing the nature of the lesion. Patients with thoracic cord disease caused by a herniated disk, intraspinal tumor, arteriovenous malformation, multiple sclerosis, or vitamin B_{12} deficiency receive different treatments. Clinical neurophysiology can assist in distinguishing these as well as in localizing the disease.

Clinical neurophysiology can help define prognosis by classifying changes as acute, subacute, chronic, or residual from an old process. Rapid, moderate, slow, stable, or improving disease progression results in different clinical neurophysiologic findings.

An *acute process* develops within seconds to a few days. A *subacute disorder* evolves over a few days to weeks, and a *chronic disorder* evolves over months to years. In progressive diseases, there is increasing damage and impairment of function. Improvement occurs when the disease process has subsided so that neural mechanisms of repair can begin to reduce the severity of damage. In a stable process, damage has occurred but remains unchanged, either because the rate of neural reparative processes is able to keep pace with the rate of neural damage in a chronic continuous disorder or because the disease process has entirely subsided and left damage that cannot be repaired. This is usually referred to as a *residual* of the disease. Clinical neurophysiology can help identify or classify disorders into these categories. Physicians and clinical neurophysiologists must be aware of these potential applications and make full use of them.

How Can Clinical Neurophysiology Help?

Clinical neurophysiologic assessment of a suspected disorder of the nervous system can help define each of the disease features described previously. This section discusses considerations important in deciding whether one or more clinical neurophysiologic techniques is warranted in a particular clinical problem. Often, after the initial medical history taking and neurologic examination, a physician has formulated a hypothesis about the neural system involved, level of involvement, type of disease, and prognosis. If these are sufficiently certain, electrophysiologic testing is not needed. However, the information a physician has is not always definitive enough to determine the best approach to treatment. Additional information that clinical neurophysiology can provide is described below.

The most common application of clinical neurophysiology is to *confirm* a tentative clinical diagnosis. Uncertainty about the diagnosis usually reflects an atypical or incomplete symptom complex, incomplete or mixed findings that do not all fit with the suspected disorder, a relatively mild stage of the disease with a minimum of symptoms and signs, or unexpected findings that are not consistent with the diagnosis.

Clinical neurophysiology can also be of value when a specific disease is strongly suspected but other diseases with similar findings have to be *excluded*. For example, a patient with a C-6 radiculopathy may have features suggesting carpal tunnel syndrome. Clinical neurophysiology can be used to help exclude the latter possibility.

In situations in which the physician cannot obtain an adequate clinical history and neurologic examination, clinical neurophysiology may provide new information. These situations include patients who are in a coma, who have dementia or psychiatric disease, or who may be unable to cooperate. A language barrier also may interfere with taking a medical history and performing a neurologic examination. When traumatic injuries such as fractures or postoperative immobilization preclude thorough neurologic examination, clinical neurophysiology may be able to assess function and provide new information.

Several electrophysiologic techniques can be used to *identify subclinical disease* by detecting an abnormality that is below threshold for clinical identification or that has no clinical accompaniments. Examples are epileptiform discharges between seizures, slowing of conduction in a hereditary neuropathy with no deficit, and fibrillation potentials in a radiculopathy with no clinical deficit.

In clinical situations in which the physician is able to identify a category of disease, clinical neurophysiology may be needed to *characterize the disease*. A patient who has brief spells that are clearly seizures may require electroencephalography (EEG) to determine whether they are absence seizures or partial complex seizures. A patient with a peripheral neuropathy may require nerve conduction studies to determine whether the neuropathy is axonal or demyelinating.

In patients in whom the disease can be clearly identified clinically, *clinical neurophysiology may help localize the disease* with a precision not otherwise possible clinically. For example, EEG may define the origin of complex partial seizures in the anterior temporal lobe or nerve conduction studies may identify compression at the aponeurosis of the flexor carpi ulnaris muscle as the cause of an ulnar neuropathy.

Clinical assessment can usually define the severity of a disease as mild, moderate, severe, or with total loss of function. Clinical neurophysiology can *quantitate* the severity with reproducible measures of the extent of abnormality. Quantitation can assist in making decisions about the best treatment approach and prognosis, especially the likelihood of improvement. As an example, nerve conduction studies can define a 30% block as the cause of a peroneal neuropathy. Such quantitation provides evidence that the process is relatively mild and that the patient can be expected to make a good recovery.

SYMPTOM COMPLEXES AND NEURAL SYSTEMS

The neurologic symptoms described by a patient will suggest involvement of one or more neural systems. A physician can be more certain about the hypothesis of involvement of a specific system if confirmatory signs are found on examination of that neural system. If further confirmation is needed, clinical neurophysiologic testing can often provide it. The symptoms that suggest disorders of specific systems are as follows:

1. Motor system: paralysis, weakness, tremor, other extraneous movements, and posture abnormalities.
2. Sensory system: sensory loss, paresthesia, pain, and impairment of vision, hearing, or balance.
3. Disorders of consciousness: confusion, coma, syncope, and seizures.
4. Cognitive disorders: dementia, confusion, and disorders of language.
5. Internal control system and autonomic disorders: perspiration abnormalities, fatigue, vascular changes, pain, and emotional disorders.

Assessment of Motor Symptoms

The presence of symptoms or signs involving movement is strong evidence that disease is affecting the motor system at some level of the nervous system. If there is atrophy, loss of power, jerking, shaking, stiffness, or any of the many manifestations of disease of the motor system, clinical neurophysiologic testing with one of the following modalities that assesses the motor system should be considered.

1. *EEG* should be considered for patients in whom the motor deficit may be due to disease at the cortical level—either a seizure or a destructive process (e.g., tumor or infarct). EEG is of little or no value in assessing movement disorders except for those involving seizure activity.
2. Transcranial *motor evoked potentials*, approved for clinical use in Canada and Europe but not in the United States, test the entire motor pathway from the level of the cerebral cortex (where the motor system is activated by a magnetic or electrical pulse) to the level of the muscle (where the response is recorded). Motor evoked potentials are most useful in distinguishing primary motor system disease from func-

tional or psychiatric disease. For example, a patient with a hysterical paralysis of the legs can be tested using this technique.
3. Disorders of movement such as tremors, jerks, and twitches cannot always be clearly classified by clinical observation. Quantitative measurements by *multichannel surface electromyography (EMG)* recordings from widespread muscle groups can often characterize the presence or absence of a motor system disorder and can sometimes define it precisely.
4. *Motor nerve conduction studies* can be critically important in identifying disease of the peripheral motor pathways through the plexus, peripheral nerve, neuromuscular junction, or muscle.
5. *Repetitive stimulation* can separate functional fatigue from defects of neuromuscular transmission due to disease at the neuromuscular junction.
6. *Amplitudes of evoked responses* can help define severity of the disease process and presence of conduction block.
7. *Conduction slowing* can define the precise location of damage.
8. *Needle EMG* can identify and characterize peripheral neurogenic and myopathic diseases.

Assessment of Sensory Symptoms

Complaints of numbness, tingling, localized pain, loss of vision, hearing impairment, and poor balance suggest involvement of sensory systems. Clinical neurophysiologic testing can assist in identifying and characterizing the disease involving the system. Sensory symptoms can arise at any level of the nervous system. The following tests should be considered in patients with sensory symptoms:

1. *EEG* can help identify sensory symptoms arising at the cortical level. For example, unilateral paresthesia due to a seizure discharge or a local destructive lesion may be associated with focal spikes or slowing on the EEG.
2. *Somatosensory evoked potentials* can help determine the presence of impairment along the sensory pathways, par-

ticularly if there is an identifiable sensory loss.

3. *Sensory nerve conduction studies* can localize disease in peripheral nerves. Slowing of sensory nerve action potentials may distinguish between primarily axonal and demyelinating disorders.

4. *Visual evoked potentials* in cases of visual impairment and *auditory evoked potentials* in cases of hearing impairment can help localize the site of abnormality and characterize the type of abnormality.

Assessing Impairments of Consciousness and Cognition

Episodic or continuous confusion, unconsciousness, and disorders of sleep can be identified and characterized by electrophysiologic testing. (1) *EEG*, often critical in defining the nature of an impairment of consciousness, can distinguish seizure disorders from metabolic disturbances, focal lesions with increased pressure, hysterical disorders, and sleep disorders. (2) *Polysomnography* provides more precise assessment and characterization of sleep disorders, often assisting in defining their specific nature and, thereby, their appropriate treatment modalities. (3) *Somatosensory evoked potentials* and *brain stem auditory evoked potentials* can determine whether central sensory pathways from the level of the brain stem to the cerebral cortex are intact and functioning normally. This helps define the status of patients in coma. Severity of impairment of somatosensory evoked potentials after head trauma is helpful in judging prognosis.

Disorders of cognition with impairment of mentation, language, and memory are primarily due to disorders at the level of the cerebral cortex. *EEG* is the most precise laboratory technique available for assessing cortical function. Only clinical testing of thinking and memory can provide more information. The wide variety of EEG changes in the many disorders that produce impaired mentation—ranging from Alzheimer's disease, chronic infection, ischemic vascular disease, and subdural hematomas to

mass lesions of the frontal lobe—produce specific types of abnormality on the EEG which can be recognized and which can be used to help define the nature of the disorder.

Assessing Impairment of Visceral Function and Sleep

Impairment of visceral function (including autonomic disorders, syncope, and disorders of reproductive organs) can be assessed by two groups of electrophysiologic testing: autonomic function testing and polysomnography. *Autonomic function testing* can assess peripheral sympathetic function in vascular disease or peripheral nerve disease. It also can assess central vascular control mechanisms that may be altered in autonomic diseases such as multisystem atrophy.

Polysomnography assesses sleep disorders by measuring both EEG activity and associated autonomic function. Disorders of the autonomic nervous system may be manifest on polysomnography. Polysomnography is particularly valuable in central disorders of the internal regulatory system, but it can also be helpful in assessing impotence and sleep apnea.

LOCALIZATION OF DISEASE

Clinical neurophysiologic testing can often be more precise than clinical evaluation in defining location of a disease process. Even in cases in which clinical assessment indicates the possibility of localized disease, clinical neurophysiologic testing is often necessary to confirm the localization. With these techniques, specific localization can be made at the supratentorial, posterior fossa, spinal, and peripheral levels.

Localization at the Supratentorial Level

In patients in whom the physician is able to identify the likelihood of disease at the supratentorial level, clinical neurophysiology can assist in localizing the area of involvement. Each of the following measures may provide evidence of specific involvement at the supratentorial level.

1. *EEG* is one of the most helpful techniques for distinguishing among supratentorial diseases because of its ability to identify localized areas of cortical involvement caused by either an epileptiform or a destructive process. EEG can also help distinguish between subcortical and cortical diseases.
2. *Evoked potentials* can be used to distinguish subcortical from cortical involvement. Somatosensory evoked potentials may increase or decrease in size in cortical disease and be reduced or delayed in subcortical processes such as multiple sclerosis. Visual evoked potentials are able to distinguish disease in the optic nerves and tracts from that in the cerebral cortex.
3. The origin of disorders with tremor, jerking, and twitching may be identified as cortical or basal ganglion by the character of their pattern of firing and distribution on *multichannel movement recordings*.

Localization at the Posterior Fossa Level

Clinical neurophysiology can be helpful in identifying the presence of a lesion at the posterior fossa level and occasionally in localizing it within that level.

1. *Auditory evoked potentials* and *auditory testing* can specifically identify involvement of the peripheral auditory pathways and distinguish it from involvement of those auditory pathways at the level of the pons or midbrain. Auditory evoked potentials can also distinguish the nature of these disorders.
2. *Posturography* can specifically identify the presence of a peripheral vestibular disorder and distinguish it from central disease of the vestibular pathways.
3. Tibial and median *somatosensory evoked potentials* recorded from the scalp in

comparison with recordings at the neck can identify involvement at the posterior fossa level.

4. *Blink reflexes* and *facial nerve conduction studies* can identify involvement of the fifth and seventh cranial nerves and distinguish their involvement from diseases of midbrain or pons.
5. *Needle EMG* of cranial muscles can provide evidence of damage to brain stem motor neurons or the peripheral motor pathways in primary neurogenic processes, myasthenia gravis, or primary myopathies. Patients with bulbar symptoms can often be separated into those with an upper motor neuron–pseudobulbar disorder and those with lower motor neuron involvement in amyotrophic lateral sclerosis or myasthenia gravis.

Localization at the Spinal Cord Level

Clinical findings can usually identify the level of spinal cord disease unless multiple disorders or multilevel involvement are present, in which case clinical neurophysiology can be particularly helpful.

1. *Somatosensory evoked potentials* recorded with median, ulnar, or tibial nerve stimulation can distinguish involvement of the peripheral and spinal nerves from direct spinal cord damage and separate lumbar and thoracic cord lesions from cervical cord lesions. Occasionally, somatosensory evoked potentials are able to determine the nature of the disorder, especially demyelinating disorders.
2. On *nerve conduction studies*, F waves, patterns of amplitude changes with motor and sensory conduction studies, and H reflexes can sometimes identify involvement of specific levels of the spinal cord.
3. Paraspinal fibrillation potentials on needle EMG provide evidence of lower motor neuron involvement at the spinal cord level; they can also help define distribution of lower motor neuron loss along the spinal cord.

For example, in amyotrophic lateral sclerosis, evidence of subclinical involvement at the thoracic level may be demonstrated by EMG.

4. *Autonomic function testing* in primary spinal cord disease, particularly localized disease with trauma, inflammatory disease, or ischemia, will show localized changes at a specific segmental level.

Localizing Peripheral Lesions

Peripheral lesions can be localized to the level of a spinal nerve, plexus, peripheral nerve, neuromuscular junction, or muscle and specifically to individual nerves and muscles. Three electrophysiologic tests assist in localizing peripheral lesions.

1. *Motor and sensory nerve conduction studies* identify localized areas of damage to individual nerves.
2. *Needle EMG* localizes lesions in cases in which nerve conduction studies are not successful, either because the damage is primarily axonal or because the nerves are not accessible for stimulation and recording. EMG can also assist in distinguishing primary neurogenic disease from neuromuscular junction and muscle diseases.
3. *Autonomic function tests* (quantitative sudomotor autonomic response testing, thermoregulatory sweat test, and skin vascular reflexes) separate peripheral sympathetic and parasympathetic disorders from spinal cord disease and involvement of central autonomic pathways. Patterns of distribution of temperature change, alteration in sweating, and vascular reflexes are combined to provide this information.

IDENTIFYING DISEASE TYPES

Clinical neurophysiologic testing can facilitate identification of specific diseases. The physician usually initially classifies a disease as vascular, inflammatory, degenerative, or neoplastic, based on the temporal profile of the patient's symptoms. Electrophysiologic testing can sometimes aid in this classification, but it does so in different ways for different tests. In many instances, only a broad category of disease can be suggested; in others, however, specific diseases can be identified. These are described in detail in the chapters on each of the techniques. The following are some examples.

EEG can distinguish epileptogenic from destructive lesions, but rarely can it categorize the changes as neoplastic, inflammatory, infectious, or degenerative. At times, specific clinical entities can be suggested, such as the Lennox-Gastaut syndrome, hepatic encephalopathy, and hypsarrhythmia of infancy. Evoked potentials are rarely able to provide evidence of a specific clinical entity. Marked increase in latency is usually evidence of a demyelinating process in patients with multiple sclerosis. Evoked potentials are particularly useful in identifying areas of subclinical involvement to confirm the presence of multiple lesions in multiple sclerosis.

Nerve conduction studies infrequently identify specific disease categories. Marked slowing and dispersion are evidence of a demyelinating neuropathy, which may be due to either an inherited or an acquired disorder. Nerve conduction studies also help distinguish demyelinating disease from axonal disease. Repetitive stimulation can demonstrate specific patterns of abnormalities seen in myasthenia gravis and distinguish them from Lambert-Eaton myasthenic syndrome.

EMG can occasionally assist in identifying specific disorders by characteristic findings such as changes with polymyositis, periodic paralysis, and radiation damage with myokymia. Autonomic function testing can provide evidence of specific disorders such as multisystem atrophy, reflex sympathetic dystrophy, or both. Peripheral audiologic testing and posturography are able to distinguish central from peripheral structural lesions. Polysomnography testing can specifically identify sleep apnea and some forms of sleep impairment such as periodic movements of sleep.

PROGNOSIS

Although each of the clinical neurophysiologic techniques can define the severity of a disease, most of them are unable to characterize the stage of evolution of the disease or to provide information about progression or prognosis. Peripheral EMG and nerve conduction studies are the exception. On the basis of the findings of these studies, specific comments can be made about prognosis.

The clinical electromyographer can not only distinguish myopathies from neuropathies and the subdivisions among them but can also characterize the severity of a disease and specify the stage of its evolution. Identification of disease type on EMG is well known, but changes in EMG findings with time are less familiar and are reviewed here. Recognition of different stages in the evolution of a disease depends on understanding the pathophysiologic changes that occur in nerve and muscle. The three types of nerve damage—conduction block, slowing of conduction, and axonal destruction—evolve over very different time courses. Secondary changes in muscle with each of these evolve over time courses that vary with severity of disease.

Electrophysiologic Classification of Nerve Injury

Conduction block is a localized area of abnormality that is unable to conduct an action potential in an axon or group of axons. The proportion of fibers that are blocked in a nerve is a direct measurement of the amount of clinical deficit. Nerve function proximal and distal to the conduction block can be entirely normal. Conduction block may be due to either metabolic or structural changes. With local anesthetics, anoxia, and some toxins, nerve function can improve over minutes to hours and may not be associated with histologic alteration. Conduction block due to distortion or loss of myelin persists for days to weeks because it requires remodeling of the histologic abnormality. Conduction block caused by either of these mechanisms may not show improvement if the offending mechanism is not eliminated.

Slowing of conduction is usually due to myelin changes and, thus, requires weeks to months for improvement to occur if the underlying cause can be eliminated. In contrast to conduction block, slowing of conduction alone may be associated with little or no clinical deficit.

Axonal disruption or degeneration is associated with a loss of axons due to wallerian degeneration. Therefore, recovery of function depends on reinnervation. Reinnervation can occur rapidly, within days to weeks, if the number of axons lost is not great and the remaining axons can provide reinnervation by local collateral sprouting. Reinnervation is much slower, over months to years, if it requires sprouting and regrowth of the damaged axons.

Electrophysiologic changes in muscle associated with nerve damage depend primarily on whether the degeneration is wallerian. Slowing of conduction is not associated with measurable changes on needle EMG or in estimates of number of motor units. Conduction block is associated with reduced recruitment of motor unit potentials on needle EMG and reduced estimates of the number of motor units proximal to the site of damage. Wallerian degeneration produces reduced recruitment of motor unit potentials, reduced estimates of the number of motor units, and muscle changes associated with denervation and reinnervation.

Denervation of muscle results in a loss of the trophic factors required to maintain normal membrane function. With the loss of innervation, a muscle fiber discharges spontaneously and contracts in a rhythmical fashion. The contractions are called *fibrillations* and the associated discharges are *fibrillation potentials*. Fibrillation potentials develop 1 to 3 weeks after acute denervation. Delay in their appearance varies with species and muscle characteristics. In humans, the delay depends most on the length of axon attached to the muscle fiber. If axonal destruction occurs close to a muscle fiber, fibrillation develops more quickly than if the damage is more proximal (i.e., the shorter the segment of axon attached to the muscle, the more quickly wallerian degeneration occurs). The corollary is that muscles closer to the lesion show fibrillation potentials sooner than muscles more distant to the damage.

Reinnervation of muscle is associated with a defined sequence of changes in the estimate of the number of motor units and in motor

Table 43–1. **COMPOUND ACTION POTENTIAL AMPLITUDE AFTER PERIPHERAL NERVE INJURY***

	AMPLITUDE		
	0–5 Days	After 5 Days	During Recovery
Conduction block			
Proximal stimulation	Low	Low	Increases
Distal stimulation	Normal	Normal	Normal
Axonal disruption			
Proximal stimulation	Low	Low	Increases
Distal stimulation	Normal	Low	Increases

*Supramaximal stimulation.

unit potentials. If reinnervation is by collateral sprouting, the estimate of the number of motor units remains low and increases only in proportion to the number of axons that regenerate and reach the muscle. With initial reinnervation, motor unit potentials consist of activity in only a small number of muscle fibers with poor synchrony of firing and unstable neuromuscular junctions. Therefore, potentials are low amplitude, polyphasic, and unstable. These have been called *nascent* motor unit potentials. As reinnervation proceeds to include more muscle fibers with better synchrony of firing, motor unit potentials become higher in amplitude, longer in duration, less polyphasic, and more stable. Therefore, late after reinnervation, motor unit potentials are of long duration and high amplitude. Reinnervation usually is completed by fewer than the normal number of axons; thus, recruitment is decreased because there are fewer motor units. These changes are summarized in Ta-

bles 43–1 and 43–2. Conduction block and axonal disruption both can have different time courses, depending on the underlying mechanism and the number of axons involved. The changes over time in compound muscle action potentials and results of needle EMG examination after a localized nerve injury are shown in Tables 43–1 and 43–2.

Focal Neuropathies and Radiculopathies

Electrophysiologic changes on nerve conduction studies in mononeuropathies vary with rapidity of development, duration of damage, severity of damage, and with the underlying pathologic condition. Localized narrowing of axons or paranodal or internodal demyelination with a chronic compressive lesion produces localized slowing of conduction. Narrowing of axons distal to a

Table 43–2. **RESULTS OF NEEDLE EXAMINATION AFTER PERIPHERAL NERVE INJURY**

	0–15 Days	After 15 Days	During Recovery
Conduction block			
Fibrillation potentials	None	None	None
Motor unit potentials	↓ number	↓ number	↑ number
Axonal disruption			
Fibrillation potentials	None	Present	Reduced
Motor unit potentials	↓ number	↓ number	Nascent

↓ = decrease; ↑ = increase.

chronic compression results in slowing of conduction along the entire length of the nerve. Telescoping of axons with intussusception of one internode into another produces distortion and obliteration of the nodes of Ranvier and, thus, a block of conduction. Moderate segmental demyelination and local metabolic alterations are often associated with conduction block. The conduction block is manifested as a lower amplitude evoked response with stimulation proximal to the site of damage. In an acute lesion with disruption of the axons, that segment of nerve distal to the lesion may continue to function normally for up to 5 days. Then, as the axons undergo wallerian degeneration, they cease to conduct and amplitude of the evoked response diminishes and finally disappears. One week after an acute injury, the amplitude of the evoked response is a rough gauge of the number of intact viable axons.

Interpretations about the duration and severity of nerve injury after focal neuropathies and radiculopathies can be made on the basis of an analysis of the combination of these changes. Examples of these interpretations are given in Tables 43–3 and 43–4.

Motor Neuron Disease

Gradual loss of anterior horn cells in motor neuron disease produces changes in the EMG findings during the course of the disease. These changes allow electromyographers to assess evolution of the disease as well as severity. In initial stages of the disease, before clinical weakness is evident, collateral sprouting of viable motor neuron axons maintains innervation of all muscle fibers; thus, few if any fibrillation potentials are evident. However, the loss of motor unit potentials can be recognized. Motor unit potential size later increases with innervation of greater numbers of muscle fibers. If a significant amount of collateral sprouting has occurred, some motor unit potentials vary in configuration.

As these changes progress to the stage where reinnervation cannot keep pace with

Table 43–3. INTERPRETATION OF ELECTROMYOGRAPHIC FINDINGS AFTER PERIPHERAL NERVE INJURY

Finding	Interpretation
0–5 days	
Motor unit potentials present	Nerve intact, functional axons
Fibrillation present	Old lesion
Low-amplitude compound action potentials	Old lesion
5–15 days	
Compound action potential, distal only	Conduction block
Low-amplitude compound action potential	Axonal disruption
Motor unit potentials present	Nerve intact
After 15 days	
Compound action potential, distal only	Conduction block
Motor unit potentials present	Nerve intact
Fibrillation potentials	Axonal disruption
Recovery	
Increasing compound action potential	Block clearing
Decreasing number of fibrillation potentials	Reinnervation
Nascent motor unit potentials	Reinnervation

Table 43–4. **EVOLUTION OF ELECTROMYOGRAPHIC CHANGES IN RADICULOPATHY**

	Acute (<7 Days)	Subacute	Chronic Progressive	Residual
Nerve conduction study				
Compound muscle action potential amplitude	Normal	Low	Low	(Low)
Motor conduction velocity	Normal	Slow <30%	Slow <30%	(Slow <30%)
Distal motor latency	Normal	<30% increase	Prolonged	(Prolonged)
F wave/H reflex	Prolonged or absent	Absent or prolonged	Absent or prolonged	(Prolonged) (absent)
Needle electromyography				
Fibrillation potentials	None	Proximal, brief	Many proximal and distal	Distal, small, few
Fasciculation potentials	Rare	Rare	Occasional	Occasional
Motor unit potentials	Reduced recruitment	Reduced recruitment, polyphasic, unstable	Long duration, high amplitude, polyphasic, unstable	Long duration, high amplitude (especially distal)
Complex repetitive discharges	None	None	Rare	Occasional

( ), May recover to normal level.

denervation, fibrillation potentials become prominent. During this time, larger numbers of regenerating fibers are present and intermittent blocking of the components of a motor unit potential, *motor unit potential variation*, becomes more evident. The potentials become increasingly polyphasic, with satellite potentials. This combination of polyphasic, varying motor unit potentials is evidence of a severe, progressing disorder. At times, it is accompanied by a decrement on slow repetitive stimulation (Table 43–5).

Myositis

Inflammatory myopathies also evolve with time, beginning with small motor unit potentials and quickly developing fibrillation potentials and polyphasic motor unit potentials. The regenerating muscle fibers and fibers that have lost their innervation because of nerve terminal damage, segmental necrosis, or fiber splitting produce a number of

fibrillation potentials which roughly parallel the degree of disease severity. As the disease subsides, fibrillation potentials become less prominent and motor unit potentials have a more normal size. The number of muscle fibers in some motor units increases, resulting in larger than normal motor unit potentials late in the disorder.

ASSESSING CLINICAL DISORDERS

Clearly, from the previous discussion, clinical neurophysiologic testing cannot be applied in a routine fashion. For each patient, the testing should be designed to answer the clinical question posed by the patient's problem. This requires giving careful thought to the selection of the testing procedures.

Because of the relatively defined type and location of neuromuscular disorders, algorithms of testing can be developed. These algorithms must take into account findings

Table 43–5. **EVOLUTION OF ELECTROMYOGRAPHY (EMG) CHANGES IN MUSCLE AND MOTOR NEURON DISEASE**

	Subacute Active	Chronic Active	Inactive
Myositis			
Nerve conduction studies and needle EMG	Normal	Low	Normal
Fibrillation potentials	Proximal, paraspinal muscles	Widespread	None
Bizarre repetitive potentials	None	Present	None
Motor unit potentials	Short duration, few polyphasic	Short duration, polyphasic, rarely long duration	Short duration, polyphasic, rarely long duration
Motor neuron disease			
Nerve conduction studies			
Compound muscle action potential	Normal or low	Low	May remain low
Motor conduction velocity	Normal	Slightly low	Normal
Repetitive stimulation	Rarely decrement	Rarely decrement	Normal
Needle EMG			
Fibrillation potentials	Few or many	Many	Few
Fasciculation potentials	Rare	Rare	Rare
Bizarre repetitive potentials	None	Rare	Occasional
Motor unit potentials	Reduced recruitment, increased duration, may be unstable	Reduced recruitment, long duration, high amplitude, polyphasic, unstable	Long duration, high amplitude

obtained during testing in determining the amount and types of testing. These algorithms are suggestions for EMG and nerve conduction studies in neuromuscular disorders and nearly always need to be modified according to the particular problem and findings in individual cases. Subsequent sections outline a set of possible algorithms for several neuromuscular problems.

EMG in the Evaluation of Cervical and Lumbar Radiculopathies

Increased understanding of nerve and muscle and the development of new testing techniques have added to the diagnostic value of EMG in assessing patients who may have radiculopathy; however, they have not displaced standard methods in answering the questions that can be asked of electromyographers: (1) Is there evidence of radiculopathy? (2) Which nerve root is involved in the radiculopathy? (3) How severe is the neural damage caused by the radiculopathy? (4) Is the radiculopathy of recent onset, is it ongoing, or is it a residual of an old lesion? (5) Is there evidence of other peripheral nerve disease? One or more of these questions should be in the mind of the physician requesting the EMG. The presence of back or neck pain without evidence of radiculopathy is not necessarily an indication for EMG. Nerve conduction studies serve primarily to answer question 5 by identifying peripheral neuropathies. However, recently introduced measurements of F-wave latencies and H-reflex latencies can measure conduction through the nerve roots. In a small proportion of patients with lesions of the C-7, C-8, L-5, or S-1 nerve roots, particularly those with recent damage, the F waves or H reflexes may be abnormal when other measurements are normal. Damage to the C-8, L-5, or S-1 roots may also cause low amplitude motor responses and mild slowing of conduction velocity in the median, ulnar, peroneal, or tibial nerves. Determining the amount of amplitude reduction helps in defining the severity of axonal destruction. Sensory nerve action potentials are generally normal in nerve root disease and are helpful in differentiating it from more peripheral disease.

Needle electrode examination of muscles is still the most useful method for identifying radiculopathy. Because EMG changes evolve over time, the age of the lesion can be judged from both distribution and type of abnormality. Well-defined fibrillation potentials are not seen until 3 weeks after nerve damage. Proximal and paraspinal muscles are the earliest to show fibrillation potentials (evidence of axonal destruction) and also the earliest to show improvement. Changes in paraspinal muscles can locate the damage proximal to the plexus. Persistent abnormalities in paraspinal muscles after neck or back surgery preclude postoperative testing of these muscles. If diagnosis is uncertain, preoperative EMG studies may be helpful. Peripheral distribution of abnormalities defines the root involved. Severity of damage can be estimated by the amount of motor fiber degeneration (fibrillation potentials and motor unit potential changes). EMG is particularly valuable in differentiating relatively recent nerve damage with abundant fibrillations (especially in proximal muscles) from the residual of old disease with scanty fibrillation potentials (mainly in distal muscles).

EMG does not define the cause of the radiculopathy. EMG signs of a localized radiculopathy could be similar whether caused by a disk, tumor, or diabetes. Because disorders of the nerve roots produce changes only if the nerve fibers are damaged, EMG can never exclude the presence of a radiculopathy and EMG findings may be normal even when the radiculopathy causes severe pain. The following algorithms suggest specific approaches to suspected radiculopathy.

CERVICAL RADICULOPATHY (ARM PAIN)

Nerve Conduction Studies to Identify and Localize Peripheral Nerve Damage

1. Median motor conduction study with F waves.
2. Ulnar motor conduction study with F waves.
3. If either of the above show unexpected abnormal findings, check technical factors, temperature, anomalies, and consider another algorithm, as appropriate.
4. If symptoms are nonspecific or suggest need for C-4–C-8 sensory conduction studies:
   a. Median sensory conduction study
   b. Ulnar sensory conduction study
5. If a plexus lesion is suspected clinically, consider evaluation of the plexus:
   a. Musculocutaneous motor conduction study
   b. Ulnar motor conduction study with supraclavicular and root stimulation
   c. Median sensory conduction study with supraclavicular stimulation
6. If the only abnormality is a prolonged F wave or low compound muscle action potential, compare with the opposite side.

Needle Examination

1. First dorsal interosseous, pronator teres, biceps, triceps, infraspinatus, cervical paraspinal muscles, and one clinically involved muscle
2. If one root is suspected clinically or if any abnormality is seen, examine two more muscles in the distribution of the suspected nerve root (proximal and distal) and demonstrate normal muscle above and below the level of the involved root.
3. Check paraspinal muscles if limb muscles are abnormal, if symptoms are of recent onset, if radiculopathy is likely, or if symptoms are only in neck (but not if there has been neck surgery).

4. If results of needle examination are abnormal, check at least one contralateral muscle.

LUMBOSACRAL RADICULOPATHY

Nerve Conduction Studies to Identify and Localize a Peripheral Nerve Disorder

1. Peroneal motor conduction study with F waves.
2. If knee amplitude is greater than ankle, check for supramaximal stimulation at ankle, excess stimulation at knee, or accessory peroneal nerve.
3. If ankle amplitude is 20% greater than knee or velocity is less than 35 m/s, consider algorithm for peroneal neuropathy.
4. Tibial motor conduction study with F waves
5. If tibial ankle amplitude is 50% greater than knee or velocity is less than 35 m/s, consider another algorithm.
6. If S-1 radiculopathy is strongly suspected, no clinical signs are absent, and motor nerve conduction studies are normal, check tibial H reflex from knee (must be compared with opposite side).
7. If peroneal amplitude is low or peroneal conduction velocity is 35 to 42 m/s, consider superficial peroneal sensory testing.
8. Sural sensory conduction study. Consider medial plantar sensory conduction study if patient is less than 55 years old.
9. If anything in steps 6 through 8 is abnormal, consider another algorithm.
10. If peroneal or tibial amplitude is low, F waves are abnormal or borderline, H reflex is abnormal or borderline, or velocity is slow, compare with the opposite side.

Needle Examination

1. With no specific root suspected: anterior tibial, peroneus longus, posterior tibial, medial gastrocnemius, lateral gastrocnemius, vastus medialis, rectus femoris, gluteus medius, gluteus maximus, sacral paraspinal, low lumbar paraspinal, midlumbar paraspinal, upper lumbar paraspinal muscles.

2. When a specific root is suspected: add muscles in that distribution. If any muscle is abnormal, compare the most abnormal muscle with the other side.

EMG in the Evaluation of Peripheral Neuropathy

EMG is a sensitive aid to the clinical evaluation of patients with suspected peripheral neuropathy. Testing for a neuropathy includes performing both nerve conduction studies and needle EMG to search for signs of muscle denervation. Diagnostically, EMG can confirm presence of peripheral nerve dysfunction and separate patients so affected from those with complaints due to spinal cord or nonorganic disease. In the presence of neuropathy, EMG may provide localizing or etiologic information. Different patterns of abnormality are seen in demyelinating and axonal neuropathies, in large-fiber and small-fiber neuropathies, polyneuropathy, polyradiculopathy, peripheral neuropathies, and mononeuritis multiplex. Normal findings do not exclude peripheral neuropathy with exclusive small-fiber involvement, for example, as in some forms of diabetic, amyloid, or hereditary sensory neuropathy with involvement primarily of pain and autonomic nerve fibers. In some patients (especially diabetics) with complaints of vague or nonspecific pain, EMG may provide evidence of nerve damage before it is evident clinically. Of 989 patients seen in 1 year who had electrodiagnostic signs of neuropathy, 30% had minimal or subclinical signs on clinical examination. Painful diabetic radiculopathies are often confirmed with EMG.

After a peripheral neuropathy is diagnosed, EMG can help classify the disease and suggest a possible cause. Findings of segmental demyelination suggest an inherited, autoimmune, or inflammatory neuropathy; axonal degeneration suggests a toxic, metabolic, or nutritional neuropathy. Of patients with neuropathy documented by EMG, 35% had an underlying cause: diabetes (21%), inherited (5%), alcoholic (3%), collagen-vascular (2%), uremia (1%), malignancy (1%), and vitamin deficiency, toxic, or others

(2%). Some neuropathies have characteristic electrodiagnostic patterns, such as prominent paraspinal fibrillation potentials with a mixed demyelinating and axonal neuropathy in diabetes or bilateral carpal tunnel syndrome superimposed on a largely axonal neuropathy in amyloidosis.

Some techniques may add to the standard assessment of neuropathy. F-wave latencies provide a measure of proximal conduction and, in some disorders, may show early abnormality. With so many nerves that can be tested, the most appropriate ones must be selected. For example, plantar nerves show earlier abnormalities in neuropathy than more proximal sensory nerves do. Physicians who have specific questions that EMG can answer will greatly assist the electromyographer in making the selection. The following are options for assessing specific peripheral nerve problems:

PERIPHERAL NEUROPATHY

1. Test the most involved extremity (arm or leg) first, if the deficit is mild or moderate; test the least involved extremity if the deficit is severe.
2. Peroneal motor conduction study with F waves.
3. Tibial motor conduction study with F waves.
4. If no response, test peroneal nerve conduction to the anterior tibial muscle.
5. Sural sensory conduction study.
   Consider: superficial peroneal sensory conduction study if no sural response.
   If age is less than 55 years and/or sural sensory conduction study is normal: plantar sensory conduction.
6. If any of the above are abnormal or upper extremities are symptomatic, test arm.
7. If arm is tested first, do at least one motor and sensory conduction study in a leg.
8. Median sensory conduction study.
   a. Palmar stimulation if antidromic response is absent or carpal tunnel syndrome is suspected.
   b. If no responses are seen in distal sensory nerves, consider proximal testing, for example, antebrachial cutaneous.
   c. Consider ulnar sensory conduction study if symptoms are appropriate or median sensory conduction is abnormal.
   d. Consider radial sensory conduction study if compression syndromes are present.
9. Ulnar motor conduction study with F waves.
   Four-point ulnar study only if there is a significant decrease in amplitude across the elbow.
10. Test additional nerves if findings are equivocal.
11. Use a specific algorithm for the nerve and test the opposite side, if only a single nerve is involved.
12. Test additional proximal conduction if there are no responses or polyradiculopathy is possible:
    a. Blink reflex: unilateral R1 and R2 responses.
    b. Musculocutaneous motor conduction study, or
    c. Somatosensory evoked potential study in arm and/or leg.
13. Needle examination: The following muscles are appropriate in each case: anterior tibial, medial gastrocnemius, lumbar paraspinal, and first dorsal interosseous muscles.
    a. If these are normal, test foot muscles.
    b. If any of these are abnormal, test the opposite extremity and proximal muscles of leg (and arm), including paraspinal muscles.

CARPAL TUNNEL SYNDROME

1. Median motor conduction study with F waves.
2. Ulnar motor conduction study with F waves.
3. Check for anomalous innervation if amplitude or configuration changes from elbow to wrist on ulnar or median motor conduction studies.
4. Use an ulnar algorithm if the ulnar motor conduction study is abnormal.
5. Median palmar sensory conduction study with wrist and elbow recording:

a. Consider antidromic median sensory conduction study if technical problems are anticipated.

b. Ipsilateral orthodromic median sensory studies may be needed to confirm sensory versus motor waveform.

6. Determine ulnar sensory conduction using same method as median sensory conduction, including conduction velocity.

7. Consider recording antidromic responses from thumb or middle finger for focal symptoms.

8. Consider using radial sensory conduction for comparison if both ulnar and median sensory conductions are abnormal.

9. Check temperature of the hand if ulnar and median sensory distal latencies are long.

10. Change algorithm if the abnormality consists of more than a median neuropathy at the wrist.

11. Contralateral median and ulnar sensory conduction studies if ipsilateral median conduction study is abnormal or symptoms are bilateral.

12. Contralateral median motor conduction study if contralateral median sensory conduction study is abnormal.

13. Needle examination of ipsilateral first dorsal and thenar muscles:

a. Examine flexor pollicis longus and pronator teres if thenar muscle EMG is abnormal.

b. Examine additional muscles, especially those innervated by C-6, if nerve conduction values are normal.

ULNAR NEUROPATHY

1. Ipsilateral median motor conduction study with F waves.

2. Ipsilateral ulnar motor conduction study with F waves.

a. If ulnar conduction velocity is less than 52 m/s (compare with median) or has elbow-wrist amplitude difference greater than 20%, add below elbow and upper arm stimulation (four points must be stimulated).

b. If abnormal, use 2-cm inching from below to above elbow.

3. If clinically indicated, do ulnar study with wrist, elbow, upper arm, supraclavicular, and/or root stimulation (arm down) for possible plexopathy.

4. If ulnar and median study distal latencies are prolonged, check for technical problems (e.g., low temperature, anode distal).

5. If amplitude drop occurs above the wrist but below the elbow, stimulate median nerve at the wrist and elbow, recording hypothenar muscles for anomaly (median-ulnar crossover).

6. If ulnar nerve conduction is slow only in forearm, consider needle stimulation to localize the site of damage.

7. If ulnar conduction velocity is normal with atrophy or only a long distal latency, consider ulnar stimulation while recording first dorsal interosseous muscle for Guyon's canal lesion or flexor carpi ulnaris recording for selective lesion of branch to that muscle. Consider using flexor carpi ulnaris muscle recording if there is no hand response.

8. Ulnar antidromic sensory conduction study. Add below elbow and upper arm stimulation if abnormal. Use palmar stimulation if wrist lesion is suspected. Near-nerve needle recording may show selective slowing in some fascicles.

9. Median antidromic sensory conduction study (palmar stimulation if carpal tunnel syndrome is also suspected).

10. Perform the same study contralaterally if motor amplitude decreases more than 20% across the elbow, motor conduction slows more than 8 m/s across the elbow (long-segment calculation), ulnar sensory conduction is absent, ulnar sensory conduction is slow, other specific abnormality is identified, or symptoms are bilateral.

11. Needle examination: first dorsal interosseous, abductor digiti quinti, and flexor carpi ulnaris.

12. If ulnar muscles are normal: flexor pollicis longus, extensor indicis proprius, pronator teres, triceps, and biceps.

13. If ulnar muscles are abnormal: add abductor pollicis brevis, extensor carpi ulnaris, extensor indicis proprius, and contralateral first dorsal interosseous to the above.

14. If other C-8 muscles are abnormal, examine paraspinal muscles.
15. If only distal muscles are abnormal, examine leg muscles.

BRACHIAL PLEXOPATHY

1. Median motor conduction study with F waves.
2. Ulnar motor conduction study with stimulation at wrist, elbow, upper arm, and supraclavicular with F waves (arm straight). If normal with a lower trunk deficit, consider nerve root stimulation.
3. If ulnar study is abnormal at elbow with the arm straight, test ulnar conduction in the standard position with the arm bent.
4. Musculocutaneous motor conduction study with supraclavicular stimulation. If normal with an upper trunk deficit, consider nerve root stimulation.
5. Median sensory conduction study with wrist, elbow, upper arm, and supraclavicular stimulation. Consider nerve root stimulation.
6. Ulnar sensory conduction study with wrist, elbow, upper arm, and supraclavicular stimulation. Consider nerve root stimulation.
7. Radial sensory conduction study.
8. Conduction to specific muscle group, for example, infraspinatus, if clinically indicated (compared with opposite side).
9. Compare abnormal results with opposite side.
10. Needle examination: first dorsal interosseous, pronator teres, biceps, triceps, deltoid, infraspinatus, rhomboid, cervical paraspinal muscles.
11. Additional muscles as needed to localize lesions to upper, middle, or lower trunk or to medial, posterior, or lateral cord.

PERONEAL NEUROPATHY

1. Ipsilateral peroneal motor conduction study with F waves.
    a. Ankle first if extensor digitorum brevis is severely atrophic (may require a large stimulus, 0.5 milliseconds and 100 mA).
    b. Use slow sweep and high gain (200

μV), if there is a twitch with no compound muscle action potential.
    c. Measure to initial negativity if there is a positive dip at the knee and not at the ankle.
2. If the knee compound muscle action potential is larger:
    a. Recheck ankle for supramaximal stimulation (slide stimulator); use long-duration high-voltage stimulus, especially with low amplitude.
    b. Recheck knee for current spread and tibial stimulation (watch twitch and move lateral).
    c. If knee compound muscle action potential remains larger, check for accessory peroneal anomaly.
3. If the only abnormality is prolonged distal latency, check temperature.
4. If ankle compound muscle action potential is 20% larger than that of the knee or velocity is not in the normal range:
    a. Stimulate below head of fibula with at least a 10-cm distance between knee site and site below head of fibula.
    b. If fibular compound muscle action potential is 20% larger than that of knee, "inch" proximally along nerve starting at head of fibula.
5. If no response from extensor digitorum brevis or if there is anterior tibial muscle weakness with normal peroneal studies:
    a. Peroneal conduction to anterior tibial muscle, stimulating fibula head and 10 cm proximal.
    b. Rotate stimulator for artifact.
6. Tibial motor conduction study with F waves.
7. If tibial study is abnormal, change algorithm.
8. Superficial peroneal sensory conduction study.
    a. If amplitude is greater than 5 μV, record above head of fibula.
    b. If no response or velocity is clearly not normal above the fibula, record below the head of the fibula.
    c. Use needle stimulation if no response.
9. Sural conduction study.
10. If sural study is abnormal, change algorithm.
11. If any part of the peroneal recording is

abnormal or borderline, repeat peroneal studies on the opposite leg.
12. Needle examination:
   a. Always test—anterior tibial, peroneus
      longus, medial gastrocnemius, and
      posterior tibial (flexor digitorum lon-
      gus) muscles.
   b. If indicated, study short head of bi-
      ceps femoris, extensor hallucis, ex-
      tensor digitorum brevis, abductor
      hallucis, quadriceps, gluteus medius,
      and lumbar paraspinal muscles.
   c. If any muscle EMG is abnormal, test
      most involved muscles on the oppo-
      site side.

EMG in the Evaluation of Weakness

Generalized weakness is a common complaint; most often, it is *not* due to neuromuscular disease. The following algorithms
can help identify the specific cause of weakness or the disorder, such as myopathy, neuromuscular junction disease, or motor neuron disease.

1. If weakness if generalized or arms are
   weaker: ulnar motor conduction study
   with F waves and 2-Hz repetitive stimulation with the hand immobilized.
2. If ulnar study is equivocal: median motor
   conduction study with F waves and 2-Hz
   repetitive stimulation.
3. If proximal muscles are clearly weaker:
   a. Musculocutaneous motor conduction
      study with 2-Hz repetitive stimulation
      with arm immobilized or
   b. Trapezius conduction study with 2-Hz
      repetitive stimulation and arm immo-
      bilized.
4. If legs are clearly weaker: peroneal motor
   conduction study with F waves, and peroneal study to anterior tibial with repetitive stimulation and leg immobilized.
5. Median sensory conduction study.
6. If legs clearly are weaker, sural sensory
   conduction study.
7. If an abnormality is found at any point,
   use the appropriate algorithm, for example, polyradiculopathy, myopathy, myasthenia gravis.
8. Needle examination:

   a. If arms are involved, examine the first
      dorsal interosseous, biceps, triceps,
      and infraspinatus muscles.
   b. If legs are involved, examine the an-
      terior tibial, gluteus medius, lumbar
      paraspinal, and other muscles as clin-
      ically indicated.

MYOPATHY/MYOSITIS

1. If the lower extremities are weaker:
   a. Peroneal motor conduction study with
      F waves and 2-Hz repetitive stimula-
      tion; brief (10 seconds) exercise if
      compound muscle action potential is
      low amplitude. Use myasthenia gravis
      algorithm if there is a decrement.
   b. Tibial motor conduction study with F
      waves if peroneal conduction not sat-
      isfactory
   c. Sural sensory conduction study
   d. Plantar sensory conduction study if pa-
      tient is less than 55 years old
   e. Needle examination if all are normal
   f. Use another protocol if any are abnor-
      mal.
2. If there is generalized weakness or the
   arms are weaker:
   a. Ulnar motor conduction study with F
      waves and 2-Hz repetitive stimulation;
      brief exercise if compound muscle ac-
      tion potential is low amplitude; neu-
      romuscular junction studies if there is
      a decrement.
   b. Median motor conduction study with
      F waves if ulnar study is equivocal or
      not satisfactory.
   c. If proximal myopathy or neuromus-
      cular junction abnormality is possible:
      1) Musculocutaneous motor conduc-
         tion study with 2-Hz repetitive stim-
         ulation or
      2) Spinal accessory motor conduction
         study with 2-Hz repetitive stimula-
         tion
   d. Median sensory conduction study
3. Needle examination: one side only (same
   as previous EMG if done before):
   a. Check to ensure that creatine kinase
      level has been determined.
   b. Test moderately weak muscles, but not
      severely weak or atrophic muscles.
   c. Examine first dorsal interosseous, bi-
      ceps, triceps, infraspinatus, deltoid,

anterior tibial, vastus lateralis, gluteus medius, and lumbar paraspinal muscles.

d. If weakness is focal or selective, sample the involved muscles, for example, brachioradialis, sternocleidomastoid, or facial muscles.

e. If the findings are uncertain or unusual, quantitate motor unit potentials.

MYASTHENIA GRAVIS

1. Patient should not take pyridostigmine (Mestinon) for at least 4 hours (preferably more) before test, if possible.
2. If there is generalized myasthenia or if arm or bulbar muscles are primarily involved, perform an ulnar motor conduction study on symptomatic side.
   a. Hand is kept warm (above 32°C) and immobilized on board.
   b. Measure amplitude, latency, conduction velocity, and F wave.
   c. If abnormal, consider another algorithm, for example, peripheral neuropathy or ulnar neuropathy. Do repetitive stimulation on an uninvolved nerve.
   d. Repetitive stimulation at the wrist.
      1) 2 Hz, 4 shocks 3 times for a reproducible response.
      2) If normal, 1 minute of exercise (if low amplitude or decrement, 10 seconds of exercise).
      3) 2 Hz, 4 shocks at 5, 30, 60, 120, and 180 seconds (longer if results are equivocal).
3. Repetitive stimulation as above until two nerves are clearly abnormal, selecting nerve by clinical findings and symptoms. Routine motor nerve conduction studies should be performed first on each nerve; findings should be normal before proceeding with repetitive stimulation.
   a. Median—thumb immobilized by hand; repetitive stimulation at wrist as above.
   b. Musculocutaneous—immobilize arm on arm board.
   c. Spinal accessory—immobilize arm with strap.
   d. Axillary—immobilize arm with strap.
   e. Facial.

f. Peroneal with anterior tibial recording—immobilize leg on leg board.
g. Femoral.
4. If leg is more involved clinically, start with step 3f.
5. Median sensory conduction study.
6. Needle examination: first dorsal interosseous, biceps, triceps, cervical paraspinal, and masseter muscles (with isolation of single motor unit potentials to check for variation).
   Examine leg muscles if clinically indicated.
7. If normal, perform needle examination of other symptomatic muscles.
8. If all test findings are normal and it is clinically indicated, perform single fiber EMG.

POLYRADICULOPATHY

1. Peroneal motor conduction study with F waves. Stimulate below fibula if amplitude is lower at knee without dispersion.
2. Tibial motor conduction study with F waves.
3. Sural sensory conduction study
   a. Consider plantar sensory conduction study if patient is less than 55 years old.
   b. Superficial peroneal sensory conduction study if only peroneal motor conduction study is abnormal.
4. Median sensory conduction study with proximal stimulation.
5. Ulnar motor conduction study with F waves.
   If normal, consider wrist, elbow, axilla, supraclavicular, and root stimulation.
   Watch for dispersion.
6. If findings in step 4 or 5 are borderline:
   a. Median motor conduction study with F waves.
   b. Ulnar sensory conduction study with proximal stimulation.
7. If above studies are normal or borderline or no responses are obtained, consider:
   a. Musculocutaneous motor conduction study with proximal stimulation.
   b. Unilateral blink reflex testing.
   c. Facial motor conduction studies.
   d. Tibial H reflex.
8. Needle examination: First dorsal interosseous, biceps, triceps, anterior tibial, abductor hallucis, vastus lateralis, gluteus

medius, and lumbar paraspinal muscles. Others as needed for focal or questionable abnormality or as dictated by neurologic deficit.

9. Consider tibial and median somatosensory evoked potential if nerve conduction studies are all normal.

MOTOR NEURON DISEASE

1. Ulnar motor conduction study with F waves.
   Consider repetitive stimulation at 2 Hz for evaluation of progression. If there is a decrement, stimulate after exercise for 10 seconds.
2. If median nerve involvement is suspected, ulnar nerve conduction is abnormal, or legs are uninvolved:
   a. Median motor conduction study with F waves.
   b. If there is a decrement in ulnar nerve, repetitive stimulation at 2 Hz.
3. If the legs are involved more than the arms, or if a motor polyradiculopathy is being considered:
   a. Peroneal motor conduction study with F waves.
   b. Consider motor unit number estimate.
4. If peroneal abnormal or if a leg F wave is needed: tibial motor conduction study with F waves.
5. If legs are tested, consider: sural sensory conduction study, medial plantar/ankle (if patient is less than 55 years), or superficial peroneal if motor response is lost.
6. Median sensory conduction study
7. If median sensory conduction is abnormal or ulnar motor conduction is low amplitude: ulnar sensory conduction study, using same method as median sensory study
8. Consider phrenic nerve conduction study.
9. Consider motor unit number estimates to quantitate severity of cell loss.
10. Needle examination must be based on clinical findings, together with the following considerations:
    a. Two to three muscles innervated by different nerves and roots must have fibrillation potentials and motor unit

potential changes at two distinct levels to confirm diagnosis (levels are brain stem and cervical, thoracic, and lumbar spinal cord).
    b. Fasciculation potentials are usually present; if not, consider another disease.
    c. First, select muscles that are most likely to be abnormal, such as weak, atrophic, or distal muscles.
    d. Do not attempt to localize damage to one nerve before looking for widespread changes.
    e. If clinically involved muscles show no definite abnormality, do not spend much time or effort examining many normal muscles in the same distribution.
    f. Examine some muscles in each limb even if there is no clinical abnormality. Paraspinal muscles, especially thoracic ones, should also be examined.
    g. Look for motor unit potential variation.
    h. Consider needle examination of diaphragm.
11. If needle EMG shows clear abnormality in only one or two extremities, consider single-fiber EMG for jitter, using standard concentric electrode (500-Hz low-frequency filter; 0.5 milliseconds/cm sweep) to look for motor unit potential variation.

POSTPOLIO SYNDROME

Postpolio syndrome is a clinical diagnosis that cannot be confirmed by EMG or nerve conduction studies. The purpose of these tests in this syndrome is to look for other superimposed new diseases in the face of known residuals of the old poliomyelitis.

1. Nerve conduction studies:
   a. These studies are used to search for carpal tunnel syndrome, ulnar neuropathy, peroneal neuropathy, peripheral neuropathy, or other neuropathy that may be suggested clinically.
   b. Extremity with greatest number of new symptoms.
   c. Motor conduction study with F waves (note size of F waves).
   d. Sensory conduction study with velocities.

e. If findings are abnormal or borderline, compare with the opposite extremity.

2. Needle EMG
   a. This test is used to search for radiculopathy or myopathy.
   b. Test most symptomatic extremity.
      1) Least atrophic muscles: Distribution of major roots. Proximal and distal.
      2) Markedly atrophic muscles are always abnormal, and will not give useful information.
   c. Always compare opposite, less symptomatic extremity.
      1) Muscles with comparable atrophy.
      2) Muscles in same distribution.
   d. Large motor unit potentials are of no significance, but motor unit potential variation may be.
   e. Fibrillation potentials are significant only if they are found in distribution of:
      1) Mononeuropathy or radiculopathy:
         Limited to distribution of one nerve or root.
         Present proximally and distally.
         Present in muscles without atrophy.
      2) Peripheral neuropathy or polyradiculopathy:
         Present bilaterally.
         Present in muscles without atrophy.

FACIAL WEAKNESS

1. Ipsilateral facial motor conduction study:
   a. 2-Hz repetitive stimulation if defect of neuromuscular junction is suspected.
   b. Reorient stimulating electrode if initial positive dip occurs because of masseter activation.
2. Opposite facial conduction study if comparison is needed.
3. Record from another facial muscle if weakness is focal.
4. Bilateral blink reflex testing.
5. Blink reflex testing with additional orbicularis oris recording if synkinesis is

suspected (hemifacial spasm, old facial neuropathy).

6. Limb nerve conduction (polyradiculopathy protocol) if two cranial nerves are slowed or if a generalized disorder is suspected.
7. Jaw jerk if abnormalities suggest multiple cranial nerve disease.
8. Lateral spread responses if hemifacial spasm is suspected.
9. Needle examination: examine ipsilateral orbicularis oris, orbicularis oculi, mentalis, and frontalis muscles.
10. If findings on ipsilateral needle examination are abnormal, check masseter, sternocleidomastoid muscles, and a contralateral facial muscle.
11. If motor unit potential variation or a decrement is found, use the protocol for myasthenia.

MYOTONIC SYNDROMES

1. Ulnar motor conduction study with F waves and hand immobilized.
   a. 2-Hz repetitive stimulation before and at 3, 15, and 30 seconds after 10 seconds of exercise
   b. Interpretation—after 10 seconds of exercise, an immediate decrease in compound muscle action potential amplitude and repair of a decrement (if present) are specific for myotonic syndromes. Decreased compound muscle action potential amplitude after 10 seconds of exercise will return to normal in less than 2 minutes (usually 30 to 40 seconds) in all forms of myotonia, except for paramyotonia, in which recovery may take up to 90 minutes.
2. Median motor conduction studies with F waves with thumb restrained: Repetitive stimulation and exercise as for ulnar nerve in step 1.
3. Median sensory conduction study
4. If legs are more involved than arms, consider peroneal/anterior tibial–knee (foot board), with repetitive stimulation and exercise as for ulnar nerve in step 1.
5. If leg motor conduction is tested, sural sensory conduction study
6. Needle examination—rested muscle:
   a. First dorsal interosseous, abductor

pollicis brevis, extensor carpi radialis, triceps, biceps, infraspinatus muscles, anterior tibial, gluteus medius, and lumbar paraspinal muscles
   b. Interpretation—motor unit potentials are of short duration in rested muscle in myotonic dystrophy and in some cases of recessive myotonia congenita. On average, myotonic discharges occur in shorter bursts, are more variable in rate and amplitude, and fire at higher rates in myotonia congenita than in other forms of myotonia. Also, myotonic discharges, on average, fire at slower rates in paramyotonia.
7. Conduct the following special needle examination on a clinically involved muscle after performing steps 1 to 6:
   a. After 20 repeated forceful contractions, examine for changes in myotonia, loss of motor unit potential, or postexercise fibrillation.
   b. After exercise in paramyotonia, myotonia increases and the motor unit potentials drop out. In all other forms of primary myotonia, myotonia subsides and motor unit potentials do not change. Postexercise fibrillation may occur in paramyotonia.
   c. After cooling a muscle to 20°C (intramuscular temperature measured with needle thermistor), examine for spontaneous discharges, change in myotonia, loss of motor unit potential, and the effect of 20 strong contractions. Observe EMG activity as the muscle warms up.
      1) After cooling, myotonia increases, except in paramyotonia, in which the muscle becomes electrically silent. Spontaneous discharges indistinguishable from fibrillation potentials occur only in paramyotonia, especially as the temperature decreases from 32°C to 28°C (intramuscular temperature).
      2) If poor recruitment develops in cooled muscle, exercise will rapidly produce normal patterns, except in paramyotonia, in which all motor unit potentials are lost.
      3) Any changes seen in cold muscle are rapidly reversed by warming, except in paramyotonia, in which

abnormalities recover slowly over several hours.
8. If the diagnosis is still unclear:
   a. Rapid repetitive stimulation (greater than 25 Hz) of immobilized ulnar nerve with hand restrained.
      1) Rapid repetitive stimulation produces a waxing and waning pattern that is most prominent in myotonia congenita, particularly the autosomal recessive form.
      2) More than 30 seconds of stimulation may be needed before this pattern appears.
      3) The waxing and waning effect disappears in exercised muscle.
   b. Cool the contralateral hypothenar muscle to 28°C and apply 2-Hz stimulation to ulnar nerve with the hand restrained on a board to check for occurrence or enhancement of a decrement. Occurrence or enhancement of a decrement at 2-Hz stimulation during cooling is characteristic of paramyotonia.
   c. If periodic paralysis is suspected, consider prolonged exercise, potassium challenge, and intra-arterial epinephrine tests.

Unexpected Findings on Nerve Conduction Studies: Cause and Action

COMMON PROBLEMS

1. If compound muscle action potential or sensory nerve action potential is low-amplitude or absent:
   a. Preamplifier input may not be turned on.
   b. Stimulator may not be over the nerve: slide the stimulating electrode without changing the stimulating voltage.
   c. If the nerve may be deep or the patient obese: firmly push in stimulating electrode, separate cathode and anode, increase voltage and duration to maximum, try monopolar stimulating electrode, or consider using needle stimulation.
   d. Anode and cathode may be reversed, causing anodal block, or they may be

too close together with current shunt. Check cathode location.

e. Recording electrode may be placed incorrectly. Check the positions of the active and reference electrodes.

f. Wrong electrodes may be plugged into the recording system. Check input plugs.

g. Gain setting may be wrong. Check amplification.

h. Sweep speed may be wrong, with the response off the oscilloscope screen. Check sweep speed.

i. Filter setting may be incorrect. Check filters.

j. There may be stimulator malfunction. Test on self.

k. The amplifier may be turned off.

l. Innervation may be anomalous. Stimulate other nerves as appropriate.

m. It may be a manifestation of disease.

2. If the compound muscle action potential amplitude differs between proximal and distal stimulations.

a. At the site of the lower amplitude response—

1) The stimulating electrode may be off the nerve. Slide the electrode medial and lateral without voltage change.

2) The nerve may be deeper (see step 1c above).

3) Stimulation may not be supramaximal for other reasons listed in step 1 above.

b. At the site of higher amplitude response—There may be excessive stimulation with current spread to activate other nerves. Slide toward other nerve while watching the configuration of the evoked response and the muscle twitch.

c. Dispersion of the action potential:

1) It may be normal variant, for instance, tibial nerve to abductor hallucis brevis muscle response.

2) It may be disease with late components or increasing duration.

d. Anomalous innervation: stimulate other nerves as above.

e. Diseases with conduction block.

3. If there is initial positivity preceding a negative M wave of the compound muscle action potential:

a. The active recording electrode may not be over the end-plate region. Check the position of the recording electrode and slide it while stimulating.

b. The wrong nerve may be stimulated. Check twitch and configuration for current spread. Check location of the stimulating electrode.

c. Active and reference input may be reversed. Check input of electrodes.

d. There may be volume-conducted response from a distant muscle, especially at high gain (e.g., peroneal with knee stimulation). Check twitch and change the stimulation site.

e. Innervation may be anomalous. Check change in the response with stimulation at different sites on different nerves.

f. The reference electrode may not be in correct location or may not be in contact with the skin. Check the electrode.

g. There may be disease with atrophy.

4. If there is excessive stimulus artifact or baseline shift (i.e., active and reference electrodes are not isopotential):

a. The contact of the recording electrode with skin may be poor. Reclean and abrade skin and reapply electrodes.

b. There may be a current bridge between the stimulating and recording or ground electrodes. Check for smeared paste and clean skin between the electrodes.

c. The active, reference, or ground electrode may be broken, not in contact, or not plugged in. Check electrodes, especially if 60-cycle artifact is present.

d. Stimulating electrode may be too close to the recording electrode. Check their relative position and distance.

e. Stimulating electrodes may be oriented incorrectly. Rotate stimulating electrodes if the artifact is not bidirectional. Skin must be cleaned if the artifact is bidirectional.

f. There may be equipment malfunction. Check the effect of voltage change.

g. The stimulus may be excessive, especially duration (see step 5 below).

h. There may be amplifier/preamplifier overload. Reduce shock artifact.

5. High threshold (excessive voltage required to achieve supramaximal stimulation) (also see step 1 above):

a. Stimulating electrode may not be over the nerve. Slide stimulator without changing the voltage.

b. The nerve may be deep: push in stimulating electrode firmly, spread cathode and anode, or consider monopolar stimulation.

c. Intervening tissue may be excessive (e.g., scar or fat) (see step b above).

d. Contact between skin and electrode may be poor. Abrade skin or add electrode paste.

e. Amplification may be incorrect or amplifier may not be turned on.

f. Anode and cathode may be too close with current shunt. Separate them.

g. There may be disease with small, regenerating, or hypertrophic axons.

6. If distal latency is long:

a. The extremity may be cold. Make sure temperature probe is working correctly.

b. Distance may be wrong. Check against normal distance.

c. Cathode and anode may be reversed. Check cathode location.

d. Stimulus delay set may be wrong. Check delay.

e. Stimulation may be submaximal.

f. Gain may be too low.

g. There may be local or diffuse disease.

7. If there are no recognizable F waves:

a. Gain may be too low. Increase amplification.

b. Sweep speed may be too fast. Try a slower sweep speed.

c. There may be poor relaxation. Manipulate extremity and stimulate only when the audio EMG recording is quiet.

d. Too few stimuli may have been given. Increase their number.

e. Voltage may not be supramaximal. Check for supramaximal voltage with the cathode proximal.

f. There may be anodal block with the cathode distal. Place cathode proximal.

g. F waves may be confused with axon reflex or satellite potentials. Check persistence and latency change with site of stimulation.

h. Lost in M wave with proximal stimulation. Stimulate distally.

i. It may be a normal variant, especially with peroneal stimulation.

j. It may be due to disease.

ANOMALOUS INNERVATION

1. If the hypothenar compound muscle action potential with ulnar nerve stimulation differs in configuration between the wrist and the elbow or if the elbow amplitude/area is less than 80% of the wrist amplitude:

a. Check for current spread at wrist.

b. Check for supramaximal stimulation at elbow.

c. Record hypothenar compound muscle action potential with median stimulation at the wrist and elbow. An amplitude or configuration difference between the elbow and wrist confirms median/ulnar nerve anastomosis.

d. If step c above gives no difference, stimulate ulnar nerve at the wrist, below elbow, at elbow, and in upper arm to check for ulnar conduction block (ulnar neuropathy protocol).

e. If further characterization is desired, record first dorsal interosseous muscle with stimulation of the median and ulnar nerves at elbow and wrist or record with a needle electrode to localize site of anomaly.

2. If thenar compound muscle action potential with median nerve stimulation differs in configuration between the elbow and wrist or if the elbow amplitude/area exceeds the wrist amplitude:

a. Check for supramaximal stimulation at wrist.

b. Check for current spread at elbow.

c. Record thenar compound muscle action potential with ulnar stimulation at the wrist and elbow. Amplitude or configuration difference between the wrist and elbow responses confirms median/ulnar nerve anastomosis.

d. If an initial positivity is present in the thenar compound muscle action potential with both wrist and elbow stimulation, reposition the active electrode.

e. An initial positivity is the usual finding with ulnar and thenar stimulation.

f. If further characterization is desired, record first dorsal interosseous muscle with stimulation of the median and ulnar nerves at the elbow and wrist.

3. If the extensor digitorum brevis response changes configuration or has a higher amplitude with knee than with ankle peroneal nerve stimulation:

a. Check for supramaximal stimulation at ankle.

b. Check for current spread to the tibial nerve at the knee. Observe for plantar flexion due to tibial nerve stimulation and move the knee stimulating electrode lateral and distal if present.

c. Stimulate behind the lateral malleolus while recording extensor digitorum brevis muscle. A response confirms presence of a deep accessory branch of the superficial peroneal nerve. Make sure that you are not stimulating the tibial nerve at the ankle.

Interpretations of Nerve Conduction Studies

DEMYELINATING NEUROPATHY (OR REGENERATION) IS LIKELY IF:

1. Conduction velocity is within 50% to 70% of normal range with:

a. Amplitude in normal range.

b. No fibrillation potentials.

2. Conduction velocity is less than 50% of normal range with:

a. Amplitude greater than 50% of normal.

b. Minimal fibrillation potentials.

3. Other strong evidence:

a. Distal latencies 50% longer than normal range with normal amplitude.

b. Progressive dispersion and amplitude reduction with more proximal sites of stimulation.

c. F-wave latency is slowed out of proportion to nerve conduction slowing.

d. Conduction velocities are less than 10% of normal.

e. Blink reflex is longer than 16 milliseconds.

AXONAL NEUROPATHY (OR WALLERIAN DEGENERATION) IS LIKELY IF:

1. Motor amplitude is 50% to 70% of normal with:

a. Normal conduction velocity and latency.

b. No change between stimulation sites.

c. Clear-cut fibrillation.

2. Motor amplitude is less than 50% of normal with:

a. Conduction velocity greater than 70% of normal.

b. No change between stimulation sites.

c. Prominent fibrillation.

3. Absent sensory potentials with:

a. Normal motor conduction.

b. Fibrillation potentials.

FOCAL LESION (UNSPECIFIED TYPE) IS LIKELY IF:

1. Amplitude reduction is greater than 20% over a 10-cm or shorter segment:

a. With or without slowing of conduction.

b. With or without fibrillation potentials.

c. May be a loss of area rather than of amplitude.

2. Slowing of conduction velocity with localized slowing greater than 10 m/s over a 10-cm segment:

a. Measured over length to distal site (or greater than 20 m/s measured over 10-cm segment).

b. Must be faster proximal and distal to site.

c. With or without amplitude decrease.

d. Without or without fibrillation potentials.

3. Distal latency is abnormal with other conduction studies normal:

a. Distal latencies are normal in other nerves in the same limb.

b. Distal latencies are normal in the opposite limb.

c. Conduction velocity is normal.

d. Amplitude is normal or both motor and sensory axons are involved.

e. Extremity temperature is appropriate.

CHAPTER SUMMARY

The major value and primary application of clinical neurophysiology is in assessment and characterization of neurologic disease. Selection of appropriate studies for the problem of an individual patient requires careful clinical evaluation of the patient to determine possible causes of the patient's symptoms. The nature of the symptoms and the conclusions of the clinical evaluation are the best guides to appropriate use of clinical neurophysiologic testing.

The approach to testing can be assisted by deciding which structures are likely to be involved. For example, motor and sensory symptoms are best assessed using the different methods of motor and sensory nerve conduction studies. Electroencephalography, autonomic function testing, and polysomnography provide distinct assessment of disturbances of consciousness, cognition, visceral function, and sleep.

The level of the nervous system that is likely to be involved by the disease process can also guide selection of the neurophysiologic methods that will be most helpful in sorting out the clinical problem. Disorders of the cerebral hemisphere are best characterized electrically by electroencephalography, somatosensory evoked potentials, polysomnography, and movement recordings. Lesions in the posterior fossa may benefit from the addition of cranial conduction studies and brain stem auditory evoked potentials. Spinal cord disease produces alterations in EMG, nerve conduction studies, and somatosensory evoked potentials. Peripheral diseases show changes on nerve conduction studies, EMG, and autonomic function testing. The multiplicity of different neurophysiologic measures that can be applied in peripheral disorders is sometimes assisted by application of guideline protocols based on the patient's clinical findings and on what is found during testing.

Although clinical neurophysiologic assessment rarely provides evidence for a specific diagnosis, it can sometimes provide valuable information about the severity, progression, and prognosis of the disease.

GLOSSARY OF ELECTROPHYSIOLOGIC TERMS

QUANTITATIVE PREFIXES	
Meg (M)	1,000,000
Kilo (K)	1000
Milli (m)	1/1000
Micro (μ)	1/1,000,000
Pico (p)	1/1,000,000,000,000

A Wave: A compound action potential evoked consistently from a muscle by submaximal electrical stimuli to the nerve; frequently abolished by supramaximal stimuli.

Absence: Use of term discouraged when describing EEG patterns. Terms suggested, whenever appropriate: 3-Hz spike-and-slow waves; atypical repetitive spike-and-slow waves.

Absolute Amplitude: In frequency analysis, the average peak height of electrical activity in a particular frequency band, irrespective of the amount of electrical activity in other frequency bands; measured in microvolt (μV) units.

Absolute Power: In frequency analysis, the average amount of electrical activity in a particular frequency band, measured in units of amplitude squared (microvolts [μV] squared).

Abundance: Use of term discouraged. Term suggested: quantity (not a synonym).

AC: Abbreviation for alternating current.

Accommodation: True accommodation in neuronal physiology is a rise in the threshold of transmembrane depolarization required to initiate a spike when depolarization is maintained. In older literature, accommodation described the observation that the final intensity of current applied in a slowly rising fashion to stimulate a nerve was greater than the intensity of a pulse of current required to stimulate the same nerve. The latter may largely be an artifact of the nerve sheath and bears little relation to true accommodation as measured intracellularly.

Action Potential: A self-propagating regenerative change in membrane potential.

Activation: Any procedure designed to enhance or elicit normal or abnormal EEG activity, especially paroxysmal activity.

Active Electrode: Use of term discouraged. *Cf.* exploring electrode (not a synonym).

Afterdischarge: Continuation of an impulse train in neurons, axons, or muscle fibers after termination of an applied stimulus.

The following annotations are used throughout the glossary:

*Cf. Report of the Committee on Methods of Clinical Examination in Electroencephalography. Electroencephalogr Clin Neurophysiol 10:370–375, 1958.

†*Cf.* Report of the Committee on EEG Instrumentation Standards of the International Federation of Societies for Electroencephalography and Clinical Neurophysiology. Electroencephalogr Clin Neurophysiol 37:549–553, 1974.

The Report of the Committee on Terminology, page 529, to which part of this glossary is an appendix, should be consulted.

Modified from Chatrian, GE (Chairman), Bergamini, L, Dondey, M, Klass, DW, Lennox-Buchthal, M, and Petersén, I. A glossary of terms most commonly used by clinical electroencephalographers. Electroencephalogr Clin Neurophysiol 37:538–548, 1974, © Elsevier Scientific Publishing Company, Amsterdam—printed in the Netherlands, and from the Nomenclature Committee, American Association of Electromyography and Electrodiagnosis (Jablecki, CK, Bolton, CF, Bradley, WG, Brown, WF, Buchthal, F, Cracco, RQ, Johnson, EW, Kraft, GH, Lambert, EH, Lüders, HO, Ma, DM, Simpson, JA, and Stålberg, EV): AAEE glossary of terms in clinical electromyography. Muscle Nerve 10 Suppl:G1–G23, 1987, with permission. (Approval for inclusion of the glossary in this book in no way implies review or endorsement by the AAEE of material contained in the book.)

The number of extra impulses and their periodicity in the train may vary.

Afterhyperpolarization: A hyperpolarizing membrane potential that follows an action potential.

Afterpolarization: A depolarizing membrane potential that follows an action potential.

Alpha Band: Frequency band of 8 to 13 Hz. Greek letter: α.

Alpha Rhythm: Rhythm at 8 to 13 Hz occurring during wakefulness over the posterior regions of the head, generally with higher voltage over the occipital areas. Blocked or attenuated by attention, especially during visual and mental effort.

Alpha Variant Rhythms: Term applies to certain characteristic EEG rhythms recorded most prominently over posterior regions of the head. They differ in frequency but resemble the alpha rhythm in reactivity. *Cf.* fast alpha variant rhythms; slow alpha variant rhythms.

Alpha Wave: Wave with duration of 1/8 to 1/13 second.

Alternating Current (AC): Direction of current flow is episodically reversed, not necessarily at a regular rate.

Amplitude: With reference to an action potential or EEG waves, the maximum voltage difference between two points, usually baseline to peak or peak to peak.

Anode: The positive terminal of a source of electrical current.

Anode Block: The focal block of nerve conduction that may occur if the anode is placed between the cathode and the recording electrodes. It can cause a study to be considered technically unsatisfactory.

Antidromic: Propagation of an impulse in the direction opposite to physiologic conduction; for example, conduction along motor nerve fibers away from the muscle and conduction along sensory fibers away from the spinal cord. *Cf.* orthodromic.

Aperiodic: Applies to: (1) Waves or complexes occurring in a sequence at an irregular rate. (2) Waves or complexes occurring intermittently at irregular intervals.

Arceau: Synonym: Mu rhythm.

Area Under the Curve: The mathematical integral between two poststimulus times under the curve; sometimes calculated after taking the absolute value of each point.

Array: Regular arrangement of electrodes over the scalp or brain or within the brain substance.

Arrhythmic Activity: Sequence of waves of inconstant period. *Cf.* rhythm.

Artifact: A voltage change generated by a biologic or nonbiologic source other than the ones of interest.

Asymmetry: Unequal amplitude or form and frequency of activities over homologous areas.

Asynchrony: Nonsimultaneous occurrence of activities over different regions.

Attenuation: Reduction in amplitude of activity.

Atypical Repetitive Spike-and-Slow Waves: Paroxysms consisting of a sequence of spike-and-slow-wave complexes that occur bilaterally synchronously but do not meet one or more of the criteria of 3-Hz spike-and-slow waves. *Cf.* 3-Hz spike-and-slow waves.

Augmentation: Increase in amplitude of electrical activity.

Average Potential Reference: Average of potentials of all or many EEG electrodes, used as a reference.

Backfiring: Discharge of an antidromically activated motor neuron.

Background Activity: Any EEG activity representing the setting in which a given normal or abnormal pattern appears and from which such a pattern is distinguished.

BAEPs: Abbreviations for brain stem auditory evoked potentials.

Balanced Amplifier: An amplifier that consists essentially of two identical single-ended amplifiers operated as a pair and in opposite phases.

Band: Portion of a frequency spectrum.

Bandwidth: Range of frequencies between which response of a channel is within stated limits. Determined by frequency response of the amplifier-writer combination and frequency filters used.

Bar Format: Display of data in colored or dot-density strips in which color or density codes for amplitude.

Basal Electrode: Any electrode located in proximity to the base of the skull. *Cf.* nasopharyngeal electrode; sphenoidal electrode.

Baseline: Obtained when an identical voltage is applied to the two input terminals of an amplifier or when the instrument is in the calibrate position but no calibration signal is applied.

Beta Band: Frequency band greater than 13 Hz. Greek letter: β.

Beta Rhythm: Any EEG rhythm greater than 13 Hz.

Biphasic Action Potential: An action potential with two phases.

Bipolar Derivation: Recording from a pair of exploring electrodes. *Cf.* bipolar montage; exploring electrode.

Bipolar Montage: Multiple bipolar derivations, with no electrode common to all derivations. In most instances, bipolar derivations are linked, that is, adjacent derivations from electrodes along the same array have one electrode in common, connected to input terminal 2 of one amplifier and input terminal 1 of the following amplifier. *Cf.* referential montage.

Bipolar Reconstruction: Transformation of referential EEG data to a bipolar montage by serial subtraction of data channels.

Bipolar Stimulation: By convention, stimulating electrodes are called bipolar if they are encased or attached. *Cf.* monopolar stimulation.

Blink Responses: Potentials evoked from orbicularis oculi muscle as a result of brief electrical or mechanical stimuli to the cutaneous area innervated by the supraorbital (or less commonly the infraorbital) branch of the trigeminal nerve.

Blocking: (1) Apparent, temporary obliteration of biologic rhythms in response to physiologic or other stimuli. *Cf.* attenuation. (2) A condition of temporary unresponsiveness of an amplifier, caused by major overload. *Cf.* clipping; overload.

Brain Stem Auditory Evoked Potentials (BAEPs): Electrical waveforms of biologic origin elicited in response to sound stimuli.

Build-up: Colloquialism. Used to describe progressive increase in voltage of the EEG or appearance of waves of increasing amplitude, frequently associated with decrease in frequency.

Burst: A group of waves that appear and disappear abruptly and are distinguished from background activity by differences in frequency, form, and/or amplitude. Comments: (1) Term does not imply abnor-

mality. (2) Not a synonym of paroxysm. *Cf.* paroxysm.

Burst-Suppression: Pattern characterized by bursts of theta and/or delta waves, at times intermixed with faster waves, and intervening periods of relative quiescence.

C: Abbreviation for capacitance.

c/sec: Abbreviation for cycles per second. Equivalent: Hz (preferred).

Calcium Spike: A depolarizing membrane potential caused by the influx of calcium into the cell.

Calibration: Procedure of testing and recording responses to voltage differences applied to the input terminals of their respective amplifiers.

Canonogram: Ratio between frequency bands shown in a topographic map or similar display.

Capacitance (C): Amount of electrical charge that can be stored on a capacitor; usually measured in microfarads (μF) and picofarads (pF).

Capacitor: Consists of two conductive surfaces held close together but separated by an insulator. It may be used to store electrical charge, to conduct alternating current while blocking direct current, or to construct filters that will allow some frequencies to pass, but not others.

Cartooning: Rapid sequence of topographic maps displayed on a video monitor.

Cathode: The negative terminal of a source of electrical current.

Central Electromyography (Central EMG): Use of electromyographic recording techniques to study reflexes and the control of movement by the spinal cord and brain.

Channel: Complete system for detection, amplification, and display of potential differences between a pair of electrodes.

Charge: A surplus or deficit of electrons. *Cf.* negative charge and positive charge.

Chopper: Device consisting of a mechanical or electronic switch used for interrupting (chopping) DC and low-frequency AC signals and converting them into square waves of relatively high frequency.

Chopper Amplifier: A DC amplifier in which a chopper interrupts DC and low-frequency AC signals and converts them into square waves of relatively high frequency.

Circumferential Bipolar Montage: Derivations from pairs of electrodes along circumferential arrays.

Clipping: Distortion of waves which makes them appear flat-topped in the write-out. Caused by overload.

CMAP: Abbreviation for compound muscle action potential.

Coherence: Degree of interrelationship or correlation between electrical activity at two sites expressed in terms of their relative power in common at a specific frequency or band; values range from 0 (no interrelationship) to 1 (maximal correlation).

Collision: In reference to nerve conduction studies, collision is interaction of two action potentials propagated toward each other from opposite directions on the same nerve fiber.

Common EEG Input Test: Procedure in which the same pair of EEG electrodes is connected to the two input terminals of all channels of the electroencephalograph. *Cf.* calibration.

Common-Mode Rejection: A characteristic of differential amplifiers whereby they provide markedly decreased amplification of common-mode signals, compared with differential signals. Expressed as common-mode rejection ratio, that is, ratio of amplifications of differential and common mode signals. Example:

$$\frac{\text{amplification, differential}}{\text{amplification, common mode}} =$$

$$\frac{20,000}{1} = 20,000:1$$

Common-Mode Signal: Common component of two signals applied to the two respective input terminals of a differential amplifier.

Common Reference Electrode: One connected to the input terminal 2 of several or all amplifiers.

Common Reference Montage: Several referential derivations sharing a single reference electrode. *Cf.* referential derivation; reference electrode.

Complex: Sequence of two or more waves having a characteristic form or recurring with a fairly consistent form, distinguished from background activity.

Complex Motor Unit Action Potential: A motor unit action potential that is polyphasic or serrated.

Complex Repetitive Discharge: Polyphasic or serrated action potentials that have a uniform frequency, shape, and amplitude, with abrupt onset, cessation, or change in configuration.

Compound Muscle Action Potential (CMAP): Summation of nearly synchronous muscle fiber action potentials recorded from a muscle. It is commonly produced by stimulating, either directly or indirectly, the nerve supplying the muscle.

Compound Nerve Action Potential: Summation of nearly synchronous nerve fiber action potentials recorded from a nerve trunk. It is commonly produced by stimulation of the nerve.

Compound Sensory Nerve Action Potential: This is considered to have been evoked from afferent fibers if the recording electrodes detect activity only in a sensory nerve or in a sensory branch of a mixed nerve, or if the electrical stimulus is applied to a sensory nerve or a dorsal nerve root, or if an adequate stimulus is applied synchronously to sensory receptors. Amplitude, latency, duration, and configuration should be noted. Generally, amplitude is measured as the maximum peak-to-peak voltage, latency as either the "latency" to the initial deflection or the "peak latency" to the negative peak, and duration as the interval from the first deflection of the waveform from baseline to final return to baseline. Compound sensory nerve action potential has been referred to as the "sensory response" or "sensory potential."

Concentric Needle Electrode: Recording electrode that measures electrical potential difference between the bare tip of an insulated wire, usually stainless steel, silver, or platinum, and the bare shaft of the steel cannula through which the wire is inserted.

Conduction Block: Failure of an action potential to be conducted past a particular point in the nervous system, whereas conduction is possible below the point of the block.

Conduction Velocity (CV): Speed of propagation of an action potential along a nerve or muscle fiber.

Coronal Bipolar Montage: A montage consisting of derivations from pairs of electrodes along coronal (transverse) arrays. Synonym: transverse bipolar montage.*

Cortical Electrode: Electrode applied directly upon or inserted into the cerebral cortex.

Covariance: Relationship between two variables; often expressed statistically.

Cramp Discharge: Involuntary repetitive firing of motor unit action potentials at a high frequency (up to 150 Hz) in a large area of muscles, usually associated with painful muscle contraction.

Current (I): Flow of ions or electrons; measured in amperes (A), milliamperes (mA), and microamperes (μA).

CV: Abbreviation for conduction velocity.

Cycle: Complete sequence of potential changes undergone by individual components of a sequence of regularly repeated waves or complexes.

Cycles per Second: Unit of frequency. Abbreviation is c/sec. Suggested synonym: hertz (Hz) (preferred).

dB: Abbreviation for decibel.

DC: Abbreviation for direct current.

Decibel (dB): Logarithmic measure of relative power. A difference between two signals of 20 dB is a power ratio of 100:1 and an amplitude ratio of 10:1. A difference of 6 dB is an amplitude ratio of 2:1.

Decrementing Response: Reproducible decline in amplitude or area of the M wave of successive responses to repetitive nerve stimulation.

DEEG: Abbreviation for depth electroencephalogram.

Deglitching: Process by which the operator indicates to the computer which epochs of electrical activity are contaminated by artifacts and should thus be removed from further analysis.

Delay: As originally used in clinical electromyography, time between beginning of the horizontal sweep of the oscilloscope and onset of an applied stimulus. Also refers to an information storage device (delay line) used to display events occurring before a trigger signal.

Delta Band: Frequency band less than 4 Hz. Greek letter: δ.

Delta Rhythm: Rhythm less than 4 Hz.

Delta Wave: Wave with duration greater than 1/4 second.

Dendritic Spike: Depolarizing potential recorded from a dendrite.

Depolarization Block: Failure of an excitable cell to respond to a stimulus because of depolarization of the cell membrane.

Depth Electrode: Electrode implanted within the brain substance.

Depth Electroencephalogram (DEEG): Record of electrical activity of the brain by means of electrodes implanted within the brain substance itself. *Cf.* stereotactic (stereotaxic) depth electroencephalogram.

Derivation: (1) Process of recording from a pair of electrodes in an EEG channel. (2) EEG record obtained by this process.

Differential Amplifier: An amplifier whose output is proportional to the voltage difference between its two input terminals.

Differential Signal: Difference between two unlike signals applied to the respective two input terminals of a differential amplifier.

Diffuse: Occurring over large areas of one or both sides of the head. *Cf.* generalized.

Diode: Electronic device that allows current to flow in only one direction.

Diphasic Wave: One consisting of two components developed on alternate sides of the baseline.

Dipole: Two adjacent current monopoles of opposite polarity.

Direct-Coupled Amplifier: One in which successive stages are connected (coupled) by devices that are not frequency-dependent.

Direct Current (DC): Current flows in only one direction.

Direct Current Amplifier: An amplifier capable of magnifying DC (zero frequency) voltages and slowly varying voltages. *Cf.* chopper amplifier; direct-coupled amplifier.

Discharge: Firing of one or more excitable elements (neurons, axons, or muscle fibers).

Discharge Frequency: Rate of repetition of potentials. When potentials occur in groups, rate of recurrence of the group and rate of repetition of individual components in the groups should be specified.

Discriminant Analysis: An algorithm used to search a large list of variables, such as EEG features, seeking those that can discriminate best between several categories, for example, disease versus normal condition.

Disk Electrode: Metal disk attached to the scalp with an adhesive such as collodion, a paste, or wax.*

Disorganization: Gross alteration in frequency, form, topography, or quantity of physiologic EEG rhythms in an individual record, relative to previous records in the same subject, or rhythms of homologous regions on the opposite side of the head.

Distortion: An instrumental alteration in waveform. *Cf.* artifact.

Double Discharge: Two action potentials (motor unit action potential, fibrillation potential) of the same form and of nearly the same amplitude, occurring consistently in the same relationship to one another at intervals of 2 to 20 milliseconds.

Doublet: Synonym for double discharge.

Driving Force: Chemical or electrical force responsible for the movement of ions across cell membranes.

Duration: (1) Interval from beginning to end of an individual wave, complex, or stimulus. (2) Length of time that a sequence of waves or complexes or any other distinguishable feature lasts.

Earth Connection: *Cf.* synonym: ground electrode.

ECoG: Abbreviation for electrocorticogram and electrocorticography.

EEG: Abbreviation for electroencephalogram and electroencephalography.

Electrical Silence: Absence of measurable electrical activity due to biologic or nonbiologic sources. The sensitivity and signal-to-noise level of the recording system should be specified.

Electrocorticography (ECoG): Technique of recording electrical activity of the brain by means of electrodes applied over or implanted in the cerebral cortex.

Electrode: A conducting device used to record an electrical potential (recording electrode) or to apply an electrical current (stimulating electrode). In addition to the ground electrode used in clinical recordings, two electrodes are always required either to record an electrical potential or to apply an electrical current.

Electrode Impedance: Opposition to flow of an AC current through interface between an electrode and the scalp or brain. Measured between pairs of electrodes or, in some electroencephalographs, between each individual electrode and all the other electrodes connected in parallel. Expressed in ohms (generally kilohms, kΩ). *Cf.* electrode resistance; input impedance.

Electrode Resistance: Opposition to flow of a DC current through the interface between an EEG electrode and the scalp or brain. Measured between pairs of electrodes or, in some electroencephalographs, between each individual electrode and all other electrodes connected in parallel. Expressed in ohms (generally kilohms, kΩ). *Cf.* electrode impedance.

Electroencephalogram: Record of electrical activity of the brain measured by electrodes placed on the surface of the head, unless otherwise specified.

Electroencephalograph: Instrument employed to record electroencephalograms.

Electroencephalography (EEG): (1) Science related to the electrical activity of the brain. (2) Technique of recording electroencephalograms.

Electromotive Force: Driving force behind the flow of charge, analogous to the head of pressure in a water pipe. For neurophysiologic applications, electromotive force is measured in volts (V), millivolts (mV), or microvolts (μV).

Electromyography (EMG): Strictly defined, recording and study of insertion, spontaneous, and voluntary electrical activity of muscle.

Electroneurography (ENG): Recording and study of the action potentials of peripheral nerves. Synonym: nerve conduction studies.

EMG: Abbreviation for electromyography.

End-Plate Activity: Spontaneous electrical activity recorded using a needle electrode close to muscle end-plates.

End-Plate Potential (EPP): Graded, nonpropagated membrane potential induced in postsynaptic membrane of the muscle fiber by the action of acetylcholine released in response to an action potential in the presynaptic axon terminal.

ENG: Abbreviation for electroneurography.

EP: Abbreviation for evoked potential.

Epidural Electrode: Electrode located over the dural covering of the cerebrum.

Epoch: A period of time measured in an EEG record. Duration of epochs is determined arbitrarily. Example: a 10-second epoch.

Epoch Length: Number of seconds in an individually recorded and analyzed piece (epoch) of electrical activity, usually between 1 and 30.

EPP: Abbreviation for end-plate potential.

EPSP: Abbreviation for excitatory postsynaptic potential.

Equilibrium Potential: Membrane potential of a cell at equilibrium. Synonym: resting potential.

Equipotential: Applies to regions of the head or electrodes that are at the same potential at a given instant in time.

Equipotential Line: Imaginary line joining a series of points that are at the same potential at a given instant in time.

Ergodicity: An event that is statistically invariant over time.

Event-Related Desynchronization: Attenuation of a rhythmic EEG feature due to an event, for example, mu (μ) suppression caused by finger movement.

Evoked Potential: Wave or complex elicited by, and time-locked to, a physiologic or other stimulus, for instance, electrical, delivered to a sensory receptor or nerve or applied directly to a discrete area of the brain.

Excitability: Capacity to be activated by, or react to, a stimulus.

Excitatory Postsynaptic Potential (EPSP): Local, graded depolarization of a neuron in response to activation by a nerve terminal of a synapse.

Exploring Electrode: Synonymous with active electrode. *Cf.* recording electrode.

Extracerebral Potential: Any potential that does not originate in the brain, referred to as an artifact in EEG. May arise from electrical interference external to the subject and recording system, the subject, the electrodes and their connections to the subject and the electroencephalograph, and the electroencephalograph itself. *Cf.* artifact.

Facilitation: Improvement of neuromuscular transmission that results in activation of previously inactive muscle fibers. Facilitation may be identified in several ways.

Far-Field Potential: Electrical activity of biologic origin generated at a considerable distance from the recording electrodes. All potentials in clinical neurophysiology are recorded at some distance from the generator. Far-field potential usually implies stationary rather than propagating signals recorded at a distance.

Fasciculation: Random, spontaneous twitching of a group of muscle fibers or a motor unit; may produce movement of the overlying skin (limb), mucous membrane (tongue), or digits.

Fasciculation Potential: The electrical potential often associated with a visible fasciculation that has the configuration of a motor unit action potential but occurs spontaneously.

Fast Activity: Activity of frequency higher than alpha, that is, beta activity.

Fast Alpha Variant Rhythms: Characteristic rhythms at 14 to 20 Hz, detected most prominently over the posterior regions of the head. May alternate or be intermixed with alpha rhythms. Blocked or attenuated by attention, especially visual, and mental effort.

Fast Wave: Wave with duration shorter than alpha waves, that is, less than 1/13 second.

Fatigue: Generally, a state of depressed responsiveness resulting from protracted activity and requiring appreciable recovery time. Muscle fatigue is a decrease in the force of contraction of muscle fibers; it follows repeated voluntary contraction or direct electrical stimulation of the muscle.

Fiber Density: Anatomically, fiber density is a measure of the number of muscle or nerve fibers per unit area.

Fibrillation: Spontaneous contractions of individual muscle fibers that are not visible through the skin. This term has been used loosely in electromyography for the preferred term, fibrillation potential.

Fibrillation Potential: Electrical activity associated with a spontaneously contracting (fibrillating) muscle fiber. Action potentials may occur spontaneously or after movement of the needle electrode.

Field Gradients: Change in a feature across space, for example, how quickly the amplitude of a spike decreases with distance from the generator.

Firing Pattern: Qualitative and quantitative descriptions of the sequence of discharge of potential waveforms recorded from muscle to nerve.

Firing Rate: Frequency of repetition of a potential. Relationship of the frequency to the occurrence of other potentials and the force of muscle contraction may be described.

Focus: A limited region of the scalp, cerebral cortex, or depth of the brain displaying a given EEG activity, whether normal or abnormal.

Form: Shape of a wave. Synonyms: morphology; waveform.

Fourteen- and Six-Hz Positive Burst: Burst of arc-shaped waves at 13 to 17 Hz or 5 to 7 Hz but most commonly at 14 and 6 Hz, seen generally over the posterior temporal and adjacent areas of one or both sides of the head during sleep.

Frequency: Number of complete cycles of repetitive waves or complexes in 1 second. Measured in hertz (Hz), a unit preferred to its equivalent, cycles per second (c/sec).

Frequency Analysis: Determination of range of frequencies composing a potential waveform, with measurement of absolute or relative amplitude of each component frequency.

Frequency Response: *Cf.* bandwidth; low-frequency response; high-frequency response.[†]

Frequency-Response Curve: A graph depicting the relationships between output pen deflection or amplifier output and input frequency in an EEG channel, for a particular setting of low- and high-frequency filters.[†]

Frequency Spectrum: Range of frequencies composing the EEG. Divided into four bands: delta, theta, alpha, and beta. *Cf.* alpha; beta; delta; theta bands.

Frontal Intermittent Rhythmic Delta Activity: Fairly regular or approximately sinusoidal waves, mostly occurring in bursts at 1.5 to 3 Hz over frontal areas of either or both sides of the head.

F Wave: Compound action potential evoked intermittently from a muscle by a supramaximal electrical stimulus to the nerve. Compared with the maximal amplitude M wave of the same muscle, the F wave has a smaller amplitude (1% to 5% of the M wave), a variable configuration, and a longer, more variable latency.

G1: Abbreviation for grid 1 (use of term discouraged).

G2: Abbreviation for grid 2 (use of term discouraged).

Gain: Ratio of output signal voltage to input signal voltage of an EEG channel. Example:

$$\text{Gain} = \frac{\text{output voltage}}{\text{input voltage}}$$

$$= \frac{10 \text{ volts}}{10 \text{ microvolts}} = 1{,}000{,}000$$

Often expressed in decibels (dB), a logarithmic unit. Example: a voltage gain of 10 = 20 dB, of 1,000 = 60 dB, of 1,000,000 = 120 dB. *Cf.* sensitivity.

Generalization: Propagation of EEG activity from limited areas to all regions of the head.

Generalized: Occurring over all regions of the head.

Ground Electrode: An electrode connected to the patient and to a large conducting body (such as the earth), used as a common return for an electrical circuit and as an arbitrary zero potential reference point.

Harness, Head: Combination of straps that are fitted over the head to hold pad electrodes in position.

Hertz (Hz): Unit of frequency. Preferred to synonym: cycles per second.

High-Frequency Filter: A circuit that decreases sensitivity to relatively high frequencies. For each position of the high-frequency filter control, this attenuation is expressed as percent reduction at a given frequency relative to frequencies unaffected by the filter, that is, in the mid-frequency band of the channel.

High-Frequency Response: Sensitivity to relatively high frequencies. Determined by the high-frequency response of the amplifier and the high-frequency filter used. Expressed as percent reduction in output at certain specific high frequencies relative to other frequencies in the mid-frequency band of the channel.[†]

High-Pass Filter: Synonym: Low-frequency filter.

H-Reflex: Abbreviation for Hoffmann reflex. *Cf.* H wave.

H Wave: A compound muscle action potential having a consistent latency evoked regularly, when present, from a muscle by an electrical stimulus to the nerve. It is regularly found only in a limited group of physiologic extensors, particularly the calf muscles.

Hypsarrhythmia: Pattern consisting of high-voltage arrhythmic slow waves interspersed with spike discharges, without consistent synchrony between the two sides of the head or different areas on the same side.

Hz: Abbreviation for hertz. Preferred to equivalent: c/sec.

I: Abbreviation for current.

Impedance (Z): Sum of all obstacles to the flow of AC in an electrical circuit. Impedance is produced by resistors, capacitors, and inductors.

Impedance Meter: Instrument used to measure impedance. *Cf.* electrode impedance.

Incrementing Response: A reproducible increase in the amplitude and area of successive responses (M wave) to repetitive nerve stimulation.

Independent (Temporally): Synonym: asynchronous.

Index: Percentage of time an EEG activity is present in an EEG sample. Example: alpha index.

Inhibitory Postsynaptic Potential (IPSP): A local graded hyperpolarization of a neuron in response to activation at a synapse by a nerve terminal.

In-Phase Signals: Waves with no phase difference between them. *Cf.* common-mode signal (not a synonym).

Input Circuit: System consisting of EEG electrodes and intervening tissues, electrode leads, jack box, input cable, and electrode selectors.

Input Impedance: Impedance that exists between the two inputs of an EEG amplifier. Measured in ohms (generally megohms, MΩ), with or without additional specification of input shunt capacitance (measured in picofarads, pF).[†]

Input Terminal 1: Wherein negativity, relative to the other input terminal, produces an upward deflection on the graphic display.

Input Terminal 2: Wherein negativity, relative to the other input terminal, produces a downward deflection on the graphic display.

Input Voltage: Potential difference between the two input terminals of a differential EEG amplifier.

Insertion Activity: Electrical activity caused by insertion or movement of a needle electrode.

Integrated Circuit: Array of transistors, capacitors, resistors, and diodes fabricated on a silicon chip and used as a single package.

Interdischarge Interval: Time between consecutive discharges of the same potential.

Interelectrode Distance: Spacing between pairs of electrodes.

Interference: Unwanted electrical activity arising outside the system being studied.

Interference Pattern: Electrical activity recorded from a muscle with a needle electrode during maximal voluntary effort.

Interhemispheric derivation: Recording between a pair of electrodes located on opposite sides of the head.

Interpeak Interval: Difference between the peak latencies of two components of a waveform.

Interpotential Interval: Time between two different potentials. Measurement should be made between the corresponding parts on each waveform.

Intracerebral Electrode: Synonym: depth electrode.

Intracerebral Electroencephalogram: *Cf.* depth electroencephalogram.

IPSP: Abbreviation for inhibitory postsynaptic potential.

Isoelectric: Record obtained from a pair of equipotential electrodes. *Cf.* equipotential.

Isolated: Occurring singly.

Isopotential Contours: Lines on a map connecting points that have the same electrical potential.

Jitter: Variability with consecutive discharges of the interpotential interval between two muscle fiber action potentials belonging to the same motor unit. Synonym: single-fiber electromyographic jitter.

Kappa Rhythm: Bursts of alpha or theta frequency occurring over temporal areas of the scalp of subjects engaged in mental activity.

K Complex: A burst of somewhat variable appearance, consisting most commonly of a high-voltage diphasic slow wave, frequently associated with a sleep spindle. Amplitude is generally maximal in proximity of the vertex. *Cf.* vertex sharp transient.

Lambda Wave: Sharp transient occurring over occipital regions of the head of waking subjects during visual exploration. Mainly positive relative to other areas. Time-locked to saccadic eye movement. Greek letter: λ.

Laplacian: Mathematical transformation involving the second spatial derivative; the laplacian of the potential may be approximated by using the average of all neighboring sites as the reference for each site.

Latency: Interval between onset of a stimulus and onset of a response.

Latency of Activation: Time required for an electrical stimulus to depolarize a nerve fiber (or bundle of fibers, as in a nerve trunk) beyond threshold and to initiate a regenerative action potential in the fiber(s).

Lateralized: Involving mainly either the right or left side of the head. *Cf.* unilateral.

Late Response: General term used to describe an evoked potential with a longer latency than the M wave.

Lead: Strictly, wire connecting an electrode to the electroencephalograph. Loosely, synonym of electrode.

Leakage: Artifact due to ordinary fast Fourier transformations used in frequency analysis, causing excess activity in frequencies adjacent to the true frequency.

Line Format: Ordinary display format of traditional electrical activity, in contrast with graphic maps, bar format, or tables of numeric data.

Linkage: Connection of a pair of electrodes to two respective input terminals of a differential EEG amplifier. *Cf.* derivation.

Longitudinal Bipolar Montage: Derivations from pairs of electrodes along longitudinal, usually anteroposterior, arrays.*

Low-Frequency Filter: A circuit that decreases sensitivity to relatively low frequencies.* For each position of the low-frequency filter control, this attenuation is expressed as percentage reduction of output pen deflection at a given stated frequency relative to frequencies unaffected by the filter, that is, in the mid-frequency band of the channel.

Low-Frequency Response: Sensitivity to relatively low frequencies. Determined by the low-frequency response of the amplifier and by the low-frequency filter (time constant) used.

Low-Pass Filter: Synonym: high-frequency filter.

Macro-EMG needle electrode: A modified single-fiber electromyography electrode insulated to within 15 mm from the tip and with a small recording surface (25 μm in diameter) 7.5 mm from the tip.

Macro Motor Unit Action Potential (Macro MUAP): Average electrical activity of that part of an anatomic motor unit that is within the recording range of a macro-EMG electrode.

Macro MUAP: Abbreviation for macro motor unit action potential.

MCD: Abbreviation for mean consecutive difference. *Cf.* jitter.

Mean Consecutive Difference (MCD): Measure of the variability of consecutive interpotential intervals on single-fiber electromyography. *Cf.* jitter.

Mean Frequency: Statistical mean of data in a frequency spectrum; sometimes restricted to just one frequency band, for instance, mean alpha frequency.

Membrane Potential: Electrical potential inside a cell subtracted from that outside a cell.

Membrane Space Constant: Distance over which the membrane potential due to a localized injection of current decreases to 37% of its maximal value.

Membrane Time Constant: Time required for membrane potential to reach 63% of its new value.

MEPP: Abbreviation for miniature end-plate potential.

Microneurography: Technique of recording peripheral nerve action potentials in humans using intraneural electrodes.

Miniature End-Plate Potential (MEPP): Postsynaptic muscle fiber potentials produced through spontaneous release of individual quanta of acetylcholine from the presynaptic axon terminals.

Monophasic Potential: Potential developed on one side of the baseline.

Monopolar Needle Electrode: A solid wire,

usually of stainless steel, coated, except at its tip, with an insulating material.

Monopolar Stimulation: Stimulating electrodes are called monopolar if they are not encased or attached.

Monopole: Isolated source (or sink) of current in a volume conductor.

Montage: The particular arrangement by which a number of derivations are displayed simultaneously in an EEG record.*

Morphology: (1) Study of the form of EEG waves. (2) Form of EEG waves.

Motor Point: The point over a muscle where its contraction may be elicited by a minimal intensity, short-duration electrical stimulus. The motor point corresponds anatomically to the location of the terminal portion of the motor nerve fibers (end-plate zone).

Motor Unit: The anatomic unit of an anterior horn cell, its axon, the neuromuscular junctions, and all of the muscle fibers innervated by the axon.

Motor Unit Action Potential (MUAP): Action potential reflecting the electrical activity of a single anatomic motor unit. It is the compound action potential of those muscle fibers within the recording range of an electrode.

Motor Unit Territory: Area in a muscle over which muscle fibers belonging to an individual motor unit are distributed.

Mu Rhythm: Rhythm at 7 to 11 Hz, composed of arc-shaped waves occurring over the central or centroparietal regions of the scalp during wakefulness. Amplitude varies but is mostly less than 50 μV. Blocked or attenuated most clearly by contralateral movement, thought of movement, readiness to move, or tactile stimulation. Greek letter: μ. Synonym: *arceau*, wicket.

MUAP: Abbreviation for motor unit action potential.

Multiple Discharge: Four or more motor unit action potentials of the same form and of nearly the same amplitude occurring consistently in the same relationship to one another and generated by the same axon or muscle fiber.

Multiple Foci: Two or more spatially separated foci.

Multiple Spike-and-Slow-Wave Complex: A sequence of two or more spikes associated with one or more slow waves. Preferred to synonym: polyspike-and-slow-wave complex.

Multiple Spike Complex: A sequence of two or more spikes. Preferred to synonym: polyspike complex.

M Wave: A compound action potential evoked from a muscle by a single electrical stimulus to its motor nerve.

Myokymia: Continuous quivering or undulating movement of surface and overlying skin and mucous membrane associated with spontaneous, repetitive discharge of motor unit potentials.

Myokymic Discharges: Motor unit action potentials that fire repetitively and may be associated with clinical myokymia in one of two firing patterns.

Myotonia: Clinical observation of delayed relaxation of muscle after its voluntary contraction or percussion.

Myotonic Discharge: Repetitive discharge at rates of 20 to 80 Hz are of two types: (1) biphasic (positive-negative) spike potentials shorter than 5 milliseconds in duration resembling fibrillation potentials, (2) positive waves of 5 to 20 milliseconds in duration resembling positive sharp waves.

NAP: Abbreviation for nerve action potential.

Nasopharyngeal Electrode: Rod electrode introduced through the nose and placed against the nasopharyngeal wall, with its tip lying near the body of the sphenoid bone.

NCV: Abbreviation for nerve conduction velocity.

Near-Field Potential: Electrical activity of biologic origin that is registered as the generated signal passes near the recording electrodes. In contrast to far-field potentials, near-field potentials propagate, showing progressive change in latency with increasing distance to the recording site.

Needle Electrode: Used for recording or stimulating, an electrode shaped like a needle.

Negative Charge: State present in an object with an excess of electrons.

Nernst Potential: Membrane potential that would occur if a membrane were permeable to only one ionic species. It depends on the ratio of the concentration of the ions inside and outside the cell.

Nerve Action Potential (NAP): Strictly defined, term refers to an action potential recorded from a single nerve fiber. It is also commonly used to refer to the compound nerve action potential.

Nerve Conduction Velocity (NCV): Loosely used to refer to the maximum nerve conduction velocity.

Neuromyotonia: Clinical syndrome of continuous muscle fiber activity manifested as continuous muscle rippling and stiffness.

Neuromyotonic Discharges: Bursts of motor unit action potentials that originate in the motor axons firing at high rates (150–300 Hz) for a few seconds and that often start and stop abruptly.

Noise: Potentials produced by electrodes, cables, amplifier, or storage media and unrelated to the potentials of biologic origin.

Notch Filter: A filter that selectively attenuates a very narrow frequency band, thus producing a sharp notch in the frequency response curve. A 60-Hz notch filter is used in some electroencephalographs to provide attenuation of 60-Hz interference under extremely unfavorable technical conditions.

Occipital Intermittent Rhythmic Delta Activity: Fairly regular or approximately sinusoidal waves, mostly occurring in bursts at 2 to 3 Hz over occipital areas of one or both sides of the head. Frequently blocked or attenuated by opening the eyes.

Ohmmeter: An instrument used to measure resistance. *Cf.* electrode resistance.

Ohm's Equations: Expressions of relationships among voltage (V), current (I), and resistance (R) in an electrical circuit: V = IR, I = V/R, R = V/I.

Onset Frequency: Lowest stable frequency of firing for a single motor unit action potential that can be voluntarily maintained by a subject.

Orthodromic: Propagation of an impulse in the same direction as physiologic conduction; for example, conduction along motor nerve fibers toward the muscle and conduction along sensory nerve fibers toward the spinal cord.

Out-of-Phase Signals: Two waves of opposite phases. *Cf.* differential signal (not a synonym).

Overload: Condition resulting from the application to the input terminals of an amplifier of voltage differences larger than the channel is designed or set to handle. Causes clipping of waves and blocking of the amplifier, depending on its magnitude.

Pad Electrode: Metal electrode covered with a cotton or felt and gauze pad, held in position by a head cap or harness.

Paired Stimuli: Two consecutive stimuli. Time interval between the two stimuli and the intensity of each stimulus should be specified.

Paper Speed: Velocity of movement of EEG paper.

Paroxysm: Phenomenon with abrupt onset, rapid attainment of a maximum, and sudden termination; distinguished from background activity. It is also commonly used to refer to epileptiform and seizure patterns.

Pattern: Any characteristic activity.

Peak: Point of maximum amplitude of a wave.

Peak Frequency: Statistical mode of data in a frequency spectrum; often, the spectrum restricted to just one frequency band, for example, alpha frequency.

Peak Latency: Interval between onset of a stimulus and a specific peak of the evoked potential.

Period: Duration of complete cycle of individual component of a sequence of regularly repeated waves or complexes.

Periodic: Term describes: (1) Waves or complexes occurring in a sequence at an approximately regular rate. (2) Waves or complexes occurring intermittently at approximately regular intervals, generally of 1 to several seconds.

Phase: (1) That portion of a wave between the departure from and return to baseline. (2) Time or polarity relationships between a point on a wave displayed in a derivation and the identical point on the same wave recorded simultaneously in another derivation. (3) Time or angular relationships between a point on a wave and the onset of the cycle of the same wave. Usually expressed in degrees or radians.[†]

Phase Cancellation: Decrease in the size of the compound action potential as the result of a latency difference that lines up the positive peaks of the fast fibers with the negative peaks of the slow fibers, canceling both.

Phase Reversal: *Cf.* true phase reversal.

Photic Driving: Physiologic response consisting of rhythmic activity elicited over posterior regions of the head by repetitive photic stimulation at frequencies of about 5 to 30 Hz.

Photic Stimulation: Delivery of intermittent flashes of light to the eyes of a subject. Used as EEG activation procedure.*

Photic Stimulator: Device for delivering intermittent flashes of light. Synonym: Stroboscope (discouraged).

Photoconvulsive Response: Synonym: photoparoxysmal response (preferred).

Photomyoclonic Response: Synonym: photomyogenic response (preferred).

Photomyogenic Response: A response to intermittent photic stimulation characterized by the appearance in the record of brief, repetitive muscular spikes over anterior regions of the head. These often increase gradually in amplitude as stimuli are continued and cease promptly when the stimulus is withdrawn.

Photoparoxysmal Response: A response to intermittent photic stimulation characterized by the appearance in the record of spike-and-slow-wave and multiple spike-and-slow-wave complexes.

Polarity Convention: International agreement whereby differential amplifiers are constructed so that negativity at input terminal 1 relative to input terminal 2 of the same amplifier produces an upward display deflection.*†

Polarity, EEG Wave: Sign of potential difference existing at a given instant between an electrode affected by a given potential change and another electrode not appreciably, or less, affected by the same change. *Cf.* polarity convention.

Polygraphic Recording: Simultaneous monitoring of multiple physiologic measures such as the EEG, respiration, electrocardiogram, electromyogram, eye movement, galvanic skin resistance, and blood pressure.

Polyphasic Action Potential: An action potential with five or more phases.

Polyphasic Wave: Wave with two or more components developed on alternating sides of the baseline. *Cf.* diphasic wave; triphasic wave.

Polyspike-and-Slow-Wave Complex: Synonym: multiple spike-and-slow-wave complex (preferred).

Polyspike Complex: Synonym: multiple spike complex (preferred).

Positive Charge: State present in an object with a deficit of electrons.

Positive Occipital Sharp Transient of Sleep: Sharp transient maximal over the occipital regions, positive relative to other areas, occurring apparently spontaneously during sleep. May be single or repetitive.

Positive Sharp Wave: Biphasic positive-negative action potential initiated by needle movement and recurring in a uniform, regular pattern at a rate of 1 to 50 Hz; discharge frequency may decrease slightly just before cessation of discharge.

Postactivation Depression: A descriptive term indicating a reduction in amplitude associated with a reduction in the area of the M wave(s) in response to a single stimulus or train of stimuli which occurs a few minutes after a 30- to 60-second, strong voluntary contraction or a period of repetitive nerve stimulation that produces tetanus.

Postactivation Exhaustion: Reduction in the safety factor (margin) of neuromuscular transmission after sustained activity of the neuromuscular junction.

Postactivation Potentiation: Increase in the force of contraction (mechanical response) after tetanus or strong voluntary contraction.

Postsynaptic Potential: Membrane potential caused by neurotransmitter-induced opening or closing of ion channels in a postsynaptic neuron or muscle fiber. It may be either excitatory or inhibitory.

Potential: (1) Strictly, voltage; "potential" usually is preferred when describing the presence of an electrical field in a volume conductor such as the human body. (2) Loosely, a synonym of wave.

Potential Field: Amplitude distribution of a wave measured at a given instant. Represented in diagrams by equipotential lines.

Power Ratio: Quotient of values from two different power spectral bands.

Projected Patterns: Abnormal EEG activities believed to result from a disturbance at a site remote from the recording electrodes. Description of specific EEG patterns preferred.

QSART: Abbreviation for quantitative sudomotor axon reflex test.

Quadrupole: Two adjacent current dipoles of opposite orientation placed end-to-end.

Quantitative Sudomotor Axon Reflex Test: A type of test used to evaluate sympathetic axons that innervate sweat glands.

Quantity: Amount of EEG activity with respect to both number and amplitude of waves.

Quasi-Periodic: EEG waves or complexes occurring at intervals only approaching regularity.

R-C Coupled Amplifier: Abbreviation for resistance-capacitance coupled amplifier.

Reactivity: Susceptibility of individual rhythms or the EEG as a whole to change after sensory stimulation or other physiologic actions.

Reactivity Ratio: Change in a variable between two states or between two different situations (e.g., alpha attenuation with eye opening), expressed as a ratio.

Record: End product of the EEG recording process.

Recording: (1) The process of obtaining a record. (2) The end product of the recording process. Synonyms: record; tracing.

Recording Electrode: Device used to record electrical potential difference. All electrical recordings require two electrodes. The recording electrode close to the source of the activity to be recorded is called the "active or exploring electrode," and the other electrode is called the "reference electrode."

Recruitment: Successive activation of the same and additional motor units with increasing strength of voluntary contraction.

Recruitment Frequency: Firing rate of a motor unit action potential (MUAP) when a different MUAP first appears with gradually increasing strength of voluntary muscle contraction.

Recruitment Pattern: A qualitative and quantitative description of the sequence of appearance of motor unit action potentials with increasing strength of voluntary muscle contraction. Recruitment frequency and recruitment interval are two commonly used quantitative measures.

Reference Electrode: (1) In general, any electrode against which the potential variations of another electrode are measured. (2) Specifically, any suitable electrode customarily connected to input terminal 2 of an EEG amplifier and placed so as to minimize likelihood of pickup of the same EEG activity recorded by an exploring electrode, usually connected to input terminal 1 of the same amplifier, or of other activities.

Referential Derivation: Recording from a pair of electrodes consisting of an exploring electrode generally connected to input terminal 1 and a reference electrode usually connected to input terminal 2 of an amplifier.* *Cf.* exploring electrode; reference electrode; referential montage; common reference montage.

Referential Montage: A montage consisting of referential derivations. Comment: a referential montage in which reference electrode is common to multiple derivations is referred to as a common reference montage.* *Cf.* referential derivation.

Reflex: A stereotyped motor response elicited by a sensory stimulus.

Refractory Period: The absolute refractory period is the period after an action potential during which no stimulus, however strong, evokes a further response. The relative refractory period is the period during which a stimulus must be abnormally large to evoke a second response.

Regular: Applies to waves or complexes of approximately constant period and relatively uniform appearance.

Relative Power: The quotient between power in one frequency band and the total power of all bands; often expressed as a percentage.

Repetitive Discharge: General term for the recurrence of an action potential with the same or nearly the same form.

Repetitive Stimulation: The technique of repeated supramaximal stimulations of a nerve while recording M waves from muscles innervated by the nerve.

Residual Latency: Refers to the calculated time difference between the measured distal latency of a motor nerve and the expected distal latency, calculated by dividing the distance between the stimulus cathode and the active recording electrode by the maximum conduction velocity measured in a more proximal segment of a nerve.

Resistance: The "friction" in a conductor that limits the flow of current and dissipates its energy as heat. Resistance is measured in ohms (Ω), kilohms ($k\Omega$), and megohms ($M\Omega$).

Resistance-Capacitance Coupled Amplifier: An amplifier in which successive stages are connected (coupled) by networks consisting of capacitors and resistors.

Resting Potential: *Cf.* equilibrium potential.

Rhythm: Activity consisting of waves of approximately constant period.

Rhythm of Alpha Frequency: (1) In general, any rhythm in the alpha band. (2) Specifically, the term should be used to designate activities in the alpha band that differ from the alpha rhythm with regard to their topography and reactivity and do not have specific appellations (such as mu rhythm). *Cf.* alpha rhythm.

Rhythmic Temporal Theta Burst of Drowsiness: Characteristic burst of 4-to-7-Hz waves frequently notched by faster waves, occurring over the temporal regions of the head during drowsiness. Synonym: psychomotor variant pattern (use discouraged).

Rise Time: Interval from onset of a change of a potential to its peak. Method of measurement should be specified.

Satellite Potential: A small action potential separated from the main motor unit action potential by an isoelectric interval and firing in a time-locked relationship to the main action potential.

Scalp Electrode: Electrode held against, attached to, or inserted into the scalp.

Scalp Electroencephalogram: Record of electrical activity of the brain using electrodes placed on the head.

Scalp Electroencephalography: Technique of recording scalp electroencephalograms.

SDEEG: Abbreviation for stereotaxic depth electroencephalography (electroencephalogram).

Sensitivity: Ratio of input voltage to output pen deflection in an EEG channel.[†] Sensitivity is measured in microvolts per millimeter ($\mu V/mm$).[*] Example:

$$\text{Sensitivity} = \frac{\text{input voltage}}{\text{output pen deflection}}$$

$$= \frac{50\ \mu V}{10\ mm} = 5\ \mu V/mm$$

Sensory Latency: Interval between onset of a stimulus and onset of the compound sensory nerve action potential.

Sensory Nerve Action Potential (SNAP): *Cf.* compound sensory nerve action potential.

SEP: Abbreviation for somatosensory evoked potential.

Serrated Action Potential: An action potential waveform with several changes in direction (turns) that do not cross the baseline.

SFEMG: Abbreviation for single-fiber electromyography.

Sharp-and-Slow-Wave Complex: A sequence of a sharp wave and a slow wave.

Sharp Wave: An EEG transient, clearly distinguished from background activity, with pointed peak at conventional paper speeds and duration of 70 to 200 milliseconds, that is, longer than 1/14 to 1/5 second, approximately. Main component is generally negative relative to other areas. Amplitude is variable.

Shock Artifact: *Cf.* artifact.

Silent Period: Pause in the electrical activity of a muscle such as that seen after rapid unloading of a muscle.

Simultaneous: Occurring at the same time. Synonym: synchronous.

Sine Wave: Wave in the form of a sine curve.

Single-Ended Amplifier: An amplifier that operates on signals that are asymmetrical with respect to ground.

Single-Fiber Electromyography (SFEMG): Technique and conditions that permit recording of a single muscle fiber action potential.

Single-Fiber Needle Electrode: Needle electrode with a small recording surface (usually 25 μm in diameter) permitting recording of single muscle fiber action potentials between the active recording surface and the cannula.

Sinusoidal: Term describes EEG waves resembling sine waves.

Six-Hz Spike-and-Slow Waves: Spike-and-slow-wave complexes at 4 to 7 Hz, but mostly at 6 Hz, occurring generally in brief bursts bilaterally synchronously, symmetrically or asymmetrically, and either confined to, or of larger amplitude over, posterior or anterior regions of the head.

Sleep Spindle: Burst at 11 to 15 Hz, but mostly at 12 to 14 Hz, generally diffuse but of higher voltage over the central regions of the head, occurring during sleep. Amplitude is variable, but is generally less than 50 μV in adults.

Sleep Stages: Distinctive phases of sleep, best demonstrated by polygraphic recordings of the EEG and other variables, including eye movements and activity of certain voluntary muscles, at a minimum.

Slope Descriptors: Three specific EEG features—activity, complexity, and mobility—derived from statistical analysis of the frequency spectrum.

Slow Activity: Activity of frequency lower than alpha, that is, theta and delta activities.

Slow Alpha Variant Rhythms: Characteristic rhythms at 3.5 to 6 Hz but mostly at 4 to 5 Hz, recorded most prominently over posterior regions of the head. Generally alternate, or are intermixed, with alpha rhythms, to which they often are harmonically related. Amplitude is variable but is frequently close to 50 μV. Blocked or attenuated by attention, especially visual, and mental effort.

Slow Wave: EEG wave with duration longer than alpha wave, that is, over 1/8 second.

Smearing: Artifactual broadening of a frequency peak; produced by techniques used to decrease variation when performing frequency analysis.

SNAP: Abbreviation for sensory nerve action potential.

Somatosensory Evoked Potentials (SEP): Electrical waveform of biologic origin elicited by electrical stimulation or physiologic activation of peripheral sensory fibers, for example, the median nerve, common peroneal nerve, or posterior tibial nerve.

Source Derivation: An EEG technique that uses the average of all neighboring sites as reference for each site. This is an approximation of the laplacian of the potential. *Cf.* Laplacian.

Spatial Interpolation: Techniques used to define topographic mapping values evenly between the real recording sites.

Sphenoidal Electrode: Needle or wire electrode inserted through the soft tissues of the face below the zygomatic arch, with its tip lying near the base of the skull in the region of the foramen ovale.

Spike: (1) An EEG transient, clearly distinguished from background activity, with pointed peak at conventional paper speeds and a duration from 20 to less than 70 milliseconds, that is, from 1/50 to 1/14 second, approximately. Main component is generally negative relative to other areas. (2) In cellular neurophysiology, a short-lived (usually in the range of 1 to 3 milliseconds), all-or-none change in membrane potential that arises when a graded response passes threshold. (3) In EMG, electrical record of a nerve impulse or similar event in muscle or elsewhere. *Cf.* sharp wave.

Spike-and-Slow-Wave Complex: A pattern consisting of a spike followed by a slow wave.

Spinal Evoked Potentials: Electrical waveforms of biologic origin recorded over the sacral, lumbar, thoracic, or cervical spine in response to electrical stimulation or physiologic activation of peripheral sensory fibers.

Spindle: Group of rhythmic waves characterized by a progressively increasing, then gradually decreasing, amplitude. *Cf.* sleep spindle.

Spontaneous Activity: Electrical activity recorded from muscle or nerve at rest after insertion activity has subsided and when no voluntary contraction or external stimulus exists.

Spread: Propagation of EEG waves from one region of the scalp and brain to another. *Cf.* generalization.

Standard Electrode: Conventional scalp electrode. *Cf.* disk electrode; needle electrode; pad electrode.

Standard Electrode Placement. Scalp electrode location(s) determined by the 10–20 system. *Cf.* Ten-twenty system.

Stationarity: Measure of the stability of the frequency spectrum over a short period; an epileptic spike is a classic nonstationarity in an EEG.

Status Epilepticus, EEG: Occurrence of virtually continuous seizure activity in an EEG.

Stereotaxic Depth Electroencephalogram

(SDEEG): Recording of electrical activity of the brain using electrodes implanted within the brain substance according to stereotaxic measurements.

Stereotaxic Depth Electroencephalography (SDEEG): Technique of recording stereotaxic depth electroencephalograms.

Sternospinal Reference: A noncephalic reference achieved by interconnecting two electrodes placed over the right sternoclavicular junction and the apophysis spinosa of the seventh cervical vertebra, respectively, and balancing the voltage between them by means of a potentiometer to decrease electrocardiographic artifact.

Stimulating Electrode: Device used to apply electrical current. All electrical stimulation requires an anode and a cathode. Electrical stimulation for nerve conduction studies generally requires application of the cathode to produce depolarization of the fibers in the nerve trunk.

Stimulus: Any external agent, state, or change that is capable of influencing the activity of a cell, tissue, or organism.

Strength-Duration Curve: Graphic presentation of the relationship between the intensity (y-axis) and various durations (x-axis) of the threshold electrical stimulus for a muscle with the stimulating cathode positioned over the motor point.

Subdural Electrode: Electrode inserted under the dural covering of the cerebrum.

Symmetry: (1) Approximately equal amplitude, frequency, and form of EEG activities over homologous areas on opposite sides of the head. (2) Approximately equal distribution of potentials of unlike polarity on either side of a zero isopotential axis. *Cf.* true phase reversal. (3) Approximately equal distribution of EEG waves about the baseline.

Synchrony: The simultaneous occurrence of EEG waves over regions on the same or opposite sides of the head.

TC: Abbreviation for time constant.

Ten-Twenty (10–20) System: System of standardized scalp electrode placement recommended by the International Federation of Societies for Electroencephalography and Clinical Neurophysiology.* In this system, electrode placements are determined by measuring the head from external landmarks and taking 10% or 20% of such measurements.

Tetanic Contraction: Contraction produced in a muscle through repetitive maximal direct or indirect stimulation at a sufficiently high frequency to produce a smooth summation of successive maximum twitches.

Tetanus: Continuous contraction of muscle caused by repetitive stimulation or discharge of nerve or muscle. *Cf.* tetany.

Tetany: A clinical syndrome manifested by muscle twitching, cramps, and carpal and pedal spasm.

Theta Band: Frequency band from 4 to less than 8 Hz. Greek letter θ.

Theta Rhythm: Rhythm with a frequency of 4 to less than 8 Hz.

Theta Wave: Wave with duration of 1/4 to longer than 1/8 second.

Three-Hz Spike-and-Slow Waves: Characteristic paroxysm consisting of a regular sequence of spike-and-slow-wave complexes that: (1) repeat at 3 to 3.5 Hz (measured during the first few seconds of the paroxysm), (2) are bilateral in their onset and termination, generalized, and usually of maximal amplitude over the frontal areas, (3) are approximately synchronous and symmetrical on the two sides of the head throughout the paroxysm. *Cf.* atypical repetitive spike-and-slow waves.

Threshold: Level at which a clear and abrupt transition occurs from one state to another. Term is generally used to refer to the voltage level at which an action potential is initiated in a single axon or a group of axons.

Time Constant (TC), EEG Channel: Product of the values of resistance (in megohms, MΩ) and capacitance (in microfarads, μF) that make up the time constant control of an EEG channel. This product represents the time required for the pen to fall to 37% of the deflection initially produced when a DC voltage difference is applied to the input terminals of the amplifier. Expressed in seconds.

Topography: Amplitude distribution of EEG activities at the surface of the head, cerebral cortex, or in the depths of the brain.

Tracé Alternant: EEG pattern of sleeping newborns, characterized by bursts of slow waves, at times intermixed with sharp waves, and intervening periods of relative quiescence.

Tracing: Synonyms: record; recording.

Transformer: A pair of closely coupled wire

coils, usually surrounding an iron core, that inductively transfer alternating current from one circuit to another. The most common use is to convert voltage of an alternating current to a higher or lower value.

Transient, EEG: Any isolated wave or complex, distinguished from background activity.

Transistor: A solid-state circuit element used to amplify voltage or current.

Transverse Bipolar Montage: Synonym: coronal bipolar montage.

Triangular Bipolar Montage: Derivations from pairs of electrodes in a group of three electrodes arranged in a triangular pattern.

Triphasic Action Potential: Action potential with three phases.

Triphasic Wave: Wave consisting of three components alternating above the baseline.

True Phase Reversal: Simultaneous pen deflections in opposite directions occurring in two referential derivations using a suitable common reference electrode and displaying the same wave.

Turn: Point of change in direction of a waveform and magnitude of the voltage change after the turning point.

Unilateral: Confined to one side of the head.

VEPs: Abbreviation for visual evoked potentials.

Vertex Sharp Transient (V Wave): Sharp potential, maximal at the vertex, negative relative to other areas, occurring apparently spontaneously during sleep or in response to a sensory stimulus during sleep or wakefulness. *Cf.* K complex.

Visual Evoked Potentials (VEPs): Electrical waveforms of biologic origin recorded over the cerebrum and elicited by light stimuli.

Voltage: Potential difference between two recording sites.

Volume Conduction: Spread of current from the potential source through a conducting medium, such as the body tissues.

V Wave: Abbreviation for vertex sharp transient.

Waning Discharge: General term referring to a repetitive discharge that decreases in frequency or amplitude before cessation.

Wave: Any change of the potential difference between pairs of electrodes in EEG recording.

Waveform: The shape of a wave.

Wicket: Synonym: Mu rhythm.

Z: Abbreviation for impedance.

INDEX

Numbers followed by an "f" indicate figures; numbers followed by a "t" indicate tabular material.

517